ESSENTIALS OF CALCULUS

FOR BUSINESS, ECONOMICS, LIFE SCIENCES, SOCIAL SCIENCES

19. $\displaystyle \int \frac{u^n\, du}{\sqrt{a + bu}} = \frac{2u^n\sqrt{a + bu}}{b(2n + 1)} - \frac{2an}{b(2n + 1)} \int \frac{u^{n-1}\, du}{\sqrt{a + bu}}$

20. $\displaystyle \int \frac{du}{u\sqrt{a + bu}} = \begin{cases} \dfrac{1}{\sqrt{a}} \ln \left| \dfrac{\sqrt{a + bu} - \sqrt{a}}{\sqrt{a + bu} + \sqrt{a}} \right| + C & \text{if } a > 0 \\[4mm] \dfrac{2}{\sqrt{-a}} \tan^{-1} \sqrt{\dfrac{a + bu}{-a}} + C & \text{if } a < 0 \end{cases}$

21. $\displaystyle \int \frac{du}{u^n\sqrt{a + bu}} = -\frac{\sqrt{a + bu}}{a(n - 1)u^{n-1}} - \frac{b(2n - 3)}{2a(n - 1)} \int \frac{du}{u^{n-1}\sqrt{a + bu}}$

22. $\displaystyle \int \frac{\sqrt{a + bu}\, du}{u} = 2\sqrt{a + bu} + a \int \frac{du}{u\sqrt{a + bu}}$

23. $\displaystyle \int \frac{\sqrt{a + bu}\, du}{u^n} = -\frac{(a + bu)^{3/2}}{a(n - 1)u^{n-1}} - \frac{b(2n - 5)}{2a(n - 1)} \int \frac{\sqrt{a + bu}\, du}{u^{n-1}}$

Forms Containing $a^2 \pm u^2$

24. $\displaystyle \int \frac{du}{a^2 + u^2} = \frac{1}{a} \tan^{-1} \frac{u}{a} + C$

25. $\displaystyle \int \frac{du}{a^2 - u^2} = \frac{1}{2a} \ln \left| \frac{u + a}{u - a} \right| + C$

26. $\displaystyle \int \frac{du}{u^2 - a^2} = \frac{1}{2a} \ln \left| \frac{u - a}{u + a} \right| + C$

Forms Containing $\sqrt{u^2 \pm a^2}$

27. $\displaystyle \int \frac{du}{\sqrt{u^2 \pm a^2}} = \ln \left| u + \sqrt{u^2 \pm a^2} \right| + C$

28. $\displaystyle \int \sqrt{u^2 \pm a^2}\, du = \frac{u}{2} \sqrt{u^2 \pm a^2} \pm \frac{a^2}{2} \ln \left| u + \sqrt{u^2 \pm a^2} \right| + C$

29. $\displaystyle \int u^2 \sqrt{u^2 \pm a^2}\, du = \frac{u}{8}(2u^2 \pm a^2)\sqrt{u^2 \pm a^2} - \frac{a^4}{8} \ln \left| u + \sqrt{u^2 \pm a^2} \right| + C$

30. $\displaystyle \int \frac{\sqrt{u^2 + a^2}\, du}{u} = \sqrt{u^2 + a^2} - a \ln \left| \frac{a + \sqrt{u^2 + a^2}}{u} \right| + C$

31. $\displaystyle \int \frac{\sqrt{u^2 - a^2}\, du}{u} = \sqrt{u^2 - a^2} - a \sec^{-1} \left| \frac{u}{a} \right| + C$

32. $\displaystyle \int \frac{\sqrt{u^2 \pm a^2}\, du}{u^2} = -\frac{\sqrt{u^2 \pm a^2}}{u} + \ln \left| u + \sqrt{u^2 \pm a^2} \right| + C$

33. $\displaystyle \int \frac{u^2\, du}{\sqrt{u^2 \pm a^2}} = \frac{u}{2} \sqrt{u^2 \pm a^2} - \frac{\pm a^2}{2} \ln \left| u + \sqrt{u^2 \pm a^2} \right| + C$

34. $\displaystyle \int \frac{du}{u\sqrt{u^2 + a^2}} = -\frac{1}{a} \ln \left| \frac{a + \sqrt{u^2 + a^2}}{u} \right| + C$

35. $\displaystyle \int \frac{du}{u\sqrt{u^2 - a^2}} = \frac{1}{a} \sec^{-1} \left| \frac{u}{a} \right| + C$

37. $\displaystyle \int (u^2 \pm a^2)^{3/2}\, du = \frac{u}{8}(2u^2 \pm 5a^2)\sqrt{u^2 \pm a^2} + \frac{3a^4}{8} \ln \left| u + \sqrt{u^2 \pm a^2} \right| + C$

36. $\displaystyle \int \frac{du}{u^2\sqrt{u^2 \pm a^2}} = -\frac{\sqrt{u^2 \pm a^2}}{\pm a^2 u} + C$

38. $\displaystyle \int \frac{du}{(u^2 \pm a^2)^{3/2}} = \frac{u}{\pm a^2 \sqrt{u^2 \pm a^2}} + C$

ESSENTIALS OF CALCULUS

FOR BUSINESS, ECONOMICS, LIFE SCIENCES, SOCIAL SCIENCES

Louis Leithold

Pepperdine University

HARPER & ROW, PUBLISHERS, New York

Cambridge, Philadelphia, San Francisco,
London, Mexico City, São Paulo, Sydney

Sponsoring Editor: Ann Trump
Project Editor: David Nickol
Designer: Robert Sugar
Production Assistant: Debi Forrest Bochner
Compositor: Syntax International Pte. Ltd.
Printer and Binder: R. R. Donnelley and Sons Company
Art Studio: J&R Art Services, Inc.

Essentials of Calculus for Business, Economics, Life Sciences, Social Sciences

Library of Congress Cataloging in Publication Data

Leithold, Louis.
Essentials of calculus for business, economics, life sciences, social sciences.

    Includes index.
1. Calculus.   I. Title.
QA303.L4295   1984   515   83-18620
ISBN 0-06-043954-8

To Farley and Jenny

CONTENTS

PREFACE

This book is designed for use as a text in a calculus course for students whose specialty is in the management, social, or life sciences. The topics included were selected to give the kind and amount of calculus and applications essential for a professional program in these fields. Numerous examples and exercises from business, economics, biology, sociology, psychology, and statistics appear, but they are presented in such a way that no previous knowledge of technical terminology in these subjects is necessary. A background in high school algebra is required, but trigonometry is not needed. In the Appendix there are two sections devoted to the calculus of trigonometric functions for those who wish an introduction to this material.

It is assumed that the students using this book are interested mainly in the applications of the calculus to their particular field. However, to appreciate fully such applications—whether it be marginal analysis in economics, optimization in business, growth of bacteria in biology, or logistic growth in sociology—it is necessary first to have an understanding of the mathematical concepts involved. Therefore, my goal has been to give a correct treatment of elementary calculus with careful statements of the basic definitions and theorems. Bearing in mind that a textbook should be written for the student, I have attempted to keep the presentation geared to a beginner's experience and maturity and to leave no step unexplained or omitted. Most of the theorems are not proved. However, the discussion of a theorem has been augmented by illustrations and examples elaborating upon its content.

In Chapter 1 I have presented basic facts about the real numbers as well as some topics from algebra that are necessary for an understanding of the calculus. Dependent upon the preparation of the students, this chapter may be covered in detail, or some of the subject matter in the first two sections may be omitted. Sections 1.3 through 1.6 should be studied because they contain material necessary for the sequel.

The concepts of limits, continuity, and the derivative in Chapter 2 are at the heart of any first course in the calculus. The notion of the limit of a function is first given a step-by-step motivation, which brings the discussion from computing the value of a function near a number through an intuitive treatment of the limiting process. I have given a formal definition of the limit of a function that avoids "epsilon-delta" terminology. A sequence of examples progressively graded in difficulty is included. In the discussion of continuity, I have used as examples and counterexamples "common, everyday" functions and have abstained from those that would have little intuitive meaning. Before giving the formal definition of a derivative, I have defined the tangent line to a curve. Theorems on differentiation are proved and illustrated by examples.

The applications of the derivative in Chapter 3 are chosen to be intuitively appealing, and they include marginal concepts in economics and rates of change involving topics such as demand, supply, population, temperature, and pressure. The problem of finding absolute extrema of a function is presented in Chapter 4, and the concept is applied to problems from management and life sciences. Also in this chapter, the derivative is used as an aid to curve sketching. Additional methods for drawing a graph are given in Chapter 5, and they are utilized for the graphs of functions in economics. Price elasticity of demand in Section 5.4 and Profit in Section 5.5 are other important applications of the derivative to economics.

The antiderivative is treated in Chapter 6. I use the term "antidifferentiation" instead of indefinite integration, but the standard notation $\int f(x)\,dx$ is retained. This notation will suggest that some relation must exist between definite integrals, introduced in Chapter 7, and antiderivatives; however, I see no harm in this as long as the presentation gives the theoretically proper view of the definite integral as the limit of sums. In Section 6.3, differential equations with variables separable are discussed, and there are examples and exercises applying them to business and biology. Applications of antidifferentiation in economics occur in Section 6.4.

The measure of the area under a curve as a limit of sums is motivated by a business situation in Section 7.2, and this discussion precedes the introduction of the definite integral. Exercises involving the evaluation of definite integrals by finding limits of sums are given in Chapter 7 to stress that this is how they are calculated. Elementary properties of the definite integral are stated, and the fundamental theorem of the calculus is presented. It is emphasized that this is a theorem, and an important one, because it provides us with an alternative to computing limits of sums. It is also emphasized that the definite integral is in no sense some special type of antiderivative. In Section 7.4, there are applications of the definite integral in business decisions, inventory control, and probability. Consumers' surplus and producers' surplus occur in Section 7.6, and they provide an application of the definite integral in economics.

The number e is introduced in Section 8.1 by considering interest on an investment at a rate compounded continuously. This discussion is followed by a treatment of exponential functions. Exponential growth is illustrated by increases in the amount of an investment and the number of bacteria present in a culture, while exponential decay is a model for decreases in the amount of a radioactive substance and the value of a piece of equipment. The learning curve is used to demonstrate bounded growth. Logistic growth is shown as describing the spread of a disease or a rumor. In Section 8.5, the natural logarithmic function is defined as the inverse of the exponential function. In Sections 8.2 through 8.7, exponential and logarithmic functions are used to give additional applications of the calculus to management, social, and life sciences. Annuities involving both compound interest and interest compounded continuously are treated in Section 8.8.

Chapter 9, on techniques of integration, involves one of the most important computational aspects of the calculus. I have limited the discussion to those techniques that are most frequently used. They include integration by parts and integration by partial fractions. The mastery of integration techniques depends upon the examples, and I have used as illustrations problems that the student will certainly meet in practice. The material on the approximation of definite integrals includes the statements of theorems for computing the bounds of the error involved in these approximations. Improper integrals are introduced in Section 9.5, and an application of them in the field of economics is presented.

An introduction to the differential calculus of functions of several variables is furnished in Chapter 10. Applications of partial derivatives to economics include partial marginal demand and partial elasticity of demand. Extrema of functions of two variables is presented along with optimization problems in business and biology. Lagrange multipliers, discussed in Section 10.7, provide a useful device for the economist. The method of least squares is treated in Section 10.8, and there are problems pertaining to medicine, statistics, and maximization of profits.

I have used the material in this book in prepublication versions for a course given at Pepperdine University. To the students in these classes, I wish to express my thanks for their suggestions and enthusiasm.

Louis Leithold

ACKNOWLEDGMENTS

REVIEWERS

James Blackburn, Tulsa Junior College
Raymond Cannon, Baylor University
John Cunningham, Keene State College
Charles Friedman, University of Texas at Austin
Bodh Gulati, Southern Connecticut State University
Darrell Horwath, John Carroll University
James Hurley, University of Connecticut
Joseph Katz, Georgia State University
Teddy C.J. Leavitt, State University of New York at Plattsburgh
Dennis Luciano, Western New England College
Marcus McWaters, University of South Florida
Eldon Miller, University of Mississippi
James Modeer, University of Colorado
Kenneth Perrin, Pepperdine University
Eric Pianka, University of Texas at Austin
Paul Spannbauer, Hudson Valley Community College
Robert Wherritt, Wichita State University

COVER AND CHAPTER OPENING ARTIST

Douglas Bond, Pasadena
Courtesy of Tortue Gallery, Santa Monica

To these people and to the staff of Harper & Row, I express my appreciation.

L.L.

CHAPTER 1

REAL NUMBERS, GRAPHS, AND FUNCTIONS

Introduction   The two fundamental mathematical operations in calculus are *differentiation* and *integration*. These operations involve the computation of the *derivative* and the *integral*, and their applications occur in many different fields. For instance, as we progress through the book, some of the problems we will deal with are as follows:

- A cardboard box manufacturer wishes to make open boxes from square pieces of cardboard of side 12 in. by cutting equal squares from the four corners and turning up the sides. Find the length of the side of the square to be cut out in order to obtain a box of the largest possible volume.
- Points A and B are opposite each other on the shores of a straight river that is 3 km wide. Point C is on the same shore as B but 2 km down the river from B. A telephone company wishes to lay a cable from A to C. If the cost per kilometer of the cable is 25 percent more under the water than it is on the land, what line of cable would be least expensive for the company?
- A wholesale distributor has a standing order for 25,000 boxes of detergent that arrive every 20 weeks. These boxes are shipped out by the distributor at a constant rate of 1250 boxes per week. If storage costs are 0.3 cent per box per week, what is the total cost of maintaining inventory for 20 weeks?
- A deposit of $1000 is made at a savings bank that advertises interest on accounts computed at 7 percent compounded daily. Find the approximate amount at the end of 1 year by taking the rate as 7 percent compounded continuously.
- An historically important abstract painting was purchased in 1922 for $200, and its value has doubled every 10 years since its purchase. Find the rate at which the value was increasing in 1982.
- The GNP (Gross National Product) of a certain country has a rate of increase that is proportional to the GNP. If the GNP on January 1, 1978, was $80 billion and on January 1, 1981, it was $96 billion, when is it expected to be $128 billion?
- A rectangular box without a top is to be made at a cost of $10 for the material. The cost of the material for the bottom of the box is 15 cents per square foot, and the material for the sides costs 30 cents per square foot. Find the dimensions of the box of greatest volume that can be made.
- The maximum number of bacteria supportable by a particular environment is 900,000, and the rate of bacterial growth is jointly proportional to the number present and the difference between 900,000 and the number present. Determine the number of bacteria present when the rate of growth is a maximum.
- In a large forest a predator feeds on prey, and the predator population at any time is a function of the number of prey in the forest at that time. Suppose that when there are x prey in the forest, the predator population is y and that $y = \frac{1}{6}x^2 + 90$. Furthermore, if t weeks have elapsed since the end of the hunting season, $x = 7t + 85$. At what rate is the predator population growing 8 weeks after the close of the hunting season?

- In a particular small town the rate at which a rumor spreads is jointly proportional to the number of people who have heard the rumor and the number of people who have not heard it. Show that the rumor is being spread at the greatest rate when half the population of the town knows the rumor.
- Suppose that a tumor in a person's body is spherical in shape. If, when the radius of the tumor is 0.5 cm, the radius is increasing at the rate of 0.001 cm per day, what is the rate of increase in the volume of the tumor at that time?
- If a thermometer is taken from a room in which the temperature is 75° into the open where the temperature is 35°, and the reading of the thermometer is 65° after 30 sec, how long after the removal will the reading be 50°?

Before we introduce the basic ideas of calculus, it is necessary that you be familiar with some facts about the real numbers as well as some topics from algebra. This subject matter is presented in Secs. 1.1 and 1.2. Dependent upon your preparation, these sections may be covered in detail or treated as a review.

1.1 Real numbers, the number plane, and graphs of equations

A **real number** is either a positive number, a negative number, or zero, and in elementary calculus we are concerned with the set of real numbers. It is assumed that you are familiar with the algebraic operations of addition, subtraction, multiplication, and division of real numbers, as well as with the algebraic concepts of solving equations, factoring, and so forth. In this section we are concerned with properties of the real numbers that are important to the study of calculus.

Any real number may be classified as a *rational number* or an *irrational number*. A **rational number** is any number that can be expressed as the quotient of two integers. That is, a rational number is a number of the form p/q, where p and q are integers and $q \neq 0$. The rational numbers consist of the following:

- The **integers** (positive, negative, and zero)

  $\ldots, -5, -4, -3, -2, -1, 0, 1, 2, 3, 4, 5, \ldots$

- The positive and negative **fractions**, such as

  $\frac{2}{7} \qquad -\frac{4}{5} \qquad \frac{83}{5}$

- The positive and negative **terminating decimals**, such as

  $2.36 = \frac{236}{100} \qquad -0.003251 = -\frac{3,251}{1,000,000}$

- The positive and negative **nonterminating repeating decimals**, such as

  $0.333\ldots = \frac{1}{3} \qquad -0.549549549\ldots = -\frac{61}{111}$

The real numbers that are not rational are called **irrational numbers**. These are positive and negative **nonterminating nonrepeating decimals**, for example,

$$\sqrt{3} = 1.732 \ldots \qquad \pi = 3.14159 \ldots$$

The set of real numbers is denoted by R^1. This set may be represented geometrically as points on a horizontal line called an **axis**. Refer to Fig. 1.1.1. A point on the axis is chosen to represent the number 0. This point is called the **origin**. A unit of distance is selected. Then each positive number x is represented by the point at a distance of x units to the right of the origin, and each negative number x is represented by the point at a distance of $-x$ units to the left of the origin (it should be noted that if x is negative, then $-x$ is positive). To each real number there corresponds a point on the axis, and with each point on the axis there is associated only one real number; hence we say that there is a one-to-one correspondence between the set of real numbers and the points on the axis. Thus the points on the axis are identified with the numbers they represent, and the same symbol is used for both the number and the point corresponding to that number on the axis. We identify R^1 with the axis and call R^1 the real number line.

There is an ordering for the set R^1 by means of a relation denoted by the symbols $<$ (read "is less than") and $>$ (read "is greater than"), which are defined as follows:

$$a < b \quad \text{means} \quad b - a \text{ is positive}$$

$$a > b \quad \text{means} \quad a - b \text{ is positive}$$

● ILLUSTRATION 1

$3 < 5$ because $5 - 3 = 2$, and 2 is positive

$-10 < -6$ because $-6 - (-10) = 4$, and 4 is positive

$7 > 2$ because $7 - 2 = 5$, and 5 is positive

$-2 > -7$ because $-2 - (-7) = 5$, and 5 is positive

$\frac{3}{4} > \frac{2}{3}$ because $\frac{3}{4} - \frac{2}{3} = \frac{1}{12}$, and $\frac{1}{12}$ is positive   ●

We see that $a < b$ if and only if the point on the real number line that represents the number a is to the left of the point representing the number b. Similarly, $a > b$ if and only if the point representing a is to the right of the point representing b. For instance, the number 3 is less than the number 5, and the point 3 is to the left of the point 5. We could also write $5 > 3$ and say that the point 5 is to the right of the point 3.

The symbols $\leq$ (read "is less than or equal to") and $\geq$ (read "is greater than or equal to") are defined as follows:

$$a \leq b \quad \text{if and only if} \quad \text{either } a < b \text{ or } a = b$$

$$a \geq b \quad \text{if and only if} \quad \text{either } a > b \text{ or } a = b$$

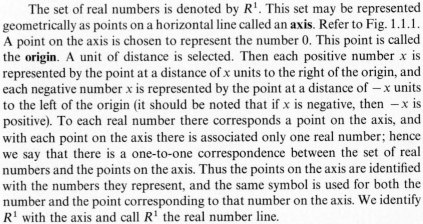

Figure 1.1.1

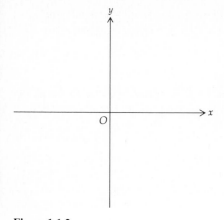

Figure 1.1.2

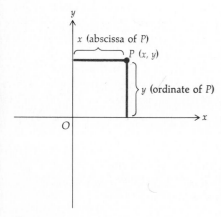

Figure 1.1.3

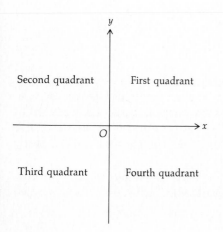

Figure 1.1.4

The statements $a < b$, $a > b$, $a \leq b$, and $a \geq b$ are called **inequalities**. In particular, $a < b$ and $a > b$ are called **strict** inequalities, whereas $a \leq b$ and $a \geq b$ are called **nonstrict** inequalities.

Ordered pairs of real numbers will now be considered. Any two real numbers form a **pair**, and when the order of the pair of real numbers is designated, we call it an **ordered pair of real numbers**. If x is the first real number and y is the second real number, we denote this ordered pair by writing them in parentheses with a comma separating them, as (x, y). Note that the ordered pair $(3, 7)$ is different from the ordered pair $(7, 3)$.

The set of all ordered pairs of real numbers is called the **number plane**, and each ordered pair (x, y) is called a **point** in the number plane. The number plane is denoted by R^2. Just as R^1 can be identified with points on an axis (a one-dimensional space), we can identify R^2 with points in a geometric plane (a two-dimensional space). The method we use with R^2 is the one attributed to the French mathematician René Descartes (1596–1650), who is credited as the originator of analytic geometry in 1637.

A horizontal line is chosen in the geometric plane and is called the *x* **axis**. A vertical line is chosen and is called the *y* **axis**. The point of intersection of the x axis and the y axis is called the **origin** and is denoted by the letter O. A unit of length is chosen (usually the unit length on each axis is the same). We establish the positive direction on the x axis to the right of the origin, and the positive direction on the y axis above the origin. See Fig. 1.1.2.

We now associate an ordered pair of real numbers (x, y) with a point P in the geometric plane. Refer to Fig. 1.1.3. The distance of P from the y axis (considered as positive if P is to the right of the y axis and negative if P is to the left of the y axis) is called the **abscissa** (or *x* **coordinate**) of P and is denoted by x. The distance of P from the x axis (considered as positive if P is above the x axis and negative if P is below the x axis) is called the **ordinate** (or *y* **coordinate**) of P and is denoted by y. The abscissa and the ordinate of a point are called the **rectangular cartesian coordinates** of the point. There is a one-to-one correspondence between the points in a geometric plane and R^2; that is, to each point there corresponds a unique ordered pair (x, y), and with each ordered pair (x, y) there is associated only one point. This one-to-one correspondence is called a **rectangular cartesian coordinate system**.

The x and y axes are called the **coordinate axes**. They divide the plane into four parts, called **quadrants**. The first quadrant is the one in which the abscissa and the ordinate are both positive, that is, the upper right quadrant. The other quadrants are numbered in the counterclockwise direction, with the fourth, for example, being the lower right quadrant. See Fig. 1.1.4.

Because of the one-to-one correspondence, R^2 is identified with the geometric plane, and for this reason an ordered pair (x, y) is called a **point**. Similarly, we refer to a **line** in R^2 as the set of all points corresponding to a line in the geometric plane, and we use other geometric terms for sets of points in R^2.

EXAMPLE 1 Plot the points $(-6, 0)$, $(-8, -6)$, $(-4, 5)$, $(0, -4)$, $(1, 2)$, $(2, 0)$, $(9, -7)$, and $(8, 5)$.

Solution Figure 1.1.5 shows a rectangular cartesian coordinate system with the given points plotted.

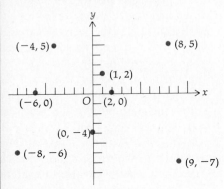

Figure 1.1.5

Consider the equation

$$y = x^2 - 2 \tag{1}$$

where (x, y) is a point in R^2. We call this an equation in R^2.

By a solution of this equation, we mean an ordered pair of numbers, one for x and one for y, that satisfies the equation. For example, if x is replaced by 3 in Eq. (1), we see that $y = 7$; thus $x = 3$ and $y = 7$ constitutes a solution of this equation. If any number is substituted for x in the right side of (1), we obtain a corresponding value for y. It is seen, then, that (1) has an unlimited number of solutions. Table 1.1.1 gives a few such solutions.

Table 1.1.1

x	0	1	2	3	4	-1	-2	-3	-4
$y = x^2 - 2$	-2	-1	2	7	14	-1	2	7	14

If we plot the points having as coordinates the number pairs (x, y) satisfying (1), we have a sketch of the graph of the equation. In Fig. 1.1.6 we have plotted points whose coordinates are the number pairs obtained from Table 1.1.1. These points are connected by a smooth curve. Any point (x, y) on this curve has coordinates satisfying (1). Also, the coordinates of any point not on this curve do not satisfy the equation. The graph of (1), shown in Fig. 1.1.6, is a **parabola**.

The **graph of an equation** in R^2 is the set of all points (x, y) in R^2 whose coordinates are numbers satisfying the equation. Such a graph is also called a **curve**. Unless otherwise stated, an equation with two unknowns, x and y, is considered an equation in R^2.

In the next example we use notation for a square root of a number. You may recall from algebra that the symbol $\sqrt{a}$, where $a \geq 0$, is defined as the

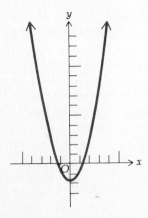

Figure 1.1.6

unique **nonnegative** number x such that $x^2 = a$. We read $\sqrt{a}$ as "the principal square root of a." For example,

$$\sqrt{4} = 2 \qquad \sqrt{0} = 0 \qquad \sqrt{\tfrac{9}{25}} = \tfrac{3}{5}$$

Note: $\sqrt{4} \neq -2$ even though $(-2)^2 = 4$, because $\sqrt{4}$ denotes only the *positive* square root of 4.

Because we are concerned only with real numbers in this book, $\sqrt{a}$ is not defined if $a < 0$.

EXAMPLE 2 Draw a sketch of the graph of the equation

$$y^2 - x - 2 = 0 \tag{2}$$

Solution Solving (2) for y we have

$$y = \pm\sqrt{x + 2} \tag{3}$$

Equations (3) are equivalent to the two equations

$$y = \sqrt{x + 2} \tag{4}$$

$$y = -\sqrt{x + 2} \tag{5}$$

The coordinates of all points that satisfy (3) will satisfy either (4) or (5), and the coordinates of any point that satisfies either (4) or (5) will satisfy (3). Table 1.1.2 gives some of these values of x and y.

Table 1.1.2

x	0	0	1	1	2	2	3	3	-1	-1	-2
y	$\sqrt{2}$	$-\sqrt{2}$	$\sqrt{3}$	$-\sqrt{3}$	2	-2	$\sqrt{5}$	$-\sqrt{5}$	1	-1	0

Note that for any value of $x < -2$ there is no real value for y. Also, for each value of $x > -2$ there are two values for y. A sketch of the graph of (2) is shown in Fig. 1.1.7. The graph is a parabola.

EXAMPLE 3 Draw sketches of the graphs of the equations

$$y = \sqrt{x + 2} \tag{6}$$

and

$$y = -\sqrt{x + 2}$$

Solution Equation (6) is the same as (4). The value of y is nonnegative; hence the graph of (6) is the upper half of the graph of (2). A sketch of this graph is shown in Fig. 1.1.8.

Similarly, the graph of the equation

$$y = -\sqrt{x + 2}$$

a sketch of which is shown in Fig. 1.1.9, is the lower half of the parabola of Fig. 1.1.7.

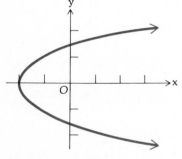

Figure 1.1.7

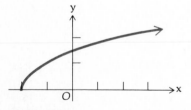

Figure 1.1.8

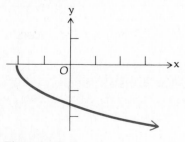

Figure 1.1.9

The concept of the *absolute value* of a number is used in some important definitions in the study of calculus. Following is the formal definition.

Definition of absolute value

The **absolute value** of x, denoted by $|x|$, is defined by

$$|x| = \quad x \quad \text{if } x > 0$$
$$|x| = -x \quad \text{if } x < 0$$
$$|0| = \quad 0$$

● ILLUSTRATION 2

$$|3| = 3 \qquad |-5| = -(-5) \qquad |8 - 14| = |-6|$$
$$= 5 \qquad\qquad = -(-6)$$
$$= 6$$

From the definition, the absolute value of a number is either a positive number or zero; that is, it is nonnegative.
Observe that

$$\sqrt{x^2} = |x|$$

For example,

$$\sqrt{5^2} = |5| = 5 \qquad \sqrt{(-3)^2} = |-3| = 3$$

Figure 1.1.10 shows a sketch of the graph of the equation

$$y = |x|$$

Table 1.1.3 gives some of the number pairs (x, y) that are coordinates of points on the graph.

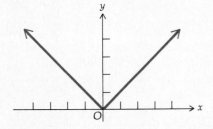

Figure 1.1.10

Table 1.1.3

x	0	1	2	3	-1	-2	-3
y	0	1	2	3	1	2	3

EXAMPLE 4   Draw a sketch of the graph of the equation

$$y = |x + 3| \tag{7}$$

Solution   From the definition of the absolute value of a number we have

$$y = x + 3 \quad \text{if} \quad x + 3 \geq 0$$

and

$$y = -(x + 3) \quad \text{if} \quad x + 3 < 0$$

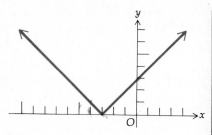

Figure 1.1.11

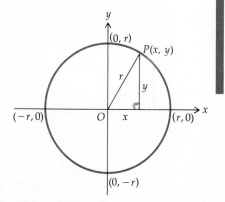

Figure 1.1.12

or, equivalently,

$$y = x + 3 \quad \text{if} \quad x \geq -3$$

and

$$y = -(x + 3) \quad \text{if} \quad x < -3$$

Table 1.1.4 gives some values of x and y satisfying (7).

Table 1.1.4

x	0	1	2	3	-1	-2	-3	-4	-5	-6	-7	-8	-9
y	3	4	5	6	2	1	0	1	2	3	4	5	6

A sketch of the graph of (7) is shown in Fig. 1.1.11.

Consider a circle having its center at the origin and radius r. See Fig. 1.1.12. From the Pythagorean theorem, involving the lengths of the sides of a right triangle, it follows that the point $P(x, y)$ is on the circle if and only if

$$x^2 + y^2 = r^2 \tag{8}$$

Thus (8) is an equation of the circle in Fig. 1.1.12.

Exercises 1.1

In Exercises 1 through 20, draw a sketch of the graph of the equation.

1. $y = 2x + 5$
2. $y = 4x - 3$
3. $y = \sqrt{x - 3}$
4. $y = -\sqrt{x - 3}$
5. $y^2 = x - 3$
6. $y = 5$
7. $x = -3$
8. $x = y^2 + 1$
9. $y = |x - 5|$
10. $y = -|x + 2|$
11. $y = |x| - 5$
12. $y = -|x| + 2$
13. $y = 4 - x^2$
14. $y = 4 + x^2$
15. $y = x^3$
16. $y = (x - 3)^2$
17. $y = (x + 3)^2$
18. $y = -x^3$
19. $x^2 + y^2 = 16$
20. $4x^2 + 4y^2 = 1$

In Exercises 21 through 26, draw a sketch of the graph of each of the equations.

21. (a) $y = \sqrt{2x}$ (b) $y = -\sqrt{2x}$ (c) $y^2 = 2x$
22. (a) $y = \sqrt{-2x}$ (b) $y = -\sqrt{-2x}$ (c) $y^2 = -2x$
23. (a) $y = \sqrt{4 - x^2}$ (b) $y = -\sqrt{4 - x^2}$ (c) $x^2 + y^2 = 4$
24. (a) $y = \sqrt{25 - x^2}$ (b) $y = -\sqrt{25 - x^2}$ (c) $x^2 + y^2 = 25$
25. (a) $x + 3y = 0$ (b) $x - 3y = 0$ (c) $x^2 - 9y^2 = 0$
26. (a) $2x - 5y = 0$ (b) $2x + 5y = 0$ (c) $4x^2 - 25y^2 = 0$

27. (a) Write an equation whose graph is the x axis. (b) Write an equation whose graph is the y axis. (c) Write an equation whose graph is the set of all points on either the x axis or the y axis.
28. (a) Write an equation whose graph is the set of all points having an abscissa of 4. (b) Write an equation whose graph is the set of all points whose ordinate is -3. (c) Write an equation whose graph is the set of all points having either an abscissa of 4 or an ordinate of -3.

1.2 Equations of a line

There are situations in which the rate of change of one quantity with respect to another is constant. For instance, suppose that it costs \$15 per unit to manufacture a particular commodity and there is a fixed daily overhead of \$400. Then if x units are produced per day and y dollars is the manufacturer's total daily cost,

$$y = 15x + 400$$

Some of the solutions of this equation are given in Table 1.2.1.

Table 1.2.1

x	0	10	20	30	40
$y = 15x + 400$	400	550	700	850	1000

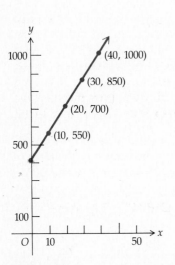

Figure 1.2.1

In Fig. 1.2.1 we have plotted points whose coordinates are the number pairs in this table, and we connected these points to obtain a *straight line*. Observe that for each 10-unit increase in x, y increases by 150 units, or, equivalently, for each 1-unit increase in x, y increases by 15 units. Thus the rate of change of y with respect to x is a constant 15. This constant rate of change is called the *slope* of the line. We proceed now to arrive at a formal definition of *slope*.

Let l be a nonvertical line and $P_1(x_1, y_1)$ and $P_2(x_2, y_2)$ be any two distinct points on l. Figure 1.2.2 shows such a line. In the figure, R is the point (x_2, y_1), and the points P_1, P_2, and R are vertices of a right triangle; furthermore, $\overline{P_1R} = x_2 - x_1$ and $\overline{RP_2} = y_2 - y_1$. The number $y_2 - y_1$ gives the measure of the change in the ordinate from P_1 to P_2, and it may be positive, negative, or zero. The number $x_2 - x_1$ gives the measure of the change in the abscissa from P_1 to P_2, and it may be positive or negative. Because the line l is not vertical, $x_2 \neq x_1$, and therefore $x_2 - x_1$ may not be zero. Let

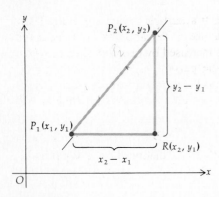

Figure 1.2.2

$$m = \frac{y_2 - y_1}{x_2 - x_1}$$

The value of m computed from this equation is independent of the choice of the two points P_1 and P_2 on l. This number m is called the *slope* of the line. Following is the formal definition.

Definition of the
slope of a line

If $P_1(x_1, y_1)$ and $P_2(x_2, y_2)$ are any two distinct points on line l, which is not parallel to the y axis, then the **slope** of l, denoted by m, is given by

$$m = \frac{y_2 - y_1}{x_2 - x_1} \tag{1}$$

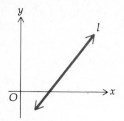

Figure 1.2.3

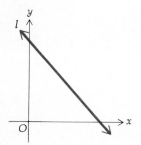

Figure 1.2.4

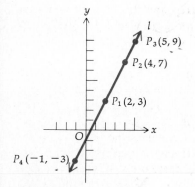

Figure 1.2.5

Multiplying on both sides of the equation in the above definition by $x_2 - x_1$ we obtain

$$y_2 - y_1 = m(x_2 - x_1) \tag{2}$$

It is seen from (2) that if we consider a particle moving along line l, the change in the ordinate of the particle is proportional to the change in the abscissa, and the constant of proportionality is the slope of the line.

If the slope of a line is positive, then as the abscissa of a point on the line increases, the ordinate increases. Such a line is shown in Fig. 1.2.3. In Fig. 1.2.4 there is a line whose slope is negative. For this line, as the abscissa of a point on the line increases, the ordinate decreases. Note that if the line is parallel to the x axis, then $y_2 = y_1$ and so $m = 0$.

If the line is parallel to the y axis, $x_2 = x_1$; thus Eq. (1) is meaningless, because we cannot divide by zero. This is the reason that lines parallel to the y axis, or vertical lines, are excluded in the definition of slope. Thus a vertical line does not have a slope.

● ILLUSTRATION 1

Let l be the line through the points $P_1(2, 3)$ and $P_2(4, 7)$. The slope of l is given by

$$m = \frac{7 - 3}{4 - 2} = 2$$

Refer to Fig. 1.2.5, which shows line l. If a particle is moving along l, the change in the ordinate is twice the change in the abscissa. That is, if the particle is at $P_2(4, 7)$ and the abscissa is increased by one unit, then the ordinate is increased by two units, and the particle is at the point $P_3(5, 9)$. Similarly, if the particle is at $P_1(2, 3)$ and the abscissa is decreased by three units, then the ordinate is decreased by six units, and the particle is at the point $P_4(-1, -3)$.   ●

It is proved in analytic geometry that two distinct lines are parallel if and only if they have the same slope.

● ILLUSTRATION 2

If L_1 is the line through the points $A(1, -2)$ and $B(-3, 6)$, and m_1 is the slope of L_1, then

$$m_1 = \frac{6 - (-2)}{-3 - 1} = \frac{8}{-4} = -2$$

If L_2 is the line through the points $C(-3, 4)$ and $D(5, -12)$, and m_2 is the

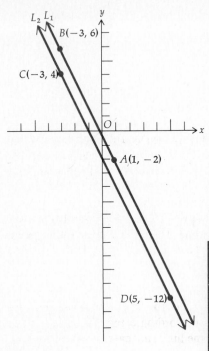

Figure 1.2.6

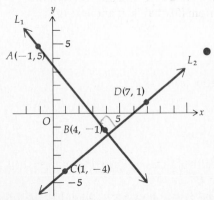

Figure 1.2.7

slope of L_2, then

$$m_2 = \frac{-12 - 4}{5 - (-3)} = \frac{-16}{8} = -2$$

Because $m_1 = m_2$, it follows that L_1 and L_2 are parallel. See Fig. 1.2.6. ●

Any two distinct points determine a line. Three distinct points may or may not lie on the same line. If three or more points lie on the same line, they are said to be **collinear**. Hence three points A, B, and C are collinear if and only if the line through the points A and B is the same as the line through the points B and C. Because the line through A and B and the line through B and C both contain the point B, they will be the same line if their slopes are equal.

EXAMPLE 1 Prove that the points $A(-2, -5)$, $B(1, -1)$, and $C(4, 3)$ are collinear.

Solution If m_1 is the slope of the line through A and B, then

$$m_1 = \frac{-1 - (-5)}{1 - (-2)} = \frac{4}{3}$$

If m_2 is the slope of the line through B and C, then

$$m_2 = \frac{3 - (-1)}{4 - 1} = \frac{4}{3}$$

Hence $m_1 = m_2$. Therefore the line through A and B and the line through B and C have the same slope and contain the common point B. Therefore they are the same line, and so A, B, and C are collinear.

Another theorem from analytic geometry states that two lines, L_1 and L_2, neither of which is vertical, are perpendicular if and only if the product of their slopes is -1. That is, if m_1 is the slope of L_1 and m_2 is the slope of L_2, then L_1 and L_2 are perpendicular if and only if

$$m_1 m_2 = -1$$

● ILLUSTRATION 3

Refer to Fig. 1.2.7. Let L_1 be the line through $A(-1, 5)$ and $B(4, -1)$. Then if m_1 is the slope of L_1,

$$m_1 = \frac{-1 - 5}{4 - (-1)} = \frac{-6}{5} = -\frac{6}{5}$$

Let L_2 be the line through $C(1, -4)$ and $D(7, 1)$. Then if m_2 is the slope of L_2,

$$m_2 = \frac{1 - (-4)}{7 - 1} = \frac{5}{6}$$

Because

$$m_1 m_2 = \left(-\frac{6}{5} \right) \frac{5}{6} = -1$$

it follows that L_1 and L_2 are perpendicular. ●

Because two points $P_1(x_1, y_1)$ and $P_2(x_2, y_2)$ determine a unique line, we should be able to obtain an equation of the line through these two points. Consider $P(x, y)$ any point on the line. We want an equation that is satisfied by x and y if and only if $P(x, y)$ is on the line through $P_1(x_1, y_1)$ and $P_2(x_2, y_2)$. We distinguish two cases.

Case 1: $x_2 = x_1$. In this case the line through P_1 and P_2 is parallel to the y axis, and all points on this line have the same abscissa. So $P(x, y)$ is any point on the line if and only if

$$x = x_1 \tag{3}$$

Equation (3) is an equation of a line parallel to the y axis. Note that this equation is independent of y; that is, the ordinate may have any value whatsoever, and the point $P(x, y)$ is on the line whenever the abscissa is x_1.

Case 2: $x_2 \neq x_1$. The slope of the line through P_1 and P_2 is given by

$$m = \frac{y_2 - y_1}{x_2 - x_1} \tag{4}$$

If $P(x, y)$ is any point on the line except (x_1, y_1), the slope is also given by

$$m = \frac{y - y_1}{x - x_1} \tag{5}$$

The point P will be on the line through P_1 and P_2 if and only if the value of m from (4) is the same as the value of m from (5), that is, if and only if

$$\frac{y - y_1}{x - x_1} = \frac{y_2 - y_1}{x_2 - x_1}$$

Multiplying on both sides of this equation by $(x - x_1)$ we obtain

$$y - y_1 = \frac{y_2 - y_1}{x_2 - x_1}(x - x_1) \tag{6}$$

Equation (6) is satisfied by the coordinates of P_1 as well as by the coordinates of any other point on the line through P_1 and P_2.

Equation (6) is called the **two-point** form of an equation of the line. It gives an equation of the line if two points on the line are known.

● ILLUSTRATION 4

An equation of the line through the two points $(6, -3)$ and $(-2, 3)$ is

$$y - (-3) = \frac{3 - (-3)}{-2 - 6}(x - 6)$$

$$y + 3 = -\tfrac{3}{4}(x - 6)$$

$$4y + 12 = -3x + 18$$

$$3x + 4y = 6$$

●

If in (6) we replace $(y_2 - y_1)/(x_2 - x_1)$ by m, we get

$$y - y_1 = m(x - x_1) \tag{7}$$

Equation (7) is called the **point-slope** form of an equation of the line. It gives an equation of the line if a point $P_1(x_1, y_1)$ on the line and the slope m of the line are known. It is recommended that you use the point-slope form even when two points are given, as shown in the following illustration.

● ILLUSTRATION 5

To find an equation of the line through the two points $Q(2, 1)$ and $R(4, 7)$, we first compute m.

$$m = \frac{7 - 1}{4 - 2} = \frac{6}{2} = 3$$

Using the point-slope form of an equation of the line, with Q as P_1, we have

$$y - 1 = 3(x - 2)$$

$$y - 1 = 3x - 6$$

$$-3x + y + 5 = 0$$

$$3x - y - 5 = 0$$

●

EXAMPLE 2 Find an equation of the line containing the point $(-4, 3)$ and having a slope $-\tfrac{2}{5}$. Draw a sketch of the line.

Solution Using the point-slope form of an equation of the line we have

$$y - 3 = -\tfrac{2}{5}[x - (-4)]$$

$$5(y - 3) = -2(x + 4)$$

$$5y - 15 = -2x - 8$$

$$2x + 5y - 7 = 0$$

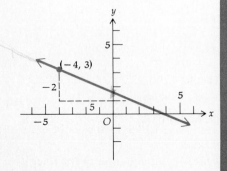

Figure 1.2.8

A sketch of the line is shown in Fig. 1.2.8.

If we choose the particular point $(0, b)$ (i.e., the point where the line intersects the y axis) for the point (x_1, y_1) in (7), we have

$$y - b = m(x - 0)$$

$$y = mx + b \qquad (8)$$

The number b, which is the ordinate of the point where the line intersects the y axis, is called the **y intercept** of the line. Consequently, (8) is called the **slope-intercept** form of an equation of the line. This form is especially important because it enables us to find the slope of a line from its equation. It is also important because it expresses the y coordinate explicitly in terms of the x coordinate.

● ILLUSTRATION 6

To find the slope of the line having the equation $3x + 4y = 7$, we solve the equation for y, and we have

$$4y = -3x + 7$$

$$y = -\tfrac{3}{4}x + \tfrac{7}{4}$$

Comparing this equation with (8) we see that $m = -\tfrac{3}{4}$ and $b = \tfrac{7}{4}$. Therefore the slope is $-\tfrac{3}{4}$.
●

EXAMPLE 3 Find the slope-intercept form of an equation of the line through the points $(-4, -1)$ and $(-7, -3)$. Draw a sketch of the line.

Solution If m is the slope of the line, then

$$m = \frac{-3 - (-1)}{-7 - (-4)} = \frac{-2}{-3} = \frac{2}{3}$$

Using the point-slope form of an equation of the line with $(-4, -1)$ as P_1, we have

$$y - (-1) = \tfrac{2}{3}[x - (-4)]$$

$$y + 1 = \tfrac{2}{3}x + \tfrac{8}{3}$$

$$y = \tfrac{2}{3}x + \tfrac{5}{3}$$

Therefore the slope of the line is $\tfrac{2}{3}$ and the y intercept is $\tfrac{5}{3}$. A sketch of the line is shown in Fig. 1.2.9.

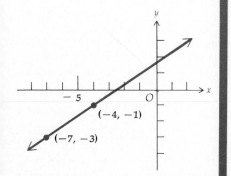

Figure 1.2.9

Consider the equation

$$Ax + By + C = 0 \qquad (9)$$

where A, B, and C are constants and where not both A and B are zero. We discuss the graph of this equation for the two cases $B \neq 0$ and $B = 0$.

Case 1: $B \neq 0$. Because $B \neq 0$, we divide on both sides of Eq. (9) by B and obtain

$$y = -\frac{A}{B}x - \frac{C}{B} \qquad (10)$$

Equation (10) is an equation of a straight line because it is in the slope-intercept form, where $m = -A/B$ and $b = -C/B$.

Case 2: $B = 0$. Because $B = 0$, we may conclude that $A \neq 0$ and thus have

$$Ax + C = 0$$

$$x = -\frac{C}{A} \qquad (11)$$

Equation (11) is in the form of (3), and so the graph is a straight line parallel to the y axis.

We have proved the following theorem.

> **Theorem 1.2.1**   The graph of the equation
>
> $$Ax + By + C = 0$$
>
> where A, B, and C are constants and where not both A and B are zero, is a straight line.

Because the graph of (9) is a straight line, it is called a **linear equation**. Equation (9) is the general equation of the first degree in x and y.

Because two points determine a line, to draw a sketch of the graph of a straight line from its equation we need only determine the coordinates of two points on the line, plot the two points, and then draw the line. Any two points will suffice, but it is usually convenient to plot the two points where the line intersects the two axes. These two points are denoted by $(a, 0)$ and $(0, b)$, where a is called the x intercept and b is called the y intercept.

EXAMPLE 4   Given line l_1, having the equation $2x - 3y = 12$, and line l_2, having the equation $4x + 3y = 6$, draw a sketch of each of the lines. Then find the coordinates of the point of intersection of l_1 and l_2.

Solution   To draw a sketch of the graph of l_1, we find the intercepts a and b. In the equation of l_1, we substitute 0 for x and get $b = -4$. In the equation of l_1, we substitute 0 for y and get $a = 6$. Similarly, we obtain the intercepts a and b for l_2, and for l_2 we have $a = \frac{3}{2}$ and $b = 2$. The two lines are plotted in Fig. 1.2.10.

To find the coordinates of the point of intersection of l_1 and l_2, we solve the two equations simultaneously. Because the point must lie on both lines, its coordinates must satisfy both equations. If both equations are put in the slope-intercept form, we have

$$y = \tfrac{2}{3}x - 4 \quad \text{and} \quad y = -\tfrac{4}{3}x + 2$$

Eliminating y gives

$$\tfrac{2}{3}x - 4 = -\tfrac{4}{3}x + 2$$

$$2x - 12 = -4x + 6$$

$$x = 3$$

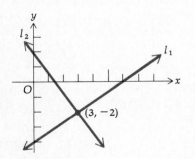

Figure 1.2.10

So

$$y = \tfrac{2}{3}(3) - 4$$
$$= -2$$

Therefore the point of intersection is $(3, -2)$.

● ILLUSTRATION 7

A company spends $1800 for equipment. The company accountant uses the straight-line method of depreciation over 10 years, which is the estimated life of the equipment; that is, the book value of the equipment decreases at a constant rate, so that at the end of 10 years the book value will be zero. Suppose y dollars is the book value of the equipment at the end of x years. Thus when $x = 0$, $y = 1800$, and when $x = 10$, $y = 0$. The graph giving the relationship between y and x is the segment in the first quadrant of the straight line through the points $(0, 1800)$ and $(10, 0)$. See Fig. 1.2.11. If m is the slope of the line, then

$$m = \frac{0 - 1800}{10 - 0} = -180$$

We use the slope-intercept form (Eq. (8)) of an equation of the line with $m = -180$ and $b = 1800$. We have

$$y = -180x + 1800 \qquad 0 \le x \le 10$$

Observe that the slope of the line is -180, and this number gives the amount by which the book value changes each year; that is, the book value decreases $180 per year.   ●

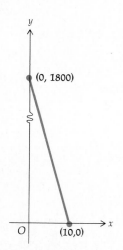

Figure 1.2.11

EXAMPLE 5 (a) Suppose that equipment is purchased for A dollars and it is depreciated by the straight-line method over n years. If its book value is y dollars at the end of x years, find an equation giving the relationship between y and x. (b) Use the result of part (a) to find the book value at the end of 5 years for equipment purchased for $3000 if the equipment is depreciated by the straight-line method over 12 years.

Solution (a) Refer to Fig. 1.2.12. The graph giving the relationship between y and x is the segment in the first quadrant of the straight line through the points $(0, A)$ and $(n, 0)$. The slope of this line is

$$\frac{0 - A}{n - 0} = -\frac{A}{n}$$

The y intercept of the line is A. Thus an equation of the line is

$$y = -\frac{A}{n}x + A \qquad 0 \le x \le n$$

(b) We use the result of part (a) with $A = 3000$ and $n = 12$. We obtain

$$y = -\frac{3000}{12}x + 3000 \qquad 0 \le x \le 12$$

$$y = -250x + 3000 \qquad 0 \le x \le 12$$

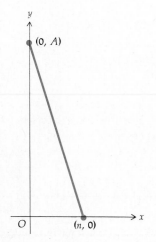

Figure 1.2.12

Let y_5 be the value of y when $x = 5$. Then

$$y_5 = -250(5) + 3000$$

$$= 1750$$

Therefore the book value of the equipment at the end of 5 years is $1750.

Exercises 1.2

In Exercises 1 through 6, find the slope of the line through the given points.

1. $(2, -3), (-4, 3)$

2. $(5, 2), (-2, -3)$

3. $(\frac{1}{3}, \frac{1}{2}), (-\frac{5}{6}, \frac{2}{3})$

4. $(\frac{3}{4}, -\frac{3}{2}), (-\frac{5}{2}, \frac{1}{4})$

5. $(-1, 4), (-6, 4)$

6. $(-2.1, 0.3), (2.3, 1.4)$

In Exercises 7 through 18, find an equation of the line satisfying the given conditions.

7. The slope is 4 and through the point $(2, -3)$.
8. Through the two points $(3, 1)$ and $(-5, 4)$.
9. Through the two points $(-3, 0)$ and $(4, 0)$.
10. Through the point $(1, 4)$ and parallel to the line whose equation is $2x - 5y + 7 = 0$.
11. Through the point $(-4, -5)$ and parallel to the line whose equation is $2x - 3y + 6 = 0$.
12. Through the point $(-2, 3)$ and perpendicular to the line whose equation is $2x - y - 2 = 0$.
13. The y intercept is -4 and perpendicular to the line whose equation is $3x - 4y + 8 = 0$.
14. Through the point $(-3, -4)$ and parallel to the y axis.
15. Through the point $(1, -7)$ and parallel to the x axis.
16. Through the point $(-2, -5)$ and having a slope of $\sqrt{3}$.
17. Through the origin and bisecting the angle between the axes in the first and third quadrants.
18. Through the origin and bisecting the angle between the axes in the second and fourth quadrants.

In Exercises 19 through 22, find the slope of the given line.

19. $4x - 6y = 5$

20. $x + 3y = 7$

21. $2y + 9 = 0$

22. $3x - 5 = 0$

23. Find an equation of the line through the points $(3, -5)$ and $(1, -2)$, and write the equation in slope-intercept form.
24. Find an equation of the line through the points $(1, 3)$ and $(2, -2)$, and write the equation in slope-intercept form.
25. Show that the lines having the equations $3x + 5y + 7 = 0$ and $5x - 3y - 2 = 0$ are perpendicular.
26. Show that the lines having the equations $3x + 5y + 7 = 0$ and $6x + 10y - 5 = 0$ are parallel.

In Exercises 27, 28, and 29, determine by means of slopes if the given points are on a line.

27. (a) $(2, 3), (-4, -7), (5, 8)$; (b) $(2, -1), (1, 1), (3, 4)$
28. (a) $(4, 6), (1, 2), (-5, -4)$; (b) $(-3, 6), (3, 2), (9, -2)$
29. (a) $(2, 5), (-1, 4), (3, -2)$; (b) $(0, 2), (-3, -1), (4, 6)$

In Exercises 30 and 31, draw a sketch of each of the given lines, and find the coordinates of their point of intersection.

30. $6x - 5y - 6 = 0; 4x - 3y - 2 = 0$

31. $3x + 2y = 2; 5x + 3y = 1$

32. Some business property was purchased in 1973 for $750,000, and the land was valued at $150,000 while the improvements were valued at $600,000. The improvements were depreciated by the straight-line method over 20 years. What was the book value of the improvements in 1981?
33. A company purchased machinery for $15,000. It is anticipated that the scrap value after 10 years will be $2000. If the straight-line method is used to depreciate the machinery from $15,000 to $2000 over 10 years, what is the book value of the machinery after 6 years?
34. This exercise is a generalization of Exercise 33. Suppose that machinery is purchased for A dollars and has a scrap value of B dollars after n years. Furthermore, the machinery is depreciated from A dollars to B dollars over n years by the straight-line

method. If the book value of the machinery is y dollars at the end of x years, find an equation giving the relationship between y and x.

35. The producer of a particular commodity has a total cost consisting of a weekly overhead of $3000 and a manufacturing cost of $25 per unit. (a) If x units are produced per week and y dollars is the total weekly cost, write an equation involving x and y. (b) Draw a sketch of the graph of the equation in part (a).

36. A producer's total cost consists of a manufacturing cost of $20 per unit and a fixed daily overhead. (a) If the total cost of producing 200 units in 1 day is $4500, determine the fixed daily overhead. (b) If x units are produced per day and y dollars is the total daily cost, write an equation involving x and y. (c) Draw a sketch of the graph of the equation in part (b).

37. Do Exercise 36 if the producer's cost is $30 per unit and the total cost of producing 200 units in 1 day is $6600.

38. The graph of an equation relating the temperature reading in Celsius degrees and the temperature reading in Fahrenheit degrees is a straight line. Water freezes at $0°$ Celsius and $32°$ Fahrenheit, and water boils at $100°$ Celsius and $212°$ Fahrenheit. (a) If y degrees Fahrenheit corresponds to x degrees Celsius, write an equation involving x and y. (b) Draw a sketch of the graph of the equation in part (a). (c) What is the Fahrenheit temperature corresponding to $20°$ Celsius? (d) What is the Celsius temperature corresponding to $86°$ Fahrenheit?

1.3 Functions and their graphs

There are many instances occurring in practice where the value of one quantity depends on the value of another. In particular, a person's salary may depend on the number of hours worked; the number of units demanded by the consumers of a certain commodity may depend on the price of the commodity; the total production at a factory may depend on the number of machines used; and so forth. A relationship between such quantities is often defined by means of a *function*. In this section we introduce the concept of a function, and in the next section we give some examples showing how a practical situation can be expressed in terms of a functional relationship. Before discussing functions, we introduce some set notation and terminology used in the development.

The idea of *set* is used extensively in mathematics and is such a basic concept that it is not given a formal definition. We can say that a **set** is a collection of objects, and the objects in a set are called the **elements** of the set. If every element of a set S is also an element of a set T, then S is a **subset** of T. Two subsets of the set R^1 of real numbers are the set N of natural numbers and the set Z of integers.

We use the symbol $\in$ to indicate that a specific element belongs to a set. Hence we may write $8 \in N$, which is read "8 is an element of N." The notation $a, b \in S$ indicates that both a and b are elements of S. The symbol $\notin$ is read "is not an element of." Thus we read $\frac{1}{2} \notin N$ as "$\frac{1}{2}$ is not an element of N."

A pair of braces $\{\ \}$ used with words or symbols can describe a set. If S is the set of natural numbers less than 6, we can write the set S as

$$\{1, 2, 3, 4, 5\}$$

We can also write the set S as

$$\{x, \text{ such that } x \text{ is a natural number less than } 6\}$$

where the symbol x is called a *variable*. A **variable** is a symbol used to represent any element of a given set. Another way of writing the above set S

is to use what is called **set-builder notation**, where a vertical bar is used in place of the words "such that." Using set-builder notation to describe the set S we have

$\{x \mid x \text{ is a natural number less than 6}\}$

which is read "the set of all x such that x is a natural number less than 6."

Two sets A and B are said to be **equal**, written $A = B$, if A and B have identical elements. The **union** of two sets A and B, denoted by $A \cup B$ and read "A union B" is the set of all elements that are in A or in B or in both A and B. The **intersection** of A and B, denoted by $A \cap B$ and read "A intersection B," is the set of all elements that are in both A and B. The set that contains no elements is called the **empty set** and is denoted by $\varnothing$.

● ILLUSTRATION 1

Suppose $A = \{2, 4, 6, 8, 10, 12\}$, $B = \{1, 4, 9, 16\}$, and $C = \{2, 10\}$. Then

$$A \cup B = \{1, 2, 4, 6, 8, 9, 10, 12, 16\} \qquad A \cap B = \{4\}$$
$$B \cup C = \{1, 2, 4, 9, 10, 16\} \qquad\qquad B \cap C = \varnothing$$ ●

A number x is between a and b if $a < x$ and $x < b$. We can write this as a continued inequality as follows:

$$a < x < b \tag{1}$$

The set of all numbers x satisfying the continued inequality (1) is called an **open interval** and is denoted by (a, b). Therefore

$(a, b) = \{x \mid a < x < b\}$

The **closed interval** from a to b is the open interval (a, b) together with the two endpoints a and b and is denoted by $[a, b]$. Thus

$[a, b] = \{x \mid a \le x \le b\}$

Figure 1.3.1

Figure 1.3.1 illustrates the open interval (a, b), and Fig. 1.3.2 illustrates the closed interval $[a, b]$.

The **interval half-open on the left** is the open interval (a, b) together with the right endpoint b. It is denoted by $(a, b]$; so

$(a, b] = \{x \mid a < x \le b\}$

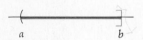

Figure 1.3.2

We define an **interval half-open on the right** in a similar way and denote it by $[a, b)$. Thus

$[a, b) = \{x \mid a \le x < b\}$

Figure 1.3.3

Figure 1.3.3 illustrates the interval $(a, b]$, and Fig. 1.3.4 illustrates the interval $[a, b)$.

We shall use the symbol $+\infty$ (positive infinity) and the symbol $-\infty$ (negative infinity); however, take care not to confuse these symbols with real numbers, for they do not obey the properties of the real numbers. We have the following intervals:

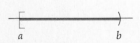

Figure 1.3.4

$$(a, +\infty) = \{x \mid x > a\}$$
$$(-\infty, b) = \{x \mid x < b\}$$
$$[a, +\infty) = \{x \mid x \geq a\}$$
$$(-\infty, b] = \{x \mid x \leq b\}$$
$$(-\infty, +\infty) = R^1$$

We intuitively consider y to be a function of x if there is some rule by which a unique value is assigned to y by a corresponding value of x. Familiar examples of such relationships are given by equations such as

$$y = 2x^2 + 5 \tag{2}$$

and

$$y = \sqrt{x^2 - 9} \tag{3}$$

The formal definition makes the concept of a function precise.

Definition of a function

> A **function** is a set of ordered pairs of numbers (x, y) in which no two distinct ordered pairs have the same first number. The set of all admissible values of x is called the **domain** of the function, and the set of all admissible values of y is called the **range** of the function.

In the above definition, the restriction that no two distinct ordered pairs can have the same first number assures that y is unique for a specific value of x.

Equation (2) defines a function. Call this function f. The equation gives the rule by which a unique value of y can be determined whenever x is given; that is, multiply the number x by itself, then multiply that product by 2, and add 5. The function f is the set of all ordered pairs (x, y) such that x and y satisfy (2); that is,

$$f = \{(x, y) \mid y = 2x^2 + 5\}$$

The numbers x and y are **variables**. Because for the function f values are assigned to x, and because the value of y is dependent on the choice of x, x is the **independent variable** and y is the **dependent variable**. The domain of the function is the set of all admissible values of the independent variable, and the range of the function is the set of all admissible values of the dependent variable. For the function f under consideration, the domain is the set of all real numbers, and it can be denoted with interval notation as $(-\infty, +\infty)$. The smallest value that y can assume is 5 (when $x = 0$). The range of f is then the set of all positive numbers greater than or equal to 5, which is $[5, +\infty)$.

● ILLUSTRATION 2

Let g be the function which is the set of all ordered pairs (x, y) defined by (3); that is,

$$g = \{(x, y) \mid y = \sqrt{x^2 - 9}\}$$

Because the numbers are confined to real numbers, y is a function of x only for $x \geq 3$ or $x \leq -3$ (or simply $|x| \geq 3$), because for any x satisfying either of these inequalities a unique value of y is determined. However, if x is in the interval $(-3, 3)$, a square root of a negative number is obtained, and hence no real number y exists. Therefore we must restrict x, and so

$$g = \{(x, y) \mid y = \sqrt{x^2 - 9} \text{ and } |x| \geq 3\}$$

The domain of g is $(-\infty, -3] \cup [3, +\infty)$, and the range of g is $[0, +\infty)$.

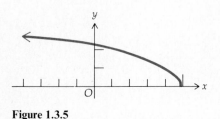

Figure 1.3.5

It should be stressed that in order to have a function there must be *exactly one value* of the dependent variable for a value of the independent variable in the domain of the function.

Definition of the graph of a function	If f is a function, then the **graph** of f is the set of all points (x, y) in R^2 for which (x, y) is an ordered pair in f.

● ILLUSTRATION 3

Let $f = \{(x, y) \mid y = \sqrt{5 - x}\}$. A sketch of the graph of f is shown in Fig. 1.3.5. The domain of f is the set of all real numbers less than or equal to 5, which is $(-\infty, 5]$, and the range of f is the set of all nonnegative real numbers, which is $[0, +\infty)$.

In the next illustration, a function is defined by more than one equation. Such a definition is permissible provided that for each number x in the domain there is a unique value for y in the range.

● ILLUSTRATION 4

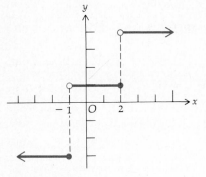

Figure 1.3.6

Let g be the function which is the set of all ordered pairs (x, y) such that

$$y = \begin{cases} -3 & \text{if } x \leq -1 \\ 1 & \text{if } -1 < x \leq 2 \\ 4 & \text{if } 2 < x \end{cases}$$

The domain of g is $(-\infty, +\infty)$, while the range of g consists of the three numbers -3, 1, and 4. A sketch of the graph is shown in Fig. 1.3.6.

Observe in Fig. 1.3.6 that there is a break at $x = -1$ and another at $x = 2$. We say that g is discontinuous at -1 and 2. Continuous and discontinuous functions are discussed in Sec. 2.3.

● ILLUSTRATION 5

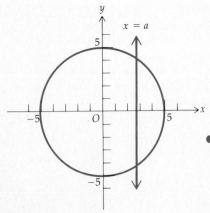

Figure 1.3.7

Consider the set

$$\{(x, y) \mid x^2 + y^2 = 25\}$$

A sketch of the graph of this set is shown in Fig. 1.3.7. This set of ordered pairs is not a function, because for any x in the interval $(-5, 5)$ there are two ordered pairs having that number as a first element. For example, both (3, 4) and (3, −4) are ordered pairs in the given set. Furthermore, observe that the graph of the given set is a circle with center at the origin and radius 5, and a vertical line having the equation $x = a$ (where $-5 < a < 5$) intersects the circle in two points. Refer to the figure.   ●

EXAMPLE 1 Let $h = \{(x, y)|y = |x|\}$. Determine the domain and range of h, and draw a sketch of the graph of h.

Solution   The domain of h is $(-\infty, +\infty)$, and the range of h is $[0, +\infty)$. A sketch of the graph of h is shown in Fig. 1.3.8.

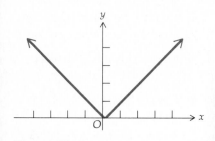

Figure 1.3.8

EXAMPLE 2 Let F be the function which is the set of all ordered pairs (x, y) such that

$$y = \begin{cases} 3x - 2 & \text{if } x < 1 \\ x^2 & \text{if } 1 \le x \end{cases}$$

Determine the domain and range of F, and draw a sketch of the graph of F.

Solution   A sketch of the graph of F is shown in Fig. 1.3.9. The domain of F is $(-\infty, +\infty)$, and the range of F is $(-\infty, +\infty)$.

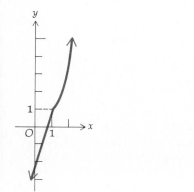

Figure 1.3.9

EXAMPLE 3 Let G be the function which is the set of all ordered pairs (x, y) such that

$$y = \frac{x^2 - 9}{x - 3}$$

Determine the domain and range of G, and draw a sketch of the graph of G.

Solution   A sketch of the graph is shown in Fig. 1.3.10. Because a value for y is determined for each value of x except $x = 3$, the domain of G consists of all real numbers except 3. When $x = 3$, both the numerator and denominator are zero, and 0/0 is undefined.

Factoring the numerator into $(x - 3)(x + 3)$ we obtain

$$y = \frac{(x - 3)(x + 3)}{(x - 3)}$$

or $y = x + 3$, provided that $x \ne 3$. In other words, the function G consists of all ordered pairs (x, y) such that

$$y = x + 3 \quad \text{and} \quad x \ne 3$$

The range of G consists of all real numbers except 6. The graph consists of all points on the line $y = x + 3$ except the point (3, 6).

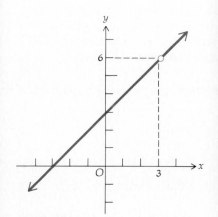

Figure 1.3.10

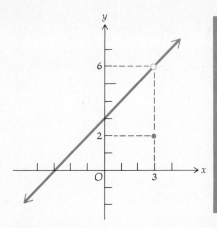

Figure 1.3.11

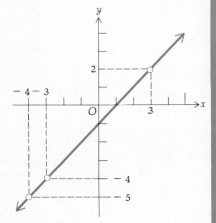

Figure 1.3.12

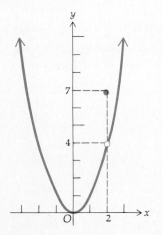

Figure 1.3.13

EXAMPLE 4 Let H be the function which is the set of all ordered pairs (x, y) such that

$$y = \begin{cases} x + 3 & \text{if } x \neq 3 \\ 2 & \text{if } x = 3 \end{cases}$$

Determine the domain and range of H, and draw a sketch of the graph of H.

Solution A sketch of the graph of this function is shown in Fig. 1.3.11. The graph consists of the point $(3, 2)$ and all points on the line $y = x + 3$ except the point $(3, 6)$. Function H is defined for all values of x, and therefore the domain of H is $(-\infty, +\infty)$. The range of H consists of all real numbers except 6.

EXAMPLE 5 Let ϕ be the function which is the set of all ordered pairs (x, y) such that

$$y = \frac{(x^2 + 3x - 4)(x^2 - 9)}{(x^2 + x - 12)(x + 3)}$$

Determine the domain and range of ϕ, and draw a sketch of the graph of ϕ.

Solution A sketch of the graph of this function is shown in Fig. 1.3.12, Factoring the numerator and denominator we obtain

$$y = \frac{(x + 4)(x - 1)(x - 3)(x + 3)}{(x + 4)(x - 3)(x + 3)}$$

The denominator is zero for $x = -4$, -3, and 3; therefore ϕ is undefined for these three values of x. For values of $x \neq -4$, -3, or 3, we may divide numerator and denominator by the common factors and obtain

$$y = x - 1 \quad \text{if } x \neq -4, -3, \text{ or } 3$$

Therefore the domain of ϕ is the set of all real numbers except -4, -3, and 3, and the range of ϕ is the set of all real numbers except those values of $x - 1$ obtained by replacing x by -4, -3, or 3, that is, all real numbers except -5, -4, and 2. The graph of this function is the straight line $y = x - 1$, with the points $(-4, -5)$, $(-3, -4)$, and $(3, 2)$ deleted.

EXAMPLE 6 Let f be the function which is the set of all ordered pairs (x, y) such that

$$y = \begin{cases} x^2 & \text{if } x \neq 2 \\ 7 & \text{if } x = 2 \end{cases}$$

Determine the domain and range of f, and draw a sketch of the graph of f,

Solution A sketch of the graph of f is shown in Fig. 1.3.13. The graph consists of the point $(2, 7)$ and all points on the parabola $y = x^2$ except the

point (2, 4). Function f is defined for all values of x, and so the domain of f is $(-\infty, +\infty)$. The range of f consists of all nonnegative real numbers.

EXAMPLE 7   Let h be the function which is the set of all ordered pairs (x, y) such that

$$y = \begin{cases} x - 1 & \text{if } x < 3 \\ 2x + 1 & \text{if } 3 \leq x \end{cases}$$

Determine the domain and range of h, and draw a sketch of the graph of h.

Solution   A sketch of the graph of h is shown in Fig. 1.3.14. The domain of h is $(-\infty, +\infty)$. The values of y are either less than 2 or greater than or equal to 7. So the range of h is $(-\infty, 2) \cup [7, +\infty)$ or, equivalently, all real numbers not in $[2, 7)$.

Figure 1.3.14

Exercises 1.3

In Exercises 1 through 14, determine the domain and range of the given function, and draw a sketch of the graph of the function.

1. $f = \{(x, y)\,|\,y = 3x - 1\}$

2. $g = \{(x, y)\,|\,y = x^2 + 2\}$

3. $F = \{(x, y)\,|\,y = 3x^2 - 6\}$

4. $G = \{(x, y)\,|\,y = 5 - x^2\}$

5. $G = \{(x, y)\,|\,y = \sqrt{x + 1}\}$

6. $f = \{(x, y)\,|\,y = \sqrt{5 - 3x}\}$

7. $h = \{(x, y)\,|\,y = \sqrt{3x - 4}\}$

8. $f = \{(x, y)\,|\,y = \sqrt{4 - x^2}\}$

9. $g = \{(x, y)\,|\,y = \sqrt{x^2 - 4}\}$.

10. $H = \{(x, y)\,|\,y = |x - 3|\}$

11. $\phi = \{(x, y)\,|\,y = |3x + 2|\}$

12. $F = \left\{(x, y)\,\middle|\,y = \dfrac{4x^2 - 1}{2x + 1}\right\}$

13. $H = \left\{(x, y)\,\middle|\,y = \dfrac{x^2 - 4x + 3}{x - 1}\right\}$

14. $g = \left\{(x, y)\,\middle|\,y = \dfrac{x^3 - 3x^2 - 4x + 12}{x^2 - x - 6}\right\}$

In Exercises 15 through 32, the function is the set of all ordered pairs (x, y) satisfying the given equation. Determine the domain and range of the function, and draw a sketch of the graph of the function.

15. $G: y = \begin{cases} -2 & \text{if } x \leq 3 \\ 2 & \text{if } 3 < x \end{cases}$

16. $h: y = \begin{cases} -4 & \text{if } x < -2 \\ -1 & \text{if } -2 \leq x \leq 2 \\ 3 & \text{if } 2 < x \end{cases}$

17. $f: y = \begin{cases} 2x - 1 & \text{if } x \neq 2 \\ 0 & \text{if } x = 2 \end{cases}$

18. $f: y = \begin{cases} x^2 - 4 & \text{if } x \neq -3 \\ -2 & \text{if } x = -3 \end{cases}$

19. $H: y = \begin{cases} x^2 - 4 & \text{if } x < 3 \\ 2x - 1 & \text{if } 3 \leq x \end{cases}$

20. $\phi: y = \begin{cases} x + 5 & \text{if } x < -5 \\ \sqrt{25 - x^2} & \text{if } -5 \leq x \leq 5 \\ x - 5 & \text{if } 5 < x \end{cases}$

21. $F: y = \begin{cases} x - 2 & \text{if } x < 0 \\ x^2 + 1 & \text{if } 0 \leq x \end{cases}$

22. $g: y = \begin{cases} 6x + 7 & \text{if } x \leq -2 \\ 4 - x & \text{if } -2 < x \end{cases}$

23. $F: y = \dfrac{(x + 1)(x^2 + 3x - 10)}{x^2 + 6x + 5}$

24. $G: y = \dfrac{(x^2 + 3x - 4)(x^2 - 5x + 6)}{(x^2 - 3x + 2)(x - 3)}$

25. $f: y = \sqrt{9 - x^2}$

26. $h: y = \sqrt{x^2 - 9}$

27. $g: y = \dfrac{x^3 - 2x^2}{x - 2}$

28. $f: y = \dfrac{x^3 + 3x^2 + x + 3}{x + 3}$

29. $h: y = \dfrac{x^3 + 5x^2 - 6x - 30}{x + 5}$

30. $F: y = \dfrac{x^4 + x^3 - 9x^2 - 3x + 18}{x^2 + x - 6}$

31. $f: y = |x| + |x - 1|$

32. $g: y = |x| \cdot |x - 1|$

1.4 Function notation, types of functions, and applications

If f is the function having as its domain variable x and as its range variable y, the symbol $f(x)$ (read "f of x") denotes the particular value of y that corresponds to the value of x.

● ILLUSTRATION 1

In Illustration 3 of Sec. 1.3,

$$f = \{(x, y) \,|\, y = \sqrt{5 - x}\}$$

Thus $f(x) = \sqrt{5 - x}$. Because when $x = 1$, $\sqrt{5 - x} = 2$, we have $f(1) = 2$. Similarly, $f(-6) = \sqrt{11}$, $f(0) = \sqrt{5}$, and so on.   ●

When defining a function, the domain of the function must be given either explicitly or implicitly. For instance, if f is defined by

$$f(x) = 3x^2 - 5x + 2$$

it is implied that x can be any real number. However, if f is defined by

$$f(x) = 3x^2 - 5x + 2 \qquad 1 \le x \le 10$$

then the domain of f consists of all real numbers between and including 1 and 10.

Similarly, if g is defined by the equation

$$g(x) = \frac{5x - 2}{x + 4}$$

it is implied that $x \ne -4$, because the quotient is undefined for $x = -4$; hence the domain of g is the set of all real numbers except -4.

If

$$h(x) = \sqrt{9 - x^2}$$

it is implied that x is in the closed interval $-3 \le x \le 3$, because $\sqrt{9 - x^2}$ is undefined (i.e., not a real number) for $x > 3$ or $x < -3$. So the domain of h is $[-3, 3]$, and the range of h is $[0, 3]$.

EXAMPLE 1   Given that f is the function defined by $f(x) = x^2 + 3x - 4$, find: (a) $f(0)$; (b) $f(2)$; (c) $f(h)$; (d) $f(2h)$; (e) $f(2x)$; (f) $f(x + h)$; (g) $f(x) + f(h)$.

Solution

(a) $f(0) = 0^2 + 3 \cdot 0 - 4 = -4$

(b) $f(2) = 2^2 + 3 \cdot 2 - 4 = 6$

(c) $f(h) = h^2 + 3h - 4$

(d) $f(2h) = (2h)^2 + 3(2h) - 4 = 4h^2 + 6h - 4$

(e) $f(2x) = (2x)^2 + 3(2x) - 4 = 4x^2 + 6x - 4$

(f) $f(x + h) = (x + h)^2 + 3(x + h) - 4$
$$= x^2 + 2hx + h^2 + 3x + 3h - 4$$
$$= x^2 + (2h + 3)x + (h^2 + 3h - 4)$$

(g) $f(x) + f(h) = (x^2 + 3x - 4) + (h^2 + 3h - 4)$
$$= x^2 + 3x + (h^2 + 3h - 8)$$

In Chapter 2 we need to simplify expressions of the form

$$\frac{f(x + h) - f(x)}{h} \qquad h \neq 0$$

as shown in the following example.

EXAMPLE 2 Find

$$\frac{f(x + h) - f(x)}{h}$$

where $h \neq 0$, if (a) $f(x) = 4x^2 - 5x + 7$; (b) $f(x) = \sqrt{x}$.

Solution

(a) $\dfrac{f(x + h) - f(x)}{h} = \dfrac{4(x + h)^2 - 5(x + h) + 7 - (4x^2 - 5x + 7)}{h}$

$$= \frac{4x^2 + 8hx + 4h^2 - 5x - 5h + 7 - 4x^2 + 5x - 7}{h}$$

$$= \frac{8hx - 5h + 4h^2}{h}$$

$$= 8x - 5 + 4h$$

(b) $\dfrac{f(x + h) - f(x)}{h} = \dfrac{\sqrt{x + h} - \sqrt{x}}{h}$

$$= \frac{(\sqrt{x + h} - \sqrt{x})(\sqrt{x + h} + \sqrt{x})}{h(\sqrt{x + h} + \sqrt{x})}$$

$$= \frac{(x + h) - x}{h(\sqrt{x + h} + \sqrt{x})}$$

$$= \frac{h}{h(\sqrt{x + h} + \sqrt{x})}$$

$$= \frac{1}{\sqrt{x + h} + \sqrt{x}}$$

In the second step of this solution, the numerator and denominator were multiplied by the conjugate of the numerator in order to rationalize the numerator, and this gave a common factor of h in the numerator and denominator.

In later discussions we refer to the *composite function* of two functions.

Definition of a composite function

Given the two functions f and g, the **composite function**, denoted by $f \circ g$, is defined by

$$(f \circ g)(x) = f(g(x))$$

and the domain of $f \circ g$ is the set of all numbers x in the domain of g such that $g(x)$ is in the domain of f.

EXAMPLE 3   Given that f is defined by $f(x) = \sqrt{x}$ and g is defined by $g(x) = 2x - 3$, (a) find $F(x)$ if $F = f \circ g$, and determine the domain of F; (b) find $G(x)$ if $G = g \circ f$, and determine the domain of G.

Solution

(a) $F(x) = (f \circ g)(x) = f(g(x))$
$$= f(2x - 3)$$
$$= \sqrt{2x - 3}$$

The domain of g is $(-\infty, +\infty)$, and the domain of f is $[0, +\infty)$. So the domain of F is the set of real numbers for which $2x - 3 \geq 0$ or, equivalently, $[\frac{3}{2}, +\infty)$.

(b) $G(x) = (g \circ f)(x) = g(f(x))$
$$= g(\sqrt{x})$$
$$= 2\sqrt{x} - 3$$

Because the domain of f is $[0, +\infty)$ and the domain of g is $(-\infty, +\infty)$, the domain of G is $[0, +\infty)$.

EXAMPLE 4   Given that f is defined by $f(x) = \sqrt{x}$ and g is defined by $g(x) = x^2 - 1$, find: (a) $f \circ f$; (b) $g \circ g$; (c) $f \circ g$; (d) $g \circ f$. Also find the domain of the composite function in each part.

Solution   The domain of f is $[0, +\infty)$, and the domain of g is $(-\infty, +\infty)$.

(a) $(f \circ f)(x) = f(f(x)) = f(\sqrt{x}) = \sqrt{\sqrt{x}} = \sqrt[4]{x}$

The domain of $f \circ f$ is $[0, +\infty)$.

(b) $(g \circ g)(x) = g(g(x)) = g(x^2 - 1) = (x^2 - 1)^2 - 1 = x^4 - 2x^2$

The domain of $g \circ g$ is $(-\infty, +\infty)$.

(c) $(f \circ g)(x) = f(g(x)) = f(x^2 - 1) = \sqrt{x^2 - 1}$

The domain of $f \circ g$ is $(-\infty, -1] \cup [1, +\infty)$ or, equivalently, all x not in $(-1, 1)$.

(d) $(g \circ f)(x) = g(f(x)) = g(\sqrt{x}) = (\sqrt{x})^2 - 1 = x - 1$

The domain of $g \circ f$ is $[0, +\infty)$. Note that even though $x - 1$ is defined for all values of x, the domain of $g \circ f$, by the definition of a composite function, is the set of all numbers x in the domain of f such that $f(x)$ is in the domain of g.

If the range of a function f consists of only one number, then f is called a **constant function**. So if $f(x) = c$, and if c is any real number, then f is a constant function and its graph is a straight line parallel to the x axis at a directed distance of c units from the x axis.

If a function f is defined by

$$f(x) = a_0 x^n + a_1 x^{n-1} + a_2 x^{n-2} + \ldots + a_{n-1} x + a_n$$

where n is a nonnegative integer, and $a_0, a_1, \ldots, a_n$ are real numbers ($a_0 \neq 0$), then f is called a **polynomial function** of degree n. Thus the function f defined by

$$f(x) = 3x^5 - x^2 + 7x - 1$$

is a polynomial function of degree 5.

If the degree of a polynomial function is 1, then the function is called a **linear function**; if the degree is 2, the function is called a **quadratic function**; and if the degree is 3, the function is called a **cubic function**.

● ILLUSTRATION 2

The function f defined by $f(x) = 3x + 4$ is a linear function.
The function g defined by $g(x) = 5x^2 - 8x + 1$ is a quadratic function.
The function h defined by $h(x) = 8x^3 - x + 4$ is a cubic function.   ●

If the degree of a polynomial function is zero, the function is a constant function.

The general linear function is defined by

$$f(x) = mx + b$$

where m and b are constants and $m \neq 0$. The graph of this function is a straight line having m as its slope and b as its y intercept. The particular linear function defined by

$$f(x) = x$$

is called the **identity function**. The general quadratic function is defined by

$$f(x) = ax^2 + bx + c$$

where a, b, and c are constants and $a \neq 0$.

If a function can be expressed as the quotient of two polynomial functions, the function is called a **rational function**. For example, the function f defined by

$$f(x) = \frac{x^3 - x^2 + 5}{x^2 - 9}$$

is a rational function for which the domain is the set of all real numbers except 3 and -3.

An **algebraic function** is a function formed by a finite number of algebraic operations on the identity function and the constant function. These algebraic operations include addition, subtraction, multiplication, division, raising to powers, and extracting roots. An example of an algebraic function is the function f defined by

$$f(x) = \frac{(x^2 - 3x + 1)^3}{\sqrt{x^4 + 1}}$$

In addition to algebraic functions, transcendental functions are considered in elementary calculus. Examples of transcendental functions are logarithmic and exponential functions, which are discussed in Chapter 8.

Applications involving the dependence of one variable on another occur in social, life, and physical sciences. The formulas used in such situations often determine functions. For instance, if y dollars is the simple interest for 1 year earned by a principal of x dollars at the rate of 12 percent per year, then

$$y = 0.12x \tag{1}$$

For a given nonnegative value of x there corresponds a unique value of y, and so the value of y depends on the value of x. If f is the function defined by $f(x) = 0.12x$ and the domain of f is the set of nonnegative real numbers, then (1) can be written as

$$y = f(x)$$

Equation (1) is an example of *direct proportion*, and y is said to be *directly proportional* to x.

Definition of directly proportional	A variable y is said to be **directly proportional** to a variable x if $$y = kx$$ where k is a nonzero constant. More generally, a variable y is said to be **directly proportional** to the nth power of x $(n > 0)$ if $$y = kx^n$$ The constant k is called the **constant of proportionality**.

EXAMPLE 5   A person's approximate brain weight is directly proportional to his or her body weight, and a person weighing 150 lb has an approximate brain weight of 4 lb. (a) Express the number of pounds in the approximate brain weight of a person as a function of the person's body weight. (b) Find the approximate brain weight of a person whose body weight is 176 lb.

Solution   (a) Let $f(x)$ lb be the approximate brain weight of a person having a body weight of x lb. Then

$$f(x) = kx \tag{2}$$

Because a person of body weight 150 lb has a brain weighing approximately 4 lb,

$$4 = k(150)$$

$$k = \tfrac{2}{75}$$

We replace k in (2) by this value and obtain

$$f(x) = \tfrac{2}{75}x \tag{3}$$

(b) From (3),

$$f(176) = \frac{2}{75}(176)$$

$$= 4.7$$

Therefore the approximate brain weight of a person weighing 176 lb is 4.7 lb.

Definition of inversely proportional

A variable y is said to be **inversely proportional** to a variable x if

$$y = \frac{k}{x}$$

where k is a nonzero constant. More generally, a variable y is said to be **inversely proportional** to the nth power of x ($n > 0$) if

$$y = \frac{k}{x^n}$$

EXAMPLE 6   The intensity of light from a given source is inversely proportional to the square of the distance from it. (a) Express the number of candlepower in the intensity of light as a function of the number of meters in the distance of the light from the source if the intensity is 225 candlepower (CP) at a distance of 5 meters (m) from the source. (b) Find the intensity at a point 12 m from the source.

Solution   (a) Let $f(x)$ candlepower be the intensity of light from a source that is x meters from it. Then

$$f(x) = \frac{k}{x^2} \tag{4}$$

Because the intensity is 225 CP at a distance of 5 m from the source,

$$225 = \frac{k}{5^2}$$

$$k = 25(225)$$

$$k = 5625$$

Substituting this value of k in (4) we obtain

$$f(x) = \frac{5625}{x^2} \tag{5}$$

(b) From (5),

$$f(12) = \frac{5625}{144}$$

$$= \frac{625}{16}$$

Therefore the intensity at a point 12 m from the source is $\frac{625}{16}$ CP.

Definition of jointly proportional

A variable z is said to be **jointly proportional** to variables x and y if

$$z = kxy$$

where k is a nonzero constant. More generally, a variable z is said to be **jointly proportional** to the nth power of x and the mth power of y ($n > 0$ and $m > 0$) if

$$z = kx^n y^m$$

EXAMPLE 7 In a limited environment where A is the maximum number of bacteria supportable by the environment, the rate of bacterial growth is jointly proportional to the number present and the difference between A and the number present. Suppose 1,000,000 bacteria is the maximum number supportable by the environment and the rate of growth is 60 bacteria per minute when there are 1000 bacteria present. (a) Express the rate of bacterial growth as a function of the number of bacteria present. (b) Find the rate of growth when there are 100,000 bacteria present.

Solution   (a) Let $f(x)$ bacteria per minute be the rate of growth when there are x bacteria present. Then

$$f(x) = kx(1,000,000 - x) \tag{6}$$

Because the rate of growth is 60 bacteria per minute when there are 1000 bacteria present,

$$60 = k(1000)(1,000,000 - 1000)$$

$$k = \frac{60}{999,000,000}$$

$$k = \frac{1}{16,650,000}$$

Replacing k in (6) by this value we have

$$f(x) = \frac{x(1,000,000 - x)}{16,650,000} \tag{7}$$

(b) From (7),

$$f(100,000) = \frac{100,000(1,000,000 - 100,000)}{16,650,000}$$

$$= \frac{100,000(900,000)}{16,650,000}$$

$$= 5405.4$$

Therefore the rate of growth is 5405.4 bacteria per minute when there are 100,000 bacteria present.

Exercises 1.4

1. Given $f(x) = 2x - 1$, find: (a) $f(3)$; (b) $f(-2)$; (c) $f(0)$; (d) $f(a + 1)$; (e) $f(x + 1)$; (f) $f(2x)$; (g) $2f(x)$; (h) $f(x + h)$; (i) $f(x) + f(h)$;
    (j) $\dfrac{f(x + h) - f(x)}{h}$, $h \neq 0$.

2. Given $f(x) = \dfrac{3}{x}$, find: (a) $f(1)$; (b) $f(-3)$; (c) $f(6)$; (d) $f(\tfrac{1}{3})$; (e) $f\left(\dfrac{3}{a}\right)$; (f) $f\left(\dfrac{3}{x}\right)$; (g) $\dfrac{f(3)}{f(x)}$; (h) $f(x - 3)$; (i) $f(x) - f(3)$;
    (j) $\dfrac{f(x + h) - f(x)}{h}$, $h \neq 0$.

3. Given $f(x) = 2x^2 + 5x - 3$, find: (a) $f(-2)$; (b) $f(-1)$; (c) $f(0)$; (d) $f(3)$; (e) $f(h + 1)$; (f) $f(2x^2)$; (g) $f(x^2 - 3)$; (h) $f(x + h)$;
    (i) $f(x) + f(h)$; (j) $\dfrac{f(x + h) - f(x)}{h}$, $h \neq 0$.

4. Given $g(x) = 3x^2 - 4$; find: (a) $g(-4)$; (b) $g(\tfrac{1}{2})$; (c) $g(x^2)$; (d) $g(3x^2 - 4)$; (e) $g(x - h)$; (f) $g(x) - g(h)$; (g) $\dfrac{g(x + h) - g(x)}{h}$, $h \neq 0$.

5. Given $g(x) = \dfrac{2}{x + 1}$, find: (a) $g(3)$; (b) $g(-3)$; (c) $g(x - 1)$; (d) $g(x) - g(1)$; (e) $g(x^2)$; (f) $[g(x)]^2$; (g) $\dfrac{g(x + h) - g(x)}{h}$, $h \neq 0$.

6. Given $f(x) = \dfrac{x + 1}{x - 1}$, find: (a) $f(-1)$; (b) $f(0)$; (c) $f(x - 1)$; (d) $f\left(\dfrac{1}{x}\right)$; (e) $\dfrac{1}{f(x)}$; (f) $\dfrac{f(-1)}{f(x)}$; (g) $\dfrac{f(x + h) - f(x)}{h}$, $h \neq 0$.

7. Given $F(x) = \sqrt{2x + 3}$, find: (a) $F(-1)$; (b) $F(4)$; (c) $F(\frac{1}{2})$; (d) $F(30)$; (e) $F(2x + 3)$; (f) $\dfrac{F(x + h) - F(x)}{h}$, $h \neq 0$.

8. Given $G(x) = \sqrt{2x^2 + 1}$, find: (a) $G(-2)$; (b) $G(0)$; (c) $G(\frac{1}{3})$; (d) $G(\frac{4}{7})$; (e) $G(2x^2 - 1)$; (f) $\dfrac{G(x + h) - G(x)}{h}$, $h \neq 0$.

In Exercises 9 through 16, define the following functions, and determine the domain of the resulting function: (a) $f \circ g$; (b) $g \circ f$.

9. $f(x) = x - 5$; $g(x) = x^2 - 1$

10. $f(x) = \sqrt{x}$; $g(x) = x^2 + 1$

11. $f(x) = \dfrac{x + 1}{x - 1}$; $g(x) = \dfrac{1}{x}$

12. $f(x) = \sqrt{x}$; $g(x) = 4 - x^2$

13. $f(x) = \sqrt{x}$; $g(x) = x^2 - 1$

14. $f(x) = \sqrt{x + 4}$; $g(x) = x^2 - 4$

15. $f(x) = \dfrac{1}{x + 1}$; $g(x) = \dfrac{x}{x - 2}$

16. $f(x) = x^2$; $g(x) = \dfrac{1}{\sqrt{x}}$

17. A company that makes electronic devices is placing a new product on the market. During the first year the fixed costs for the new production run are \$140,000 and the variable costs for producing each unit are \$25. During the first year the selling price is to be \$65 per unit. (a) If x units are sold during the first year, express the number of dollars in the first year's profit as a function of x. (b) It is estimated that 23,000 units will be sold during the first year. Use the result of part (a) to determine the first year's profit if this sales figure is reached. (c) How many units must be sold during the first year for the company to break even (that is, with no profit or loss)?

18. The monthly fixed costs for a company that manufactures skis are \$4200, and the variable costs are \$55 per pair of skis. The selling price per pair of skis is \$105. (a) If x pairs of skis are sold during a month, express the number of dollars in the monthly profit as a function of x. (b) Use the result of part (a) to determine the December profit if 600 pairs of skis are sold during that month. (c) How many pairs of skis must be sold during a month for the company to break even (no profit or loss) for the month?

19. (a) If x degrees is the Fahrenheit temperature, use the information in Exercise 38 of Exercises 1.2 to express the number of degrees in the Celsius temperature as a function of x. (b) Find the Celsius temperature when the Fahrenheit temperature is 95°.

20. The approximate weight of a person's muscles is directly proportional to his or her body weight. (a) Express the number of pounds in the approximate muscle weight of a person as a function of the person's body weight if a person weighing 150 lb has muscles weighing approximately 60 lb. (b) Find the approximate muscle weight of a person weighing 130 lb.

21. The daily payroll for a work crew is directly proportional to the number of workers, and a crew of 12 workers earns a payroll of \$540. (a) Express the number of dollars in the daily payroll as a function of the number of workers. (b) What is the daily payroll for a crew of 15 workers?

22. The volume of a gas having a constant pressure is directly proportional to the absolute temperature of the gas, and at a temperature of 180° the gas occupies 100 m³. (a) Express the number of cubic meters in the volume of the gas as a function of the number of degrees in the absolute temperature. (b) What is the volume of the gas at a temperature of 150°?

23. The period of a pendulum (the time for one complete oscillation) is directly proportional to the square root of the length of the pendulum, and a pendulum of length 8 ft has a period of 3 sec. (a) Express the number of seconds in the period of a pendulum as a function of the number of feet in its length. (a) Find the period of a pendulum of length 2 ft.

24. For a vibrating string, the rate of vibrations is directly proportional to the square root of the tension on the string. (a) If a particular string vibrates 864 times per second under a tension of 24 kg, express the number of vibrations per second as a function of the number of kilograms in the tension. (b) Find the number of vibrations per second under a tension of 6 kg.

25. The weight of a body is inversely proportional to the square of its distance from the center of the earth. (a) If a body weighs 200 lb on the earth's surface, express the number of pounds in its weight as a function of the number of miles from the center of the earth. Assume that the radius of the earth is 4000 mi. (b) How much does the body weigh at a distance of 400 mi above the earth's surface?

26. For an electrical cable of fixed length, the resistance is inversely proportional to the square of the diameter of the cable. (a) If a cable having the fixed length is $\frac{1}{2}$ cm in diameter and has a resistance of 0.1 ohm, express the number of ohms in the resistance as a function of the number of centimeters in the diameter. (b) What is the resistance of a cable having the fixed length and a diameter of $\frac{2}{3}$ cm?

27. In a small town of population 5000, the rate of growth of an epidemic (the rate of change of the number of infected persons) is jointly proportional to the number of people infected and the number of people not infected. (a) If the epidemic is growing at the rate of 9 people per day when 100 people are infected, express the rate of growth of the epidemic as a function of the number of infected people. (b) How fast is the epidemic growing when 200 people are infected?

28. In a community of 8000 people, the rate at which a rumor spreads is jointly proportional to the number of people who have heard the rumor and the number of people who have not heard it. (a) If the rumor is spreading at the rate of 20 people per hour when 200 people have heard it, express the rate at which the rumor is spreading as a function of the number of people who have heard it. (b) How fast is the rumor spreading when 500 people have heard it?

1.5 Functions as mathematical models

In applications of calculus we shall be dealing with functions, and it will be necessary to express a practical situation in terms of a functional relationship. The function obtained gives a *mathematical model* of the situation. We now give examples showing the procedure involved in obtaining some mathematical models.

EXAMPLE 1 A clock manufacturer can produce a particular clock at a cost of $15 per clock. It is estimated that if the selling price of the clock is x dollars each, then the number of clocks sold per week is $125 - x$. (a) Express the number of dollars in the manufacturer's weekly profit as a function of x. (b) Use the result of part (a) to determine the weekly profit if the selling price is $45 per clock.

Solution (a) The profit can be obtained by subtracting the total cost from the total revenue. Let R dollars be the weekly revenue. Because the revenue is the product of the cost of each clock and the number of clocks sold,

$$R = x(125 - x) \tag{1}$$

Let C dollars be the total cost of the clocks sold each week. Because the total cost is the product of the cost of each clock and the number of clocks sold,

$$C = 15(125 - x) \tag{2}$$

If $P(x)$ dollars is the weekly profit, then

$$P(x) = R - C \tag{3}$$

Substituting from (1) and (2) into (3) we obtain

$$P(x) = x(125 - x) - 15(125 - x)$$
$$= (125 - x)(x - 15) \tag{4}$$

(b) If the selling price is $45, the number of dollars in the weekly profit is $P(45)$. From (4),

$$P(45) = (125 - 45)(45 - 15)$$
$$= 80 \cdot 30$$
$$= 2400$$

Therefore the weekly profit is $2400 when the clocks are sold at $45 each.

The function of Example 1 is discussed again in Sec. 2.4, where we determine the selling price of the clock in order for the manufacturer's weekly profit to be a maximum.

EXAMPLE 2   A cardboard box manufacturer wishes to make open boxes from square pieces of cardboard of side 12 in. by cutting equal squares from the four corners and turning up the sides. (a) If x in. is the length of the side of the square to be cut out, express the number of cubic inches in the volume of the box as a function of x. (b) What is the domain of the resulting function?

Solution   (a) Figure 1.5.1 represents a given piece of cardboard, and Fig. 1.5.2 represents the box obtained from the cardboard. The number of inches in the dimensions of the box are then x, $(12 - 2x)$, and $(12 - 2x)$. The volume of the box is the product of the three dimensions. Therefore, if $V(x)$ cubic inches is the volume of the box,

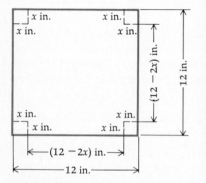

$$V(x) = x(12 - 2x)(12 - 2x)$$
$$= 144x - 48x^2 + 4x^3 \tag{5}$$

(b) From (5) we note that $V(0) = 0$ and $V(6) = 0$. From the conditions of the problem we see that x cannot be negative and x cannot be greater than 6. Thus the domain of V is the closed interval $[0, 6]$.

Figure 1.5.1

In Sec. 4.2 we return to the function of Example 2 and learn a method for determining the value of x that will give a box having the largest possible volume.

In the next two examples we first obtain two equations involving a dependent variable and two independent variables. We then express the dependent variable as a function of a single independent variable by eliminating the other independent variable from the pair of simultaneous equations.

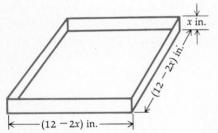

Figure 1.5.2

EXAMPLE 3   A rectangular field is to be fenced off along the bank of a river, and no fence is required along the river. The material for the fence costs \$8 per running foot for the two ends and \$12 per running foot for the side parallel to the river; \$3600 worth of fence is to be used. (a) If x ft is the length of an end, express the number of square feet in the area of the field as a function of x. (b) What is the domain of the resulting function?

Solution   (a) From the statement of the problem, x is the number of feet in the length of an end of the field. Let y be the number of feet in the length of the side parallel to the river and A ft^2 be the area of the field. See Fig. 1.5.3. Then

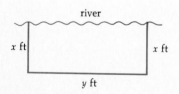

$$A = xy \tag{6}$$

Figure 1.5.3

Because the cost of the material for each end is \$8 per running foot and the length of an end is x ft, the total cost of the fence for each end is $8x$ dollars. Similarly, the total cost of the fence for the third side is $12y$ dollars.

We have, then,

$$8x + 8x + 12y = 3600 \qquad (7)$$

To express A in terms of a single variable we first solve (7) for y in terms of x. We have

$$12y = 3600 - 16x$$

$$y = 300 - \tfrac{4}{3}x$$

We substitute this value of y into (6), yielding A as a function of x, and

$$A(x) = x(300 - \tfrac{4}{3}x)$$

(b) Both x and y must be nonnegative. The smallest value that x can attain is 0. The smallest value that y can attain is 0, and when $y = 0$, we obtain, from (7), $x = 225$. Thus 225 is the largest value that x can assume. Hence x must be in the closed interval $[0, 225]$, and this closed interval is the domain of A.

In Sec. 4.2 we return to the function of Example 3 and learn how to determine the dimensions of the field of largest possible area that can be enclosed by the $3600 worth of fence.

EXAMPLE 4 A closed box with a square base is to have a volume of 2000 in.^{3}. The material for the top and bottom of the box is to cost $3 per square inch, and the material for the sides is to cost $1.50 per square inch. (a) If x in. is the length of a side of the square base, express the number of dollars in the cost of the material as a function of x. (b) What is the domain of the resulting function?

Solution (a) From the statement of the problem, x in. is the length of a side of the square base. Let y in. be the depth of the box and C dollars be the cost of the material. See Fig. 1.5.4. The total number of square inches in the combined area of the top and bottom is $2x^2$, and for the sides it is $4xy$; so

$$C = 3(2x^2) + \tfrac{3}{2}(4xy) \qquad (8)$$

Because the volume of the box is the product of the area of the base and the depth,

$$x^2y = 2000 \qquad (9)$$

Solving (9) for y in terms of x and substituting into (8) we obtain C as a function of x:

$$C(x) = 6x^2 + \frac{12{,}000}{x}$$

(b) Observe that x cannot be zero, because x is the denominator of the second term in the right member of the equation defining $C(x)$. However, x can be any positive number. Therefore the domain of C is the interval $(0, +\infty)$.

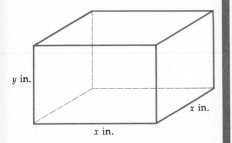

y in.

x in.

x in.

Figure 1.5.4

We return to the function of Example 4 in Sec. 4.5, where we determine the dimensions of the box so that the cost of the material is least.

EXAMPLE 5   In the planning of a coffee shop it is estimated that if there are places for 40 to 80 people, the daily profit will be $8 per place. However, if the seating capacity is above 80 places, the daily profit on each place will be decreased by 4 cents times the number of places over 80. If x is the number of places in the seating capacity, express the number of dollars in the daily profit as a function of x. Assume that the profit is to be nonnegative.

Solution   From the statement of the problem, x is the number of places in the seating capacity. Let $P(x)$ be the number of dollars in the daily profit.

We obtain $P(x)$ by multiplying x by the number of dollars in the profit per place. When $40 \leq x \leq 80$, $8 is the profit per place, and so $P(x) = 8x$. However, when $x > 80$, the number of dollars in the profit per place is $[8 - 0.04(x - 80)]$, thus giving

$$P(x) = x[8 - 0.04(x - 80)] = 11.20x - 0.04x^2$$

So we have

$$P(x) = \begin{cases} 8x & \text{if } 40 \leq x \leq 80 \\ 11.20x - 0.04x^2 & \text{if } 80 < x \leq 280 \end{cases}$$

The upper bound of 280 for x is obtained by noting that $11.20x - 0.04x^2 = 0$ when $x = 280$; and when $x > 280$, $11.20x - 0.04x^2$ is negative.

By definition, x is an integer; therefore x is any integer in the closed interval $[40, 280]$.

The function in Example 5 is discussed in Sec. 4.2, where we find the seating capacity necessary to yield the greatest daily profit.

Exercises 1.5

The functions obtained in these exercises will be used as mathematical models for the applications of extreme values in Secs. 2.4, 4.2, and 4.5.

1. A carpenter can construct bookcases at a cost of $40 each. If the carpenter sells the bookcases for x dollars each, it is estimated that $300 - 2x$ bookcases will be sold per month. (a) Express the number of dollars in the carpenter's monthly profit as a function of x. (b) Use the result of part (a) to determine the monthly profit if the selling price is $110 per bookcase.
2. A toy manufacturer can produce a particular toy at a cost of $10 per toy. It is estimated that if the selling price of the toy is x dollars, then the number of toys sold each day is $45 - x$. (a) Express the number of dollars in the manufacturer's daily profit as a function of x. (b) Use the result of part (a) to determine the daily profit if the selling price is $30 per toy.
3. A manufacturer of open tin boxes wishes to make use of pieces of tin with dimensions 8 in. by 15 in. by cutting equal squares from the four corners and turning up the sides. (a) If x in. is the length of the side of the square cut out, express the number of cubic inches in the volume of the box as a function of x. (b) What is the domain of the resulting function?
4. Suppose the manufacturer of Exercise 3 makes the open boxes from square pieces of tin that measure k cm on a side. (a) If x cm is the length of the side of the square cut out, express the number of cubic centimeters in the volume of the box as a function of x. Remember that k is a constant. (b) What is the domain of the resulting function?
5. A rectangular field is to be enclosed with 240 m of fence. (a) If x meters is the length of the field, express the number of square meters in the area of the field as a function of x. (b) What is the domain of the resulting function?

6. A rectangular garden is to be fenced off with 100 ft of fencing material. (a) If x ft is the length of the garden, express the number of square feet in the area of the garden as a function of x. (b) What is the domain of the resulting function?

7. Do Exercise 5 if one side of the field is to have a river as a natural boundary and the fencing material is to be used for the other three sides. Let x meters be the length of the side of the field that is parallel to the river.

8. Do Exercise 6 if the garden is to be placed so that a side of a house serves as a boundary and the fencing material is to be used for the other three sides. Let x feet be the length of the side of the garden that is parallel to the house.

9. A rectangular plot of ground is to be enclosed by a fence and then divided down the middle by another fence. The fence down the middle costs \$2 per running foot and the other fence costs \$5 per running foot, and \$960 worth of fencing material is to be used. (a) If x ft is the length of the fence down the middle, express the number of square feet in the area of the plot as a function of x. (b) What is the domain of the resulting function?

10. A package in the shape of a rectangular box with a square cross section is to have the sum of its length and girth (the perimeter of a cross section) equal to 100 in. (a) If x in. is the length of the package, express the volume of the box as a function of x. (b) What is the domain of the resulting function?

11. A rectangular field having an area of 2700 m^2 is to be enclosed by a fence, and an additional fence is to be used to divide the field down the middle. The cost of the fence down the middle is \$12 per running meter, and the fence along the sides costs \$18 per running meter. (a) If x meters is the length of the fence down the middle, express the number of dollars in the total cost of the fence as a function of x. (b) What is the domain of the resulting function?

12. A rectangular open tank is to have a square base, and its volume is to be 125 m^3. The cost per square meter for the bottom is \$24 and for the sides is \$12. (a) If x meters is the length of a side of the base, express the number of dollars in the total cost of the material as a function of x. (b) What is the domain of the resulting function?

13. A box manufacturer is to produce a closed box of volume 288 in.^{3}, where the base is a rectangle having a length that is three times its width. (a) If x in. is the width of the base, express the number of square inches in the total surface area of the box as a function of x. (b) What is the domain of the resulting function?

14. Do Exercise 13 if the box is to have an open top.

15. A page of print is to contain 24 in.^{2} of printed material, a margin of $1\frac{1}{2}$ in. at the top and bottom, and a margin of 1 in. at the sides. (a) If x in. is the length of the top and bottom of the printed material, express the number of square inches in the total area of the page as a function of x. (b) What is the domain of the resulting function?

16. A one-story building having a rectangular floor space of 13,200 ft^2 is to be constructed where a walkway 22 ft wide is required in the front and back and a walkway 15 ft wide is required on each side. (a) If x ft is the length of the front and back of the building, express as a function of x the number of square feet in the area of the lot on which the building and the walkways will be located. (b) What is the domain of the resulting function?

17. Orange trees grown in California produce 600 oranges per year if not more than 20 trees are planted per acre. For each additional tree planted per acre, the yield decreases by 15 oranges. (a) If x trees are planted per acre, express as a function of x the number of oranges produced per year. (b) What is the domain of the resulting function?

18. A manufacturer can make a profit of \$20 on each item if not more than 800 items are produced each week. The profit on each item decreases 2 cents for every item over 800. (a) If x items are produced each week, express the number of dollars in the manufacturer's weekly profit as a function of x. Assume that the profit is nonnegative. (b) What is the domain of the resulting function?

19. A private club charges each member annual membership dues of \$100 less 50 cents for each member over 600 and plus 50 cents for each member under 600. (a) If the club has x members, express the number of dollars in the revenue from annual dues as a function of x. (b) What is the domain of the resulting function?

20. A charity theatrical performance will cost each person \$15 if not more than 150 tickets are sold. However, the cost per ticket will be reduced 7 cents for each ticket in excess of 150. (a) If x tickets are sold, express the number of dollars in the charity's receipts as a function of x. (b) What is the domain of the resulting function?

21. The maximum number of bacteria supportable by a particular environment is 900,000, and the rate of bacterial growth is jointly proportional to the number present and the difference between 900,000 and the number present. (a) If $f(x)$ bacteria per minute is the rate of growth when x bacteria are present, write an equation defining $f(x)$. (b) What is the domain of the function f in part (a)?

22. A particular lake can support a maximum of 14,000 fish, and the rate of growth of the fish population is jointly proportional to the number of fish present and the difference between 14,000 and the number present. (a) If $f(x)$ fish per day is the rate of growth when x fish are present, write an equation defining $f(x)$. (b) What is the domain of the function f in part (a)?

1.6 Demand and supply equations

Consider circumstances affecting a manufacturer in which the only variables are the price and quantity of the commodity demanded. Let p dollars be the price of one unit of the commodity, and let x be the number of units of the commodity demanded.

Upon reflection, it should seem reasonable that the amount of the commodity demanded in the marketplace by consumers will depend on the price of the commodity. As the price falls, consumers generally demand more of the commodity. Should the price rise, the opposite will occur: Consumers will demand less.

An equation giving the relationship between the amount, given by x, of a commodity demanded and the price, given by p, is called a **demand equation**. Such an equation is arrived at by the application of statistical methods to economic data, and it may be written in one of the following two forms:

$$p = f(x) \tag{1}$$

$$x = g(p) \tag{2}$$

The function f in (1) is called the **price function**, and $f(x)$ dollars is the price of one unit of the commodity when x units are demanded. The function g in (2) is called the **demand function**, and $g(p)$ is the number of units of the commodity that will be demanded if p dollars is the price per unit. In normal economic situations the domains of the price and demand functions consist, as you would expect, of nonnegative numbers.

The graph of the demand equation is called the **demand curve**. When drawing a sketch of the demand curve, it is customary in economics to use the vertical axis to represent the price and the horizontal axis to represent the demand. Because the given demand equation may apply for only certain values of x and p, it is often necessary to restrict them to closed intervals; that is, $x \in [0, a]$ and $p \in [0, b]$. Even though in actual practice quantities and prices usually take on rational values, we allow x and p to be any real numbers within these closed intervals.

● ILLUSTRATION 1

Consider the following demand equation:

$$p^2 + 2x - 16 = 0 \tag{3}$$

Because in normal economic situations the variables x and p will be nonnegative, when (3) is solved for p we reject the negative values of p and get

$$p = \sqrt{16 - 2x} \tag{4}$$

which is of the form of (1). Hence the price function for the demand equation (3) is the function f for which $f(x) = \sqrt{16 - 2x}$. Solving (3) for x we get

$$x = 8 - \tfrac{1}{2}p^2 \tag{5}$$

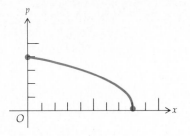

Figure 1.6.1

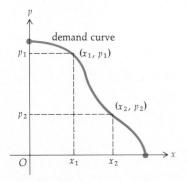

Figure 1.6.2

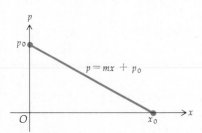

Figure 1.6.3

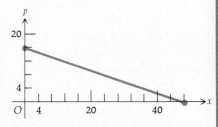

Figure 1.6.4

which expresses x as a function of p as in (2), and so the demand function is the function g for which $g(p) = 8 - \frac{1}{2}p^2$. A sketch of the demand curve is shown in Fig. 1.6.1. The graph is restricted to the first quadrant because of the requirement that x and p be nonnegative. From (4) we see that $p \leq 4$ and $16 - 2x \geq 0$ or, equivalently, $x \leq 8$. Therefore $x \in [0, 8]$ and $p \in [0, 4]$. ●

In addition to the restriction that x and p be nonnegative under normal circumstances, we impose the condition that as the price per unit decreases, the demand for a commodity increases, and as the price per unit increases, the demand for a commodity decreases; that is, if p_1 is the number of dollars in the price of x_1 units of a commodity and p_2 is the number of dollars in the price per unit of x_2 units, then $x_2 > x_1$ if and only if $p_2 < p_1$. This condition, in turn, reflects nothing more than "common economic sense" and is illustrated in Fig. 1.6.2.

The simplest demand equation is linear, and it can be written in the form

$$p = mx + p_0 \tag{6}$$

where $m < 0$. The graph of this equation is the segment in the first quadrant of the straight line having slope m and p_0 as intercept on the p axis. Refer to Fig. 1.6.3. Observe that p_0 is the number of dollars in the highest unit price anyone would pay according to the demand equation (6). If (6) is solved for x, we obtain an equation in the form

$$x = kp + x_0$$

where $k < 0$. Because $x = x_0$ when $p = 0$, x_0 is the number of units in the quantity demanded when the commodity is "free." As $k < 0$ implies, the amount demanded decreases as the price increases from zero and the commodity loses its free status.

EXAMPLE 1 A tour company learned that when the price of a sightseeing trip is \$6, the average number of tickets sold per trip is 30, and when the price is \$10, the average number of tickets sold is only 18. If the demand equation is linear, find it and draw a sketch of the demand curve.

Solution Let x equal the number of tickets demanded, and let p equal the number of dollars in the price per ticket. Because $x = 30$ when $p = 6$, and $x = 18$ when $p = 10$, the points $(30, 6)$ and $(18, 10)$ will lie on the line segment that is the graph of the required equation. Using the point-slope form of an equation of a line we have

$$p - 6 = \frac{10 - 6}{18 - 30}(x - 30)$$

$$x + 3p = 48$$

Because $x \geq 0$ and $p \geq 0$, the demand curve is restricted to the first quadrant. A sketch of the demand curve is shown in Fig. 1.6.4.

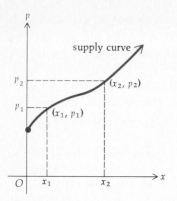

Figure 1.6.5

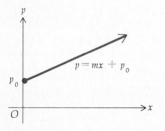

Figure 1.6.6

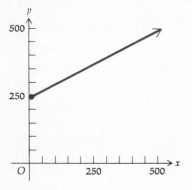

Figure 1.6.7

Suppose now that x is the number of units of a certain commodity to be supplied by a producer and, as above, p dollars is the price of one unit of the commodity. Assume that these are the only two variables. A **supply equation** is an equation involving these two variables. In a normal economic situation, x and p are nonnegative, and $x_2 > x_1$ if and only if $p_2 > p_1$; that is, as the price of the commodity received by the producer increases, the producer will of course increase the supply to take advantage of the higher price. By the same token, there will be a tendency to decrease the amount produced as the price received decreases. The trivial case, when the output is constant regardless of price, is an exception to this statement. The graph of the supply equation is called a **supply curve,** and Fig. 1.6.5 shows a sketch of a supply curve in normal circumstances. When $x = 0$, $p = p_0$, and this is the number of dollars in the unit price at which none of the commodity will be available for market. When the unit price is large, the producer supplies a large quantity of the commodity for market.

The simplest supply equation is linear and can be written in the form

$$p = mx + p_0$$

where $m > 0$. Figure 1.6.6 shows a sketch of the graph of such an equation. The graph is the portion in the first quadrant of the line having p intercept p_0 and slope m. Nothing is produced until $p > p_0$.

EXAMPLE 2 Unless the price of a desk of a specific type is more than $250, no desk is available for market. However, when the price is $350, 200 desks are available for market. Find the supply equation, if it is linear, and draw a sketch of the supply curve.

Solution Let x be the number of desks supplied when p dollars is the price per desk. When $p = 250$, $x = 0$, and when $p = 350$, $x = 200$. So the points $(0, 250)$ and $(200, 350)$ are on the supply curve. Using the point-slope form of the equation of a line we have

$$p - 250 = \frac{350 - 250}{200 - 0}(x - 0)$$

$$p - 250 = \tfrac{1}{2}x$$

$$p = \tfrac{1}{2}x + 250$$

A sketch of the supply curve is shown in Fig. 1.6.7.

We shall call the totality of all companies producing the same commodity an **industry**. The **market** for a particular commodity is made up of the industry and the consumers of the commodity (which may include business and government consumers as well as individual consumers). The market's supply equation is determined from the supply equations of all the companies in the industry, and the market's demand equation is determined from the demand equations of all the consumers. We shall now show how to determine the *equilibrium price* and *equilibrium amount* of a market commodity.

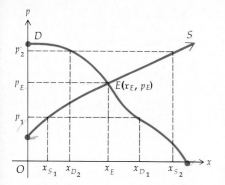

Figure 1.6.8

Market equilibrium is said to occur when the quantity of the commodity demanded, at a given price, is equal to the quantity of the commodity supplied at that price. That is, market equilibrium occurs when everything offered for sale at the price in question is purchased. When market equilibrium occurs, the quantity of the commodity produced is called the **equilibrium amount**, and the price of the commodity is called the **equilibrium price**. The equilibrium amount and the equilibrium price are found by solving simultaneously the market's demand and supply equations. Refer to Fig. 1.6.8, showing sketches of a market's demand and supply curves, labeled D and S, respectively. The point E is the **point of equilibrium**, and its coordinates are x_E and p_E, where x_E units is the equilibrium amount and p_E dollars is the equilibrium price. Still referring to Fig. 1.6.8, suppose the price of the commodity were p_1 dollars; then the industry would plan to sell x_{S_1} units and the consumers would plan to purchase x_{D_1} units, and hence the consumers would suffer a shortage of $(x_{D_1} - x_{S_1})$ units. This would force the price to rise to p_E dollars, and the quantity supplied would increase to x_E units. However, if the price were p_2 dollars, then the consumers would plan to purchase only x_{D_2} units and the industry would plan to sell x_{S_2} units. Consequently, the industry would be left with $(x_{S_2} - x_{D_2})$ units unsold, and so the price would be forced down to p_E dollars, and the quantity supplied would be reduced to x_E units.

EXAMPLE 3 The market's demand and supply equations are, respectively,

$$x^2 + p^2 - 25 = 0$$

and

$$2x - p + 2 = 0$$

where p dollars is the price and $100x$ units is the quantity. Determine the equilibrium amount and the equilibrium price. Draw sketches of the demand and supply curves on the same set of axes, and show the point of equilibrium.

Solution To find the point of equilibrium we solve the two equations simultaneously by solving the second equation for p and substituting into the first equation. We have

$$x^2 + (2x + 2)^2 - 25 = 0$$
$$x^2 + 4x^2 + 8x + 4 - 25 = 0$$
$$5x^2 + 8x - 21 = 0$$
$$(5x - 7)(x + 3) = 0$$
$$5x - 7 = 0 \qquad x + 3 = 0$$
$$x = 1.4 \qquad x = -3$$

Because $x \geq 0$, the negative value is rejected, and we obtain $x = 1.4$. Substituting into the second equation we obtain $p = 4.8$. Hence the equilibrium price is \$4.80 and the equilibrium amount is 140 units (recall that the quantity

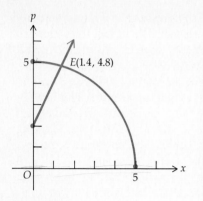

Figure 1.6.9

is 100x units). Writing the first equation in the form

$$x^2 + p^2 = 25$$

we see that its graph is a circle with its center at the origin and a radius of 5. Because $x \geq 0$ and $p \geq 0$, the demand curve is the portion of the circle in the first quadrant. Solving the supply equation for p we have

$$p = 2x + 2$$

Thus the supply curve is the portion in the first quadrant of the line having slope 2 and intercept on the p axis of 2. The required sketch is shown in Fig. 1.6.9.

It should be noted that if the demand curve and the supply curve do not intersect in the first quadrant, we say that equilibrium is not meaningful. For instance, if the curves intersect in the second quadrant, it is implied that the equilibrium amount is negative, and it is obviously meaningless to speak of producing a negative quantity.

Exercises 1.6

In Exercises 1 through 10, a linear equation is given. Draw the segment of the straight line in the first quadrant. Determine if this line segment is a demand curve, a supply curve, or neither.

1. $2x - 3p + 6 = 0$ **2.** $5p + 4x - 10 = 0$ **3.** $4p + x = 7$ **4.** $3x - 4p + 24 = 0$
5. $3x + 5p + 12 = 0$ **6.** $3p = 2$ **7.** $4p - 5 = 0$ **8.** $4x - 3p = 0$
9. $5p - 6x = 0$ **10.** $2x + 6p + 3 = 0$

In Exercises 11 through 14, the demand equation for a particular commodity is given. In each exercise do the following: (a) Draw a sketch of the demand curve, (b) find the highest price anyone would pay for the commodity, and (c) find the demand if the commodity is free.

11. $3x + 2p - 15 = 0$ **12.** $x^2 + p^2 = 36$ **13.** $p^2 + 4p + 2x - 10 = 0$ **14.** $x^2 + 2x + 3p - 23 = 0$

In Exercises 15 through 18, the supply equation for a particular commodity is given. In each exercise do the following: (a) Draw a sketch of the supply curve, and (b) find the lowest price at which the commodity would be supplied.

15. $x^2 - 4p + 12 = 0$ **16.** $2x - 6p + 9 = 0$
17. $p^2 + 8p - 6x - 20 = 0$ **18.** $2x^2 + 12x - 3p + 24 = 0$

In Exercises 19 through 26, a market's demand and supply equations are given. (a) Determine the equilibrium amount and the equilibrium price; (b) draw sketches of the demand and supply curves on the same set of axes, and show the point of equilibrium.

19. $x + 2p - 15 = 0; x - 3p + 3 = 0$ **20.** $3x + p - 21 = 0; 3x - 4p + 9 = 0$
21. $x^2 + p - 9 = 0; x - p + 3 = 0$ **22.** $3x^2 - 6x + p - 8 = 0; x^2 - p + 4 = 0$
23. $3x^2 + p - 10 = 0; x^2 + 2x - p + 4 = 0$ **24.** $p^2 + p + x - 12 = 0; 2p^2 - 2p - x - 4 = 0$
25. $x^2 + p^2 = 25; x^2 - 8p + 8 = 0$ **26.** $x^2 + p^2 = 100; x^2 - 6p + 12 = 0$

27. A company sells 20,000 units of a commodity when the unit price is \$14, and the company has determined that it can sell 2000 more units with each \$2 reduction in the unit price. Find the demand equation, if it is linear, and draw a sketch of the demand curve.

28. A company that sells office equipment can sell 1000 filing cabinets when the price is \$600. Furthermore, it is determined that with each \$30 reduction in price the company can sell 150 more cabinets. If the demand equation is linear, find it, and draw a sketch of the demand curve.

29. When the price is \$80, there are 10,000 lamps of a certain type available for market. For each \$10 increase in price, 8000 more lamps are available for market. If the supply equation is linear, find it, and draw a sketch of the supply curve.

30. A producer supplies 500 units of a commodity when the price is \$20 per unit. For each \$1 increase in price, 60 more units are supplied. Find the supply equation, if it is linear, and draw a sketch of the supply curve.

Review Exercises for Chapter 1

In Exercises 1 through 10, draw a sketch of the graph of the equation.

1. $y = 3x + 4$
2. $y = 5 - 2x$
3. $y = x^2 - 2$
4. $y = |x - 4|$

5. $y^2 = x - 4$
6. $y = \sqrt{x - 4}$
7. $y = |x + 5|$
8. $9x^2 - 25y^2 = 0$

9. $x^2 + y^2 = 16$
10. $x^2 + 4y^2 = 16$

11. Draw a sketch of the graph of each of the equations: (a) $y = 2\sqrt{x}$; (b) $y = -2\sqrt{x}$; (c) $y^2 = 4x$.

12. Draw a sketch of the graph of each of the equations: (a) $y = \sqrt{9 - x^2}$; (b) $y = -\sqrt{9 - x^2}$; (c) $x^2 + y^2 = 9$.

13. (a) Find the slope of the line through the points $(4, -5)$ and $(-2, 7)$. (b) Write an equation of the line in part (a).

14. Find an equation of the line through the points $(2, -4)$ and $(7, 3)$, and write the equation in slope-intercept form.

15. Find the slope of the line $3x - 4y = 12$.

16. Show that the lines $2x - 3y + 8 = 0$ and $4x - 6y + 5 = 0$ are parallel.

17. Find an equation of the line through the point $(5, -3)$ and perpendicular to the line $2x - 5y = 10$.

18. Determine by means of slopes if the given points are on a line: (a) $(3, 4), (-1, -2), (5, 7)$; (b) $(-6, -1), (1, 5), (5, 8)$.

In Exercises 19 through 22, the function is the set of ordered pairs (x, y) satisfying the given equation. Determine the domain and range of the function, and draw a sketch of its graph.

19. $f: y = \sqrt{2x + 5}$

20. $g: y = \dfrac{x^2 - 16}{x + 4}$

21. $h: y = \dfrac{x^2 + x - 6}{x - 2}$

22. $F: y = \begin{cases} x^2 - 1 & \text{if } x < 1 \\ x - 1 & \text{if } 1 \le x \end{cases}$

23. Given $f(x) = 3x^2 - x + 5$, find: (a) $f(1)$; (b) $f(-3)$; (c) $f(-x^2)$; (d) $-[f(x)]^2$ (e) $\dfrac{f(x + h) - f(x)}{h}$, $h \neq 0$.

24. Given $g(x) = \sqrt{x + 3}$, find: (a) $g(-3)$; (b) $g(1)$; (c) $g(x^2)$; (d) $[g(x)]^2$; (e) $\dfrac{g(x + h) - g(x)}{h}$, $h \neq 0$.

In Exercises 25 through 27, define the following functions, and determine the domain of the resulting function: (a) $f \circ g$; (b) $g \circ f$.

25. $f(x) = x^2 - 4$; $g(x) = 4x - 3$

26. $f(x) = \sqrt{x + 2}$; $g(x) = x^2 + 4$

27. $f(x) = x^2 - 9$; $g(x) = \sqrt{x + 5}$

28. The demand equation for a particular commodity is $p^2 + 2p + 2x - 24 = 0$. (a) Draw a sketch of the demand curve, (b) find the highest price anyone would pay for the commodity, and (c) find the demand if the commodity is free.

29. The supply equation for a particular commodity is $x^2 + 4x - 4p + 20 = 0$. (a) Draw a sketch of the supply curve, and (b) find the lowest price at which the commodity would be supplied.

In Exercises 30 and 31, a market's demand and supply equations are given. (a) Determine the equilibrium amount and the equilibrium price; (b) draw sketches of the demand and supply curves on the same set of axes, and show the point of equilibrium.

30. $2x + p - 12 = 0$; $x^2 - p + 4 = 0$

31. $x^2 + p^2 = 169$; $p - 2x = 2$

32. Some equipment was purchased for $20,000, and the scrap value after 10 years is expected to be $1500. If the straight-line method is used to depreciate the equipment from $20,000 to $1500 over 10 years, what is the book value of the equipment after 5 years?

33. The construction costs of an apartment house were $1,500,000, and this amount was depreciated from 1975 by the straight-line method over 15 years. What was the book value of the building in 1983?

34. A manufacturer's cost is $12 per unit, and the total cost of producing 400 units in 1 day is $5400. (a) Determine the fixed daily overhead. (b) If x units are produced per day and y dollars is the total daily cost, write an equation involving x and y. (c) Draw a sketch of the graph of the equation in part (b).

35. The distance a body falls from rest is directly proportional to the square of the time it has been falling, and a body falls 64 ft in 2 sec. (a) Express the number of feet in the distance a body falls from rest as a function of the number of seconds in the time it has been falling. (b) How far will a body fall from rest in $2\frac{1}{2}$ sec?

36. Boyle's law states that at a constant temperature the volume of a gas is inversely proportional to the pressure of the gas, and a gas occupies 100 m^3 at a pressure of 24 kg per square centimeter. (a) Express the number of cubic meters occupied by a gas as a function of the number of kilograms per square centimeter in its pressure. (b) What is the volume of a gas when its pressure is 16 kg per square centimeter?

37. Square pieces of metal of side 20 in. are used to construct open boxes by cutting equal squares from the four corners and turning up the sides. (a) If x in. is the length of the side of the square cut out, express the number of cubic inches in the volume of the box as a function of x. (b) What is the domain of the resulting function?

38. A wholesaler offers to deliver to a dealer 300 chairs at $90 per chair and to reduce the price per chair on the entire order by 25 cents for each additional chair over 300. (a) If x chairs are ordered, express the number of dollars in the dealer's cost as a function of x. (b) What is the domain of the resulting function?

39. A school-sponsored trip that can accommodate up to 250 students will cost each student $15 if not more than 150 students make the trip; however, the cost per student will be reduced 5 cents for each student in excess of 150 until the cost reaches $10 per student. (a) If x students make the trip, express the number of dollars in the gross income as a function of x. (b) What is the domain of the resulting function?

40. In a town of population 11,000, the growth rate of an epidemic is jointly proportional to the number of people infected and the number of people not infected. (a) If the epidemic is growing at the rate of $f(x)$ people per day when x people are infected, write an equation defining $f(x)$. (b) What is the domain of the function f in part (a)?

41. An open box having a square base is to have a volume of 32 ft^3. (a) If x in. is the length of a side of the base, express the number of square inches in the total surface area of the box as a function of x. (b) What is the domain of the resulting function?

42. Do Exercise 41 if the box is to be closed.

43. A sign, for which margins of 4 m at the top and bottom and 2 m on the sides are required, is to contain 50 m^2 of printed material. (a) If x meters is the length of the top and bottom of the printed material, express the number of square meters in the total area of the sign as a function of x. (b) What is the domain of the resulting function?

44. When the price is $140, there are 6000 radios available for market. For each $20 increase in price, 3000 more radios are available for market. If the supply equation is linear, find it, and draw a sketch of the supply curve.

45. A company can sell 10,000 units of a particular commodity when the unit price is $30, and the company has determined that it can sell 1000 more units with each $2 reduction in price. If the demand equation is linear, find it, and draw a sketch of the demand curve.

46. If a pond can support a maximum of 10,000 fish, the rate of growth of the fish population is jointly proportional to the number of fish present and the difference between 10,000 and the number present. (a) If the rate of growth is 90 fish per week when 1000 fish are present, express the rate of population growth as a function of the number present. (b) Find the rate of population growth when 2000 fish are present.

CHAPTER 2

THE DERIVATIVE

2.1 The limit of a function

The concept of the limit of a function is basic to the study of calculus. In this section the notion of the limit of a function is first given a step-by-step motivation, which brings the discussion from computing the value of a function near a number through an intuitive treatment of the limiting process.

Consider the function f defined by the equation

$$f(x) = \frac{(2x + 3)(x - 1)}{x - 1} \tag{1}$$

f is defined for all values of x except $x = 1$. Furthermore, if $x \neq 1$, the numerator and denominator can be divided by $x - 1$ to obtain

$$f(x) = 2x + 3 \qquad x \neq 1 \tag{2}$$

We shall investigate the function values, $f(x)$, when x is close to 1 but not equal to 1. First, let x take on the values 0, 0.25, 0.5, 0.75, 0.9, 0.99, 0.999, 0.9999, and so on. We are taking values of x closer and closer to 1 but less than 1; in other words, the variable x is approaching 1 through values that are less than 1. We illustrate this in Table 2.1.1.

Table 2.1.1

x	0	0.25	0.5	0.75	0.9	0.99	0.999	0.9999	0.99999
$f(x) = 2x + 3$ $(x \neq 1)$	3	3.5	4	4.5	4.8	4.98	4.998	4.9998	4.99998

Now let the variable x approach 1 through values that are greater than 1; that is, let x take on the values 2, 1.75, 1.5, 1.25, 1.1, 1.01, 1.001, 1.0001, 1.00001, and so on. Refer to Table 2.1.2.

Table 2.1.2

x	2	1.75	1.5	1.25	1.1	1.01	1.001	1.0001	1.00001
$f(x) = 2x + 3$ $(x \neq 1)$	7	6.5	6	5.5	5.2	5.02	5.002	5.0002	5.00002

We see from both tables that as x gets closer and closer to 1, $f(x)$ gets closer and closer to 5; and the closer x is to 1, the closer $f(x)$ is to 5. For instance, from Table 2.1.1, when $x = 0.9$, $f(x) = 4.8$; that is, when x is 0.1 less than 1, $f(x)$ is 0.2 less than 5. When $x = 0.999$, $f(x) = 4.998$; that is, when x is 0.001 less than 1, $f(x)$ is 0.002 less than 5. Furthermore, when $x = 0.9999$, $f(x) = 4.9998$; that is, when x is 0.0001 less than 1, $f(x)$ is 0.0002 less than 5.

Table 2.1.2 shows that when $x = 1.1$, $f(x) = 5.2$, that is, when x is 0.1 greater than 1, $f(x)$ is 0.2 greater than 5. When $x = 1.001$, $f(x) = 5.002$; that is, when x is 0.001 greater than 1, $f(x)$ is 0.002 greater than 5. When $x =$

1.0001, $f(x) = 5.0002$; that is, when x is 0.0001 greater than 1, $f(x)$ is 0.0002 greater than 5.

Therefore, from the two tables, we see that when x differs from 1 by ± 0.001 (i.e., $x = 0.999$ or $x = 1.001$), $f(x)$ differs from 5 by ± 0.002 [i.e., $f(x) = 4.998$ or $f(x) = 5.002$]. And when x differs from 1 by ± 0.0001, $f(x)$ differs from 5 by ± 0.0002.

Now looking at the situation another way, we consider the values of $f(x)$ first. We see from the two tables that $|f(x) - 5| = 0.2$ when $|x - 1| = 0.1$, and that

$$|f(x) - 5| < 0.2 \qquad \text{whenever} \quad 0 < |x - 1| < 0.1$$

Note that we impose the condition $0 < |x - 1|$ because we are concerned only with values of $f(x)$ for x close to 1, but not for $x = 1$ (as a matter of fact, this function is not defined for $x = 1$).

Furthermore, $|f(x) - 5| = 0.002$ when $|x - 1| = 0.001$, and

$$|f(x) - 5| < 0.002 \qquad \text{whenever} \quad 0 < |x - 1| < 0.001$$

Similarly, $|f(x) - 5| = 0.0002$ when $|x - 1| = 0.0001$, and

$$|f(x) - 5| < 0.0002 \qquad \text{whenever} \quad 0 < |x - 1| < 0.0001$$

We could go on and make the value of $f(x)$ as close to 5 as we please by taking x close enough to 1 but not equal to 1. Another way of saying this is that we can make the absolute value of the difference between $f(x)$ and 5 as small as we please by making the absolute value of the difference between x and 1 small enough, but not zero. That is, $|f(x) - 5|$ can be made as small as we please by making $|x - 1|$ small enough but $|x - 1| > 0$. We are now in a position to define the limit of a function in general.

Definition of the limit of a function

> Let f be a function which is defined at every number in some open interval I containing a, except possibly at the number a itself. The **limit of $f(x)$ as x approaches a is L**, written as
>
> $$\lim_{x \to a} f(x) = L$$
>
> if $|f(x) - L|$ can be made as small as we please by making $|x - a|$ small enough, but $|x - a| > 0$.

Another way of stating the above definition is as follows: "The function values $f(x)$ approach a limit L as x approaches a number a if the absolute value of the difference between $f(x)$ and L can be made as small as we please by taking x sufficiently near a, but not equal to a."

A geometric interpretation of the definition of the limit of a function for the one defined by Eq. (1) is illustrated in Fig. 2.1.1. Observe in the figure that $f(x)$ on the vertical axis will be between $5 - e$ and $5 + e$ (that is, $f(x)$ will be within e units of 5) whenever x on the horizontal axis lies between $1 - d$ and $1 + d$ (that is, whenever x lies within d units of 1). Another way

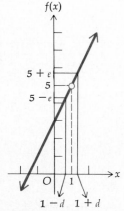

Figure 2.1.1

of stating this is that $f(x)$ on the vertical axis can be restricted to lie between $5 - e$ and $5 + e$ by restricting x on the horizontal axis to lie between $1 - d$ and $1 + d$.

It is important to realize that in the definition nothing is mentioned about the value of the function when $x = a$. That is, it is not necessary that the function be defined for $x = a$ in order for $\lim\limits_{x \to a} f(x)$ to exist. In particular, in the example,

$$\lim_{x \to 1} \frac{(2x + 3)(x - 1)}{x - 1} = 5$$

but

$$\frac{(2x + 3)(x - 1)}{x - 1}$$

is not defined for $x = 1$. However, the first sentence in the definition requires that the function of our example be defined at all numbers except 1 in some open interval containing 1.

To find limits of functions in a straightforward manner we need some theorems. The proofs of the theorems are based on the definition of the limit of a function. These proofs are advanced for this text and are not given here. The statements of the theorems are probably intuitively apparent to you if you think of the concept of a limit in the sense of "$f(x)$ being close to L when x is sufficiently near a." However, you should realize that each of the theorems can be given a rigorous proof. These theorems are labeled "limit theorems."

Limit Theorem 1   If m and b are any constants,

$$\lim_{x \to a} (mx + b) = ma + b$$

● ILLUSTRATION 1

From Limit theorem 1 it follows that

$$\lim_{x \to 2} (3x + 5) = 3 \cdot 2 + 5$$

$$= 11$$

●

Limit Theorem 2   If c is a constant, then for any number a,

$$\lim_{x \to a} c = c$$

Limit theorem 2 is the special case of Limit theorem 1 where $m = 0$ and $b = c$.

Limit Theorem 3

$$\lim_{x \to a} x = a$$

Limit theorem 3 is the special case of Limit theorem 1 where $m = 1$ and $b = 0$.

● ILLUSTRATION 2

From Limit theorem 2,

$$\lim_{x \to 5} 7 = 7$$

and from Limit theorem 3,

$$\lim_{x \to -6} x = -6$$

●

Limit Theorem 4 If $\lim_{x \to a} f(x) = L$ and $\lim_{x \to a} g(x) = M$, then

$$\lim_{x \to a} [f(x) \pm g(x)] = L \pm M$$

Limit theorem 4 can be extended to any finite number of functions.

Limit Theorem 5 If $\lim_{x \to a} f_1(x) = L_1, \lim_{x \to a} f_2(x) = L_2, \ldots,$ and $\lim_{x \to a} f_n(x) = L_n$, then

$$\lim_{x \to a} [f_1(x) \pm f_2(x) \pm \ldots \pm f_n(x)] = L_1 \pm L_2 \pm \ldots \pm L_n$$

Limit Theorem 6 If $\lim_{x \to a} f(x) = L$ and $\lim_{x \to a} g(x) = M$, then

$$\lim_{x \to a} f(x) \cdot g(x) = L \cdot M$$

● ILLUSTRATION 3

From Limit theorem 3, $\lim\limits_{x \to 3} x = 3$, and from Limit theorem 1, $\lim\limits_{x \to 3}(2x + 1) = 7$. Thus from Limit theorem 6 we have

$$\lim_{x \to 3} x(2x + 1) = \lim_{x \to 3} x \cdot \lim_{x \to 3} (2x + 1)$$

$$= 3 \cdot 7$$

$$= 21$$

●

Limit theorem 6 also can be extended to any finite number of functions.

Limit Theorem 7   If $\lim\limits_{x \to a} f_1(x) = L_1, \lim\limits_{x \to a} f_2(x) = L_2, \ldots$, and $\lim\limits_{x \to a} f_n(x) = L_n$, then

$$\lim_{x \to a} [f_1(x)f_2(x) \ldots f_n(x)] = L_1 L_2 \ldots L_n$$

If in Limit theorem 7

$$f(x) = f_1(x) = f_2(x) = \ldots = f_n(x) \text{ and } L = L_1 = L_2 = \ldots = L_n$$

we have the following theorem.

Limit Theorem 8   If $\lim\limits_{x \to a} f(x) = L$ and n is any positive integer, then

$$\lim_{x \to a} [f(x)]^n = L^n$$

● ILLUSTRATION 4

From Limit theorem 1, $\lim\limits_{x \to -2} (5x + 7) = -3$. Therefore, from Limit theorem 8, it follows that

$$\lim_{x \to -2} (5x + 7)^4 = \left[\lim_{x \to -2} (5x + 7) \right]^4$$

$$= (-3)^4$$

$$= 81$$

●

Limit Theorem 9   If $\lim\limits_{x \to a} f(x) = L$ and $\lim\limits_{x \to a} g(x) = M$, and $M \neq 0$, then

$$\lim_{x \to a} \frac{f(x)}{g(x)} = \frac{L}{M}$$

● ILLUSTRATION 5

From Limit theorem 3, $\lim\limits_{x \to 4} x = 4$, and from Limit theorem 1, $\lim\limits_{x \to 4} (-7x + 1) = -27$. Therefore, from Limit theorem 9,

$$\lim_{x \to 4} \frac{x}{-7x + 1} = \frac{\lim\limits_{x \to 4} x}{\lim\limits_{x \to 4} (-7x + 1)}$$

$$= \frac{4}{-27}$$

$$= -\frac{4}{27}$$

●

Limit Theorem 10   If $\lim\limits_{x \to a} f(x) = L$, then

$$\lim_{x \to a} \sqrt[n]{f(x)} = \sqrt[n]{L}$$

if $L > 0$ and n is any positive integer, or if $L \leq 0$ and n is a positive odd integer.

● ILLUSTRATION 6

From the result of Illustration 5 and Limit theorem 10 it follows that

$$\lim_{x \to 4} \sqrt[3]{\frac{x}{-7x + 1}} = \sqrt[3]{\lim_{x \to 4} \frac{x}{-7x + 1}}$$

$$= \sqrt[3]{-\frac{4}{27}}$$

$$= -\frac{\sqrt[3]{4}}{3}$$

●

Following are some examples illustrating the application of the above theorems. To indicate the limit theorem being used, we use the abbreviation "L.T." followed by the theorem number; for example, "L.T. 2" refers to Limit theorem 2.

EXAMPLE 1   Find $\lim\limits_{x \to 3} (x^2 + 7x - 5)$, and, when applicable, indicate the limit theorems being used.

Solution

$$\lim_{x \to 3} (x^2 + 7x - 5) = \lim_{x \to 3} x^2 + \lim_{x \to 3} 7x - \lim_{x \to 3} 5 \qquad \text{(L.T. 5)}$$

$$= \lim_{x \to 3} x \cdot \lim_{x \to 3} x + \lim_{x \to 3} 7 \cdot \lim_{x \to 3} x - \lim_{x \to 3} 5 \qquad \text{(L.T. 6)}$$

$$= 3 \cdot 3 + 7 \cdot 3 - 5 \qquad \text{(L.T. 3 and L.T. 2)}$$

$$= 9 + 21 - 5$$

$$= 25$$

It is important at this point to realize that the limit in Example 1 was evaluated by direct application of the theorems on limits. For the function f defined by $f(x) = x^2 + 7x - 5$, we see that $f(3) = 3^2 + 7 \cdot 3 - 5 = 25$, which is the same as $\lim_{x \to 3} (x^2 + 7x - 5)$. It is not always true that we have $\lim_{x \to a} f(x) = f(a)$ (see Example 4). In Example 1, $\lim_{x \to 3} f(x) = f(3)$ because the function f is continuous at $x = 3$. We discuss the meaning of continuous functions in Sec. 2.3.

EXAMPLE 2 Find

$$\lim_{x \to 2} \sqrt{\frac{x^3 + 2x + 3}{x^2 + 5}}$$

and, when applicable, indicate the limit theorems being used.

Solution

$$\lim_{x \to 2} \sqrt{\frac{x^3 + 2x + 3}{x^2 + 5}} = \sqrt{\lim_{x \to 2} \frac{x^3 + 2x + 3}{x^2 + 5}} \qquad \text{(L.T. 10)}$$

$$= \sqrt{\frac{\lim_{x \to 2} (x^3 + 2x + 3)}{\lim_{x \to 2} (x^2 + 5)}} \qquad \text{(L.T. 9)}$$

$$= \sqrt{\frac{\lim_{x \to 2} x^3 + \lim_{x \to 2} (2x + 3)}{\lim_{x \to 2} x^2 + \lim_{x \to 2} 5}} \qquad \text{(L.T. 4)}$$

$$= \sqrt{\frac{\left(\lim_{x \to 2} x\right)^3 + \lim_{x \to 2} (2x + 3)}{\left(\lim_{x \to 2} x\right)^2 + \lim_{x \to 2} 5}} \qquad \text{(L.T. 8)}$$

$$= \sqrt{\frac{2^3 + (2 \cdot 2 + 3)}{2^2 + 5}} \qquad \text{(L.T. 1, L.T. 2, and L.T. 3)}$$

$$= \sqrt{\frac{8 + 7}{4 + 5}}$$

$$= \frac{\sqrt{15}}{3}$$

EXAMPLE 3 Find

$$\lim_{x \to 5} \frac{x^2 - 25}{x - 5}$$

and, when applicable, indicate the limit theorems being used.

Solution   Here we have a more difficult problem. Limit theorem 9 cannot be applied to the quotient $(x^2 - 25)/(x - 5)$ because $\lim_{x \to 5} (x - 5) = 0$. However, factoring the numerator we obtain

$$\frac{x^2 - 25}{x - 5} = \frac{(x - 5)(x + 5)}{x - 5}$$

This quotient is $x + 5$ if $x \neq 5$, because then we can divide the numerator and denominator by $x - 5$.

When evaluating $\lim_{x \to 5} [(x^2 - 25)/(x - 5)]$, we are considering values of x close to 5 but not equal to 5. Therefore it is possible to divide the numerator and denominator by $x - 5$. The solution to this problem takes the following form:

$$\lim_{x \to 5} \frac{x^2 - 25}{x - 5} = \lim_{x \to 5} \frac{(x - 5)(x + 5)}{x - 5}$$

$$= \lim_{x \to 5} (x + 5) \qquad \text{dividing numerator and denominator by} \ x - 5 \ \text{because} \ x \neq 5$$

$$= 10 \qquad\qquad\qquad \text{(L. T. 1)}$$

Note that in Example 3, $(x^2 - 25)/(x - 5)$ is not defined when $x = 5$, but $\lim_{x \to 5} [(x^2 - 25)/(x - 5)]$ exists and is equal to 10.

EXAMPLE 4   Given

$$f(x) = \begin{cases} x - 3 & \text{if } x \neq 4 \\ 5 & \text{if } x = 4 \end{cases}$$

find $\lim_{x \to 4} f(x)$.

Solution   When evaluating $\lim_{x \to 4} f(x)$, we are considering values of x close to 4 but not equal to 4. Thus we have

$$\lim_{x \to 4} f(x) = \lim_{x \to 4} (x - 3)$$

$$= 1 \qquad\qquad\qquad \text{(L.T.1)}$$

In Example 4, $\lim_{x \to 4} f(x) = 1$ but $f(4) = 5$; therefore $\lim_{x \to 4} f(x) \neq f(4)$. This is an example of a function which is discontinuous at $x = 4$. In terms of geometry this means that there is a break in the graph of the function at the point where $x = 4$ (see Fig. 2.1.2). The graph of the function consists of the isolated point $(4, 5)$ and the straight line whose equation is $y = x - 3$, with the point $(4, 1)$ deleted.

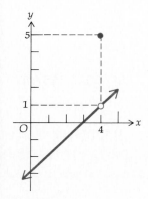

Figure 2.1.2

EXAMPLE 5 Find

$$\lim_{x \to 4} \frac{\sqrt{x} - 2}{x - 4}$$

and, when applicable, indicate the limit theorems being used.

Solution As in Example 3, Limit theorem 9 cannot be applied to the quotient $(\sqrt{x} - 2)/(x - 4)$ because $\lim\limits_{x \to 4} (x - 4) = 0$. To simplify the quotient we rationalize the numerator by multiplying the numerator and denominator by $\sqrt{x} + 2$. We have, then,

$$\frac{\sqrt{x} - 2}{x - 4} = \frac{(\sqrt{x} - 2)(\sqrt{x} + 2)}{(x - 4)(\sqrt{x} + 2)}$$

$$= \frac{x - 4}{(x - 4)(\sqrt{x} + 2)}$$

Because we are evaluating the limit as x approaches 4, we are considering values of x close to 4 but not equal to 4. Hence we can divide the numerator and denominator by $x - 4$, and we have

$$\frac{\sqrt{x} - 2}{x - 4} = \frac{1}{\sqrt{x} + 2} \quad \text{if } x \neq 4$$

The solution is as follows:

$$\lim_{x \to 4} \frac{\sqrt{x} - 2}{x - 4} = \lim_{x \to 4} \frac{(\sqrt{x} - 2)(\sqrt{x} + 2)}{(x - 4)(\sqrt{x} + 2)}$$

$$= \lim_{x \to 4} \frac{x - 4}{(x - 4)(\sqrt{x} + 2)}$$

$$= \lim_{x \to 4} \frac{1}{\sqrt{x} + 2} \qquad \text{dividing numerator and denominator} \\ \text{by } x - 4 \text{ because } x \neq 4$$

$$= \frac{\lim\limits_{x \to 4} 1}{\lim\limits_{x \to 4} (\sqrt{x} + 2)} \qquad \text{(L.T. 9)}$$

$$= \frac{1}{\lim\limits_{x \to 4} \sqrt{x} + \lim\limits_{x \to 4} 2} \qquad \text{(L.T. 2 and L.T. 4)}$$

$$= \frac{1}{\sqrt{\lim\limits_{x \to 4} x} + 2} \qquad \text{(L.T. 10 and L.T. 2)}$$

$$= \frac{1}{\sqrt{4} + 2} \qquad \text{(L.T. 3)}$$

$$= \tfrac{1}{4}$$

Exercises 2.1

In Exercises 1 through 30, find the value of the limit, and, when applicable, indicate the limit theorems being used.

1. $\lim_{x \to 5} (2x + 4)$

2. $\lim_{x \to -3} (4x + 7)$

3. $\lim_{x \to -1} (3x^2 - 5)$

4. $\lim_{x \to 2} (10 - x^2)$

5. $\lim_{y \to 2} (y^2 + 2y - 1)$

6. $\lim_{t \to 4} (t^2 - 3t + 6)$

7. $\lim_{z \to -3} (2z^3 + z^2 + 5z - 2)$

8. $\lim_{y \to -1} (y^3 - 2y^2 + 3y - 4)$

9. $\lim_{x \to 1/2} \dfrac{3x + 1}{5x - 2}$

10. $\lim_{x \to 1/4} \dfrac{2x - 3}{6x + 1}$

11. $\lim_{t \to 2} \dfrac{t^2 - 5}{2t^3 + 6}$

12. $\lim_{x \to -1} \dfrac{2x + 1}{x^2 - 3x + 4}$

13. $\lim_{x \to -1} \dfrac{x^2 - 1}{x + 1}$

14. $\lim_{x \to 1/2} \dfrac{2x - 1}{4x^2 - 1}$

15. $\lim_{x \to 2} \dfrac{x - 2}{x^2 - 4}$

16. $\lim_{x \to -3} \dfrac{9 - x^2}{3 + x}$

17. $\lim_{y \to 3} \dfrac{y^2 - y - 6}{y^2 - 4y + 3}$

18. $\lim_{t \to -3} \dfrac{t^2 + 5t + 6}{t^2 - t - 12}$

19. $\lim_{x \to 4} \dfrac{3x^2 - 17x + 20}{4x^2 - 25x + 36}$

20. $\lim_{x \to 5} \dfrac{2x^2 - 9x - 5}{4x^2 - 23x + 15}$

21. $\lim_{s \to 1} \dfrac{s^3 - 1}{s - 1}$

22. $\lim_{y \to -2} \dfrac{y^3 + 8}{y + 2}$

23. $\lim_{r \to 1} \sqrt{\dfrac{8r + 1}{r + 3}}$

24. $\lim_{s \to 2} \sqrt{\dfrac{s^2 + 3s + 4}{s^3 + 1}}$

25. $\lim_{y \to -3} \sqrt{\dfrac{y^2 - 9}{2y^2 + 7y + 3}}$

26. $\lim_{t \to 3/2} \sqrt{\dfrac{8t^3 - 27}{4t^2 - 9}}$

27. $\lim_{x \to 9} \dfrac{3 - \sqrt{x}}{9 - x}$

28. $\lim_{x \to 1} \dfrac{1 - x}{1 - \sqrt{x}}$

29. $\lim_{t \to 0} \dfrac{2 - \sqrt{4 - t}}{t}$

30. $\lim_{x \to 0} \dfrac{\sqrt{x + 2} - \sqrt{2}}{x}$

31. If $f(x) = x^2 + 5x - 3$, show that $\lim_{x \to 2} f(x) = f(2)$.

32. If $F(x) = 2x^3 + 7x - 1$, show that $\lim_{x \to -1} F(x) = F(-1)$.

33. If $g(x) = \dfrac{x^2 - 16}{x - 4}$, show that $\lim_{x \to 4} g(x) = 8$ but that $g(4)$ is not defined.

34. If $h(x) = \dfrac{x + 1}{x^3 + 1}$, show that $\lim_{x \to -1} h(x) = \frac{1}{3}$, but that $h(-1)$ is not defined.

35. If $G(x) = \dfrac{\sqrt{x + 9} - 3}{x}$, show that $\lim_{x \to 0} G(x) = \frac{1}{6}$, but that $G(0)$ is not defined.

36. Given

$$f(x) = \begin{cases} 2x - 1 & \text{if } x \neq 2 \\ 1 & \text{if } x = 2 \end{cases}$$

(a) Find $\lim_{x \to 2} f(x)$, and show that $\lim_{x \to 2} f(x) \neq f(2)$. (b) Draw a sketch of the graph of f.

37. Given

$$f(x) = \begin{cases} x^2 - 9 & \text{if } x \neq -3 \\ 4 & \text{if } x = -3 \end{cases}$$

(a) Find $\lim_{x \to -3} f(x)$, and show that $\lim_{x \to -3} f(x) \neq f(-3)$. (b) Draw a sketch of the graph of f.

2.2 One-sided limits and infinite limits

Suppose a wholesaler sells a product by the pound (or fraction of a pound), and if not more than 10 pounds are ordered, the wholesaler charges $1 per pound. However, to invite large orders the wholesaler charges only 90 cents per pound if more than 10 pounds are purchased. Thus if x pounds of the product are purchased and $C(x)$ dollars is the total cost of the order, then

$$C(x) = \begin{cases} x & \text{if } 0 \leq x \leq 10 \\ 0.9x & \text{if } 10 < x \end{cases} \qquad (1)$$

A sketch of the graph of C is shown in Fig. 2.2.1. Observe that $C(x)$ is obtained from the equation $C(x) = x$ when $0 \leq x \leq 10$ and from the equation $C(x) = 0.9x$ when $10 < x$. Because of this situation, when considering the limit of $C(x)$ as x approaches 10 we must distinguish between a *left-hand limit* at 10 and a *right-hand limit* at 10. If x approaches 10 from the left (that is, x approaches 10 through values less than 10), we write $x \to 10^-$. If x approaches 10 from the right (that is, x approaches 10 through values greater than 10), we write $x \to 10^+$. For the function C defined by (1) we have

$$\lim_{x \to 10^-} C(x) = \lim_{x \to 10^-} x = 10 \qquad (2)$$

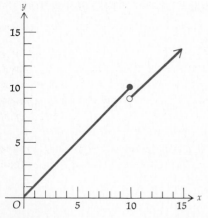

Figure 2.2.1

and

$$\lim_{x \to 10^+} C(x) = \lim_{x \to 10^+} 0.9x = 9 \qquad (3)$$

Because $\lim_{x \to 10^-} C(x) \neq \lim_{x \to 10^+} C(x)$, we say that $\lim_{x \to 10} C(x)$ does not exist. Note in Fig. 2.2.1 that at $x = 10$ there is a break in the graph of the function C. In Sec. 2.3 we return to this function and show that C is *discontinuous* at $x = 10$.

Following are the formal definitions of the two one-sided limits, called the **right-hand limit** and **left-hand limit**.

Definition of one-sided limits

(i) Let f be a function that is defined at every number in some open interval (a, c). Then the **limit of $f(x)$ as x approaches a from the right is L**, written

$$\lim_{x \to a^+} f(x) = L$$

if $\left| f(x) - L \right|$ can be made as small as we please by making $x - a$ small enough but $x - a > 0$.

(ii) Let f be a function that is defined at every number in some open interval (d, a). Then the **limit of $f(x)$ as x approaches a from the left is L**, written

$$\lim_{x \to a^-} f(x) = L$$

if $\left| f(x) - L \right|$ can be made as small as we please by making $a - x$ small enough but $a - x > 0$.

Observe in the statement of part (i) of the above definition that there are no absolute value bars around $x - a$ because $x - a > 0$ when x approaches a from the right. Similarly, in part (ii) of the definition $a - x > 0$ when x approaches a from the left.

Limit theorems 1–10 given in Sec. 2.1 remain unchanged when "$x \to a$" is replaced by "$x \to a^+$" or "$x \to a^-$."

We refer to $\lim_{x \to a} f(x)$ as the two-sided limit to distinguish it from the one-sided limits. When considering the two-sided limit, $\lim_{x \to a} f(x)$, we are concerned with values of x in an open interval containing a but not at a itself, that is, at values of x close to a and either greater than a or less than a.

Suppose that we have the function f for which $f(x) = \sqrt{x - 4}$. Because $f(x)$ does not exist if $x < 4$, f is not defined on any open interval containing 4. Hence we cannot consider the two-sided limit $\lim_{x \to 4} \sqrt{x - 4}$. However, if x is restricted to values greater than 4, the value of $\sqrt{x - 4}$ can be made as close to 0 as we please by taking x sufficiently close to 4 but greater than 4. Therefore the right-hand limit at 4 exists, and

$$\lim_{x \to 4^+} \sqrt{x - 4} = 0$$

EXAMPLE 1 Let f be defined by

$$f(x) = \begin{cases} -1 & \text{if } x < 0 \\ 0 & \text{if } x = 0 \\ 1 & \text{if } 0 < x \end{cases}$$

(a) Draw a sketch of the graph of f. (b) Determine $\lim_{x \to 0^-} f(x)$ if it exists.

(c) Determine $\lim_{x \to 0^+} f(x)$ if it exists

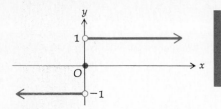

Figure 2.2.2

Solution A sketch of the graph is shown in Fig. 2.2.2.

$$\lim_{x \to 0^-} f(x) = \lim_{x \to 0^-} (-1) = -1$$

$$\lim_{x \to 0^+} f(x) = \lim_{x \to 0^+} 1 = 1$$

For the function f in Example 1, the two-sided limit, $\lim_{x \to 0} f(x)$, does not exist because $\lim_{x \to 0^-} f(x) \neq \lim_{x \to 0^+} f(x)$. The concept of the two-sided limit failing to exist because the two one-sided limits are unequal is stated in the following theorem.

Theorem 2.2.1 $\lim_{x \to a} f(x)$ exists and is equal to L if and only if $\lim_{x \to a^-} f(x)$ and $\lim_{x \to a^+} f(x)$ both exist and both are equal to L.

EXAMPLE 2 Let g be defined by

$$g(x) = \begin{cases} |x| & \text{if } x \neq 0 \\ 2 & \text{if } x = 0 \end{cases}$$

(a) Draw a sketch of the graph of g. (b) Find $\lim_{x \to 0} g(x)$ if it exists.

Solution (a) A sketch of the graph of g is shown in Fig. 2.2.3.

(b) $$\lim_{x \to 0^-} g(x) = \lim_{x \to 0^-} (-x) = 0$$

$$\lim_{x \to 0^+} g(x) = \lim_{x \to 0^+} x = 0$$

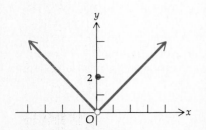

Figure 2.2.3

Because $\lim_{x \to 0^-} g(x) = \lim_{x \to 0^+} g(x) = 0$, it follows from Theorem 2.2.1 that $\lim_{x \to 0} g(x)$ exists and is equal to 0. Note that $g(0) = 2$, which has no effect on $\lim_{x \to 0} g(x)$.

EXAMPLE 3 Let h be defined by

$$h(x) = \begin{cases} 4 - x^2 & \text{if } x \leq 1 \\ 2 + x^2 & \text{if } 1 < x \end{cases}$$

(a) Draw a sketch of the graph of h. (b) Find each of the following limits if they exist: $\lim_{x \to 1^-} h(x)$, $\lim_{x \to 1^+} h(x)$, $\lim_{x \to 1} h(x)$.

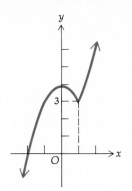

Figure 2.2.4

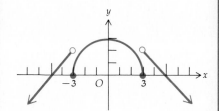

Figure 2.2.5

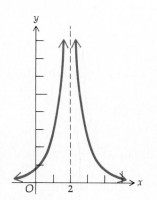

Figure 2.2.6

Solution (a) A sketch of the graph of h is shown in Fig. 2.2.4.

(b) $\lim\limits_{x\to1^-} h(x) = \lim\limits_{x\to1^-} (4 - x^2) = 3$

$\lim\limits_{x\to1^+} h(x) = \lim\limits_{x\to1^+} (2 + x^2) = 3$

Because $\lim\limits_{x\to1^-} h(x) = \lim\limits_{x\to1^+} h(x) = 3$, then from Theorem 2.2.1 $\lim\limits_{x\to1} h(x)$ exists and is equal to 3. Observe that $h(1) = 3$.

EXAMPLE 4 Let f be defined by

$$f(x) = \begin{cases} x + 5 & \text{if } x < -3 \\ \sqrt{9 - x^2} & \text{if } -3 \le x \le 3 \\ 5 - x & \text{if } 3 < x \end{cases}$$

Draw a sketch of the graph of f, and find, if they exist, each of the following limits: (a) $\lim\limits_{x\to-3^-} f(x)$; (b) $\lim\limits_{x\to-3^+} f(x)$; (c) $\lim\limits_{x\to-3} f(x)$; (d) $\lim\limits_{x\to3^-} f(x)$; (e) $\lim\limits_{x\to3^+} f(x)$; (f) $\lim\limits_{x\to3} f(x)$

Solution A sketch of the graph of f is shown in Fig. 2.2.5.

(a) $\lim\limits_{x\to-3^-} f(x) = \lim\limits_{x\to-3^-} (x + 5) = 2$

(b) $\lim\limits_{x\to-3^+} f(x) = \lim\limits_{x\to-3^+} \sqrt{9 - x^2} = 0$

(c) Because $\lim\limits_{x\to-3^-} f(x) \ne \lim\limits_{x\to-3^+} f(x)$, it follows from Theorem 2.2.1 that $\lim\limits_{x\to-3} f(x)$ does not exist.

(d) $\lim\limits_{x\to3^-} f(x) = \lim\limits_{x\to3^-} \sqrt{9 - x^2} = 0$

(e) $\lim\limits_{x\to3^+} f(x) = \lim\limits_{x\to3^+} (5 - x) = 2$

(f) Because $\lim\limits_{x\to3^-} f(x) \ne \lim\limits_{x\to3^+} f(x)$, then $\lim\limits_{x\to3} f(x)$ does not exist.

Let f be the function defined by

$$f(x) = \frac{3}{(x - 2)^2}$$

A sketch of the graph of this function is in Fig. 2.2.6.

We investigate the function values of f when x is close to 2. Letting x approach 2 from the right, we have the values of $f(x)$ given in Table 2.2.1. From this table we see intuitively that as x gets closer and closer to 2 through values greater than 2, $f(x)$ increases without bound. In other words, we can make $f(x)$ greater than any preassigned positive number (that is, we can make $f(x)$ as large as we please) by taking x close enough to 2 and x greater than 2.

Table 2.2.1

x	3	$\frac{5}{2}$	$\frac{7}{3}$	$\frac{9}{4}$	$\frac{21}{10}$	$\frac{201}{100}$	$\frac{2001}{1000}$
$f(x) = \dfrac{3}{(x-2)^2}$	3	12	27	48	300	30,000	3,000,000

To indicate that $f(x)$ increases without bound as x approaches 2 through values greater than 2, we write

$$\lim_{x \to 2^+} \frac{3}{(x-2)^2} = +\infty$$

If x approaches 2 from the left, we have the values of $f(x)$ given in Table 2.2.2. We see intuitively from this table that as x gets closer and closer to 2, through values less than 2, $f(x)$ increases without bound; so we write

$$\lim_{x \to 2^-} \frac{3}{(x-2)^2} = +\infty$$

Table 2.2.2

x	1	$\frac{3}{2}$	$\frac{5}{3}$	$\frac{7}{4}$	$\frac{19}{10}$	$\frac{199}{100}$	$\frac{1999}{1000}$
$f(x) = \dfrac{3}{(x-2)^2}$	3	12	27	48	300	30,000	3,000,000

Therefore, as x approaches 2 from either the right or the left, $f(x)$ increases without bound, and we write

$$\lim_{x \to 2} \frac{3}{(x-2)^2} = +\infty$$

We have the following definition.

Definition of $\lim\limits_{x \to a} f(x) = +\infty$

> Let f be a function which is defined at every number in some open interval containing a, except possibly at the number a itself. **As x approaches a, $f(x)$ increases without bound**, which is written
>
> $$\lim_{x \to a} f(x) = +\infty \tag{4}$$
>
> if $f(x)$ can be made larger than any preassigned positive number by making $|x - a|$ small enough and $|x - a| > 0$.

Another way of stating the above definition is as follows: "The function values $f(x)$ increase without bound as x approaches a number a if $f(x)$ can be made as large as we please by taking x sufficiently close to a but not equal to a."

Note: It should be stressed again (as in Sec. 1.3) that $+\infty$ is not a real number; hence, when we write $\lim\limits_{x \to a} f(x) = +\infty$, it does not have the same meaning as $\lim\limits_{x \to a} f(x) = L$, where L is a real number. Equation (4) can be read as "the limit of $f(x)$ as x approaches a is positive infinity." In such a case the limit does not exist, but the symbolism "$+\infty$" indicates the behavior of the function values $f(x)$ as x gets closer and closer to a.

In an analogous manner we can indicate the behavior of a function whose function values decrease without bound. To lead up to this, consider the function g defined by the equation

$$g(x) = \frac{-3}{(x-2)^2}$$

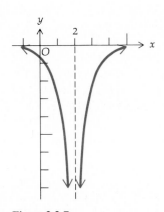

Figure 2.2.7

A sketch of the graph of this function is in Fig. 2.2.7.

The function values given by $g(x) = -3/(x-2)^2$ are the negatives of the function values given by $f(x) = 3/(x-2)^2$. So for the function g as x approaches 2, either from the right or the left, $g(x)$ decreases without bound, and we write

$$\lim_{x \to 2} \frac{-3}{(x-2)^2} = -\infty$$

In general, we have the following definition.

Definition of $\lim\limits_{x \to a} f(x) = -\infty$

> Let f be a function which is defined at every number in some open interval containing a, except possibly at the number a itself. **As x approaches a, $f(x)$ decreases without bound**, which is written
>
> $$\lim_{x \to a} f(x) = -\infty \tag{5}$$
>
> if $f(x)$ can be made less than any preassigned negative number by making $|x - a|$ small enough and $|x - a| > 0$.

Note: Equation (5) can be read as "the limit of $f(x)$ as x approaches a is negative infinity," observing again that the limit does not exist and the symbolism "$-\infty$" indicates only the behavior of the function values as x approaches a.

We can consider one-sided limits that are *infinite*. In particular, $\lim\limits_{x \to a^+} f(x) = +\infty$ if f is defined at every number in some open interval (a, c)

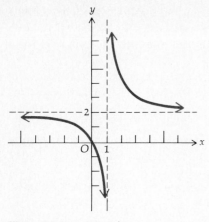

Figure 2.2.8

and if we can make $f(x)$ greater than any preassigned positive number by making $x - a$ small enough and $x - a > 0$. Similar definitions can be given for $\lim\limits_{x \to a^-} f(x) = +\infty$, $\lim\limits_{x \to a^+} f(x) = -\infty$, and $\lim\limits_{x \to a^-} f(x) = -\infty$.

Now suppose that h is the function defined by the equation

$$h(x) = \frac{2x}{x - 1} \qquad (6)$$

A sketch of the graph of this function is in Fig. 2.2.8. Referring to Figs. 2.2.6, 2.2.7, and 2.2.8, observe the difference in the behavior of the function whose graph is sketched in Fig. 2.2.8 from the functions of the other two figures. We see that

$$\lim_{x \to 1^-} \frac{2x}{x - 1} = -\infty \qquad (7)$$

and

$$\lim_{x \to 1^+} \frac{2x}{x - 1} = +\infty \qquad (8)$$

That is, for the function defined by (6), as x approaches 1 through values less than 1, the function values decrease without bound, and as x approaches 1 through values greater than 1, the function values increase without bound.

We have the following limit theorem involving *infinite* limits.

Limit Theorem 11   If r is any positive integer, then

(i) $\lim\limits_{x \to 0^+} \dfrac{1}{x^r} = +\infty$

(ii) $\lim\limits_{x \to 0^-} \dfrac{1}{x^r} = \begin{cases} -\infty & \text{if } r \text{ is odd} \\ +\infty & \text{if } r \text{ is even} \end{cases}$

● ILLUSTRATION 1

From Limit theorem 11(i) it follows that

$$\lim_{x \to 0^+} \frac{1}{x^3} = +\infty \quad \text{and} \quad \lim_{x \to 0^+} \frac{1}{x^4} = +\infty$$

From Limit theorem 11(ii) we have

$$\lim_{x \to 0^-} \frac{1}{x^3} = -\infty \quad \text{and} \quad \lim_{x \to 0^-} \frac{1}{x^4} = +\infty \qquad \qquad ●$$

Limit theorem 12, which follows, involves the limit of a rational function for which the limit of the denominator is zero and the limit of the numerator is a nonzero constant. Such a situation occurs in (7) and (8).

Limit Theorem 12   If $\lim\limits_{x \to a} f(x) = 0$ and $\lim\limits_{x \to a} g(x) = c$, where c is a nonzero constant, then

(i) if $c > 0$ and if $f(x) \to 0$ through positive values of $f(x)$,

$$\lim_{x \to a} \frac{g(x)}{f(x)} = +\infty$$

(ii) if $c > 0$ and if $f(x) \to 0$ through negative values of $f(x)$,

$$\lim_{x \to a} \frac{g(x)}{f(x)} = -\infty$$

(iii) if $c < 0$ and if $f(x) \to 0$ through positive values of $f(x)$,

$$\lim_{x \to a} \frac{g(x)}{f(x)} = -\infty$$

(iv) if $c < 0$ and if $f(x) \to 0$ through negative values of $f(x)$,

$$\lim_{x \to a} \frac{g(x)}{f(x)} = +\infty$$

The theorem is also valid if "$x \to a$" is replaced by "$x \to a^+$" or "$x \to a^-$."

When Limit theorem 12 is applied, we can determine whether the result is $+\infty$ or $-\infty$ by taking a suitable value of x near a to ascertain if the quotient is positive or negative, as shown in the following illustration and example.

● ILLUSTRATION 2

In (7) we have

$$\lim_{x \to 1^-} \frac{2x}{x - 1}$$

Limit theorem 12 is applicable because $\lim\limits_{x \to 1^-} 2x = 2$ and $\lim\limits_{x \to 1^-} (x - 1) = 0$. We wish to determine if we have $+\infty$ or $-\infty$. Because $x \to 1^-$, take a value of x near 1 and less than 1; for instance, take $x = 0.9$. Then

$$\frac{2x}{x - 1} = \frac{2(0.9)}{0.9 - 1} = \frac{1.8}{-0.1} = -18$$

Therefore

$$\lim_{x \to 1^-} \frac{2x}{x - 1} = -\infty$$

This result also follows from part (ii) of Limit theorem 12.
For the limit in (8), because $x \to 1^+$, take $x = 1.1$. Then

$$\frac{2x}{x - 1} = \frac{2(1.1)}{1.1 - 1} = \frac{2.2}{0.1} = 22$$

Therefore

$$\lim_{x \to 1^+} \frac{2x}{x - 1} = +\infty$$

This result also follows from part (i) of Limit theorem 12. ●

EXAMPLE 5   Find: (a) $\lim_{x \to 4^+} \dfrac{2x - 1}{x - 4}$; (b) $\lim_{x \to 4^-} \dfrac{2x - 1}{x - 4}$; (c) $\lim_{x \to 4^+} \dfrac{1 - 2x}{x - 4}$;

(d) $\lim_{x \to 4^-} \dfrac{1 - 2x}{x - 4}$.

Solution   For each of the limits, the limit of the numerator is 7 or -7, and the limit of the denominator is zero. Thus we apply Limit theorem 12.
(a) Because $x \to 4^+$, let $x = 4.1$. Then

$$\frac{2x - 1}{x - 4} = \frac{8.2 - 1}{4.1 - 4} = \frac{7.2}{0.1} = 72$$

Hence

$$\lim_{x \to 4^+} \frac{2x - 1}{x - 4} = +\infty$$

(b) Because $x \to 4^-$, let $x = 3.9$, and we have

$$\frac{2x - 1}{x - 4} = \frac{7.8 - 1}{3.9 - 4} = \frac{6.8}{-0.1} = -68$$

Therefore

$$\lim_{x \to 4^-} \frac{2x - 1}{x - 4} = -\infty$$

(c) Because $x \to 4^+$, let $x = 4.1$, and we have

$$\frac{1 - 2x}{x - 4} = \frac{1 - 8.2}{4.1 - 4} = \frac{-7.2}{0.1} = -72$$

Thus

$$\lim_{x \to 4^+} \frac{1 - 2x}{x - 4} = -\infty$$

(d) Because $x \to 4^-$, let $x = 3.9$. Then

$$\frac{1 - 2x}{x - 4} = \frac{1 - 7.8}{3.9 - 4} = \frac{-6.8}{-0.1} = 68$$

Therefore

$$\lim_{x \to 4^-} \frac{1 - 2x}{x - 4} = +\infty$$

Exercises 2.2

In Exercises 1 through 16, draw a sketch of the graph, and find the indicated limit if it exists. If the limit does not exist, give the reason.

1. $f(x) = \begin{cases} 2 & \text{if } x < 1 \\ -1 & \text{if } x = 1 \\ -3 & \text{if } 1 < x \end{cases}$; (a) $\lim\limits_{x \to 1^+} f(x)$; (b) $\lim\limits_{x \to 1^-} f(x)$; (c) $\lim\limits_{x \to 1} f(x)$

2. $f(x) = \begin{cases} -2 & \text{if } x < 0 \\ 2 & \text{if } 0 \le x \end{cases}$; (a) $\lim\limits_{x \to 0^+} f(x)$; (b) $\lim\limits_{x \to 0^-} f(x)$; (c) $\lim\limits_{x \to 0} f(x)$

3. $f(t) = \begin{cases} t + 4 & \text{if } t \le -4 \\ 4 - t & \text{if } -4 < t \end{cases}$; (a) $\lim\limits_{t \to -4^+} f(t)$; (b) $\lim\limits_{t \to -4^-} f(t)$; (c) $\lim\limits_{t \to -4} f(t)$

4. $g(s) = \begin{cases} s + 3 & \text{if } s \le -2 \\ 3 - s & \text{if } -2 < s \end{cases}$; (a) $\lim\limits_{s \to -2^+} g(s)$; (b) $\lim\limits_{s \to -2^-} g(s)$; (c) $\lim\limits_{s \to -2} g(s)$

5. $F(x) = \begin{cases} x^2 & \text{if } x \le 2 \\ 8 - 2x & \text{if } 2 < x \end{cases}$; (a) $\lim\limits_{x \to 2^+} F(x)$; (b) $\lim\limits_{x \to 2^-} F(x)$; (c) $\lim\limits_{x \to 2} F(x)$

6. $h(x) = \begin{cases} 2x + 1 & \text{if } x < 3 \\ 10 - x & \text{if } 3 \le x \end{cases}$; (a) $\lim\limits_{x \to 3^+} h(x)$; (b) $\lim\limits_{x \to 3^-} h(x)$; (c) $\lim\limits_{x \to 3} h(x)$

7. $g(r) = \begin{cases} 2r + 3 & \text{if } r < 1 \\ 2 & \text{if } r = 1 \\ 7 - 2r & \text{if } 1 < r \end{cases}$; (a) $\lim\limits_{r \to 1^+} g(r)$; (b) $\lim\limits_{r \to 1^-} g(r)$; (c) $\lim\limits_{r \to 1} g(r)$

8. $g(t) = \begin{cases} 3 + t^2 & \text{if } t < -2 \\ 0 & \text{if } t = -2 \\ 11 - t^2 & \text{if } -2 < t \end{cases}$; (a) $\lim\limits_{t \to -2^+} g(t)$; (b) $\lim\limits_{t \to -2^-} g(t)$; (c) $\lim\limits_{t \to -2} g(t)$

9. $f(x) = \begin{cases} x^2 - 4 & \text{if } x < 2 \\ 4 & \text{if } x = 2 \\ 4 - x^2 & \text{if } 2 < x \end{cases}$; (a) $\lim\limits_{x \to 2^+} f(x)$; (b) $\lim\limits_{x \to 2^-} f(x)$; (c) $\lim\limits_{x \to 2} f(x)$

10. $f(x) = \begin{cases} 2x + 3 & \text{if } x < 1 \\ 4 & \text{if } x = 1 \\ x^2 + 2 & \text{if } 1 < x \end{cases}$; (a) $\lim\limits_{x \to 1^+} f(x)$; (b) $\lim\limits_{x \to 1^-} f(x)$; (c) $\lim\limits_{x \to 1} f(x)$

11. $F(x) = |x - 5|$; (a) $\lim\limits_{x \to 5^+} F(x)$; (b) $\lim\limits_{x \to 5^-} F(x)$; (c) $\lim\limits_{x \to 5} F(x)$

12. $F(x) = \begin{cases} |x - 1| & \text{if } x < -1 \\ 0 & \text{if } x = -1 \\ |1 - x| & \text{if } -1 < x \end{cases}$; (a) $\lim\limits_{x \to -1^+} F(x)$; (b) $\lim\limits_{x \to -1^-} F(x)$; (c) $\lim\limits_{x \to -1} F(x)$

13. $f(x) = \dfrac{|x|}{x}$; (a) $\lim\limits_{x \to 0^+} f(x)$; (b) $\lim\limits_{x \to 0^-} f(x)$; (c) $\lim\limits_{x \to 0} f(x)$

14. $f(t) = \begin{cases} \sqrt[3]{t} & \text{if } t < 0 \\ \sqrt{t} & \text{if } 0 \le t \end{cases}$; (a) $\lim\limits_{t \to 0^+} f(t)$; (b) $\lim\limits_{t \to 0^-} f(t)$; (c) $\lim\limits_{t \to 0} f(t)$

15. $f(x) = \begin{cases} x + 1 & \text{if } x < -1 \\ x^2 & \text{if } -1 \le x \le 1 \\ 1 - x & \text{if } 1 < x \end{cases}$; (a) $\lim\limits_{x \to -1^-} f(x)$; (b) $\lim\limits_{x \to -1^+} f(x)$; (c) $\lim\limits_{x \to -1} f(x)$; (d) $\lim\limits_{x \to 1^-} f(x)$; (e) $\lim\limits_{x \to 1^+} f(x)$; (f) $\lim\limits_{x \to 1} f(x)$

16. $F(x) = \begin{cases} \sqrt{x^2 - 9} & \text{if } x \le -3 \\ \sqrt{9 - x^2} & \text{if } -3 < x < 3 \\ \sqrt{x^2 - 9} & \text{if } 3 \le x \end{cases}$; (a) $\lim\limits_{x \to -3^-} F(x)$; (b) $\lim\limits_{x \to -3^+} F(x)$; (c) $\lim\limits_{x \to -3} F(x)$; (d) $\lim\limits_{x \to 3^-} F(x)$; (e) $\lim\limits_{x \to 3^+} F(x)$; (f) $\lim\limits_{x \to 3} F(x)$

In Exercises 17 through 36, evaluate the limit.

17. $\lim\limits_{x \to 5^+} \dfrac{1}{x - 5}$

18. $\lim\limits_{x \to 5^-} \dfrac{1}{x - 5}$

19. $\lim\limits_{x \to 5} \dfrac{1}{(x - 5)^2}$

20. $\lim\limits_{x \to 1^-} \dfrac{x + 2}{1 - x}$

21. $\lim\limits_{x \to 1^+} \dfrac{x + 2}{1 - x}$

22. $\lim\limits_{x \to 1} \dfrac{x + 2}{(x - 1)^2}$

23. $\lim\limits_{x \to -1^+} \dfrac{x - 2}{x + 1}$

24. $\lim\limits_{x \to -1^-} \dfrac{x - 2}{x + 1}$

25. $\lim\limits_{x \to -4^-} \dfrac{x}{x + 4}$

26. $\lim\limits_{x \to 4^+} \dfrac{x}{x - 4}$

27. $\lim\limits_{x \to -3^-} \dfrac{4x}{9 - x^2}$

28. $\lim\limits_{x \to 3^+} \dfrac{4x^2}{9 - x^2}$

19. $\lim\limits_{t \to 2^+} \dfrac{t + 2}{t^2 - 4}$

30. $\lim\limits_{t \to 2^-} \dfrac{-t + 2}{(t - 2)^2}$

31. $\lim\limits_{t \to 2^-} \dfrac{t + 2}{t^2 - 4}$

32. $\lim\limits_{x \to 0^+} \dfrac{\sqrt{3 + x^2}}{x}$

33. $\lim\limits_{x \to 0^-} \dfrac{\sqrt{3 + x^2}}{x}$

34. $\lim\limits_{x \to 0} \dfrac{\sqrt{3 + x^2}}{x^2}$

35. $\lim\limits_{x \to 3^+} \dfrac{\sqrt{x^2 - 9}}{x - 3}$

36. $\lim\limits_{x \to 4^-} \dfrac{\sqrt{16 - x^2}}{x - 4}$

37. Shipping charges are often based on a formula that offers a lower charge per pound as the size of the shipment is increased. Suppose x pounds is the weight of a shipment, $C(x)$ dollars is the total cost of the shipment, and

$$C(x) = \begin{cases} 0.80x & \text{if } 0 < x \le 50 \\ 0.70x & \text{if } 50 < x \le 200 \\ 0.65x & \text{if } 200 < x \end{cases}$$

(a) Draw a sketch of the graph of C. Find each of the following limits: (b) $\lim\limits_{x \to 50^-} C(x)$; (c) $\lim\limits_{x \to 50^+} C(x)$; (d) $\lim\limits_{x \to 200^-} C(x)$; (e) $\lim\limits_{x \to 200^+} C(x)$.

2.3 Continuity of a function

At the beginning of Sec. 2.2 we discussed the function C defined by

$$C(x) = \begin{cases} x & \text{if } 0 \le x \le 10 \\ 0.9x & \text{if } 10 < x \end{cases} \tag{1}$$

where $C(x)$ dollars is the total cost of x pounds of a product. We showed that $\lim\limits_{x \to 10} C(x)$ does not exist because $\lim\limits_{x \to 10^-} C(x) \ne \lim\limits_{x \to 10^+} C(x)$. A sketch of the graph of C is shown in Fig. 2.3.1. Observe that there is a break in the graph of C where $x = 10$. We state that C is *discontinuous* at 10. This discontinuity is caused by the fact that $\lim\limits_{x \to 10} C(x)$ does not exist. We refer to this function C again in Illustration 1.

In Sec. 2.1 we considered the function f defined by

$$f(x) = \frac{(2x + 3)(x - 1)}{x - 1} \tag{2}$$

We noted that f is defined for all values of x except 1. A sketch of the graph consisting of all points on the line $y = 2x + 3$ except $(1, 5)$ is shown in Fig. 2.3.2. There is a break in the graph at the point $(1, 5)$, and we state that the function f is discontinuous at the number 1. This discontinuity occurs because $f(1)$ does not exist.

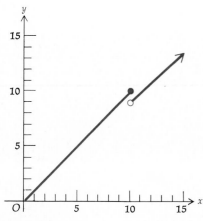

Figure 2.3.1

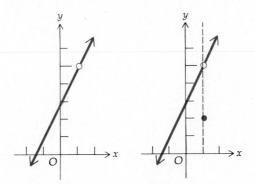

Figure 2.3.2   **Figure 2.3.3**

If f is the function defined by (2) when $x \ne 1$, and if we define $f(1) = 2$, for instance, the function is defined for all values of x, but there is still a break in the graph (see Fig. 2.3.3), and the function is still discontinuous at 1. If we define $f(1) = 5$, however, there is no break in the graph, and the function f is said to be *continuous* at all values of x. We have the following definition.

Definition of a function continuous at a number

> The function f is said to be **continuous at the number a** if the following three conditions are satisfied:
>
> (i) $f(a)$ exists.
>
> (ii) $\lim\limits_{x \to a} f(x)$ exists.
>
> (iii) $\lim\limits_{x \to a} f(x) = f(a)$.
>
> If one or more of these three conditions fails to hold at a, the function f is said to be **discontinuous** at a.

● ILLUSTRATION 1

The function C defined by (1) has the graph shown in Fig. 2.3.1. Because there is a break in the graph at the point where $x = 10$, we will investigate the conditions of the above definition at 10.

Because $C(10) = 10$, condition (i) is satisfied.

Because $\lim\limits_{x \to 10} C(x)$ does not exist, condition (ii) fails to hold at 10.

We conclude that C is discontinuous at 10.

Observe that because of the discontinuity of C, it would be advantageous to increase the size of some orders to take advantage of a lower total cost. In particular, it would be unwise to purchase $9\frac{1}{2}$ pounds for $9.50 when $10\frac{1}{2}$ pounds can be bought for $9.45. ●

In Illustration 2 there is another situation in which the formula for computing the cost of more than 10 pounds of a product is different from the formula for computing the cost of 10 pounds or less. However, here the cost function is continuous at 10.

● ILLUSTRATION 2

A wholesaler who sells a product by the pound (or fraction of a pound) charges $1 per pound if 10 pounds or less are ordered. However, if more than 10 pounds are ordered, the wholesaler charges $10 plus 70 cents for each pound in excess of 10 pounds. Therefore, if x pounds of the product are purchased and $C(x)$ dollars is the total cost, then $C(x) = x$ if $0 \le x \le 10$, and $C(x) = 10 + 0.7(x - 10) = 0.7x + 3$ if $10 < x$. Therefore,

$$C(x) = \begin{cases} x & \text{if } 0 \le x \le 10 \\ 0.7x + 3 & \text{if } 10 < x \end{cases}$$

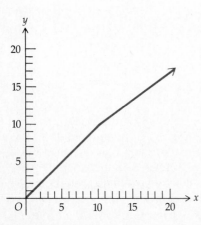

Figure 2.3.4

A sketch of the graph of C is shown in Fig. 2.3.4. For this function, $C(10) = 10$, $\lim\limits_{x \to 10^-} C(x) = \lim\limits_{x \to 10^-} x = 10$ and $\lim\limits_{x \to 10^+} C(x) = \lim\limits_{x \to 10^+} (0.7x + 3) = 10$. Thus $\lim\limits_{x \to 10} C(x) = C(10)$, and C is continuous at 10. ●

We now consider some illustrations of discontinuous functions. For each illustration there is a sketch of the graph of the function. We determine the points where there is a break in the graph and show which of the three conditions in the definition of continuity fails to hold at each discontinuity.

● ILLUSTRATION 3

Let f be defined by

$$f(x) = \begin{cases} 2x + 3 & \text{if } x \neq 1 \\ 2 & \text{if } x = 1 \end{cases}$$

A sketch of the graph of this function is given in Fig. 2.3.3. We see that there is a break in the graph at the point where $x = 1$, and so we investigate there the conditions of the definition of a continuous function.

$f(1) = 2$; therefore, condition (i) is satisfied.

$\lim_{x \to 1} f(x) = \lim_{x \to 1} (2x + 3) = 5$; therefore condition (ii) is satisfied.

$\lim_{x \to 1} f(x) = 5$, but $f(1) = 2$; therefore condition (iii) is not satisfied.

We conclude that f is discontinuous at 1. ●

Note, in Illustration 3, that if $f(1)$ is defined to be 5, then $\lim_{x \to 1} f(x) = f(1)$, and f would be continuous at 1.

● ILLUSTRATION 4

The function F is defined by

$$F(x) = \frac{1}{x - 2}$$

A sketch of the graph of F appears in Fig. 2.3.5. There is a break in the graph at the point where $x = 2$, and so we investigate there the three conditions.

$F(2)$ is not defined; therefore condition (i) is not satisfied, and so F is discontinuous at 2. ●

● ILLUSTRATION 5

Let g be defined by

$$g(x) = \begin{cases} \dfrac{1}{x - 2} & \text{if } x \neq 2 \\ 3 & \text{if } x = 2 \end{cases}$$

A sketch of the graph of g is shown in Fig. 2.3.6. Investigating the three conditions at 2, we have the following.

$g(2) = 3$; therefore, condition (i) is satisfied.

$\lim_{x \to 2^-} g(x) = -\infty$, and $\lim_{x \to 2^+} g(x) = +\infty$; therefore condition (ii) is not satisfied.

Thus g is discontinuous at 2. ●

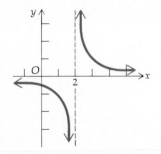

Figure 2.3.5

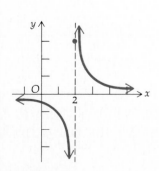

Figure 2.3.6

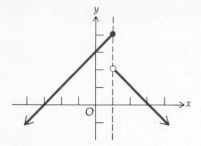

Figure 2.3.7

• ILLUSTRATION 6

Let h be defined by

$$h(x) = \begin{cases} 3 + x & \text{if } x \le 1 \\ 3 - x & \text{if } 1 < x \end{cases}$$

A sketch of the graph of h is shown in Fig. 2.3.7. Because there is a break in the graph at the point where $x = 1$, we investigate the three conditions there. We have the following.

$h(1) = 4$; therefore condition (i) is satisfied.

$$\lim_{x \to 1^-} h(x) = \lim_{x \to 1^-} (3 + x) = 4$$

$$\lim_{x \to 1^+} h(x) = \lim_{x + 1^+} (3 - x) = 2$$

Because $\lim_{x \to 1^-} h(x) \ne \lim_{x \to 1^+} h(x)$, we conclude that $\lim_{x \to 1} h(x)$ does not exist; therefore condition (ii) fails to hold at 1.

Hence h is discontinuous at 1. •

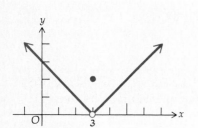

Figure 2.3.8

• ILLUSTRATION 7

Let f be defined by

$$f(x) = \begin{cases} |x - 3| & \text{if } x \ne 3 \\ 2 & \text{if } x = 3 \end{cases}$$

A sketch of the graph of f is shown in Fig. 2.3.8. We investigate the three conditions at the point where $x = 3$.

Because $f(3) = 2$, condition (i) is satisfied.

$$\lim_{x \to 3^+} f(x) = \lim_{x \to 3^+} (x - 3) = 0$$

and

$$\lim_{x \to 3^-} f(x) = \lim_{x \to 3^-} (3 - x) = 0$$

Thus $\lim_{x \to 3} f(x)$ exists and is 0; therefore condition (ii) is satisfied.

$\lim_{x \to 3} f(x) = 0$ but $f(3) = 2$; hence condition (iii) is not satisfied.

Therefore f is discontinuous at 3. •

It should be apparent that the geometric notion of a break in the graph at a certain point is synonymous with the analytic concept of a function being discontinuous at a certain value of the independent variable.

Consider the polynomial function f defined by

$$f(x) = b_0 x^n + b_1 x^{n-1} + b_2 x^{n-2} + \ldots + b_{n-1} x + b_n \qquad b_0 \neq 0$$

where n is a nonnegative integer and $b_0, b_1, \ldots, b_n$ are real numbers. By successive applications of limit theorems we can show that if a is any number,

$$\lim_{x \to a} f(x) = b_0 a^n + b_1 a^{n-1} + b_2 a^{n-2} + \ldots + b_{n-1} a + b_n$$

from which it follows that

$$\lim_{x \to a} f(x) = f(a)$$

thus establishing the following theorem.

Theorem 2.3.1 A polynomial function is continuous at every number.

● ILLUSTRATION 8

If

$$f(x) = x^3 - 2x^2 + 5x + 1$$

then f is a polynomial function and therefore, by Theorem 2.3.1, is continuous at every number. In particular, because f is continuous at 3, then $\lim_{x \to 3} f(x) = f(3)$; thus

$$\lim_{x \to 3} (x^3 - 2x^2 + 5x + 1) = 3^3 - 2(3)^2 + 5(3) + 1$$

$$= 27 - 18 + 15 + 1$$

$$= 25$$

●

If f is a rational function, it can be expressed as the quotient of two polynomial functions. So f can be defined by

$$f(x) = \frac{g(x)}{h(x)}$$

where g and h are two polynomial functions, and the domain of f consists of all numbers except those for which $h(x) = 0$.

If a is any number in the domain of f, then $h(a) \neq 0$; and so by Limit theorem 9,

$$\lim_{x \to a} f(x) = \frac{\lim_{x \to a} g(x)}{\lim_{x \to a} h(x)} \qquad (3)$$

Because g and h are polynomial functions, by Theorem 2.3.1 they are continuous at a, and so $\lim\limits_{x\to a} g(x) = g(a)$ and $\lim\limits_{x\to a} h(x) = h(a)$. Consequently, from (3),

$$\lim_{x\to a} f(x) = \frac{g(a)}{h(a)}$$

Thus we can conclude that f is continuous at every number in its domain. We state this result as a theorem.

> **Theorem 2.3.2**   A rational function is continuous at every number in its domain.

EXAMPLE 1   Given

$$f(x) = \frac{x^3 + 1}{x^2 - 9}$$

Determine all values of x for which f is continuous.

Solution   The domain of f is the set of all real numbers except those for which $x^2 - 9 = 0$. Because $x^2 - 9 = 0$ when $x = \pm 3$, it follows that the domain of f is the set of all real numbers except 3 and -3.
    The function f is a rational function. Therefore, by Theorem 2.3.2, f is continuous at all real numbers except 3 and -3.

We shall need the following theorem in subsequent discussions.

> **Theorem 2.3.3**   If $g(x) = \sqrt[n]{x}$, where n is any positive integer, then g is continuous at a if either
>
>   (i) a is any positive number, or
>   (ii) a is a negative number or zero, and n is odd.

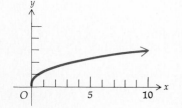

Figure 2.3.9

● ILLUSTRATION 9

(a) If $g(x) = \sqrt{x}$, then from Theorem 2.3.3(i) it follows that g is continuous at every positive number. In Fig. 2.3.9 there is a sketch of the graph of g.
    (b) If $h(x) = \sqrt[3]{x}$, it follows from Theorem 2.3.3(i) and (ii) that h is continuous at every real number. A sketch of the graph of h is shown in Fig. 2.3.10.
                                                                                                        ●

Figure 2.3.10

The next theorem states that a *continuous function of a continuous function is continuous.*

Theorem 2.3.4 If the function g is continuous at a and the function f is continuous at $g(a)$, then the composite function $f \circ g$ is continuous at a.

EXAMPLE 2 Given $h(x) = \sqrt{4 - x^2}$. Determine all values of x for which h is continuous.

Solution If $g(x) = 4 - x^2$ and $f(x) = \sqrt{x}$, then h is the composite function $f \circ g$, and

$$h(x) = (f \circ g)(x) = \sqrt{4 - x^2}$$

Because g is a polynomial function, g is continuous everywhere. Furthermore, f is continuous at every positive number by Theorem 2.3.3(i). Thus, by Theorem 2.3.4, h is continuous at every number x for which $g(x) > 0$, that is, when

$$4 - x^2 > 0$$

or, equivalently,

$$x^2 < 4$$

or, equivalently,

$$-2 < x < 2$$

Hence h is continuous at every number in the open interval $(-2, 2)$.

Because the function h of Example 2 is continuous at every number in the open interval $(-2, 2)$, we say that h is *continuous on the open interval* $(-2, 2)$.

Definition of a function continuous on an open interval

A function is said to be **continuous on an open interval** if it is continuous at every number in the open interval.

We refer again to the function h of Example 2. Because h is not defined on any open interval containing either -2 or 2, we cannot consider $\lim\limits_{x \to -2} h(x)$ or $\lim\limits_{x \to 2} h(x)$. Hence, to discuss the question of the continuity of h on the closed interval $[-2, 2]$, we must extend the concept of continuity to include continuity at an endpoint of a closed interval. We do this by first defining *right-hand continuity* and *left-hand continuity*.

Definition of right-hand continuity

> The function f is said to be **continuous from the right at the number** a if and only if the following three conditions are satisfied:
>
> (i) $f(a)$ exists.
>
> (ii) $\lim\limits_{x \to a^+} f(x)$ exists.
>
> (iii) $\lim\limits_{x \to a^+} f(x) = f(a)$.

Definition of left-hand continuity

> The function f is said to be **continuous from the left at the number** a if and only if the following three conditions are satisfied:
>
> (i) $f(a)$ exists.
>
> (ii) $\lim\limits_{x \to a^-} f(x)$ exists.
>
> (iii) $\lim\limits_{x \to a^-} f(x) = f(a)$.

The concepts of right-hand and left-hand continuity are used to define *continuity on a closed interval*.

Definition of a function continuous on a closed interval

> A function is said to be **continuous on the closed interval** $[a, b]$ if it is continuous on the open interval (a, b) as well as continuous from the right at a and continuous from the left at b.

EXAMPLE 3   If h is the function for which $h(x) = \sqrt{4 - x^2}$, show that h is continuous on the closed interval $[-2, 2]$. Draw a sketch of the graph of h.

Solution   The function h is continuous on the open interval $(-2, 2)$, and

$$\lim_{x \to -2^+} h(x) = \lim_{x \to -2^+} \sqrt{4 - x^2} = 0 = h(-2)$$
$$\lim_{x \to 2^-} h(x) = \lim_{x \to 2^-} \sqrt{4 - x^2} = 0 = h(2)$$

Therefore h is continuous from the right at -2 and continuous from the left at 2. Thus h is continuous on the closed interval $[-2, 2]$. The graph of h is a semicircle. A sketch is shown in Fig. 2.3.11.

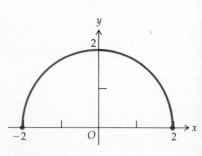

Figure 2.3.11

The following two illustrations involve the applications discussed in Examples 2 and 5 of Sec. 1.5. In each illustration there is a function whose domain is a closed interval, and we show that the function is continuous on that closed interval.

● ILLUSTRATION 10

Refer to Example 2 of Sec. 1.5. If x in. is the length of the side of the square cut out from each of the four corners of the piece of cardboard and $V(x)$ in.[3] is the volume of the box, then

$$V(x) = 144x - 48x^2 + 4x^3 \qquad x \in [0, 6]$$

Because V is a polynomial function, V is continuous everywhere. Thus V is continuous on the closed interval $[0, 6]$.   ●

● ILLUSTRATION 11

In Example 5 of Sec. 1.5, x is the number of places in the seating capacity of the coffee shop, $P(x)$ is the number of dollars in the daily gross profit, and

$$P(x) = \begin{cases} 8x & \text{if } 40 \le x \le 80 \\ 11.20x - 0.04x^2 & \text{if } 80 < x \le 280 \end{cases}$$

Even though x, by definition, is an integer, to have a continuous function we let x be any real number in the interval $[40, 280]$. Because $8x$ and $11.20x - 0.04x^2$ are polynomials, it follows that P is continuous on the interval $[40, 80)$ and the interval $(80, 280]$. We now wish to determine if P is continuous at 80. $P(80) = 8(80) = 640$.

$$\lim_{x \to 80^-} P(x) = \lim_{x \to 80^-} 8x = 640$$

and

$$\lim_{x \to 80^+} P(x) = \lim_{x \to 80^+} (11.20x - 0.04x^2) = 640$$

Because $\lim\limits_{x \to 80^-} P(x) = \lim\limits_{x \to 80^+} P(x) = 640$, it follows that the two-sided limit $\lim\limits_{x \to 80} P(x) = 640 = P(80)$. Thus P is continuous at 80. Hence P is continuous on the closed interval $[40, 280]$.   ●

Exercises 2.3

In Exercises 1 through 20, draw a sketch of the graph of the function; then, by observing where there are breaks in the graph, determine the values of the independent variable at which the function is discontinuous, and show why the definition of a function continuous at a number is not satisfied at each discontinuity.

1. $f(x) = \dfrac{x^2 + x - 6}{x + 3}$

2. $F(x) = \dfrac{x^2 - 3x - 4}{x - 4}$

3. $g(x) = \begin{cases} \dfrac{x^2 + x - 6}{x + 3} & \text{if } x \ne -3 \\ 1 & \text{if } x = -3 \end{cases}$

4. $G(x) = \begin{cases} \dfrac{x^2 - 3x - 4}{x - 4} & \text{if } x \ne 4 \\ 2 & \text{if } x = 4 \end{cases}$

5. $h(x) = \dfrac{5}{x - 4}$

6. $H(x) = \dfrac{1}{x + 2}$

7. $f(x) = \begin{cases} \dfrac{5}{x - 4} & \text{if } x \ne 4 \\ 2 & \text{if } x = 4 \end{cases}$

8. $g(x) = \begin{cases} \dfrac{1}{x + 2} & \text{if } x \ne -2 \\ 0 & \text{if } x = -2 \end{cases}$

9. $F(x) = \dfrac{x^4 - 16}{x^2 - 4}$

10. $h(x) = \dfrac{(x-1)(x^2 - x - 12)}{x^2 - 5x + 4}$

11. $G(x) = \dfrac{x^2 - 4}{x^4 - 16}$

12. $H(x) = \dfrac{x^2 - 5x + 4}{(x-1)(x^2 - x - 12)}$

13. $f(x) = \begin{cases} -1 & \text{if } x < 0 \\ 0 & \text{if } x = 0 \\ \sqrt{x} & \text{if } 0 < x \end{cases}$

14. $f(x) = \begin{cases} x - 1 & \text{if } x < 1 \\ 1 & \text{if } x = 1 \\ 1 - x & \text{if } 1 < x \end{cases}$

15. $g(t) = \begin{cases} t^2 - 4 & \text{if } t < 2 \\ 4 & \text{if } t = 2 \\ 4 - t^2 & \text{if } 2 < t \end{cases}$

16. $H(x) = \begin{cases} 1 + x & \text{if } x \le -2 \\ 2 - x & \text{if } -2 < x \le 2 \\ 2x - 1 & \text{if } 2 < x \end{cases}$

17. $g(x) = \begin{cases} \sqrt{-x} & \text{if } x < 0 \\ \sqrt[3]{x+1} & \text{if } 0 \le x \end{cases}$

18. $f(x) = \begin{cases} |x + 2| & \text{if } x \ne -2 \\ 3 & \text{if } x = -2 \end{cases}$

19. $f(x) = \dfrac{|x|}{x}$

20. $g(x) = \begin{cases} \dfrac{|x|}{x} & \text{if } x \ne 0 \\ 1 & \text{if } x = 0 \end{cases}$

In Exercises 21 through 30, determine all values of x for which the given function is continuous.

21. $f(x) = 3x^2 - 8x + 1$

22. $f(x) = (x^2 - 2)^4$

23. $f(x) = x^2(x + 3)^2$

24. $f(x) = (x - 5)^3(x^2 + 4)^5$

25. $g(x) = \dfrac{x}{x - 3}$

26. $h(x) = \dfrac{x + 1}{2x + 5}$

27. $F(x) = \dfrac{x + 1}{x^2 - 1}$

28. $G(x) = \dfrac{x - 2}{x^2 + 2x - 8}$

29. $f(x) = \dfrac{x^3 + 7}{x^2 - 4}$

30. $g(x) = \dfrac{x^2 - x - 2}{x^3 - 2x^2 + x - 2}$

31. If $f(x) = \sqrt{25 - x^2}$, draw a sketch of the graph of f, and show that f is continuous on the closed interval $[-5, 5]$.
32. If $g(x) = \sqrt{9 - x^2}$, draw a sketch of the graph of g, and show that g is continuous on the closed interval $[-3, 3]$.
33. If $F(x) = \sqrt{x^2 - 25}$, draw a sketch of the graph of F, and show that F is continuous on $(-\infty, -5]$ and $[5, +\infty)$.
34. If $G(x) = \sqrt{x^2 - 9}$, draw a sketch of the graph of G, and show that G is continuous on $(-\infty, -3]$ and $[3, +\infty)$.
35. The function C of Exercise 37 in Exercises 2.2 is defined by

$$C(x) = \begin{cases} 0.80x & \text{if } 0 < x \le 50 \\ 0.70x & \text{if } 50 < x \le 200 \\ 0.65x & \text{if } 200 < x \end{cases}$$

Draw a sketch of the graph of C. Determine where C is discontinuous, and show why the definition of a continuous function is not satisfied at each discontinuity.

36. Suppose the postage of a letter is computed as follows: 20 cents for each of the first 7 oz (or fractional part of an ounce), and then \$1.60 if the weight is greater than 7 oz and less than or equal to 1 lb. If x ounces represents the weight of a letter, express the number of cents in the postage as a function of x. Assume $0 \le x \le 16$. Draw a sketch of the graph of this function. Where is the function discontinuous? Show why the definition of a continuous function is not satisfied at each discontinuity.

37. Refer to Exercise 3 in Exercises 1.5. A manufacturer of open tin boxes wishes to make use of pieces of tin with dimensions 8 in. by 15 in. by cutting equal squares from the four corners and turning up the sides. (a) If x in. is the length of the side of the square cut out, express the number of cubic inches in the volume of the box as a function of x. (b) Show that the function in part (a) is continuous on its domain.

38. Refer to Exercise 4 in Exercises 1.5. Suppose the manufacturer of Exercise 37 makes the open boxes from square pieces of tin that measure k cm on a side. (a) If x cm is the length of the side of the square cut out, express the number of cubic centimeters in the volume of the box as a function of x. (b) Show that the function in part (a) is continuous on its domain.

39. Refer to Exercise 5 in Exercises 1.5. A rectangular field is to be enclosed with 240 m of fence. (a) If x meters is the length of the field, express the number of square meters in the area of the field as a function of x. (b) Show that the function in part (a) is continuous on its domain.

40. Refer to Exercise 8 in Exercises 1.5. A rectangular garden is to be placed so that a side of a house serves as a boundary and 100 ft of fencing material is to be used for the other three sides. (a) If x ft is the length of the side of the garden that is parallel to the house, express the number of square feet in the area of the garden as a function of x. (b) Show that the function in part (a) is continuous on its domain.

41. Refer to Exercise 17 in Exercises 1.5. Orange trees grown in California produce 600 oranges per year if not more than 20 trees are planted per acre. For each additional tree planted per acre, the yield decreases by 15 oranges. (a) If x trees are planted per acre, express the number of oranges produced per year as a function of x. (b) Show that the function in part (a) is continuous at 20 and hence continuous on its domain.

42. Refer to Exercise 18 in Exercises 1.5. A manufacturer can make a profit of $20 on each item if not more than 800 items are produced each week. The profit on each item decreases 2 cents for every item over 800. (a) If x items are produced each week, express the number of dollars in the manufacturer's weekly profit as a function of x. Assume that the profit is nonnegative. (b) Show that the function in part (a) is continuous at 800 and hence continuous on its domain.

43. Refer to Exercise 21 in Exercises 1.5. The maximum number of bacteria supportable by a particular environment is 900,000, and the rate of bacterial growth is jointly proportional to the number present and the difference between 900,000 and the number present. (a) If $f(x)$ bacteria per minute is the rate of growth when x bacteria are present, write an equation defining $f(x)$, (b) Show that the function f in part (a) is continuous on its domain.

44. Refer to Exercise 22 in Exercises 1.5. A particular lake can support a maximum of 14,000 fish, and the rate of growth of the fish population is jointly proportional to the number of fish present and the difference between 14,000 and the number present. (a) If $f(x)$ fish per day is the rate of growth when x fish are present, write an equation defining $f(x)$. (b) Show that the function f in part (a) is continuous on its domain.

2.4 The tangent line

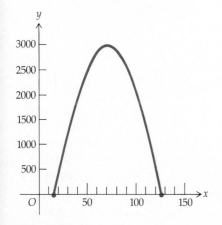

Figure 2.4.1

In Example 1 of Sec. 1.5 we had the following problem: "A clock manufacturer can produce a particular clock at a cost of $15 per clock. It is estimated that if the selling price of the clock is x dollars each, then the number of clocks sold per week is $125 - x$." In the solution of this example we found that if $P(x)$ is the number of dollars in the manufacturer's weekly profit, then

$$P(x) = (125 - x)(x - 15) \qquad (1)$$

Observe from (1) that $P(15) = 0$ and $P(125) = 0$ and that $P(x) > 0$ when x is in the open interval $(15, 125)$. A sketch of the graph of P is shown in Fig. 2.4.1.

Suppose we wish to determine what the selling price should be for the manufacturer's weekly profit to be a maximum. From the graph in Fig. 2.4.1 it is apparent that if $P(x)$ is to have a maximum value, it must occur for some number in the open interval $(15, 125)$. In Chapter 4 we learn that if a polynomial function has a maximum value, then it must occur at a point where the tangent line is horizontal, that is, at a point where the

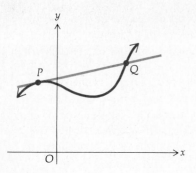

Figure 2.4.2

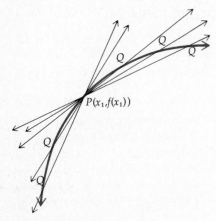

Figure 2.4.3

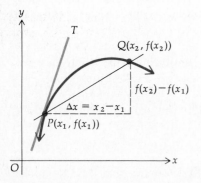

Figure 2.4.4

slope of the tangent line is zero. Therefore we can determine the value of x that gives a maximum value for $P(x)$ if we have a method for computing the slope of the tangent line to the graph of the function P. We now proceed to develop this method, and then we will return to this example.

We first consider how to define a tangent line. For a circle, we know from plane geometry that the tangent line at a point on the circle is the line intersecting the circle at only that point. This definition does not suffice for a curve in general. For example, in Fig. 2.4.2 the line that we wish to be the tangent line to the curve at the point P intersects the curve at another point Q. To arrive at a suitable definition of the tangent line to the graph of a function at a point on the graph, we proceed by considering how to define the slope of the tangent line at the point. Then the tangent line is determined by its slope and the point of tangency.

Consider the function f continuous at x_1. We wish to define the slope of the tangent line to the graph of f at $P(x_1, f(x_1))$. Let I be an open interval that contains x_1 and on which f is defined. Let $Q(x_2, f(x_2))$ be another point on the graph of f such that x_2 is also in I. Draw a line through P and Q. Any line through two points on a curve is called a **secant line**; therefore the line through P and Q is a secant line. In Fig. 2.4.3 the secant line is shown for various values of x_2. Figure 2.4.4 shows one particular secant line. In this figure Q is to the right of P. However, Q may be on either the right or the left side of P, as seen in Fig. 2.4.3.

Denote the difference of the abscissas of Q and P by Δx so that

$$\Delta x = x_2 - x_1$$

Δx may be either positive or negative. The slope of the secant line PQ then is given by

$$m_{PQ} = \frac{f(x_2) - f(x_1)}{\Delta x}$$

provided that line PQ is not vertical. Because $x_2 = x_1 + \Delta x$, the above equation can be written as

$$m_{PQ} = \frac{f(x_1 + \Delta x) - f(x_1)}{\Delta x}$$

Now think of point P as being fixed, and move point Q along the curve toward P; that is, Q approaches P. This is equivalent to stating that Δx approaches zero. As this occurs, the secant line turns about the fixed point P. If this secant line has a limiting position, it is this limiting position that we wish to be the tangent line to the graph at P. So we want the slope of the tangent line to the graph at P to be the limit of m_{PQ} as Δx approaches zero, if this limit exists. If $\lim_{\Delta x \to 0} m_{PQ} = +\infty$ or $-\infty$, then as Δx approaches zero, the line PQ approaches the line through P that is parallel to the y axis. In this case we would want the tangent line to the graph at P to be the line $x = x_1$. The preceding discussion leads to the following definition.

Definition of the tangent line to the graph of a function at a point

Suppose the function f is continuous at x_1. The **tangent line** to the graph of f at the point $P(x_1, f(x_1))$ is

(i) the line through P having slope $m(x_1)$, given by

$$m(x_1) = \lim_{\Delta x \to 0} \frac{f(x_1 + \Delta x) - f(x_1)}{\Delta x} \qquad (2)$$

if this limit exists;

(ii) the line $x = x_1$ if

$$\lim_{\Delta x \to 0} \frac{f(x_1 + \Delta x) - f(x_1)}{\Delta x} = +\infty \text{ or } -\infty$$

If neither (i) nor (ii) of the above definition holds, then there is no tangent line to the graph of f at the point $P(x_1, f(x_1))$.

EXAMPLE 1 Given the parabola $y = x^2$. In parts (a) through (c) find the slope of the secant line through the two points: (a) $(2, 4)$, $(3, 9)$; (b) $(2, 4)$, $(2.1, 4.41)$; (c) $(2,4)$, $(2.01, 4.0401)$. (d) Find the slope of the tangent line to the parabola at the point $(2, 4)$. (e) Draw a sketch of the graph of the parabola and a piece of the tangent line at $(2, 4)$.

Solution Let m_a, m_b, and m_c be the slopes of the secant lines for (a), (b), and (c), respectively.

(a) $m_a = \dfrac{9 - 4}{3 - 2}$ \qquad (b) $m_b = \dfrac{4.41 - 4}{2.1 - 2}$ \qquad (c) $m_c = \dfrac{4.0401 - 4}{2.01 - 2}$

$\qquad\quad = 5 \qquad\qquad\qquad = \dfrac{0.41}{0.1} \qquad\qquad\quad = \dfrac{0.0401}{0.01}$

$\qquad\qquad\qquad\qquad\qquad\quad = 4.1 \qquad\qquad\qquad = 4.01$

(d) $f(x) = x^2$. From (2) we have

$$m(2) = \lim_{\Delta x \to 0} \frac{f(2 + \Delta x) - f(2)}{\Delta x}$$

$$= \lim_{\Delta x \to 0} \frac{(2 + \Delta x)^2 - 4}{\Delta x}$$

$$= \lim_{\Delta x \to 0} \frac{4 + 4\,\Delta x + (\Delta x)^2 - 4}{\Delta x}$$

$$= \lim_{\Delta x \to 0} \frac{4\,\Delta x + (\Delta x)^2}{\Delta x}$$

$$= \lim_{\Delta x \to 0} (4 + \Delta x)$$

$$= 4$$

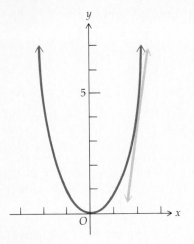

Figure 2.4.5

Table 2.4.1

x	y	m
2	-1	0
1	0	-2
0	3	-4
-1	8	-6
3	0	2
4	3	4
5	8	6

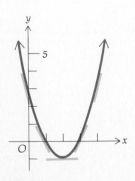

Figure 2.4.6

(e)   A sketch of the graph of the parabola and a piece of the tangent line at (2, 4) are shown in Fig. 2.4.5.

EXAMPLE 2   Find the slope of the tangent line to the curve $y = x^2 - 4x + 3$ at the point (x_1, y_1).

Solution   We let $f(x) = x^2 - 4x + 3$; therefore $f(x_1) = x_1{}^2 - 4x_1 + 3$, and $f(x_1 + \Delta x) = (x_1 + \Delta x)^2 - 4(x_1 + \Delta x) + 3$. From (2),

$$m(x_1) = \lim_{\Delta x \to 0} \frac{f(x_1 + \Delta x) - f(x_1)}{\Delta x}$$

$$= \lim_{\Delta x \to 0} \frac{\left[(x_1 + \Delta x)^2 - 4(x_1 + \Delta x) + 3\right] - \left[x_1{}^2 - 4x_1 + 3\right]}{\Delta x}$$

$$= \lim_{\Delta x \to 0} \frac{x_1{}^2 + 2x_1\,\Delta x + (\Delta x)^2 - 4x_1 - 4\,\Delta x + 3 - x_1{}^2 + 4x_1 - 3}{\Delta x}$$

$$= \lim_{\Delta x \to 0} \frac{2x_1\,\Delta x + (\Delta x)^2 - 4\,\Delta x}{\Delta x}$$

Because $\Delta x \neq 0$, the numerator and the denominator can be divided by Δx to obtain

$$m(x_1) = \lim_{\Delta x \to 0} (2x_1 + \Delta x - 4)$$

$$= 2x_1 - 4 \tag{3}$$

To draw a sketch of the graph of the equation in Example 2 we plot some points and a segment of the tangent line at some points. Values of x are taken arbitrarily, and the corresponding value of y is computed from the given equation, as well as the value of m from (3). The results are given in Table 2.4.1, and a sketch of the graph is shown in Fig. 2.4.6. It is important to determine the points where the graph has a horizontal tangent. Because a horizontal line has a slope of zero, these points are found by setting $m(x_1) = 0$ and solving for x_1. Doing this calculation for this example we have $2x_1 - 4 = 0$, which gives $x_1 = 2$. Therefore, at the point having an abscissa of 2 the tangent line is parallel to the x axis.

EXAMPLE 3   Find an equation of the tangent line to the curve of Example 2 at the point (5, 8).

Solution   Because the slope of the tangent line at any point (x_1, y_1) is given by

$$m(x_1) = 2x_1 - 4$$

the slope of the tangent line at the point (5, 8) is $m(5) = 2(5) - 4 = 6$. Therefore an equation of the desired line in the point-slope form is

$$y - 8 = 6(x - 5)$$
$$y = 6x - 22$$

In the following illustration we return to the example discussed at the beginning of this section.

● ILLUSTRATION 1

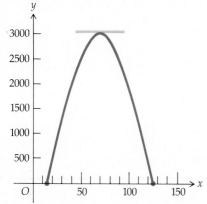

Figure 2.4.7

A clock manufacturer can produce a particular clock at a cost of $15 per clock. It is estimated that if the selling price of the clock is x dollars each, then the number of clocks sold per week is $125 - x$. If $P(x)$ dollars is the number of dollars in the manufacturer's weekly profit, then

$$P(x) = (125 - x)(x - 15)$$
$$= -x^2 + 140x - 1875$$

We wish to find where $P(x)$ is a maximum. This maximum value of $P(x)$ occurs at the point on the graph of P where the tangent line is horizontal, that is, where the slope of the tangent line is zero. Figure 2.4.7 shows a sketch of the graph of P and a piece of the horizontal tangent line. We apply formula (2) to find the slope of the tangent line:

$$m(x_1) = \lim_{\Delta x \to 0} \frac{P(x_1 + \Delta x) - P(x_1)}{\Delta x}$$

$$= \lim_{\Delta x \to 0} \frac{[-(x_1 + \Delta x)^2 + 140(x_1 + \Delta x) - 1875] - (-x_1^2 + 140x_1 - 1875)}{\Delta x}$$

$$= \lim_{\Delta x \to 0} \frac{-x_1^2 - 2x_1 \Delta x - (\Delta x)^2 + 140x_1 + 140 \Delta x - 1875 + x_1^2 - 140x_1 + 1875}{\Delta x}$$

$$= \lim_{\Delta x \to 0} \frac{-2x_1 \Delta x - (\Delta x)^2 + 140 \Delta x}{\Delta x}$$

$$= \lim_{\Delta x \to 0} (-2x_1 - \Delta x + 140)$$

$$= -2x_1 + 140$$

To determine the value of x_1 for which the slope of the tangent line is zero, we set $m(x_1) = 0$. We have

$$-2x_1 + 140 = 0$$
$$-2x_1 = -140$$
$$x_1 = 70$$

We therefore conclude that if there is a maximum weekly profit, it will occur when the selling price of the clock is $70. Furthermore, because $P(70) = 3025$, the maximum weekly profit will then be $3025.

●

Exercises 2.4

In Exercises 1 through 6, find the slope of the tangent line to the graph at the point (x_1, y_1). Make a table of values of x, y, and m for integer values of x in the closed interval $[a, b]$, and include in the table all points where the graph has a horizontal tangent. Draw a sketch of the graph, and show a segment of the tangent line at each of the plotted points.

1. $y = 9 - x^2$; $[a, b] = [-3, 3]$
2. $y = x^2 + 4$; $[a, b] = [-2, 2]$
3. $y = -2x^2 + 4x$; $[a, b] = [-1, 3]$
4. $y = x^2 - 6x + 9$; $[a, b] = [1, 5]$
5. $y = x^3 + 1$; $[a, b] = [-2, 2]$
6. $y = 1 - x^3$; $[a, b] = [-2, 2]$

In Exercises 7 through 10, find the slope of the tangent line to the graph at the point (x_1, y_1). Make a table of values of x, y, and m at various points on the graph, and include in the table all points where the graph has a horizontal tangent. Draw a sketch of the graph.

7. $y = x^3 - 3x$
8. $y = 7 - 6x - x^2$
9. $y = x^3 - 4x^2 + 4x - 2$
10. $y = x^3 - x^2 - x + 10$

In Exercises 11 through 16, find an equation of the tangent line to the given curve at the indicated point.

11. $y = x^2 - 4x - 5$; $(-2, 7)$
12. $y = x^2 - x + 2$; $(2, 4)$
13. $y = \sqrt{4 + x}$; $(5, 3)$
14. $y = \sqrt{9 - 4x}$; $(-4, 5)$
15. $y = \dfrac{6}{x}$; $(3, 2)$
16. $y = -\dfrac{8}{\sqrt{x}}$; $(4, -4)$

17. Find an equation of the tangent line to the curve $y = 2x^2 + 3$ that is parallel to the line $8x - y + 3 = 0$.
18. Find an equation of the tangent line to the curve $y = 3x^2 - 4$ that is parallel to the line $3x + y = 4$.
19. Find an equation of the tangent line to the curve $y = 2 - \frac{1}{3}x^2$ that is perpendicular to the line $x - y = 0$.
20. Find an equation of the tangent line to the curve $y = \sqrt{4x - 3} - 1$ that is perpendicular to the line $x + 2y - 11 = 0$.
21. Find an equation of each of the tangent lines to the curve $y = 1/x$ that is parallel to the line $x + 4y = 0$.
22. Find an equation of each of the tangent lines to the curve $y = x^3 - 3x$ that is perpendicular to the line $2x + 18y - 9 = 0$.
23. Refer to Exercise 1 in Exercises 1.5. A carpenter can construct bookcases at a cost of $40 each. If the carpenter sells the bookcases for x dollars each, it is estimated that $300 - 2x$ bookcases will be sold per month. (a) If $P(x)$ dollars is the carpenter's monthly profit, write an equation defining $P(x)$. (b) Use the method of Illustration 1 to determine the selling price of each bookcase for the carpenter's monthly profit to be a maximum. Assume that there is a maximum profit.
24. Refer to Exercise 2 in Exercises 1.5. A toy manufacturer can produce a particular toy at a cost of $10 per toy. It is estimated that if the selling price of the toy is x dollars, then the number of toys sold each day is $45 - x$. (a) If $P(x)$ dollars is the manufacturer's daily profit, write an equation defining $P(x)$. (b) Under the assumption that there is a maximum daily profit, use the method of Illustration 1 to determine what the selling price of each toy should be for the manufacturer to realize the maximum daily profit.

2.5 The derivative

In Sec. 2.4 the slope of the tangent line to the graph of $y = f(x)$ at the point $(x_1, f(x_1))$ is defined by

$$m(x_1) = \lim_{\Delta x \to 0} \frac{f(x_1 + \Delta x) - f(x_1)}{\Delta x} \tag{1}$$

if this limit exists.

This type of limit occurs in other problems, too, and has a specific name.

Definition of the derivative
of a function

> The **derivative** of the function f is that function, denoted by f', such that its value at any number x in the domain of f is given by
>
> $$f'(x) = \lim_{\Delta x \to 0} \frac{f(x + \Delta x) - f(x)}{\Delta x} \qquad (2)$$
>
> if this limit exists.

Another symbol used instead of $f'(x)$ is $D_x f(x)$, which is read "the derivative of f of x with respect to x."

If x_1 is a particular number in the domain of f, then

$$f'(x_1) = \lim_{\Delta x \to 0} \frac{f(x_1 + \Delta x) - f(x_1)}{\Delta x} \qquad (3)$$

if this limit exists. Comparing (1) and (3), note that the slope of the tangent line to the graph of $y = f(x)$ at the point $(x_1, f(x_1))$ is precisely the derivative of f evaluated at x_1.

EXAMPLE 1 Given $f(x) = 3x^2 + 12$, find the derivative of f.

Solution If x is any number in the domain of f, then from (2),

$$f'(x) = \lim_{\Delta x \to 0} \frac{f(x + \Delta x) - f(x)}{\Delta x}$$

$$= \lim_{\Delta x \to 0} \frac{[3(x + \Delta x)^2 + 12] - (3x^2 + 12)}{\Delta x}$$

$$= \lim_{\Delta x \to 0} \frac{3x^2 + 6x\,\Delta x + 3(\Delta x)^2 + 12 - 3x^2 - 12}{\Delta x}$$

$$= \lim_{\Delta x \to 0} \frac{6x\,\Delta x + 3(\Delta x)^2}{\Delta x}$$

$$= \lim_{\Delta x \to 0} (6x + 3\,\Delta x)$$

$$= 6x$$

Therefore the derivative of f is the function f' defined by $f'(x) = 6x$. The domain of f' is the set of all real numbers, which is the same as the domain of f.

EXAMPLE 2 For the function of Example 1, find the derivative of f at 2 in two ways: (a) Apply formula (3); (b) substitute 2 for x in the expression for $f'(x)$ found in Example 1.

Solution   (a) $f(x) = 3x^2 + 12$. From (3),

$$f'(2) = \lim_{\Delta x \to 0} \frac{f(2 + \Delta x) - f(2)}{\Delta x}$$

$$= \lim_{\Delta x \to 0} \frac{[3(2 + \Delta x)^2 + 12] - [3(2)^2 + 12]}{\Delta x}$$

$$= \lim_{\Delta x \to 0} \frac{12 + 12\,\Delta x + 3(\Delta x)^2 + 12 - 12 - 12}{\Delta x}$$

$$= \lim_{\Delta x \to 0} \frac{12\,\Delta x + 3(\Delta x)^2}{\Delta x}$$

$$= \lim_{\Delta x \to 0} (12 + 3\,\Delta x)$$

$$= 12$$

(b) Because from Example 1, $f'(x) = 6x$, then $f'(2) = 12$.

EXAMPLE 3   Given $f(x) = \sqrt{x - 3}$, find $D_x f(x)$.

Solution

$$D_x f(x) = \lim_{\Delta x \to 0} \frac{f(x + \Delta x) - f(x)}{\Delta x} = \lim_{\Delta x \to 0} \frac{\sqrt{(x + \Delta x) - 3} - \sqrt{x - 3}}{\Delta x}$$

To obtain a common factor of Δx in the numerator and denominator, we rationalize the numerator by multiplying numerator and denominator by $(\sqrt{x + \Delta x - 3} + \sqrt{x - 3})$. We have

$$D_x f(x) = \lim_{\Delta x \to 0} \frac{(\sqrt{x + \Delta x - 3} - \sqrt{x - 3})(\sqrt{x + \Delta x - 3} + \sqrt{x - 3})}{\Delta x(\sqrt{x + \Delta x - 3} + \sqrt{x - 3})}$$

$$= \lim_{\Delta x \to 0} \frac{x + \Delta x - 3 - (x - 3)}{\Delta x(\sqrt{x + \Delta x - 3} + \sqrt{x - 3})}$$

$$= \lim_{\Delta x \to 0} \frac{\Delta x}{\Delta x(\sqrt{x + \Delta x - 3} + \sqrt{x - 3})}$$

$$= \lim_{\Delta x \to 0} \frac{1}{\sqrt{x + \Delta x - 3} + \sqrt{x - 3}}$$

$$= \frac{1}{\sqrt{x - 3} + \sqrt{x - 3}}$$

$$= \frac{1}{2\sqrt{x - 3}}$$

If the function f is given by the equation $y = f(x)$, we can let

$$\Delta y = f(x + \Delta x) - f(x)$$

and write $\dfrac{dy}{dx}$ in place of $f'(x)$ so that from formula (2) we have

$$\frac{dy}{dx} = \lim_{\Delta x \to 0} \frac{\Delta y}{\Delta x}$$

Remember that when using $\dfrac{dy}{dx}$ as a notation for a derivative, dy and dx have so far not been given independent meaning.

EXAMPLE 4   Given $y = \dfrac{2 + x}{3 - x}$, find $\dfrac{dy}{dx}$.

Solution

$$\frac{dy}{dx} = \lim_{\Delta x \to 0} \frac{\Delta y}{\Delta x}$$

$$= \lim_{\Delta x \to 0} \frac{f(x + \Delta x) - f(x)}{\Delta x}$$

$$= \lim_{\Delta x \to 0} \frac{\dfrac{2 + x + \Delta x}{3 - x - \Delta x} - \dfrac{2 + x}{3 - x}}{\Delta x}$$

$$= \lim_{\Delta x \to 0} \frac{(3 - x)(2 + x + \Delta x) - (2 + x)(3 - x - \Delta x)}{\Delta x(3 - x - \Delta x)(3 - x)}$$

$$= \lim_{\Delta x \to 0} \frac{(6 + x - x^2 + 3\,\Delta x - x\,\Delta x) - (6 + x - x^2 - 2\,\Delta x - x\,\Delta x)}{\Delta x(3 - x - \Delta x)(3 - x)}$$

$$= \lim_{\Delta x \to 0} \frac{5\,\Delta x}{\Delta x(3 - x - \Delta x)(3 - x)}$$

$$= \lim_{\Delta x \to 0} \frac{5}{(3 - x - \Delta x)(3 - x)}$$

$$= \frac{5}{(3 - x)^2}$$

The German mathematician Gottfried Wilhelm Leibniz (1646–1716) was the first to use the notation $\dfrac{dy}{dx}$ for the derivative of y with respect to x. The concept of a derivative was introduced in the seventeenth century, almost simultaneously by Leibniz and Sir Isaac Newton (1642–1727), who were working independently. Leibniz probably thought of dx and dy as small changes in the variables x and y and of the derivative of y with respect to x as the ratio of dy to dx as dy and dx become small. The concept of a limit as we know it today was not known to Leibniz.

If $y = f(x)$, sometimes the notation $D_x y$ is used for the derivative of f. The notation y' is also used for the derivative of y with respect to an independent variable if the independent variable is understood.

EXAMPLE 5   Given $f(x) = x^{1/3}$. (a) Find $f'(x)$. (b) Show that $f'(0)$ does not exist even though f is continuous at 0. (c) Draw a sketch of the graph of f.

Solution

(a) $f'(x) = \lim\limits_{\Delta x \to 0} \dfrac{f(x + \Delta x) - f(x)}{\Delta x} = \lim\limits_{\Delta x \to 0} \dfrac{(x + \Delta x)^{1/3} - x^{1/3}}{\Delta x}$

We rationalize the numerator in order to obtain a common factor of Δx in the numerator and the denominator; this yields

$$f'(x) = \lim_{\Delta x \to 0} \frac{[(x + \Delta x)^{1/3} - x^{1/3}][(x + \Delta x)^{2/3} + (x + \Delta x)^{1/3}x^{1/3} + x^{2/3}]}{\Delta x[(x + \Delta x)^{2/3} + (x + \Delta x)^{1/3}x^{1/3} + x^{2/3}]}$$

$$= \lim_{\Delta x \to 0} \frac{(x + \Delta x) - x}{\Delta x[(x + \Delta x)^{2/3} + (x + \Delta x)^{1/3}x^{1/3} + x^{2/3}]}$$

$$= \lim_{\Delta x \to 0} \frac{1}{(x + \Delta x)^{2/3} + (x + \Delta x)^{1/3}x^{1/3} + x^{2/3}}$$

$$= \frac{1}{x^{2/3} + x^{1/3}x^{1/3} + x^{2/3}}$$

$$= \frac{1}{3x^{2/3}}$$

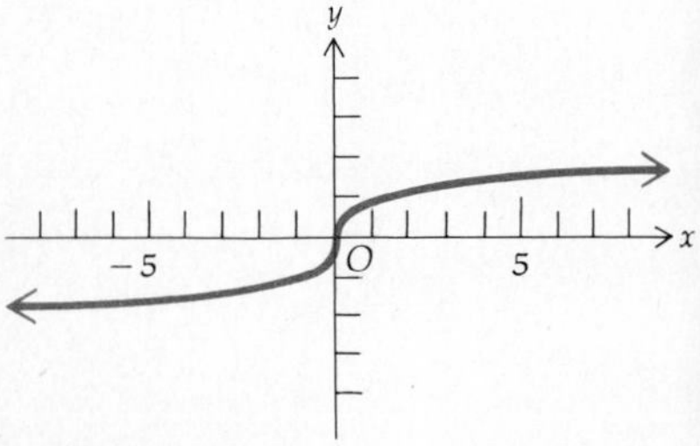

(b) $f'(0)$ does not exist because $\dfrac{1}{3x^{2/3}}$ is not defined when $x = 0$. However, the function f is continuous at 0 from Theorem 2.3.3(ii).

(c) A sketch of the graph of f is shown in Fig. 2.5.1.

Figure 2.5.1

● ILLUSTRATION 1

For the function f of Example 5, consider

$$\lim_{\Delta x \to 0} \frac{f(x_1 + \Delta x) - f(x_1)}{\Delta x} \quad \text{at } x_1 = 0$$

We have

$$\lim_{\Delta x \to 0} \frac{f(0 + \Delta x) - f(0)}{\Delta x} = \lim_{\Delta x \to 0} \frac{(\Delta x)^{1/3} - 0}{\Delta x}$$

$$= \lim_{\Delta x \to 0} \frac{1}{(\Delta x)^{2/3}}$$

$$= +\infty$$

From part (ii) of the definition of a tangent line given in Sec. 2.4, it follows that the line $x = 0$ (the y axis) is the tangent line to the graph of f at the origin. ●

Example 5 and Illustration 1 show that $f'(x)$ can exist for some values of x in the domain of f but fail to exist for other values of x in the domain of f. We have the following definition.

Definition of a function differentiable at a number

> The function f is said to be **differentiable** at x_1 if $f'(x_1)$ exists.

● ILLUSTRATION 2

From the above definition it follows that the function of Example 5 and Illustration 1 is differentiable at every number except 0. ●

● ILLUSTRATION 3

In Example 1, $f(x) = 3x^2 + 12$, and the domain of f is the set of all real numbers. Because $f'(x) = 6x$, and $6x$ exists for all real numbers, it follows that f is differentiable everywhere. ●

● ILLUSTRATION 4

Let f be the absolute value function. Thus

$$f(x) = |x|$$

A sketch of the graph of this function was obtained in Example 1 of Sec. 1.3. It is shown here in Fig. 2.5.2. From formula (3),

$$f'(0) = \lim_{\Delta x \to 0} \frac{f(0 + \Delta x) - f(0)}{\Delta x}$$

if this limit exists. Because $f(0 + \Delta x) = |\Delta x|$ and $f(0) = 0$,

$$\lim_{\Delta x \to 0} \frac{f(0 + \Delta x) - f(0)}{\Delta x} = \lim_{\Delta x \to 0} \frac{|\Delta x|}{\Delta x}$$

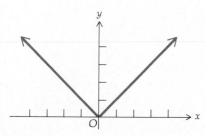

Figure 2.5.2

Because $|\Delta x| = \Delta x$ if $\Delta x > 0$ and $|\Delta x| = -\Delta x$ if $\Delta x < 0$, one-sided limits at 0 must be considered.

$$\lim_{\Delta x \to 0^+} \frac{|\Delta x|}{\Delta x} = \lim_{\Delta x \to 0^+} \frac{\Delta x}{\Delta x} = \lim_{\Delta x \to 0^+} 1 = 1$$

and

$$\lim_{\Delta x \to 0^-} \frac{|\Delta x|}{\Delta x} = \lim_{\Delta x \to 0^-} \frac{-\Delta x}{\Delta x} = \lim_{\Delta x \to 0^-} (-1) = -1$$

Because $\lim\limits_{\Delta x \to 0^+} \dfrac{|\Delta x|}{\Delta x} \neq \lim\limits_{\Delta x \to 0^-} \dfrac{|\Delta x|}{\Delta x}$, it follows that the two-sided limit $\lim\limits_{\Delta x \to 0} \dfrac{|\Delta x|}{\Delta x}$ does not exist. Therefore $f'(0)$ does not exist, and so f is not differentiable at 0.

Because $f'(0)$ does not exist and is neither $+\infty$ nor $-\infty$, there is no tangent line at the origin for the graph of the absolute value function. ●

Because the functions of Illustration 4 and Example 5 are continuous at a number but not differentiable there, it may be concluded that continuity of a function at a number does not imply differentiability of the function at that number. However, differentiability *does* imply continuity, as given by the next theorem, which is stated without proof.

Theorem 2.5.1   If a function f is differentiable at x_1, then f is continuous at x_1.

A function f can fail to be differentiable at a number c for one of the following reasons:

1. The function f is discontinuous at c. This follows from Theorem 2.5.1. Refer to Fig. 2.5.3 for a sketch of the graph of such a function.
2. The function f is continuous at c, and the graph of f has a vertical tangent line at the point where $x = c$. See Fig. 2.5.4 for a sketch of the graph of a function having this property. This situation also occurs in Example 5.
3. The function f is continuous at c, and the graph of f does not have a tangent line at the point where $x = c$. Figure 2.5.5 shows a sketch of the graph of a function satisfying this condition. Observe that there is a "sharp turn" in the graph at $x = c$. The absolute value function, discussed in Illustration 4, is another such function.

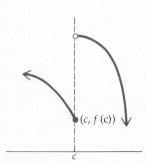

$(c, f(c))$

c

Figure 2.5.3

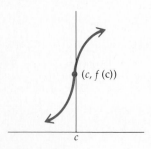

$(c, f(c))$

c

Figure 2.5.4

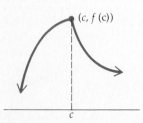

$(c, f(c))$

c

Figure 2.5.5

Exercises 2.5

In Exercises 1 through 10, find $f'(x)$ for the given function by applying formula (2) of this section.

1. $f(x) = 7x + 3$
2. $f(x) = 8 - 5x$
3. $f(x) = -4$
4. $f(x) = 3x^2 + 4$
5. $f(x) = 4 - 2x^2$
6. $f(x) = 3x^2 - 2x + 1$
7. $f(x) = 4x^2 + 5x + 3$
8. $f(x) = 6 - 3x - x^2$
9. $f(x) = 8 - x^3$
10. $f(x) = x^3$

In Exercises 11 through 16, find $D_x f(x)$.

11. $f(x) = \sqrt{x}$
12. $f(x) = \sqrt{3x + 5}$
13. $f(x) = \dfrac{1}{x + 1}$

14. $f(x) = \dfrac{2x + 3}{3x - 2}$
15. $f(x) = \dfrac{1}{\sqrt{2x + 1}}$
16. $f(x) = \dfrac{1}{\sqrt{3x}}$

In Exercises 17 through 22, find $f'(a)$ by applying formula (3) of this section.

17. $f(x) = 1 - x^2; a = 3$
18. $f(x) = \dfrac{4}{5x}; a = 2$
19. $f(x) = \dfrac{2}{x^3}; a = 4$

20. $f(x) = \dfrac{2}{\sqrt{x}} - 1; a = 4$
21. $f(x) = \sqrt{10 - x}; a = 6$
22. $f(x) = \dfrac{1}{x} + x + x^2; a = -3$

In Exercises 23 through 26, find $\dfrac{dy}{dx}$.

23. $y = x^2 + x^{-2}$
24. $y = \dfrac{1}{x^2} - x$
25. $y = \dfrac{x}{2x - 3}$
26. $y = \dfrac{4}{2x - 1}$

In Exercises 27 through 30, find $D_x y$.

27. $y = \dfrac{4}{x^2} + 3x - x^2$
28. $y = 4 - 5x + \dfrac{3}{x^2}$
29. $y = \sqrt{4 - x^2}$
30. $y = \dfrac{6}{\sqrt{2x + 3}}$

31. Given $f(x) = \sqrt[3]{x - 1}$. (a) Find $f'(x)$. (b) Is f differentiable at 1? (c) Draw a sketch of the graph of f.
32. Given $f(x) = x^{2/3}$. (a) Find $f'(x)$. (b) Is f differentiable at 0? (c) Draw a sketch of the graph of f.
33. Given $f(x) = |x - 2|$. (a) Draw a sketch of the graph of f. (b) Show that f is not differentiable at 2.
34. Given $f(x) = |x + 1|$. (a) Draw a sketch of the graph of f. (b) Show that f is not differentiable at -1.

2.6 Techniques of differentiation

The operation of finding the derivative of a function is called **differentiation**, which can be performed by applying the definition of a derivative given in Sec. 2.5. However, because this process is usually quite lengthy, we need some theorems that enable us to find the derivative of certain functions more easily. These theorems are proved by applying the definition of a derivative. Along with the statement of each theorem, we give the corresponding formula of differentiation.

The first theorem gives the derivative of a constant function. Suppose

$$f(x) = c$$

where c is a constant. Then

$$f'(x) = \lim_{\Delta x \to 0} \frac{f(x + \Delta x) - f(x)}{\Delta x}$$

$$= \lim_{\Delta x \to 0} \frac{c - c}{\Delta x}$$

$$= \lim_{\Delta x \to 0} 0$$

$$= 0$$

Thus the derivative of a constant is zero. This fact agrees with the geometric interpretation of the derivative because the graph of $f(x) = c$ is a horizontal straight line having a slope of zero. The result is stated formally as a theorem.

The derivative of a constant

> **Theorem 2.6.1** If c is a constant, and if $f(x) = c$ for all x, then
>
> $$f'(x) = 0$$
>
> $$\boxed{D_x(c) = 0}$$
>
> *The derivative of a constant is zero.*

● ILLUSTRATION 1

(a) If $f(x) = 5$, then

$$f'(x) = 0$$

(b) If $g(x) = -\pi$, then

$$g'(x) = 0$$ ●

We proceed now to obtain the formula for the derivative of a power function, which is a function f defined by $f(x) = x^n$, where n is any real number. Examples of power functions are $f(x) = x^3$, $g(x) = x^{-2}$, $h(x) = x^{1/2}$, and $F(x) = x^{-2/3}$.

Consider a power function where n is any positive integer; that is,

$$f(x) = x^n$$

where n is a positive integer. Then

$$f'(x) = \lim_{\Delta x \to 0} \frac{f(x + \Delta x) - f(x)}{\Delta x}$$

$$= \lim_{\Delta x \to 0} \frac{(x + \Delta x)^n - x^n}{\Delta x}$$

Applying the binomial theorem to $(x + \Delta x)^n$ we have

$$f'(x) = \lim_{\Delta x \to 0} \frac{\left[x^n + nx^{n-1}\,\Delta x + \dfrac{n(n-1)}{2!}x^{n-2}(\Delta x)^2 + \ldots + nx(\Delta x)^{n-1} + (\Delta x)^n\right] - x^n}{\Delta x}$$

$$= \lim_{\Delta x \to 0} \frac{nx^{n-1}\,\Delta x + \dfrac{n(n-1)}{2!}x^{n-2}(\Delta x)^2 + \ldots + nx(\Delta x)^{n-1} + (\Delta x)^n}{\Delta x}$$

Dividing the numerator and the denominator by Δx we have

$$f'(x) = \lim_{\Delta x \to 0}\left[nx^{n-1} + \frac{n(n-1)}{2!}x^{n-2}(\Delta x) + \ldots + nx(\Delta x)^{n-2} + (\Delta x)^{n-1}\right]$$

Every term except the first has a factor of Δx; therefore every term except the first approaches zero as Δx approaches zero. So we obtain

$$f'(x) = nx^{n-1}$$

We have proved the following theorem for n any positive integer. In Illustration 4, the theorem is proved for n any negative integer. The proof for any real number n is deferred until Sec. 8.5.

The derivative of a
power function

Theorem 2.6.2   If n is any real number, and if $f(x) = x^n$, then

$$f'(x) = nx^{n-1}$$

$$D_x(x^n) = nx^{n-1}$$

● ILLUSTRATION 2

(a) If $f(x) = x^8$, then

$$f'(x) = 8x^7$$

(b) If $f(x) = x$, then

$$f'(x) = 1 \cdot x^0$$
$$= 1 \cdot 1$$
$$= 1$$

(c) If $f(x) = \sqrt{x}$, then

$$f(x) = x^{1/2}$$

Thus

$$f'(x) = \tfrac{1}{2}x^{-1/2}$$

$$= \frac{1}{2\sqrt{x}}$$

●

The next theorem involves the derivative of a constant times a function. If

$$g(x) = c \cdot f(x)$$

where c is a constant, then

$$g'(x) = \lim_{\Delta x \to 0} \frac{g(x + \Delta x) - g(x)}{\Delta x}$$

$$= \lim_{\Delta x \to 0} \frac{cf(x + \Delta x) - cf(x)}{\Delta x}$$

$$= \lim_{\Delta x \to 0} c \cdot \left[\frac{f(x + \Delta x) - f(x)}{\Delta x} \right]$$

$$= c \cdot \lim_{\Delta x \to 0} \frac{f(x + \Delta x) - f(x)}{\Delta x}$$

$$= cf'(x)$$

We have proved the following theorem.

The derivative of a constant
times a function

> **Theorem 2.6.3** If f is a function, c is a constant, and g is the function defined by
>
> $$g(x) = c \cdot f(x)$$
>
> then if $f'(x)$ exists,
>
> $$g'(x) = c \cdot f'(x)$$
>
> $$D_x[c \cdot f(x)] = c \cdot D_x f(x)$$
>
> *The derivative of a constant times a function is the constant times the derivative of the function if this derivative exists.*

By combining Theorems 2.6.2 and 2.6.3 we obtain the following result.

The derivative of a constant times
a power function

> If $f(x) = cx^n$, where n is any real number and c is a constant,
>
> $$f'(x) = cnx^{n-1}$$
>
> $$D_x(cx^n) = cnx^{n-1}$$

● ILLUSTRATION 3

(a) If $f(x) = 5x^7$, then

$$f'(x) = 5 \cdot 7x^6$$

$$= 35x^6$$

(b) If $f(x) = 9x^{2/3}$, then

$$f'(x) = 9 \cdot \tfrac{2}{3} x^{-1/3}$$

$$= 6 \frac{1}{x^{1/3}}$$

$$= \frac{6}{\sqrt[3]{x}}$$

To obtain a formula for the derivative of the sum of two functions let

$$h(x) = f(x) + g(x)$$

where $f'(x)$ and $g'(x)$ exist. Then

$$h'(x) = \lim_{\Delta x \to 0} \frac{h(x + \Delta x) - h(x)}{\Delta x}$$

$$= \lim_{\Delta x \to 0} \frac{[f(x + \Delta x) + g(x + \Delta x)] - [f(x) + g(x)]}{\Delta x}$$

$$= \lim_{\Delta x \to 0} \frac{[f(x + \Delta x) - f(x)] + [g(x + \Delta x) - g(x)]}{\Delta x}$$

$$= \lim_{\Delta x \to 0} \frac{f(x + \Delta x) - f(x)}{\Delta x} + \lim_{\Delta x \to 0} \frac{g(x + \Delta x) - g(x)}{\Delta x}$$

$$= f'(x) + g'(x)$$

Therefore the sum of two differentiable functions is the sum of their derivatives, and we have the following theorem.

The derivative of the sum of two functions

> **Theorem 2.6.4**   If f and g are functions, and if h is the function defined by
>
> $$h(x) = f(x) + g(x)$$
>
> then if $f'(x)$ and $g'(x)$ exist,
>
> $$h'(x) = f'(x) + g'(x)$$
>
> $$D_x[f(x) + g(x)] = D_x f(x) + D_x g(x)$$
>
> *The derivative of the sum of two functions is the sum of their derivatives if these derivatives exist.*

The result of the preceding theorem can be extended to any finite number of functions; that is, the derivative of the sum of a finite number of functions is equal to the sum of their derivatives, if these derivatives exist. The derivative of any polynomial function can be found by applying this fact, as shown in the following example.

EXAMPLE 1 Given

$$f(x) = 7x^4 - 2x^3 + 8x + 5$$

find $f'(x)$.

Solution

$$\begin{aligned}
f'(x) &= D_x(7x^4 - 2x^3 + 8x + 5) \\
&= D_x(7x^4) + D_x(-2x^3) + D_x(8x) + D_x(5) \\
&= 28x^3 - 6x^2 + 8
\end{aligned}$$

The next theorem gives a formula for the derivative of the product of two functions. To obtain it let

$$h(x) = f(x)g(x)$$

where $f'(x)$ and $g'(x)$ exist. Then

$$h'(x) = \lim_{\Delta x \to 0} \frac{h(x + \Delta x) - h(x)}{\Delta x}$$

$$= \lim_{\Delta x \to 0} \frac{[f(x + \Delta x) \cdot g(x + \Delta x)] - [f(x) \cdot g(x)]}{\Delta x}$$

If $f(x + \Delta x) \cdot g(x)$ is subtracted and added in the numerator, then

$$h'(x) = \lim_{\Delta x \to 0} \frac{f(x + \Delta x) \cdot g(x + \Delta x) - f(x + \Delta x) \cdot g(x) + f(x + \Delta x) \cdot g(x) - f(x) \cdot g(x)}{\Delta x}$$

$$= \lim_{\Delta x \to 0} \left[f(x + \Delta x) \cdot \frac{g(x + \Delta x - g(x)}{\Delta x} + g(x) \cdot \frac{f(x + \Delta x) - f(x)}{\Delta x} \right]$$

$$= \lim_{\Delta x \to 0} \left[f(x + \Delta x) \cdot \frac{g(x + \Delta x) - g(x)}{\Delta x} \right] + \lim_{\Delta x \to 0} \left[g(x) \cdot \frac{f(x + \Delta x - f(x)}{\Delta x} \right]$$

$$= \lim_{\Delta x \to 0} f(x + \Delta x) \cdot \lim_{\Delta x \to 0} \frac{g(x + \Delta x) - g(x)}{\Delta x} + \lim_{\Delta x \to 0} g(x) \cdot \lim_{\Delta x \to 0} \frac{f(x + \Delta x) - f(x)}{\Delta x}$$

Because f is differentiable at x, by Theorem 2.5.1 f is continuous at x; therefore $\lim_{\Delta x \to 0} f(x + \Delta x) = f(x)$. Also,

$$\lim_{\Delta x \to 0} \frac{g(x + \Delta x) - g(x)}{\Delta x} = g'(x)$$

$$\lim_{\Delta x \to 0} \frac{f(x + \Delta x) - f(x)}{\Delta x} = f'(x)$$

and

$$\lim_{\Delta x \to 0} g(x) = g(x)$$

thus giving

$$h'(x) = f(x)g'(x) + g(x)f'(x)$$

The following theorem has been proved.

The derivative of the product
of two functions

> **Theorem 2.6.5** If f and g are functions, and if h is the function defined by
>
> $$h(x) = f(x)g(x)$$
>
> then if $f'(x)$ and $g'(x)$ exist,
>
> $$h'(x) = f(x)g'(x) + g(x)f'(x)$$
>
> $$D_x[f(x)g(x)] = f(x) \cdot D_x g(x) + g(x) \cdot D_x f(x)$$
>
> *The derivative of the product of two functions is the first function times the derivative of the second function plus the second function times the derivative of the first function if these derivatives exist.*

EXAMPLE 2 Given

$$h(x) = (2x^3 - 4x^2)(3x^5 + x^2)$$

find $h'(x)$.

Solution

$$\begin{aligned} h'(x) &= (2x^3 - 4x^2)(15x^4 + 2x) + (3x^5 + x^2)(6x^2 - 8x) \\ &= (30x^7 - 60x^6 + 4x^4 - 8x^3) + (18x^7 - 24x^6 + 6x^4 - 8x^3) \\ &= 48x^7 - 84x^6 + 10x^4 - 16x^3 \end{aligned}$$

In Example 2, note that if we multiply first and then perform the differentiation, the same result is obtained. Doing this we have

$$h(x) = 6x^8 - 12x^7 + 2x^5 - 4x^4$$

Thus

$$h'(x) = 48x^7 - 84x^6 + 10x^4 - 16x^3$$

To derive a formula for the derivative of the quotient of two functions let

$$h(x) = \frac{f(x)}{g(x)} \qquad g(x) \neq 0$$

where $f'(x)$ and $g'(x)$ exist. Then

$$h'(x) = \lim_{\Delta x \to 0} \frac{h(x + \Delta x) - h(x)}{\Delta x}$$

$$= \lim_{\Delta x \to 0} \frac{\dfrac{f(x + \Delta x)}{g(x + \Delta x)} - \dfrac{f(x)}{g(x)}}{\Delta x}$$

$$= \lim_{\Delta x \to 0} \frac{f(x + \Delta x) \cdot g(x) - f(x) \cdot g(x + \Delta x)}{\Delta x \cdot g(x) \cdot g(x + \Delta x)}$$

Subtracting and adding $f(x) \cdot g(x)$ in the numerator we obtain

$$h'(x) = \lim_{\Delta x \to 0} \frac{f(x + \Delta x) \cdot g(x) - f(x) \cdot g(x) - f(x) \cdot g(x + \Delta x) + f(x) \cdot g(x)}{\Delta x \cdot g(x) \cdot g(x + \Delta x)}$$

$$= \lim_{\Delta x \to 0} \frac{\left[g(x) \cdot \dfrac{f(x + \Delta x) - f(x)}{\Delta x} \right] - \left[f(x) \cdot \dfrac{g(x + \Delta x) - g(x)}{\Delta x} \right]}{g(x) \cdot g(x + \Delta x)}$$

$$= \frac{\displaystyle\lim_{\Delta x \to 0} g(x) \cdot \lim_{\Delta x \to 0} \frac{f(x + \Delta x) - f(x)}{\Delta x} - \lim_{\Delta x \to 0} f(x) \cdot \lim_{\Delta x \to 0} \frac{g(x + \Delta x) - g(x)}{\Delta x}}{\displaystyle\lim_{\Delta x \to 0} g(x) \cdot \lim_{\Delta x \to 0} g(x + \Delta x)}$$

$$= \frac{g(x) \cdot f'(x) - f(x) \cdot g'(x)}{g(x) \cdot g(x)}$$

$$= \frac{g(x)f'(x) - f(x)g'(x)}{[g(x)]^2}$$

We have proved the following theorem.

The derivative of the quotient
of two functions

Theorem 2.6.6 If f and g are functions, and if h is the function defined by

$$h(x) = \frac{f(x)}{g(x)} \qquad g(x) \neq 0$$

then if $f'(x)$ and $g'(x)$ exist,

$$h'(x) = \frac{g(x)f'(x) - f(x)g'(x)}{[g(x)]^2}$$

$$D_x \left[\frac{f(x)}{g(x)} \right] = \frac{g(x)D_x f(x) - f(x)D_x g(x)}{[g(x)]^2}$$

The derivative of the quotient of two functions is the fraction having as its denominator the square of the original denominator, and as its numerator the denominator times the derivative of the numerator minus the numerator times the derivative of the denominator if these derivatives exist.

EXAMPLE 3   Given

$$h(x) = \frac{2x^3 + 4}{x^2 - 4x + 1}$$

find $h'(x)$.

Solution

$$h'(x) = \frac{(x^2 - 4x + 1)(6x^2) - (2x^3 + 4)(2x - 4)}{(x^2 - 4x + 1)^2}$$

$$= \frac{6x^4 - 24x^3 + 6x^2 - 4x^4 + 8x^3 - 8x + 16}{(x^2 - 4x + 1)^2}$$

$$= \frac{2x^4 - 16x^3 + 6x^2 - 8x + 16}{(x^2 - 4x + 1)^2}$$

● ILLUSTRATION 4

We can prove Theorem 2.6.2, regarding the derivative of a power function, if the exponent is any negative integer, by applying Theorem 2.6.6. We make use of the fact that the theorem is true for any positive integer. Suppose that $f(x) = x^{-n}$, where $-n$ is a negative integer and $x \neq 0$. Then because $-n$ is a negative integer, n is a positive integer. We write

$$f(x) = \frac{1}{x^n}$$

From Theorem 2.6.6,

$$f'(x) = \frac{x^n \cdot 0 - 1 \cdot nx^{n-1}}{(x^n)^2}$$

$$= \frac{-nx^{n-1}}{x^{2n}}$$

$$= -nx^{n-1-2n}$$

$$= -nx^{-n-1}$$

●

EXAMPLE 4   Given

$$f(x) = \frac{3}{x^5} + 4\sqrt[4]{x^3}$$

find $f'(x)$.

Solution

$$f(x) = 3x^{-5} + 4x^{3/4}$$

$$f'(x) = 3(-5x^{-6}) + 4(\tfrac{3}{4}x^{-1/4})$$

$$= -15x^{-6} + 3x^{-1/4}$$

$$= -\frac{15}{x^6} + \frac{3}{\sqrt[4]{x}}$$

Exercises 2.6

In Exercises 1 through 42, differentiate the given function by applying the theorems of this section.

1. $f(x) = 7x - 5$ **2.** $g(x) = 8 - 3x$ **3.** $g(x) = 1 - 2x - x^2$

4. $f(x) = 4x^2 + 4x + 1$ **5.** $f(x) = x^3 - 3x^2 + 5x - 2$ **6.** $f(x) = 3x^4 - 5x^2 + 1$

7. $f(x) = \frac{1}{8}x^8 - x^4$ **8.** $g(x) = x^7 - 2x^5 + 5x^3 - 7x$ **9.** $F(t) = \frac{1}{4}t^4 - \frac{1}{2}t^2$

10. $H(x) = \frac{1}{3}x^3 - x + 2$ **11.** $v(r) = \frac{4}{3}\pi r^3$ **12.** $G(y) = y^{10} + 7y^5 - y^3 + 1$

13. $F(x) = x^2 + 3x + \dfrac{1}{x^2}$ **14.** $f(x) = \dfrac{x^3}{3} + \dfrac{3}{x^3}$ **15.** $g(x) = 4x^4 - \dfrac{1}{4x^4}$

16. $f(x) = x^4 - 5 + x^{-2} + 4x^{-4}$ **17.** $g(x) = \dfrac{3}{x^2} + \dfrac{5}{x^4}$ **18.** $H(x) = \dfrac{5}{6x^5}$

19. $f(x) = 4x^{1/2} + 6x^{-1/2}$ **20.** $f(x) = 6x^{2/3} - 9x^{1/3}$ **21.** $g(x) = \sqrt[3]{t} - 3\sqrt{\dfrac{1}{t}}$

22. $h(x) = 2\sqrt{x} + \dfrac{1}{2}\sqrt{\dfrac{1}{x}}$ **23.** $f(s) = \sqrt{3}(s^3 - s^2)$ **24.** $g(x) = (2x^2 + 5)(4x - 1)$

25. $f(x) = (2x^4 - 1)(5x^3 + 6x)$ **26.** $f(x) = (4x^2 + 3)^2$ **27.** $G(y) = (7 - 3y^3)^2$

28. $F(t) = (t^3 - 2t + 1)(2t^2 + 3t)$ **29.** $f(x) = (x^2 - 3x + 2)(2x^3 + 1)$ **30.** $g(x) = \dfrac{2x}{x + 3}$

31. $f(x) = \dfrac{x}{x - 1}$ **32.** $F(y) = \dfrac{2y + 1}{3y + 4}$ **33.** $H(x) = \dfrac{x^2 + 2x + 1}{x^2 - 2x + 1}$

34. $f(x) = \dfrac{4 - 3x - x^2}{x - 2}$ **35.** $h(x) = \dfrac{5x}{1 + 2x^2}$ **36.** $g(x) = \dfrac{x^4 - 2x^2 + 5x + 1}{x^4}$

37. $f(x) = \dfrac{x^3 - 8}{x^3 + 8}$ **38.** $f(x) = \dfrac{x^2 - a^2}{x^2 + a^2}$ **39.** $f(x) = \dfrac{2x + 1}{x + 5}(3x - 1)$

40. $g(x) = \dfrac{x^3 + 1}{x^2 + 3}(x^2 - 2x^{-1} + 1)$ **41.** $f(x) = \dfrac{\sqrt{x} - 1}{\sqrt{x} + 1}$ **42.** $f(t) = \dfrac{1}{4t} + 4t^{3/2} - \dfrac{2}{\sqrt[3]{t}}$

43. Find an equation of the tangent line to the curve $y = x^{1/2} - 3x$ at the point $(1, -2)$.

44. Find an equation of the tangent line to the curve $y = \dfrac{8}{x^2 + 4}$ at the point $(2, 1)$.

45. Given $f(x) = x^2 - 2x - 1$. (a) Find the point on the graph of f at which the tangent line is horizontal. (b) Draw a sketch of the graph of f, and show the horizontal tangent line.

46. Given $f(x) = -x^2 + 6x - 4$. (a) Find the point on the graph of f at which the tangent line is horizontal. (b) Draw a sketch of the graph of f, and show the horizontal tangent line.

47. Given $f(x) = x + 2 + \dfrac{4}{x}$. (a) Find all the points on the graph of f for which the tangent line is horizontal. (b) Draw a sketch of the graph of f, and show the horizontal tangent lines.

Review Exercises for Chapter 2

In Exercises 1 through 10, evaluate the limit if it exists.

1. $\lim\limits_{x \to 2} (3x^2 - 4x + 5)$

2. $\lim\limits_{y \to 1} \dfrac{y^2 - 4}{3y^3 + 6}$

3. $\lim\limits_{z \to -3} \dfrac{z^2 - 9}{z + 3}$

4. $\lim\limits_{x \to -2} \dfrac{x^2 - x - 6}{x^2 - 5x - 14}$

5. $\lim\limits_{x \to 1} \dfrac{2x^2 + x - 3}{3x^2 + 2x - 5}$

6. $\lim\limits_{t \to 4} \sqrt{\dfrac{t - 4}{t^2 - 16}}$

7. $\lim\limits_{x \to 4^+} \dfrac{2x}{16 - x^2}$

8. $\lim\limits_{x \to 0^-} \dfrac{x^2 - 5}{2x^3 - 3x^2}$

9. $\lim\limits_{t \to 2} \dfrac{\sqrt{t} - \sqrt{2}}{t^2 - 4}$

10. $\lim\limits_{s \to 0} \dfrac{\sqrt{9 - s} - 3}{s}$

In Exercises 11 through 14, draw a sketch of the graph of the function; then by observing where there are breaks in the graph, determine the values of the independent variable at which the function is discontinuous and show why the definition of a continuous function is not satisfied at each discontinuity.

11. $f(x) = \dfrac{x^2 + x - 2}{x + 2}$

12. $g(x) = \dfrac{x^4 - 1}{x^2 - 1}$

13. $g(x) = \begin{cases} 2x + 1 & \text{if } x \leq -2 \\ x - 2 & \text{if } -2 < x \leq 2 \\ 2 - x & \text{if } 2 < x \end{cases}$

14. $F(x) = \begin{cases} |4 - x| & \text{if } x \neq 4 \\ -2 & \text{if } x = 4 \end{cases}$

15. If $f(x) = \sqrt{16 - x^2}$, draw a sketch of the graph of f, and show that f is continuous on the closed interval $[-4, 4]$.

16. If $g(x) = \sqrt{x^2 - 16}$, draw a sketch of the graph of g, and show that g is continuous on $(-\infty, -4]$ and $[4, +\infty)$.

In Exercises 17 through 24, find the derivative of the given function.

17. $f(x) = 5x^3 - 7x^2 + 2x - 3$

18. $g(x) = 5(x^4 + 3x^7)$

19. $F(x) = \dfrac{4}{x^2} - \dfrac{3}{x^4}$

20. $f(x) = 2x^{1/2} - \tfrac{1}{2}x^{-2}$

21. $g(t) = (3t^2 - 4)(4t^3 + t - 1)$

22. $h(r) = (r^4 - 2r)(4r^2 + 2r + 5)$

23. $f(x) = \dfrac{x^3 + 1}{x^3 - 1}$

24. $f(y) = \dfrac{y^2}{y^3 + 8}$

In Exercises 25 through 28, find $\dfrac{dy}{dx}$.

25. $y = 2x^3 - 3 - 2x^{-3}$

26. $y = \dfrac{x^2}{4} + \dfrac{4}{x^2}$

27. $y = \dfrac{x^2 - 4x + 4}{x - 1}$

28. $y = \dfrac{3x + 1}{x - 5}(2x - 1)$

29. Use only the definition of a derivative to find $f'(x)$ if $f(x) = 4x^2 - 2x + 1$.

30. Use only the definition of a derivative to find $f'(x)$ if $f(x) = \dfrac{5}{x^2}$.

31. Use only the definition of a derivative to find $f'(5)$ if $f(x) = \dfrac{3}{x + 2}$.

32. Use only the definition of a derivative to find $f'(2)$ if $f(x) = \sqrt{4x + 1}$.

33. Find an equation of the tangent line to the curve $y = x^2 - 5x + 6$ at the point $(1, 2)$.

34. Find equations of the tangent lines to the curve $y = \dfrac{1}{x+1}$ that are perpendicular to the line $4x - y + 2 = 0$.

35. Refer to Exercise 37 in Review Exercises for Chapter 1. Square pieces of metal of side 20 in. are used to construct open boxes by cutting equal squares from the four corners and turning up the sides. (a) If x in. is the length of the side of the square cut out, express the number of cubic inches in the volume of the box as a function of x. (b) Show that the function in part (a) is continuous on its domain.

36. Refer to Exercise 38 in Review Exercises for Chapter 1. A wholesaler offers to deliver to a dealer 300 chairs at $90 per chair and to reduce the price per chair on the entire order by 25 cents for each additional chair over 300. (a) If x chairs are ordered, express the number of dollars in the dealer's cost as a function of x. (b) Show that the function in part (a) is continuous on its domain.

37. Refer to Exercise 39 in Review Exercises for Chapter 1. A school-sponsored trip that can accommodate up to 250 students will cost each student $15 if not more than 150 students make the trip; however, the cost per student will be reduced 5 cents for each student in excess of 150 until the cost reaches $10 per student. (a) If x students make the trip, express the number of dollars in the gross income as a function of x. (b) Show that the function in part (a) is continuous at 150 and hence continuous on its domain.

38. Refer to Exercise 40 in Review Exercises for Chapter 1. In a town of population 11,000, the growth rate of an epidemic is jointly proportional to the number of people infected and the number of people not infected. (a) If the epidemic is growing at the rate of $f(x)$ people per day when x people are infected, write an equation defining $f(x)$. (b) Show that the function f in part (a) is continuous on its domain.

39. A manufacturer can produce a certain commodity at a cost of $20 per unit. It is estimated that if the selling price is x dollars, the number of units sold per day is $200 - x$. (a) If $P(x)$ dollars is the manufacturer's daily profit, write an equation defining $P(x)$. (b) Use the method of Illustration 1 in Sec. 2.4 to determine the selling price of the commodity for the manufacturer's daily profit to be a maximum. Assume that there is a maximum profit.

40. Given $f(x) = |x - 3|$. (a) Draw a sketch of the graph of f. (b) Show that f is not differentiable at 3.

41. Given $f(x) = \sqrt[3]{x - 3}$. (a) Find $f'(x)$. (b) Is f differentiable at 3? (c) Draw a sketch of the graph of f.

CHAPTER 3

APPLICATIONS OF THE DERIVATIVE AND THE CHAIN RULE

3.1 Marginal cost, elasticity of cost, and marginal revenue

In economics the variation of one quantity with respect to another quantity may be described by either an *average* concept or a *marginal* concept. The average concept expresses the variation of one quantity over a specified range of values of a second quantity, whereas the marginal concept is the instantaneous change in the first quantity that results from a very small unit change in the second quantity. We begin our examples in economics with the definitions of average cost and marginal cost. To define a marginal concept precisely we must use the notion of a limit, and this will lead to the derivative.

Suppose $C(x)$ is the number of dollars in the total cost of producing x units of a commodity. The function C is called a **total cost function**. In normal circumstances x and $C(x)$ are positive. Note that since x represents the number of units of a commodity, x must be a nonnegative integer. However, to apply the calculus we shall assume that x is a nonnegative real number to give us the continuity requirements for the function C.

The **average cost** of producing each unit of a commodity is obtained by dividing the total cost by the number of units produced. Letting $Q(x)$ be the number of dollars in the average cost we have

$$Q(x) = \frac{C(x)}{x}$$

and Q is called an **average cost function**.

Now let us suppose that the number of units in a particular output is x_1, and this is changed by Δx. Then the change in the total cost is given by $C(x_1 + \Delta x) - C(x_1)$, and the average change in the total cost with respect to the change in the number of units produced is given by

$$\frac{C(x_1 + \Delta x) - C(x_1)}{\Delta x} \tag{1}$$

Economists use the term *marginal cost* for the limit of the quotient (1) as Δx approaches zero, provided the limit exists. This limit, being the derivative of C at x_1, gives us the following definition.

Definition of marginal cost

> If $C(x)$ is the number of dollars in the total cost of producing x units of a commodity, then the **marginal cost**, when $x = x_1$, is given by $C'(x_1)$, if it exists. The function C' is called the **marginal cost function**.

In the above definition, $C'(x_1)$ may be interpreted as the rate of change of the total cost when x_1 units are produced.

● ILLUSTRATION 1

Suppose that $C(x)$ is the number of dollars in the total cost of manufacturing x toys, and

$$C(x) = 110 + 4x + 0.02x^2$$

(a) The marginal cost function is C', and

$C'(x) = 4 + 0.04x$

(b) The marginal cost when $x = 50$ is given by $C'(50)$, and

$C'(50) = 4 + 0.04(50)$

$= 6$

Therefore the rate of change of the total cost, when 50 toys are manufactured, is $6 per toy.

(c) The number of dollars in the actual cost of manufacturing the fifty-first toy is $C(51) - C(50)$, and

$C(51) - C(50) = \left[110 + 4(51) + 0.02(51)^2\right] - \left[110 + 4(50) + 0.02(50)^2\right]$

$= 366.02 - 360$

$= 6.02$

Note that the answers in (b) and (c) differ by 0.02. This discrepancy occurs because the marginal cost is the instantaneous rate of change of $C(x)$ with respect to a unit change in x. Hence $C'(50)$ is the approximate number of dollars in the cost of producing the fifty-first toy.   ●

Observe that the computation of $C'(50)$ in Illustration 1 is simpler than computing $C(51) - C(50)$. Economists frequently approximate the cost of producing one additional unit by using the marginal cost function. Specifically, $C'(k)$ dollars is the approximate cost of the $(k + 1)$st unit after the first k units have been produced.

The graphs of the total cost function, the marginal cost function, and the average cost function are called the **total cost curve** (labeled TC), the **marginal cost curve** (labeled MC), and the **average cost curve** (labeled AC), respectively. A more thorough discussion of these graphs is given in Sec. 5.3, after we apply the derivative to drawing sketches of graphs.

EXAMPLE 1 Suppose that $C(x)$ dollars is the total cost of producing x units of a commodity, and $C(x) = 2x^2 + x + 8$. Find the function giving (a) the average cost and (b) the marginal cost. Draw sketches of the total cost curve, the marginal cost curve, and the average cost curve on the same set of axes.

Solution We are given

$C(x) = 2x^2 + x + 8$

(a) Let $Q(x)$ be the number of dollars in the average cost. Then

$Q(x) = \dfrac{C(x)}{x}$

$= \dfrac{2x^2 + x + 8}{x}$

$= 2x + 1 + \dfrac{8}{x}$

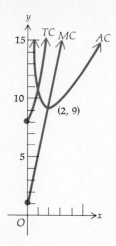

Figure 3.1.1

(b) $C'(x)$ is the number of dollars in the marginal cost, and

$$C'(x) = D_x(2x^2 + x + 8)$$
$$= 4x + 1$$

The graphs of the functions C, Q, and C' are sketched in Fig. 3.1.1.

Observe in Fig. 3.1.1 that the lowest point on the graph of Q occurs at the point of intersection $(2, 9)$ of the graphs of Q and C'. That is, the average cost is smallest when the average cost and the marginal cost are equal. This fact is proved in general and discussed further in Sec. 5.3.

● ILLUSTRATION 2

In Illustration 1 we had the cost function C for which

$$C(x) = 110 + 4x + 0.02x^2$$

where $C(x)$ is the number of dollars in the total cost of manufacturing x toys. The average cost function is Q, where

$$Q(x) = \frac{C(x)}{x}$$

$$= \frac{110}{x} + 4 + 0.02x$$

Because

$$Q(50) = \frac{110}{50} + 4 + 0.02(50)$$

$$= 7.20$$

it follows that when 50 toys have been manufactured, the average cost of producing a toy is \$7.20. In Illustration 1 we obtained

$$C'(50) = 6$$

Thus the approximate cost of producing one additional toy after 50 toys have been manufactured is \$6. This amount is less than the average cost of producing the first 50 toys. If we compute the ratio $C'(50)/Q(50)$, we have

$$\frac{C'(50)}{Q(50)} = \frac{6}{7.20} = \frac{5}{6}$$

This ratio indicates that the cost of the fifty-first toy is five-sixths of the average cost of the first 50 toys.   ●

The ratio $C'(x)/Q(x)$, computed in Illustration 2 for $x = 50$, is called the *elasticity of cost* and is denoted by the Greek letter kappa, κ.

Definition of elasticity
of cost

If $C(x)$ is the number of dollars in the total cost of producing x units of a commodity, and $Q(x)$ dollars is the average cost of producing each unit, then the **elasticity of cost** is given by the function κ for which

$$\kappa(x) = \frac{C'(x)}{Q(x)}$$

If the elasticity of cost is less than 1, then the cost of producing the next unit will be less than the average cost of the units already produced. This situation occurs in Illustration 2, where $\kappa(50) = \frac{5}{6}$. If the elasticity of cost is greater than 1, then the average cost per unit will increase when an additional unit is produced.

EXAMPLE 2   Suppose that $C(x)$ dollars is the total cost of producing x picture frames, and

$$C(x) = 50 + 8x - \frac{x^2}{100}$$

Find the average cost, the marginal cost, and the elasticity of cost when $x = 60$, and give the economic interpretation of these results.

Solution   If Q is the average cost function, then

$$Q(x) = \frac{C(x)}{x}$$

$$= \frac{50 + 8x - \dfrac{x^2}{100}}{x}$$

$$= \frac{50}{x} + 8 - \frac{x}{100}$$

Thus

$$Q(60) = \tfrac{50}{60} + 8 - \tfrac{60}{100}$$

$$= 0.83 + 8 - 0.60$$

$$= 8.23$$

Therefore the average cost of producing each of the first 60 frames is $8.23. The marginal cost function is C', and

$$C'(x) = D_x \left(50 + 8x - \frac{x^2}{100} \right)$$

$$= 8 - \frac{x}{50}$$

Hence

$$C'(60) = 8 - \tfrac{60}{50}$$

$$= 8 - 1.20$$

$$= 6.80$$

Therefore the approximate cost of producing the sixty-first frame is $6.80. The elasticity of cost when $x = 60$ is $\kappa(60)$, and

$$\kappa(60) = \frac{C'(60)}{Q(60)}$$

$$= \frac{6.80}{8.23}$$

$$= 0.83$$

Therefore the cost of producing the sixty-first frame is eighty-three one-hundredths of the average cost of the first 60 frames.

In Sec. 1.6 we stated that a demand equation is one that gives the relationship between p and x, where x units of a quantity are demanded when p dollars is the price per unit. If the demand equation is solved for p, we obtain the price function f given by

$$p = f(x)$$

We assume that x is a nonnegative real number and that the price function is continuous.

Another function important in economics is the **total revenue function**, which we denote by R, and

$$R(x) = px$$

Because p and x are nonnegative under normal circumstances, so is $R(x)$. When $x \neq 0$, from the above equation we obtain

$$\frac{R(x)}{x} = p$$

which shows that the revenue per unit (the average revenue) and the price per unit are equal.

Definition of marginal
revenue

If $R(x)$ is the number of dollars in the total revenue obtained when x units of a commodity are demanded, then the **marginal revenue**, when $x = x_1$, is given by $R'(x_1)$, if it exists. The function R' is called the **marginal revenue function**.

$R'(x_1)$ may be positive, negative, or zero, and it may be interpreted as the rate of change of the total revenue when x_1 units are demanded. Just as $C'(k)$ dollars is the approximate cost of the $(k + 1)$st unit after the first k units have been produced, $R'(k)$ is the approximate revenue from the sale of the $(k + 1)$st unit after the first k units have been sold.

● ILLUSTRATION 3

Suppose that $R(x)$ dollars is the total revenue received from the sale of x tables, and

$$R(x) = 300x - \frac{x^2}{2}$$

(a) The marginal revenue function is R' and

$$R'(x) = 300 - x$$

(b) The marginal revenue when $x = 40$ is given by $R'(40)$, and

$$R'(40) = 300 - 40$$
$$= 260$$

Thus the rate of change of the total revenue when 40 tables are sold is $260 per table.

(c) The number of dollars in the actual revenue from the sale of the forty-first table is $R(41) - R(40)$, and

$$R(41) - R(40) = \left[300(41) - \frac{(41)^2}{2} \right] - \left[300(40) - \frac{(40)^2}{2} \right]$$
$$= [12{,}300 - 840.50] - [12{,}000 - 800]$$
$$= 11{,}459.50 - 11{,}200$$
$$= 259.50$$

Hence the actual revenue from the sale of the forty-first table is $259.50. In part (b) we obtained $R'(40) = 260$, and $260 is an approximation of the revenue received from the sale of the forty-first table. ●

The graphs of the total revenue function and the marginal revenue function are called the **total revenue curve** (labeled TR) and the **marginal revenue curve** (labeled MR), respectively.

EXAMPLE 3 The demand equation for a particular commodity is

$$5x + 3p = 15$$

Find the total revenue function and the marginal revenue function. Draw sketches of the demand curve, the total revenue curve, and the marginal revenue curve on the same set of axes.

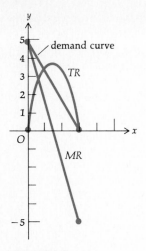

Figure 3.1.2

Solution   We solve the demand equation for p and obtain

$$p = -\tfrac{5}{3}x + 5$$

Because p and x are nonnegative, $0 \leq x \leq 3$. Thus, if R is the total revenue function and R' is the marginal revenue function, we have

$$R(x) = px$$
$$= -\tfrac{5}{3}x^2 + 5x \qquad x \in [0, 3]$$
$$R'(x) = D_x(-\tfrac{5}{3}x^2 + 5x)$$
$$= -\tfrac{10}{3}x + 5 \qquad x \in [0, 3]$$

Sketches of the demand curve and the graphs of R and R' are shown in Fig. 3.1.2.

Exercises 3.1

1. The number of dollars in the total cost of manufacturing x watches in a certain plant is given by $C(x) = 1500 + 30x + x^2$. Find: (a) the marginal cost function; (b) the marginal cost when $x = 40$; (c) the actual cost of manufacturing the forty-first watch.

2. If $C(x)$ dollars is the total cost of manufacturing x paperweights, and

$$C(x) = 200 + \frac{50}{x} + \frac{x^2}{5}$$

find: (a) the marginal cost function; (b) the marginal cost when $x = 10$; (c) the actual cost of manufacturing the eleventh paperweight.

3. Suppose that a liquid is produced by a certain chemical process and that the total cost function C is given by $C(x) = 6 + 4\sqrt{x}$, where $C(x)$ dollars is the total cost of producing x liters of the liquid. Find (a) the marginal cost when 16 liters are produced and (b) the number of liters produced when the marginal cost is 40 cents per liter.

4. The number of dollars in the total cost of producing x units of a certain commodity is given by $C(x) = 40 + 3x + 9\sqrt{2}\sqrt{x}$. Find (a) the marginal cost when 32 units are produced and (b) the number of units produced when the marginal cost is $4.50.

5. For the total cost function of Exercise 3, find each of the following, and give the economic interpretation of the result: (a) the average cost when $x = 100$; (b) the marginal cost when $x = 100$; (c) the elasticity of cost when $x = 100$.

6. For the total cost function of Exercise 4, find each of the following, and give the economic interpretation of the result: (a) the average cost when $x = 50$; (b) the marginal cost when $x = 50$; (c) the elasticity of cost when $x = 50$.

7. The number of dollars in the total cost of producing x units of a particular commodity is given by $C(x) = 20 + 5x + 2\sqrt{x}$. Find each of the following, and give the economic interpretation of the result: (a) the average cost when $x = 25$; (b) the marginal cost when $x = 25$; (c) the elasticity of cost when $x = 25$.

8. The number of dollars in the total cost of producing x units of a commodity is given by $C(x) = 3x^2 + x + 3$. Find (a) the average cost function and (b) the marginal cost function. (c) Draw sketches of the total cost, the average cost, and the marginal cost curves on the same set of axes. Observe that the average cost and marginal cost are equal when the average cost has its least value.

9. Follow the instructions of Exercise 8 if $C(x) = x^2 + 6x + 12$.

10. The total revenue received from the sale of x bookcases is $R(x)$ dollars, and

$$R(x) = 150x - \frac{x^2}{4}$$

Find: (a) the marginal revenue function; (b) the marginal revenue when $x = 20$; (c) the actual revenue from the sale of the twenty-first bookcase.

11. The total revenue received from the sale of x desks is $R(x)$ dollars, and

$$R(x) = 200x - \frac{x^2}{3}$$

Find: (a) the marginal revenue function; (b) the marginal revenue when $x = 30$; (c) the actual revenue from the sale of the thirty-first desk.

12. If $R(x)$ dollars is the total revenue received from the sale of x television sets, and

$$R(x) = 600x - \frac{x^3}{20}$$

find: (a) the marginal revenue function; (b) the marginal revenue when $x = 20$; (c) the actual revenue from the sale of the twenty-first television set.

13. If the demand equation for a particular commodity is $3x + 4p = 12$, find (a) the total revenue function and (b) the marginal revenue function. Draw sketches of the demand, total revenue, and marginal revenue curves on the same set of axes. Observe that the demand equation is linear and that the marginal revenue curve intersects the x axis at the point whose abscissa is the value of x for which the total revenue is greatest and that the demand curve intersects the x axis at the point whose abscissa is twice that.

14. Follow the instructions of Exercise 13 if the demand equation is $5x + 4p = 20$.

15. The total revenue function R for a particular commodity is given by $R(x) = 3x - \frac{2}{3}x^2$. Find (a) the demand equation and (b) the marginal revenue function. (c) Draw sketches of the demand, total revenue, and marginal revenue curves on the same set of axes.

16. Follow the instructions of Exercise 15 if the total revenue function is given by $R(x) = 6x - \frac{3}{2}x^2$.

3.2 The derivative as a rate of change

The concept of marginal variation in economics corresponds to the more general concept of instantaneous rate of change. For example, if the total cost of producing x units of a commodity is given by $C(x)$ dollars, then the marginal cost is given by $C'(x)$, which is the rate of change of $C(x)$.

In a similar way, if a quantity y is a function of a quantity x, we may express the rate of change of y with respect to x. The discussion is analogous to the discussion of marginal variation given in Sec. 3.1.

If the functional relationship between y and x is given by

$$y = f(x)$$

and if x changes from the value x_1 to $x_1 + \Delta x$, then y changes from $f(x_1)$ to $f(x_1 + \Delta x)$. So the change in y, which we may denote by Δy, is

$f(x_1 + \Delta x) - f(x_1)$ when the change in x is Δx. The average rate of change of y, with respect to x, as x changes from x_1 to $x_1 + \Delta x$ is then

$$\frac{f(x_1 + \Delta x) - f(x_1)}{\Delta x} = \frac{\Delta y}{\Delta x} \qquad (1)$$

If the limit of this quotient exists as $\Delta x \to 0$, this limit is what we intuitively think of as the instantaneous rate of change of y with respect to x at x_1. Accordingly, we have the following definition.

Definition of instantaneous rate of change

> If $y = f(x)$, the **instantaneous rate of change of y with respect to x** at x_1 is $f'(x_1)$ or, equivalently, the derivative of y with respect to x at x_1, if it exists there.

The instantaneous rate of change of y with respect to x may be interpreted as the change in y caused by a change of one unit in x if the rate of change remains constant. To illustrate this geometrically, let $f'(x_1)$ be the instantaneous rate of change of y with respect to x at x_1. Then if we multiply $f'(x_1)$ by Δx (the change in x), we have the change that would occur in y if the point (x, y) were to move along the tangent line at (x_1, y_1) of the graph of $y = f(x)$. See Fig. 3.2.1. The average rate of change of y with respect to x is given by the fraction in (1), and if this is multiplied by Δx, we have

$$\frac{\Delta y}{\Delta x} \cdot \Delta x = \Delta y$$

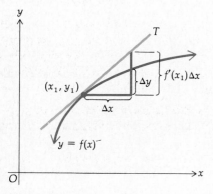

Figure 3.2.1

which is the actual change in y caused by a change of Δx in x when the point (x, y) moves along the graph.

EXAMPLE 1 Let V cubic inches be the volume of a cube having an edge of length e inches. Find the average rate of change of the volume with respect to e as e changes from (a) 3.00 to 3.20; (b) 3.00 to 3.10; (c) 3.00 to 3.01. (d) What is the instantaneous rate of change of the volume with respect to e when $e = 3$?

Solution   Because the formula for finding the volume of a cube is $V = e^3$, let f be the function defined by $f(e) = e^3$. Then the average rate of change of V with respect to e as e changes from e_1 to $e_1 + \Delta e$ is

$$\frac{f(e_1 + \Delta e) - f(e_1)}{\Delta e}$$

(a) $e_1 = 3$, $\Delta e = 0.2$, and $\dfrac{f(3.2) - f(3)}{0.2} = \dfrac{(3.2)^3 - 3^3}{0.2} = \dfrac{5.77}{0.2} = 28.8$

(b) $e_1 = 3$, $\Delta e = 0.1$, and $\dfrac{f(3.1) - f(3)}{0.1} = \dfrac{(3.1)^3 - 3^3}{0.1} = \dfrac{2.79}{0.1} = 27.9$

(c) $e_1 = 3$, $\Delta e = 0.01$, and $\dfrac{f(3.01) - f(3)}{0.01} = \dfrac{(3.01)^3 - 3^3}{0.01} = \dfrac{0.271}{0.01} = 27.1$

In part (a) we see that as the length of the edge of the cube changes from 3.00 inches to 3.20 inches, the change in the volume is 5.77 cubic inches and the average rate of change of the volume is 28.8 cubic inches per inch change in the length of the edge. There are similar interpretations of parts (b) and (c).

(d) The instantaneous rate of change of V with respect to e at 3 is $f'(3)$.

$$f'(e) = 3e^2$$

Hence

$$f'(3) = 27$$

Therefore, when the length of the edge of the cube is 3 inches, the instantaneous rate of change of the volume is 27 cubic inches per inch change in the length of the edge.

Notice in Example 1 from the results of parts (a), (b), and (c) that as Δe gets smaller, the average rate of change of the volume with respect to e gets closer to the instantaneous rate of change of the volume obtained in part (d).

EXAMPLE 2   A company estimates that if $1000x$ dollars is spent on advertising, it will sell y units of a commodity, where

$$y = 5 + 400x - 2x^2$$

(a) Find the average rate of change of y with respect to x when the advertising budget is increased from \$10,000 to \$11,000. (b) Find the instantaneous (or marginal) rate of change of y with respect to x when the advertising budget is \$10,000.

Solution   Let f be the function defined by

$$f(x) = 5 + 400x - 2x^2$$

Then the average rate of change of y with respect to x as x changes from x_1 to $x_1 + \Delta x$ is

$$\frac{f(x_1 + \Delta x) - f(x_1)}{\Delta x} \tag{2}$$

(a) We wish to find the average rate of change of y with respect to x as x changes from 10 to $10 + 1$. Thus we use quotient (2) with $x_1 = 10$ and $\Delta x = 1$. We have

$$\frac{f(10 + 1) - f(10)}{1} = \frac{f(11) - f(10)}{1}$$

$$= [5 + 400(11) - 2(11)^2] - [5 + 400(10) - 2(10)^2]$$

$$= 4163 - 3805$$

$$= 358$$

Therefore, when the advertising budget is increased from \$10,000 to \$11,000, the average rate of change of the number of units sold is 358 units per 1000-dollar increase in the advertising budget.

(b) The instantaneous rate of change of y with respect to x at 10 is $f'(10)$.

$$f'(x) = 400 - 4x$$

Therefore

$$f'(10) = 360$$

Hence, when the advertising budget is \$10,000, the instantaneous rate of change of the number of units sold is 360 units per 1000-dollar increase in the budget.

Observe in Example 2 that the independent variable is measured in units of \$1000. Thus in part (a) the increase of the advertising budget from \$10,000 to \$11,000 results in a one-unit increase in x. Notice further that $f'(10) = 360$, found in part (b), gives an approximation to the result in part (a) of 358, which is the number of items sold as a result of a one-unit increase in the size of the advertising budget. Hence we have a further example of marginal variation in economics. Recall from Sec. 3.1 that $C'(k)$, the marginal cost at k, is the approximate cost of the $(k + 1)$st unit after the first k units have been produced, and $R'(k)$, the marginal revenue at k, is the approximate revenue from the sale of the $(k + 1)$st unit after the first k units have been sold. More generally, economists apply the derivative of a function to estimate the change in the dependent variable caused by a one-unit change in the independent variable.

EXAMPLE 3   Suppose that $h(x)$ units of a commodity are produced daily when x machines are used, and

$$h(x) = 2000x + 40x^2 - x^3$$

Apply the derivative to estimate the change in the daily production if the number of machines used is increased from 20 to 21.

Solution   The value of $h'(x)$ when $x = 20$ is the approximate change in the daily production if the number of machines used is increased from 20 to 21.

$$h'(x) = 2000 + 80x - 3x^2$$
$$h'(20) = 2000 + 80(20) - 3(20)^2$$
$$= 2000 + 1600 - 1200$$
$$= 2400$$

Therefore approximately 2400 more units are produced when the number of machines is increased from 20 to 21.

For the production function in Example 3, the exact change in the daily production can be found by computing $h(21) - h(20)$, which is 2379.

EXAMPLE 4   The annual gross earnings of a particular corporation t years after January 1, 1981, is p millions of dollars, and

$$p = \tfrac{2}{5}t^2 + 2t + 10$$

Find: (a) the rate at which the gross earnings were growing on January 1, 1983; (b) the rate at which the gross earnings should be growing on January 1, 1987.

Solution   (a) On January 1, 1983, $t = 2$; hence we find $\dfrac{dp}{dt}$ when $t = 2$.

$$\frac{dp}{dt} = \tfrac{4}{5}t + 2 \qquad \frac{dp}{dt}\bigg]_{t=2} = \tfrac{8}{5} + 2 = 3.6$$

So on January 1, 1983, the gross earnings were growing at the rate of \$3.6 million per year.
(b) On January 1, 1987, $t = 6$ and

$$\frac{dp}{dt}\bigg]_{t=6} = \tfrac{24}{5} + 2 = 6.8$$

Therefore on January 1, 1987, the gross earnings should be growing at the rate of \$6.8 million per year.

The results of Example 4 are meaningful only if they are compared with the actual earnings of the corporation. For example, if on January 1, 1983, it was found that the earnings of the corporation for the year 1982 had been \$3 million, then the rate of growth on January 1, 1983, of \$3.6 million annually would have been excellent. However, if the earnings in 1982 had been \$300 million, then the growth rate on January 1, 1983, would have been poor. The measure used to compare the rate of change with the amount of the quantity that is being changed is called the *relative rate*.

Definition of relative rate of change

> If $y = f(x)$, the **relative rate of change of y with respect to x at x_1** is given by $\dfrac{f'(x_1)}{f(x_1)}$ or, equivalently, $\dfrac{dy/dx}{y}$ evaluated at $x = x_1$.

If the relative rate is multiplied by 100, we obtain the percent rate of change.

EXAMPLE 5   Find the relative rate of growth of the gross earnings on January 1, 1983, and January 1, 1987, for the corporation of Example 4.

Solution   (a) When $t = 2$, $p = \frac{2}{5}(4) + 2(2) + 10 = 15.6$. Hence on January 1, 1983, the relative rate of growth of the corporation's annual gross earnings was

$$\left. \frac{dp/dt}{p} \right]_{t=2} = \frac{3.6}{15.6} = 0.231 = 23.1 \text{ percent}$$

(b) When $t = 6$, $p = \frac{2}{5}(36) + 2(6) + 10 = 36.4$. Therefore on January 1, 1987, the relative rate of growth of the corporation's annual gross earnings should be

$$\left. \frac{dp/dt}{p} \right]_{t=6} = \frac{6.8}{36.4} = 0.187 = 18.7 \text{ percent}$$

Observe that the growth rate of \$6.8 million for January 1, 1987, is greater than the growth rate of \$3.6 million for January 1, 1983; however, the relative growth rate of 18.7 percent for January 1, 1987, is less than the relative growth rate of 23.1 percent for January 1, 1983.

EXAMPLE 6   The population of a certain town, t years after January 1, 1982, is expected to be $f(t)$, where

$$f(t) = 30t^2 + 100t + 5000$$

(a) Find the rate at which the population is expected to be growing on January 1, 1990. (b) Find the relative rate of growth of the population on January 1, 1990.

Solution   (a) On January 1, 1990, $t = 8$. Therefore we wish to find $f'(8)$.

$$f'(t) = 60t + 100$$

$$f'(8) = 580$$

Thus on January 1, 1990, the population is expected to be growing at the rate of 580 people per year.

(b) We compute the population on January 1, 1990, by finding $f(8)$.

$$f(8) = 30(8)^2 + 100(8) + 5000$$
$$= 30(64) + 800 + 5000$$
$$= 7720$$

Therefore on January 1, 1990, the relative rate of growth of the population should be

$$\frac{f'(8)}{f(8)} = \frac{580}{7720} = 0.075 = 7.5 \text{ percent}$$

Exercises 3.2

1. If A in.^{2} is the area of a square and s in. is the length of a side of the square, find the average rate of change of A with respect to s as s changes from (a) 4.00 to 4.60; (b) 4.00 to 4.30; (c) 4.00 to 4.10. (d) What is the instantaneous rate of change of A with respect to s when s is 4.00?

2. Suppose that a right-circular cylinder has a constant height of 10.00 in. If V in.^{3} is the volume of the right-circular cylinder and r in. is the radius of its base, find the average rate of change of V with respect to r as r changes from (a) 5.00 to 5.40; (b) 5.00 to 5.10; (c) 5.00 to 5.01. (d) Find the instantaneous rate of change of V with respect to r when r is 5.00. *Hint:* The formula for finding the volume of a right-circular cylinder is $V = \pi r^2 h$, where h inches is the height of the cylinder.

3. The demand equation for a certain kind of costume jewelry is

    $$x = 100 - 3p - 2p^2$$

    where x units are demanded when p dollars is the price per unit. (a) Find the average rate of change of the demand with respect to the price when the price is increased from \$4 to \$4.50. (b) Find the instantaneous rate of change of the demand with respect to the price when the price is \$4.

4. The demand equation for a certain detergent is

    $$x = 1000(50 - 5p - p^2)$$

    where x boxes are demanded when p dollars is the price per box. (a) Find the average rate of change of the demand with respect to the price when the price is increased from \$2 to \$2.20. (b) Find the instantaneous rate of change of the demand with respect to the price when the price is \$2.

5. The supply equation for a particular kind of light bulb is

    $$x = 1000(4 + 3p + 2p^2)$$

    where x bulbs are supplied when p cents is the price per bulb. (a) Find the average rate of change of the supply with respect to the price when the price is increased from 90 cents to 93 cents. (b) Find the instantaneous rate of change of the supply with respect to the price when the price is 90 cents.

6. For a particular kind of pencil, the supply equation is

    $$x = 1000(3p^2 + 2p)$$

    where x pencils are supplied when p cents is the price of each pencil. Suppose that the price per pencil was 20 cents and is increased to 21 cents. (a) Use the derivative to estimate the change in the supply. (b) Find the exact change in the supply.

7. The weekly production at a certain factory is $H(x)$ units when the factory has x employees, and $H(x) = 1800x - 2x^2$. There are currently 40 employees at the factory. (a) Apply the derivative to estimate the change in the weekly production if one additional employee is hired. (b) Find the exact change in the weekly production caused by the hiring of one additional employee.

8. A company that manufactures stoves can produce $G(x)$ stoves per day when the company's capitalization is x millions of dollars, and

$$G(x) = 400 + 280x - \frac{300}{x}$$

The current capitalization is \$5 million. If the capitalization is increased to \$6 million, (a) use the derivative to estimate the change in the daily production, and (b) find the exact change in the daily production.

9. The daily production at a particular plant is $F(x)$ units when the capital investment is x thousands of dollars, and $F(x) = 300x^{1/2}$. If the current capitalization is \$900,000, use the derivative to estimate the change in the daily production if the capital investment is increased by \$1000.

10. A particular company started doing business on April 1, 1978. The annual gross earnings of the company after t years of operation are p dollars, where $p = 50,000 + 18,000t + 600t^2$. (a) Find the rate at which the gross earnings were growing on April 1, 1980. (b) Find the relative rate of growth of the gross earnings on April 1, 1980.

11. For the company of Exercise 10, find (a) the rate at which the gross earnings should be growing on April 1, 1988, and (b) the relative rate of growth of the gross earnings on April 1, 1988.

12. It is estimated that a worker in a shop that makes picture frames can paint y frames x hours after starting work at 8 A.M., and

$$y = 3x + 8x^2 - x^3 \qquad 0 \le x \le 4$$

(a) Find the rate at which the worker is painting at 10 A.M. (b) Find the number of frames that the worker paints between 10 A.M. and 11 A.M.

13. A retail store will sell y units of a particular product when $100x$ dollars are spent on advertising the product, and

$$y = 50x - x^2 \qquad 10 \le x \le 25$$

Find the rate at which sales are increasing per \$100 increase in the advertising budget if the advertising budget is (a) \$1500 and (b) \$2000.

14. Suppose that t years after January 1, 1976, the Gross National Product (GNP) of a particular country was $f(t)$ billions of dollars, where $f(t) = 2t^2 + 3t + 80$. (a) Find the rate at which the GNP was growing on January 1, 1982. (b) What was the relative rate of growth of the GNP on January 1, 1982?

15. Suppose that the population of a particular city t years after July 1, 1980, will be $40t^2 + 200t + 10,000$. (a) Find the rate at which the population will be growing on July 1, 1989. (b) Find the rate at which the population will be growing on July 1, 1995. (c) Find the relative rate of growth of the population on July 1, 1989. (d) Find the relative rate of growth of the population on July 1, 1995.

16. The population of a certain town t years after January 1, 1984, is expected to be $f(t)$, where

$$f(t) = 10,000 - \frac{4000}{t + 1}$$

(a) Use the derivative to estimate the expected change in population from January 1, 1988, to January 1, 1989. (b) Find the exact expected change in the population from January 1, 1988, to January 1, 1989.

17. A cold front approaches the college campus. The temperature is z degrees t hours after midnight, and

$$z = 0.1(400 - 40t + t^2) \qquad 0 \le t \le 12$$

(a) Find the average rate of change of z with respect to t between 5 A.M. and 6 A.M. (b) Find the instantaneous rate of change of z with respect to t at 5 A.M.

18. If water is being drained from a swimming pool and V gal is the volume of water in the pool t min after the draining starts, where $V = 250(1600 - 80t + t^2)$ find (a) the average rate at which the water leaves the pool during the first 5 min, and (b) how fast the water is flowing out of the pool 5 min after the draining starts.

3.3 THE CHAIN RULE

19. Suppose that a person can learn $f(t)$ nonsense words in t hours and $f(t) = 15t^{2/3}$, where $0 \le t \le 9$. Find the person's rate of learning after (a) 1 hour and (b) 8 hours.

20. The profit of a retail store is $100y$ dollars when x dollars is spent daily on advertising, and $y = 2500 + 36x - 0.2x^2$. Use the derivative to determine if it would be profitable for the daily advertising budget to be increased if the daily advertising budget is (a) \$60 and (b) \$300. (c) What is the maximum value for x below which it is profitable to increase the advertising budget?

21. Boyle's law for the expansion of a gas is $PV = C$, where P is the number of pounds per square unit of pressure, V is the number of cubic units in the volume of the gas, and C is a constant. Find the instantaneous rate of change of V with respect to P when $P = 4$ and $V = 8$.

22. In an electric circuit, if E volts is the electromotive force, R ohms is the resistance, and I amperes is the current, Ohm's law states that $IR = E$. Assuming that E is constant, show that R decreases at a rate proportional to the inverse square of I.

23. A rocket is fired vertically upward, and it is s ft above the ground t sec after being fired, where $s = 560t - 16t^2$ and the positive direction is upward. If v ft/sec is the velocity of the rocket, then v is the rate of change of s with respect to t. (a) Find the velocity of the rocket 2 sec after being fired. (b) If the maximum height is reached when the velocity is zero, find how long it takes the rocket to reach its maximum height.

24. A billiard ball is hit and travels in a straight line. If s cm is the distance of the ball from its initial position at t sec, then $s = 100t^2 + 100t$. If v cm/sec is the velocity of the ball, then v is the rate of change of s with respect to t. If the ball hits a cushion that is 39 cm from its initial position, at what velocity does it hit the cushion?

3.3 The chain rule

Suppose that at a particular factory C dollars is the total cost of producing s units, and

$$C = f(s) \tag{1}$$

Furthermore, suppose that s units are produced during the t hours since production began, and

$$s = g(t) \tag{2}$$

If we know $\dfrac{ds}{dt}$, the rate of change of the number of units produced at time t hours, it appears that we could determine $\dfrac{dC}{dt}$, the rate of change of the total production cost at that time. This computation can be done by applying a very important theorem in calculus, called the **chain rule**. We now discuss the chain rule, and then in Example 1 we apply it to solving a particular problem like that described above by (1) and (2).

The chain rule

Theorem 3.3.1 If y is a function of u and $\dfrac{dy}{du}$ exists, and if u is a function of x and $\dfrac{du}{dx}$ exists, then y is a function of x and $\dfrac{dy}{dx}$ exists and is given by

$$\frac{dy}{dx} = \frac{dy}{du} \cdot \frac{du}{dx} \tag{3}$$

Observe from (3) the convenient form for remembering the chain rule. The formal statement suggests a symbolic "division" of du in the numerator and denominator of the right-hand side. However, remember in Sec. 2.5 when we introduced the Leibniz notation $\dfrac{dy}{dx}$, it was emphasized that neither dy nor dx has been given independent meaning. In Sec. 6.1 we will, however, give separate meaning to these symbols, but until then you should consider (3) an equation involving formal differentiation notation.

● ILLUSTRATION 1

Let

$$y = f(u) = u^5 \tag{4}$$

and

$$u = g(x) = 2x^3 - 5x^2 + 4 \tag{5}$$

Equations (4) and (5) together define y as a function of x, because if in (4) u is replaced by the right-hand side of (5), then

$$y = h(x) = f(g(x)) = (2x^3 - 5x^2 + 4)^5 \tag{6}$$

where h is a composite function that was defined in Sec. 1.4.

We apply the chain rule to find $\dfrac{dy}{dx}$ for the function defined by (6). Considering y as a function of u, where u is a function of x, we have

$$y = u^5 \qquad \text{where } u = 2x^3 - 5x^2 + 4$$

Therefore, from the chain rule,

$$\frac{dy}{dx} = \frac{dy}{du} \cdot \frac{du}{dx}$$

$$= 5u^4(6x^2 - 10x)$$
$$= 5(2x^3 - 5x^2 + 4)^4(6x^2 - 10x) \qquad\qquad ●$$

A rigorous proof of the chain rule is complicated and is omitted. However, following is an argument that is valid for some functions.

Suppose that an increment of x, Δx, where $\Delta x \neq 0$, causes a change in u, Δu. That is,

$$u + \Delta u = g(x + \Delta x)$$

Thus, because $u = g(x)$

$$\Delta u = g(x + \Delta x) - g(x) \tag{7}$$

Furthermore, because $y = f(u)$, Δu causes a change in y, Δy. Then

$$\frac{\Delta y}{\Delta x} = \frac{\Delta y}{\Delta u} \cdot \frac{\Delta u}{\Delta x} \qquad \text{if } \Delta u \neq 0$$

Therefore

$$\lim_{\Delta x \to 0} \frac{\Delta y}{\Delta x} = \lim_{\Delta x \to 0} \frac{\Delta y}{\Delta u} \cdot \lim_{\Delta x \to 0} \frac{\Delta u}{\Delta x} \qquad \text{if } \Delta u \neq 0 \tag{8}$$

Because $u = g(x)$ and g is differentiable, then g is also continuous. Hence, as $\Delta x \to 0$, $\Delta u \to 0$. Thus, in the first limit on the right side of (8) we replace $\Delta x \to 0$ by $\Delta u \to 0$, and we have

$$\lim_{\Delta x \to 0} \frac{\Delta y}{\Delta x} = \lim_{\Delta u \to 0} \frac{\Delta y}{\Delta u} \cdot \lim_{\Delta x \to 0} \frac{\Delta u}{\Delta x} \qquad \text{if } \Delta u \neq 0$$

Therefore

$$\frac{dy}{dx} = \frac{dy}{du} \cdot \frac{du}{dx}$$

This proof fails when $\Delta u = 0$. Because Δu depends on Δx, as shown by (7), it is possible that $\Delta u = 0$ for some values of Δx.

EXAMPLE 1 At a particular factory, if C dollars is the total cost of producing s units, then

$$C = \tfrac{1}{4}s^2 + 2s + 1000 \tag{9}$$

Furthermore, if s units are produced during the t hours since production began, then

$$s = 3t^2 + 50t \tag{10}$$

Determine the rate of change of the total cost with respect to time 2 hours after the start of production.

Solution We wish to find $\dfrac{dC}{dt}$ when $t = 2$. From the chain rule,

$$\frac{dC}{dt} = \frac{dC}{ds} \cdot \frac{ds}{dt} \tag{11}$$

From (9),

$$\frac{dC}{ds} = \tfrac{1}{2}s + 2 \tag{12}$$

From (10),

$$\frac{ds}{dt} = 6t + 50 \tag{13}$$

Substituting from (12) and (13) into (11) we have

$$\frac{dC}{dt} = (\tfrac{1}{2}s + 2)(6t + 50) \tag{14}$$

From (10), when $t = 2$,

$$s = 3(4) + 50(2) = 112$$

Therefore, from (14),

$$\frac{dC}{dt}\Bigg]_{t=2} = [\tfrac{1}{2}(112) + 2][6(2) + 50]$$

$$= (58)(62)$$

$$= 3596$$

Thus 2 hours after the start of production the total cost is increasing at the rate of $3596 per hour.

Theorem 2.6.2 states that if n is any real number, and if $f(x) = x^n$, then $f'(x) = nx^{n-1}$. An immediate consequence of this theorem and the chain rule is the following theorem.

The chain rule for powers

> **Theorem 3.3.2**   If f and g are functions such that $f(x) = [g(x)]^n$, where n is any real number, and if $g'(x)$ exists, then
>
> $$f'(x) = n[g(x)]^{n-1}g'(x)$$

EXAMPLE 2   Given

$$f(x) = \frac{1}{4x^3 + 5x^2 - 7x + 8}$$

find $f'(x)$.

Solution   Write $f(x) = (4x^3 + 5x^2 - 7x + 8)^{-1}$, and apply the chain rule for powers to obtain

$$f'(x) = -1(4x^3 + 5x^2 - 7x + 8)^{-2}(12x^2 + 10x - 7)$$

$$= \frac{-12x^2 - 10x + 7}{(4x^3 + 5x^2 - 7x + 8)^2}$$

EXAMPLE 3   Given

$$h(x) = \sqrt{2x^3 - 4x + 5}$$

find $h'(x)$.

Solution $h(x) = (2x^3 - 4x + 5)^{1/2}$. Applying the chain rule for powers we obtain

$$h'(x) = \tfrac{1}{2}(2x^3 - 4x + 5)^{-1/2}(6x^2 - 4)$$

$$= \frac{3x^2 - 2}{\sqrt{2x^3 - 4x + 5}}$$

EXAMPLE 4 Given

$$f(x) = \left(\frac{2x + 1}{3x - 1}\right)^4$$

find $f'(x)$.

Solution Applying the chain rule for powers we have

$$f'(x) = 4\left(\frac{2x + 1}{3x - 1}\right)^3 \frac{(3x - 1)(2) - (2x + 1)(3)}{(3x - 1)^2}$$

$$= \frac{4(2x + 1)^3(-5)}{(3x - 1)^5}$$

$$= -\frac{20(2x + 1)^3}{(3x - 1)^5}$$

EXAMPLE 5 Given

$$f(x) = (3x^2 + 2)^2(x^2 - 5x)^3$$

find $f'(x)$.

Solution Consider f as the product of the two functions g and h, where

$$g(x) = (3x^2 + 2)^2 \quad \text{and} \quad h(x) = (x^2 - 5x)^3$$

From Theorem 2.6.5 for the derivative of the product of two functions,

$$f'(x) = g(x)h'(x) + h(x)g'(x)$$

We find $h'(x)$ and $g'(x)$ by the chain rule, thus giving

$$f'(x) = (3x^2 + 2)^2[3(x^2 - 5x)^2(2x - 5)] + (x^2 - 5x)^3[2(3x^2 + 2)(6x)]$$

$$= 3(3x^2 + 2)(x^2 - 5x)^2[(3x^2 + 2)(2x - 5) + 4x(x^2 - 5x)]$$

$$= 3(3x^2 + 2)(x^2 - 5x)^2[6x^3 - 15x^2 + 4x - 10 + 4x^3 - 20x^2]$$

$$= 3(3x^2 + 2)(x^2 - 5x)^2(10x^3 - 35x^2 + 4x - 10)$$

The chain rule can be extended to composite functions involving any finite number of functions. The following theorem extends the chain rule to a composite function involving three functions.

The extended chain rule

> **Theorem 3.3.3** If y is a function of u, u is a function of v, and v is a function of x, and if $\dfrac{dy}{du}, \dfrac{du}{dv}$, and $\dfrac{dv}{dx}$ exist, then y is a function of x, and $\dfrac{dy}{dx}$ exists and is given by
>
> $$\frac{dy}{dx} = \frac{dy}{du} \cdot \frac{du}{dv} \cdot \frac{dv}{dx} \tag{15}$$

Notice how the Leibniz notation for the derivative makes Eq. (15) easy to remember and apply.

EXAMPLE 6   Given $f(x) = \sqrt{2 + (x^2 - 3x + 2)^3}$, find $f'(x)$.

Solution   Let $y = f(x)$. Furthermore, let

$$y = \sqrt{u} \quad \text{where} \quad u = 2 + v^3 \quad \text{and} \quad v = x^2 - 3x + 2$$

Then by Theorem 3.3.3,

$$f'(x) = \frac{dy}{dx}$$

$$= \frac{dy}{du} \cdot \frac{du}{dv} \cdot \frac{dv}{dx}$$

$$= \tfrac{1}{2}u^{-1/2}(3v^2)(2x - 3)$$

$$= \frac{3(x^2 - 3x + 2)^2(2x - 3)}{2\sqrt{2 + (x^2 - 3x + 2)^3}}$$

Exercises 3.3

In Exercises 1 through 32, find the derivative of the given function.

1. $f(x) = (2x + 1)^3$

2. $f(x) = (10 - 5x)^4$

3. $F(x) = (x^2 + 4x - 5)^4$

4. $g(r) = (2r^4 + 8r^2 + 1)^5$

5. $f(t) = (2t^4 - 7t^3 + 2t - 1)^2$

6. $H(z) = (z^3 - 3z^2 + 1)^{-3}$

7. $g(x) = \sqrt{1 + 4x^2}$

8. $f(s) = \sqrt{2 - 3s^2}$

9. $f(x) = (5 - 3x)^{2/3}$

10. $g(x) = \sqrt[3]{4x^2 - 1}$

11. $g(y) = \dfrac{1}{\sqrt{25 - y^2}}$

12. $f(x) = (5 - 2x^2)^{-1/3}$

13. $F(r) = \sqrt{\dfrac{2r - 5}{3r + 1}}$

14. $G(t) = \sqrt{\dfrac{5t + 6}{5t - 4}}$

15. $g(t) = \sqrt{2t} + \sqrt{\dfrac{2}{t}}$

16. $f(y) = \left(\dfrac{y - 7}{y + 2}\right)^2$

17. $f(x) = \left(\dfrac{2x^2 + 1}{3x^3 + 1}\right)^2$

18. $h(u) = (3u^2 + 5)^3(3u - 1)^2$

19. $f(s) = (s^2 + 1)^3(2s + 5)^2$

20. $f(x) = (4x^2 + 7)^2(2x^3 + 1)^4$

21. $g(x) = (2x - 5)^{-1}(4x + 3)^{-2}$

22. $f(z) = \dfrac{(z^2 - 5)^3}{(z^2 + 4)^2}$

23. $f(x) = (2x - 9)^2(x^3 + 4x - 5)^3$

24. $g(y) = (y^2 + 3)^{1/3}(y^3 - 1)^{1/2}$

25. $F(x) = \dfrac{\sqrt{x^2 - 1}}{x}$

26. $h(x) = \dfrac{\sqrt{x - 1}}{\sqrt[3]{x + 1}}$

27. $f(y) = \dfrac{3y}{\sqrt{2y - 3}}$

28. $f(x) = \dfrac{x}{\sqrt{x^2 - 1}}$

29. $f(x) = \sqrt{9 + \sqrt{9 - x}}$

30. $g(t) = \sqrt[3]{1 - \sqrt{1 - 2t}}$

31. $g(w) = \sqrt[3]{(1 - 3w)^4 + 4}$

32. $f(x) = \sqrt{(x^2 + 9)^4 + x^2}$

In Exercises 33 through 36, find an equation of the tangent line to the given curve at the indicated point.

33. $y = \sqrt{x^2 + 9};\ (4, 5)$

34. $y = \dfrac{1}{(3x - 2)^2};\ (1, 1)$

35. $y = \left(\dfrac{2 - x}{4 + x}\right)^3 ; (-2, 8)$ **36.** $y = (1 - x^2)^2(3 - 2x)^3 ; (2, -9)$

37. If $C(x)$ dollars is the total cost of producing x units of a certain commodity, and $C(x) = 20 + 4x + \sqrt{3x^2 + 24}$, find (a) the marginal cost function and (b) the marginal cost when 10 units are produced.

38. The total cost of producing x units of a certain commodity is $C(x)$ dollars, where $C(x) = \sqrt{3x^2 + 25} + 2x + 50$. Find (a) the marginal cost function and (b) the marginal cost when 20 units are produced.

39. The demand equation for a particular commodity is

$$p^2 + 4x^2 + 80x - 15{,}000 = 0$$

where x units are demanded when p dollars is the price per unit. Find the marginal revenue when 30 units are demanded.

40. Find the marginal revenue function for a commodity whose demand equation is $px = 5\sqrt{10x + 1}$, where x is in the interval $[1, 8]$.

41. A property development company rents each apartment at p dollars per month when x apartments are rented, and $p = 30\sqrt{300 - 2x}$. How many apartments must be rented before the marginal revenue is zero?

42. The daily production at a particular factory is $f(x)$ units when the capital investment is x thousands of dollars, and $f(x) = 200\sqrt{2x + 1}$. If the current capitalization is $760{,}000$, use the derivative to estimate the change in the daily production if the capital investment is increased by $1000.

43. The demand equation for a certain commodity is $px = 36{,}000$, where x units are demanded per week when p dollars is the price per unit. It is expected that in t weeks, where $t \in [0, 10]$, the price of the commodity will be p dollars, where $30p = 146 + 2t^{1/3}$. What is the anticipated rate of change of the demand with respect to time in 8 weeks?

44. The demand equation for a particular toy is $p^2 x = 5000$, where x toys are demanded per month when p dollars is the price per toy. It is expected that in t months, where $t \in [0, 6]$, the price of the toy will be p dollars, where $20p = t^2 + 7t + 100$. What is the anticipated rate of change of the demand with respect to time in 5 months?

45. In a forest a predator feeds on prey, and the predator population at any time is a function of the number of prey in the forest at that time. Suppose that when there are x prey in the forest, the predator population is y, and $y = \frac{1}{6}x^2 + 90$. Furthermore, if t weeks have elapsed since the end of the hunting season, $x = 7t + 85$. At what rate is the population of the predator growing 8 weeks after the close of the hunting season? Do not express y in terms of t, but use the chain rule.

3.4 Implicit differentiation

If $f = \{(x, y) | y = 3x^2 + 5x + 1\}$, then the equation

$$y = 3x^2 + 5x + 1 \tag{1}$$

defines the function f explicitly. However, not all functions are defined explicitly. For example, for the equation

$$x^6 - 2x = 3y^6 + y^5 - y^2 \tag{2}$$

we cannot solve for y in terms of x; however, there may exist one or more functions f such that if $y = f(x)$, Eq. (2) is satisfied, that is, such that the equation

$$x^6 - 2x = 3[f(x)]^6 + [f(x)]^5 - [f(x)]^2$$

is true for all values of x in the domain of f. In this case the function f is defined *implicitly* by the given equation.

With the assumption that (2) defines y as at least one differentiable function of x, the derivative of y with respect to x can be found by the process called **implicit differentiation**, which we now do.

The left side of (2) is a function of x, and the right side is a function of y. Let F be the function defined by the left side of (2), and let G be the function defined by the right side of (2). Thus

$$F(x) = x^6 - 2x \tag{3}$$

and

$$G(y) = 3y^6 + y^5 - y^2 \tag{4}$$

where y is a function of x, say,

$$y = f(x)$$

So (2) can be written as

$$F(x) = G(f(x)) \tag{5}$$

Equation (5) is satisfied by all values of x in the domain of f for which $G(f(x))$ exists.

Then for all values of x for which f is differentiable,

$$D_x[x^6 - 2x] = D_x[3y^6 + y^5 - y^2] \tag{6}$$

The derivative on the left side of (6) is easily found, and

$$D_x[x^6 - 2x] = 6x^5 - 2 \tag{7}$$

We find the derivative on the right side of (6) by the chain rule.

$$D_x[3y^6 + y^5 - y^2] = 18y^5 \cdot \frac{dy}{dx} + 5y^4 \cdot \frac{dy}{dx} - 2y \cdot \frac{dy}{dx} \tag{8}$$

Substituting the values from (7) and (8) into (6) we obtain

$$6x^5 - 2 = (18y^5 + 5y^4 - 2y)\frac{dy}{dx}$$

$$\frac{dy}{dx} = \frac{6x^5 - 2}{18y^5 + 5y^4 - 2y}$$

Equation (2) is a special type of equation involving x and y because it can be written such that all the terms involving x are on one side of the equation and all the terms involving y are on the other side.

In the following illustration the method of implicit differentiation is used to find $\dfrac{dy}{dx}$ from a more general type of equation.

● ILLUSTRATION 1

Consider the equation

$$3x^4y^2 - 7xy^3 = 4 - 8y \tag{9}$$

and assume that there exists at least one differentiable function f such that if $y = f(x)$, Eq. (9) is satisfied. Differentiating on both sides of (9) (bearing in mind that y is a differentiable function of x), and applying the theorems for the derivative of a product, the derivative of a power, and the chain rule, we obtain

$$12x^3y^2 + 3x^4(2y)\frac{dy}{dx} - 7y^3 - 7x(3y^2)\frac{dy}{dx} = 0 - 8\frac{dy}{dx}$$

$$\frac{dy}{dx}(6x^4y - 21xy^2 + 8) = 7y^3 - 12x^3y^2$$

$$\frac{dy}{dx} = \frac{7y^3 - 12x^3y^2}{6x^4y - 21xy^2 + 8} \qquad \bullet$$

Remember that we assumed that both (2) and (9) define y as at least one differentiable function of x. It may be that an equation in x and y does not imply the existence of any real-valued function, as is the case for the equation

$$x^2 + y^2 + 4 = 0$$

which is not satisfied by any real values of x and y. Furthermore, it is possible that an equation in x and y may be satisfied by many different functions, some of which are differentiable and some of which are not. A general discussion is beyond the scope of this book but can be found in an advanced calculus text. In subsequent discussions when we state that an equation in x and y defines y implicitly as a function of x, it is assumed that one or more of these functions is differentiable. Example 3, which follows, illustrates the fact that implicit differentiation gives the derivative of two differentiable functions defined by the given equation.

EXAMPLE 1   Given $(x + y)^2 - (x - y)^2 = x^4 + y^4$, find $\frac{dy}{dx}$.

Solution   Differentiating implicitly with respect to x we have

$$2(x + y)\left(1 + \frac{dy}{dx}\right) - 2(x - y)\left(1 - \frac{dy}{dx}\right) = 4x^3 + 4y^3\frac{dy}{dx}$$

$$2x + 2y + (2x + 2y)\frac{dy}{dx} - 2x + 2y + (2x - 2y)\frac{dy}{dx} = 4x^3 + 4y^3\frac{dy}{dx}$$

$$\frac{dy}{dx}(4x - 4y^3) = 4x^3 - 4y$$

$$\frac{dy}{dx} = \frac{x^3 - y}{x - y^3}$$

EXAMPLE 2   Find an equation of the tangent line to the curve $x^3 + y^3 = 9$ at the point $(1, 2)$.

Solution   Differentiating implicitly with respect to x we obtain

$$3x^2 + 3y^2\frac{dy}{dx} = 0$$

Hence

$$\frac{dy}{dx} = -\frac{x^2}{y^2}$$

Therefore, at the point $(1, 2)$, $\frac{dy}{dx} = -\frac{1}{4}$. An equation of the tangent line is then

$$y - 2 = -\tfrac{1}{4}(x - 1)$$

$$x + 4y = 9$$

EXAMPLE 3 Given the equation $x^2 + y^2 = 9$, find: (a) $\frac{dy}{dx}$ by implicit differentiation; (b) two functions defined by the equation; (c) the derivative of each of the functions obtained in part (b) by explicit differentiation. (d) Verify that the result obtained in part (a) agrees with the results obtained in part (c).

Solution   (a) Differentiating implicitly we find

$$2x + 2y \frac{dy}{dx} = 0 \quad \text{and so} \quad \frac{dy}{dx} = -\frac{x}{y}$$

(b) Solving the given equation for y we obtain

$$y = \sqrt{9 - x^2} \quad \text{and} \quad y = -\sqrt{9 - x^2}$$

Let f_1 be the function for which

$$f_1(x) = \sqrt{9 - x^2}$$

and f_2 be the function for which

$$f_2(x) = -\sqrt{9 - x^2}$$

(c) Because $f_1(x) = (9 - x^2)^{1/2}$, by using the chain rule we obtain

$$f_1'(x) = \tfrac{1}{2}(9 - x^2)^{-1/2}(^-2x)$$

$$= -\frac{x}{\sqrt{9 - x^2}}$$

Similarly,

$$f_2'(x) = \frac{x}{\sqrt{9 - x^2}}$$

(d) For $y = f_1(x)$, where $f_1(x) = \sqrt{9 - x^2}$, it follows from part (c) that

$$f_1'(x) = -\frac{x}{\sqrt{9 - x^2}} = -\frac{x}{y}$$

which agrees with the answer in part (a). For $y = f_2(x)$, where $f_2(x) = -\sqrt{9 - x^2}$, we have from part (c),

$$f_2'(x) = \frac{x}{\sqrt{9 - x^2}}$$

$$= -\frac{x}{-\sqrt{9 - x^2}}$$

$$= -\frac{x}{y}$$

which also agrees with the answer in part (a).

Exercises 3.4

In Exercises 1 through 16, find $\dfrac{dy}{dx}$ by implicit differentiation.

1. $x^2 + y^2 = 16$
2. $4x^2 - 9y^2 = 1$
3. $x^3 + y^3 = 8xy$

4. $x^2 + y^2 = 7xy$
5. $\dfrac{1}{x} + \dfrac{1}{y} = 1$
6. $\dfrac{3}{x} - \dfrac{3}{y} = 2x$

7. $\sqrt{x} + \sqrt{y} = 4$
8. $2x^3y + 3xy^3 = 5$
9. $x^2y^2 = x^2 + y^2$

10. $(2x + 3)^4 = 3y^4$
11. $\sqrt{xy} + 2x = \sqrt{y}$
12. $y + \sqrt{xy} = 3x^3$

13. $\sqrt[3]{x} + \sqrt[3]{xy} = 4y^2$
14. $\sqrt{y} + \sqrt[3]{y} + \sqrt[4]{y} = x$
15. $(x + y)^2 - (x - y)^2 = x^3 + y^3$

16. $y\sqrt{2 + 3x} + x\sqrt{1 + y} = x$

In Exercises 17 through 20, consider y as the independent variable, and find $\dfrac{dx}{dy}$.

17. $x^4 + y^4 = 12x^2y$
18. $y = 2x^3 - 5x$
19. $x^3y + 2y^4 - x^4 = 0$
20. $y\sqrt{x} - x\sqrt{y} = 9$

21. Find an equation of the tangent line to the curve $16x^4 + y^4 = 32$ at the point $(1, 2)$.
22. Find an equation of the tangent line to the curve $\sqrt[3]{xy} = 14x + y$ at the point $(2, -32)$.

In Exercises 23 through 26, a demand equation is given where x units are demanded when p dollars is the price per unit. In each exercise find the rate of change of p with respect to x by implicit differentiation.

23. $p^2 + 4p + 2x - 10 = 0$
24. $x^2 + p^2 = 36$
25. $x^3 + p^3 = 1000$
26. $p^2 + 4x^2 + 24x = 108$

In each of Exercises 27 through 30 an equation is given. Do the following in each of these problems: (a) Find two functions defined by the equation, and state their domains. (b) Draw a sketch of the graph of each of the functions obtained in part (a). (c) Draw a sketch of the graph of the equation. (d) Find the derivative of each of the functions obtained in part (a), and state the domains of the derivatives. (e) Find $\dfrac{dy}{dx}$ by implicit differentiation from the given equation, and verify that the result so obtained agrees with the results in part (d). (f) Find an equation of each tangent line at the given value of x_1.

27. $y^2 = 4x - 8$; $x_1 = 3$
28. $x^2 + y^2 = 25$; $x_1 = 4$
29. $x^2 - y^2 = 9$; $x_1 = -5$
30. $y^2 - x^2 = 16$; $x_1 = -3$

3.5 Related rates

There are many problems concerned with the rate of change of two or more related variables with respect to time in which it is not necessary to express each of these variables directly as functions of time. For example, suppose that we are given an equation involving the variables x and y, and that both x and y are functions of a third variable t, where t sec denotes time. Then, because the rate of change of x with respect to t and the rate of change of y with respect to t are given by $\dfrac{dx}{dt}$ and $\dfrac{dy}{dt}$, respectively, we differentiate implicitly with respect to t and proceed as in the following example.

EXAMPLE 1 Suppose that $5x + 3xy = 4$, x and y are functions of a third variable t, and $\dfrac{dy}{dt} = 3$. Find $\dfrac{dx}{dt}$ when $x = 2$.

Solution The given equation is

$$5x + 3xy = 4 \tag{1}$$

We differentiate on both sides of the equation with respect to t.

$$5\frac{dx}{dt} + 3y\frac{dx}{dt} + 3x\frac{dy}{dt} = 0$$

$$\frac{dx}{dt} = -\frac{3x}{3y + 5} \cdot \frac{dy}{dt} \tag{2}$$

To find the value of y when $x = 2$ we replace x by 2 in (1) and obtain $y = -1$. From (2), with $x = 2$, $y = -1$, and $\dfrac{dy}{dt} = 3$,

$$\left.\frac{dx}{dt}\right]_{x=2} = -\tfrac{6}{2}(3)$$
$$= -9$$

EXAMPLE 2 A ladder 25 ft long is leaning against a vertical wall. If the bottom of the ladder is pulled horizontally away from the wall at 3 ft/sec, how fast is the top of the ladder sliding down the wall when the bottom is 15 ft from the wall?

Solution Let t be the number of seconds in the time that has elapsed since the ladder started to slide down the wall, y be the number of feet in the distance from the ground to the top of the ladder at t sec, and x be the number of feet in the distance from the bottom of the ladder to the wall at t sec. See Fig. 3.5.1. Because the bottom of the ladder is pulled horizontally away from the wall at 3 ft/sec, $\dfrac{dx}{dt} = 3$. We wish to find $\dfrac{dy}{dt}$ when $x = 15$. From the Pythagorean theorem,

$$y^2 = 625 - x^2 \tag{3}$$

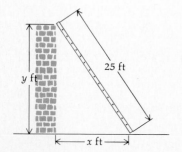

Figure 3.5.1

Because x and y are functions of t, we differentiate on both sides of (3) with respect to t and obtain

$$2y\frac{dy}{dt} = -2x\frac{dx}{dt}$$

$$\frac{dy}{dt} = -\frac{x}{y} \cdot \frac{dx}{dt} \tag{4}$$

When $x = 15$, it follows from (3) that $y = 20$. Because $\dfrac{dx}{dt} = 3$, we get from (4),

$$\left.\frac{dy}{dt}\right]_{y=20} = -\tfrac{15}{20} \cdot 3 = -\tfrac{9}{4}$$

Therefore the top of the ladder is sliding down the wall at the rate of $2\tfrac{1}{4}$ ft/sec when the bottom is 15 ft from the wall. The significance of the minus sign is that y is decreasing as t is increasing.

Example 2 involves a problem with rates of change of related variables. It is called a problem in related rates. In such problems the variables have a specific relationship for values of t, where t is a measure of time. This relationship is usually expressed in the form of an equation, as in Example 2 with Eq. (3). Values of the variables and rates of change of the variables with respect to t are often given at a particular instant. In Example 2, at the instant when $x = 15$, then $y = 20$ and $\dfrac{dx}{dt} = 3$, and we wish to find $\dfrac{dy}{dt}$.

The following steps represent a possible procedure for solving a problem involving related rates.

1. Draw a figure if it is feasible to do so.
2. Define the variables. Generally define t first, because the other variables usually depend on t.
3. Write down any numerical facts known about the variables and their derivatives with respect to t.
4. Obtain an equation involving the variables that depend on t.
5. Differentiate with respect to t on both sides of the equation found in step 4.
6. Substitute values of known quantities in the equation of step 5, and solve for the desired quantity.

EXAMPLE 3   Suppose in a certain market that p dollars is the price of a crate of oranges, x is the number of thousands of crates of oranges supplied daily, and the supply equation is $px - 20p - 3x + 105 = 0$. If the daily supply is decreasing at the rate of 250 crates per day, at what rate is the price changing when the daily supply is 5000 crates?

Solution   Let t days be the time that has elapsed since the daily supply of oranges started to decrease. Then p and x are both functions of t. Since the daily supply is decreasing at the rate of 250 crates per day, $\dfrac{dx}{dt} = -\dfrac{250}{1000} = -\dfrac{1}{4}$.

We wish to find $\dfrac{dp}{dt}$ when $x = 5$. From the given supply equation we differentiate implicitly with respect to t and obtain

$$p\,\frac{dx}{dt} + x\,\frac{dp}{dt} - 20\,\frac{dp}{dt} - 3\,\frac{dx}{dt} = 0$$

$$\frac{dp}{dt} = \frac{3 - p}{x - 20} \cdot \frac{dx}{dt} \tag{5}$$

When $x = 5$, it follows from the supply equation that $p = 6$. Because $\dfrac{dx}{dt} = -\dfrac{1}{4}$, from (5),

$$\left.\frac{dp}{dt}\right]_{p=6} = \frac{3 - 6}{5 - 20}\left(-\frac{1}{4}\right) = \frac{-3}{-15}\left(-\frac{1}{4}\right) = -\frac{1}{20}$$

Hence the price of a crate of oranges is decreasing at the rate of $0.05 per day when the daily supply is 5000 crates.

EXAMPLE 4 When a furniture manufacturer produces x chairs per week, the total weekly cost and total weekly revenue are C dollars and R dollars, respectively, and

$$C = 3000 + 40x \tag{6}$$

$$R = 150x - \tfrac{1}{4}x^2 \tag{7}$$

If the current weekly output is 200 chairs and it is increasing at the rate of 10 chairs per week, find the rate of change of each of the following: (a) the total weekly cost; (b) the total weekly revenue; (c) the total weekly profit.

Solution If t weeks represents the time, then C, R, and x are functions of t. Because the weekly output is increasing at the rate of 10 chairs per week, $\dfrac{dx}{dt} = 10$.

(a) We wish to find $\dfrac{dC}{dt}$. We differentiate implicitly with respect to t on both sides of (6) and obtain

$$\frac{dC}{dt} = 40\frac{dx}{dt}$$

Substituting 10 for $\dfrac{dx}{dt}$ we obtain

$$\frac{dC}{dt} = 40(10) = 400$$

Therefore the total weekly cost is increasing at the rate of $400 per week.

(b) To find $\dfrac{dR}{dt}$ we differentiate implicitly with respect to t on both sides of (7). We have

$$\frac{dR}{dt} = 150\frac{dx}{dt} - \frac{1}{2}x\frac{dx}{dt}$$

Because $x = 200$ and $\dfrac{dx}{dt} = 10$,

$$\frac{dR}{dt}\bigg]_{x=200} = 150(10) - \tfrac{1}{2}(200)(10) = 1500 - 1000 = 500$$

Thus the total weekly revenue is increasing at the rate of \$500 per week.

(c) Because profit equals revenue minus cost, then if P dollars is the total weekly profit,

$$P = R - C$$

Thus

$$\frac{dP}{dt} = \frac{dR}{dt} - \frac{dC}{dt}$$

From parts (a) and (b), $\dfrac{dC}{dt} = 400$ and $\dfrac{dR}{dt} = 500$. Therefore

$$\frac{dP}{dt} = 500 - 400 = 100$$

Hence the total weekly profit is increasing at the rate of \$100 per week.

EXAMPLE 5 A tank is in the form of an inverted cone having an altitude of 16 m and a base radius of 4 m. Water is flowing into the tank at the rate of 2 m³/min. How fast is the water level rising when the water is 5 m deep?

Solution Let t be the number of minutes in the time that has elapsed since water started to flow into the tank, h the number of meters in the height of the water level at t min, r the number of meters in the radius of the surface of the water at t min, and V the number of cubic meters in the volume of water in the tank at t min.

At any time, the volume of water in the tank may be expressed in terms of the volume of a cone (see Fig. 3.5.2).

$$V = \tfrac{1}{3}\pi r^2 h \tag{8}$$

V, r, and h are all functions of t. Because water is flowing into the tank at the rate of 2 m³/min, $\dfrac{dV}{dt} = 2$. We wish to find $\dfrac{dh}{dt}$ when $h = 5$. To express r in terms of h we have from similar triangles

$$\frac{r}{h} = \frac{4}{16} \qquad r = \frac{1}{4}h$$

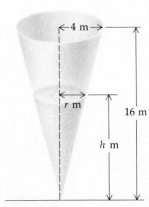

4 m

r m

16 m

h m

Figure 3.5.2

Substituting this value of r into (8) we obtain

$$V = \tfrac{1}{3}\pi(\tfrac{1}{4}h)^2(h) \quad \text{or} \quad V = \tfrac{1}{48}\pi h^3 \tag{9}$$

By differentiating on both sides of (9) with respect to t,

$$\frac{dV}{dt} = \frac{1}{16}\pi h^2 \frac{dh}{dt}$$

Substituting 2 for $\dfrac{dV}{dt}$ and solving for $\dfrac{dh}{dt}$ we get

$$\frac{dh}{dt} = \frac{32}{\pi h^2}$$

Therefore

$$\frac{dh}{dt}\bigg]_{h=5} = \frac{32}{25\pi}$$

Thus the water level is rising at the rate of $\dfrac{32}{25\pi}$ m/min when the water is 5 m deep.

Exercises 3.5

In Exercises 1 through 8, x and y are functions of a third variable t.

1. If $2x + 3y = 8$ and $\dfrac{dy}{dt} = 2$, find $\dfrac{dx}{dt}$.

2. If $\dfrac{x}{y} = 10$ and $\dfrac{dx}{dt} = -5$, find $\dfrac{dy}{dt}$.

3. If $xy = 20$ and $\dfrac{dy}{dt} = 10$, find $\dfrac{dx}{dt}$ when $x = 4$.

4. If $2xy - 3x - 4y = 3$ and $\dfrac{dy}{dt} = 4$, find $\dfrac{dx}{dt}$ when $x = 5$.

5. If $9x^2 - 4y^2 = 45$ and $\dfrac{dx}{dt} = 6$, find $\dfrac{dy}{dt}$ when $y = -3$ and $x > 0$.

6. If $x^2 + y^2 = 25$ and $\dfrac{dx}{dt} = 5$, find $\dfrac{dy}{dt}$ when $y = 4$ and $x > 0$.

7. If $\sqrt{x} + \sqrt{y} = 5$ and $\dfrac{dy}{dt} = 3$, find $\dfrac{dx}{dt}$ when $x = 1$.

8. If $2\sqrt{x} - 3\sqrt{y} + 11 = 0$ and $\dfrac{dy}{dt} = -10$, find $\dfrac{dx}{dt}$ when $x = 4$.

9. This week a factory is producing 50 units of a particular commodity, and the amount being produced is increasing at the rate of 2 units per week. If C dollars is the cost of producing x units, and $C = 0.08x^3 - x^2 + 10x + 48$, find the current rate at which the production cost is increasing.

10. The demand in a certain market for a particular kind of breakfast cereal is given by the demand equation $px + 25p - 4000 = 0$, where x thousands of boxes is the quantity demanded per week when p cents is the price of one box. If the current price of

the cereal is 80 cents per box and the price per box is increasing at the rate of 0.2 cent each week, find the rate of change of the demand.

11. The demand equation for a particular shirt is $2px + 65p - 4950 = 0$, where x hundreds of shirts are demanded per week when p dollars is the price of a shirt. If the shirt is selling this week at $30 and the price is increasing at the rate of 20 cents per week, find the rate of change of the demand.

12. The supply equation for a certain commodity is $x = 1000\sqrt{3p^2 + 20p}$, where x units are supplied per month when p dollars is the price per unit. Find the rate of change of the supply if the current price is $20 per unit and the price is increasing at the rate of 50 cents per month.

13. A producer supplies x units of a commodity per week when the price is p dollars per unit. If the supply equation is $10{,}000p = x^2 + 200x + 40{,}000$, find the rate of change of the supply if the current price is $12 per unit and the price is increasing at the rate of 12 cents per week.

14. Suppose that y is the number of workers in the labor force needed to produce x units of a certain commodity, and $x = 4y^2$. If the production of the commodity this year is 250,000 units and the production is increasing at the rate of 18,000 units per year, what is the current rate at which the labor force should be increased?

15. When a clothier produces x sweaters per week, the total weekly cost and total weekly revenue are given by C dollars and R dollars, respectively, where

$$C = 1500 + 20x \qquad R = 90x - \frac{x^3}{10{,}000}$$

If the current weekly output is 400 sweaters and it is increasing at the rate of 25 sweaters per week, find the rate of change of each of the following: (a) the total weekly cost; (b) the total weekly revenue; (c) the total weekly profit.

16. The demand equation for a certain commodity is $100p = \sqrt{250{,}000 - x^2}$, where x units are demanded per day when p dollars is the price per unit. The total daily cost of producing x units is C dollars, and $C = 200 + x + 0.001x^2$. If the current daily output is 300 units and it is increasing at the rate of 16 units per day, find the rate of change of each of the following: (a) the total daily cost; (b) the total daily revenue; (c) the total daily profit.

In Exercises 17 through 20, you need to use the formula for the volume of a sphere. If r units is the radius of a sphere and V cubic units is the volume, then $V = \frac{4}{3}\pi r^3$.

17. A spherical snowball is being made so that its volume is increasing at a rate of 8 ft^3/min. Find the rate at which the radius is increasing when the snowball has a diameter of 4 ft.

18. Suppose that when the diameter is 6 ft, the snowball in Exercise 17 started to melt at a rate of $\frac{1}{4}$ ft^3/min. Find the rate at which the radius is changing when the radius is 2 ft.

19. Suppose that a tumor in a person's body is spherical in shape. If, when the radius of the tumor is 0.5 cm, the radius is increasing at the rate of 0.001 cm per day, what is the rate of increase of the volume of the tumor at that time?

20. A bacterial cell is spherical in shape. If the radius of the cell is increasing at the rate of 0.01 micrometers per day when it is 1.5 micrometers, what is the rate of increase of the volume of the cell at that time?

In Exercises 21 through 23, you need to use the formula for the surface area of a sphere. If r units is the radius of a sphere and S . square units is the surface area, then $S = 4\pi r^2$.

21. For the tumor in Exercise 19, what is the rate of increase of the surface area when its radius is 0.5 cm?

22. For the cell of Exercise 20, what is the rate of increase of the surface area when its radius is 1.5 micrometers?

23. The volume of a spherical balloon is decreasing at a rate proportional to its surface area. Show that the radius of the balloon shrinks at a constant rate.

24. A child is flying a kite that is 40 ft high and moving horizontally at a rate of 3 ft/sec. If the string is taut, at what rate is the string being played out when the length of the string released is 50 ft?

25. If the bottom of the ladder in Example 2 is pushed horizontally toward the wall at 2 ft/sec, how fast is the top of the ladder sliding up the wall when the bottom is 20 ft from the wall?

26. A woman on a dock is pulling in a boat at the rate of 50 ft/min by means of a rope attached to the boat at water level. If the woman's hands are 16 ft above the water level, how fast is the boat approaching the dock when the amount of rope out is 20 ft?

27. A light is hung 15 ft above a straight horizontal path. If a man 6 ft tall is walking away from the light at the rate of 5 ft/sec, how fast is his shadow lengthening?

28. In Exercise 27, at what rate is the tip of the man's shadow moving?

29. Boyle's law for the expansion of gas is $PV = C$, where P is the number of pounds per square unit of pressure, V is the number of cubic units of volume of the gas, and C is a constant. At a certain instant the pressure is 3000 lb/ft^2, the volume is 5 ft^3, and the volume is increasing at the rate of 3 ft^3/min. Find the rate of change of the pressure at this instant.

30. The adiabatic law (no gain or loss of heat) for the expansion of air is $PV^{1.4} = C$, where P is the number of pounds per square unit of pressure, V is the number of cubic units of volume, and C is a constant. At a specific instant the pressure is 40 lb/in.^{2} and is increasing at the rate of 8 lb/in.^{2} each second. What is the rate of change of the volume at this instant?

31. A stone is dropped into a still pond. Concentric circular ripples spread out, and the radius of the disturbed region increases at the rate of 16 cm/sec. At what rate does the area of the disturbed region increase when its radius is 4 cm?

32. Oil is running into an inverted conical tank at the rate of 3π m^3 per minute. If the tank has a radius of 2.5 m at the top and a depth of 10 m, how fast is the depth of the oil changing when it is 8 m?

33. A water tank in the form of an inverted cone is being emptied at the rate of 6 m^3/min. The altitude of the cone is 24 m, and the base radius is 12 m. Find how fast the water level is lowering when the water is 10 m deep.

34. An automobile traveling at a rate of 30 ft/sec is approaching an intersection. When the automobile is 120 ft from the intersection, a truck traveling at the rate of 40 ft/sec crosses the intersection. The automobile and the truck are on roads that are at right angles to each other. How fast are the automobile and the truck separating 2 sec after the truck leaves the intersection?

Review Exercises for Chapter 3

In Exercises 1 through 8, find the derivative of the given function.

1. $f(x) = (3x^2 - 2x + 1)^3$

2. $f(s) = (2s^3 - 3s + 7)^4$

3. $g(t) = \sqrt{t^2 + 4t - 3}$

4. $F(x) = (4x^4 - 4x^2 + 1)^{-1/3}$

5. $f(x) = \left(\dfrac{3x^2 + 4}{x^7 + 1}\right)^{10}$

6. $f(x) = \sqrt[3]{\dfrac{x}{x^3 + 1}}$

7. $F(x) = (x^2 - 1)^{3/2}(x^2 - 4)^{1/2}$

8. $g(x) = (x^4 - x)^{-3}(5 - x^2)^{-1}$

In Exercises 9 through 14, find $\dfrac{dy}{dx}$.

9. $4x^2 + 4y^2 - y^3 = 0$

10. $xy^2 + 2y^3 = x - 2y$

11. $x^{2/3} + y^{2/3} = 4$

12. $x\sqrt{y} - y\sqrt{x} = 2$

13. $y = x^2 + [x^3 + (x^4 + x)^2]^3$

14. $y = \dfrac{x\sqrt{3 + 2x}}{4x - 1}$

15. Find an equation of the tangent line to the curve $2x^3 + 2y^3 - 9xy = 0$ at the point $(2, 1)$.

16. Find an equation of the tangent line to the curve $x - y = \sqrt{x + y}$ at the point $(3, 1)$.

17. If $C(x)$ dollars is the total cost of manufacturing x chairs, and $C(x) = x^2 + 40x + 800$, find: (a) the marginal cost function; (b) the marginal cost when 20 chairs are manufactured; (c) the actual cost of manufacturing the twenty-first chair.

18. If $C(x)$ dollars is the total cost of producing x units of a particular commodity, and $C(x) = 4x^{3/2} + 20x + 100$, find: (a) the marginal cost when 25 units are produced; (b) the number of units produced when the marginal cost is $48.

19. For the total cost function of Exercise 17, find each of the following, and give the economic interpretation of the result: (a) the average cost when $x = 10$; (b) the marginal cost when $x = 10$; (c) the elasticity of cost when $x = 10$.

20. For the total cost function of Exercise 18, find the elasticity of cost when $x = 100$, and give the economic interpretation of the result.

21. The total revenue received from the sale of x watches is $R(x)$ dollars, and $R(x) = 100x - \frac{1}{6}x^2$. Find: (a) the marginal revenue function; (b) the marginal revenue when $x = 15$; (c) the actual revenue from the sale of the sixteenth watch.

22. The total revenue function R for a particular commodity is given by $R(x) = 4x - \frac{3}{4}x^2$. Find: (a) the demand equation; (b) the marginal revenue function. (c) Draw sketches of the demand, total revenue, and marginal revenue curves on the same set of axes.

23. The supply equation for a calculator is $y = m^2 + \sqrt{m}$, where $100y$ calculators are supplied when m dollars is the price per calculator. Find: (a) the average rate of change of the supply with respect to the price when the price is increased from \$16 to \$17; (b) the instantaneous (or marginal) rate of change of the supply with respect to the price when the price is \$16.

24. If A square units is the area of an isosceles right triangle for which each leg has a length of x units, find: (a) the average rate of change of A with respect to x as x changes from 8.00 to 8.01; (b) the instantaneous rate of change of A with respect to x when x is 8.00.

25. The population of a certain city t years after January 1, 1983, will be $P(t)$, where $P(t) = 30t^2 + 100t + 20,000$. (a) Find the rate at which the population will be growing on January 1, 1987. (b) Find the relative rate of growth of the population on January 1, 1987.

26. The demand equation for a certain cracker is

$$x = 800(400 - 3p - p^2)$$

where x boxes are demanded when p dollars is the price per box. Find: (a) the average rate of change of the demand with respect to the price when the price is increased from \$1 to \$1.05; (b) the instantaneous rate of change of the demand with respect to the price when the price is \$1.

27. The demand equation for a particular brand of soap is

$$px + x + 20p = 3000$$

where $1000x$ bars of soap are demanded per week when p cents is the price per bar. If the current price of the soap is 49 cents per bar and the price per bar is increasing at the rate of 0.2 cent each week, find the rate of change in the demand.

28. The supply equation of a particular commodity is

$$8000p = 20,000 + 80x + x^2$$

where x units are supplied per week when the price is p dollars per unit. If the current price is \$8 per unit and the price is increasing at the rate of 4 cents per week, find the rate of change of the supply.

29. Suppose C millions of dollars is the capitalization of a certain corporation, P millions of dollars is the corporation's annual profit, and $P = 0.05C - 0.004C^2$. If the capitalization is increasing at the rate of \$400,000 per year, find the rate of change of the corporation's annual profit if its current capitalization is (a) \$3 million and (b) \$6 million.

30. The total weekly cost and total weekly revenue of a manufacturing company are C dollars and R dollars, respectively, when x units are produced per week, and

$$C = 50x + 4000 \qquad R = 350x - \frac{1}{3}x^2$$

If the current weekly output is 300 units and it is increasing at the rate of 20 units per week, find the rate of change of each of the following: (a) the total weekly cost; (b) the total weekly revenue; (c) the total weekly profit.

31. The demand equation for a particular article of merchandise is $px = 8000$, where x units are demanded per week when p dollars is the price per unit. It is expected that in t weeks the price of the commodity will be p dollars, where $p = 18 + \sqrt{t}$, and $t \in [0, 5]$. What is the anticipated rate of change of the demand with respect to time in 4 weeks?

32. Stefan's law states that a body emits radiant energy according to the formula $R = kT^4$, where R is the measure of the rate of emission of the radiant energy per square unit of area, T is the measure of the Kelvin temperature of the surface, and k is a constant. Find: (a) the average rate of change of R with respect to T as T increases from 200 to 300; (b) the instantaneous rate of change of R with respect to T when T is 200.

33. A burn on a person's skin is in the shape of a circle. If the radius of the burn is decreasing at the rate of 0.05 cm per day when it is 1.0 cm, what is the rate of decrease of the area of the burn at that instant?

34. A funnel in the form of a cone is 10 in. across the top and 8 in. deep. Water is flowing into the funnel at the rate of 12 in.^{3}/sec and out at the rate of 4 in.^{3}/sec. How fast is the surface of the water rising when it is 5 in. deep?

35. A man 6 ft tall is walking toward a building at the rate of 4 ft/sec. If there is a light on the ground 40 ft from the building, how fast is the man's shadow on the building growing shorter when he is 30 ft from the building?

36. A ladder 7 m long is leaning against a wall. If the bottom of the ladder is pushed horizontally toward the wall at 1.5 m/sec, how fast is the top of the ladder sliding up the wall when the bottom is 2 m from the wall?

37. In a large lake a predator fish feeds on a smaller fish, and the predator population at any time is a function of the number of small fish in the lake at that time. Suppose that when there are x small fish in the lake, the predator population is y, and $y = \frac{1}{4}x^2 + 80$; if the fishing season ended t weeks ago, $x = 8t + 90$. At what rate is the population of the predator fish growing 9 weeks after the close of the fishing season? Do not express y in terms of t, but use the chain rule.

CHAPTER 4

EXTREME FUNCTION VALUES
AND APPLICATIONS

4.1 Maximum and minimum function values

We have seen that the geometrical interpretation of the derivative of a function is the slope of the tangent line to the graph of the function at a point. This fact enables us to apply derivatives as an aid in sketching graphs. For example, the derivative may be used to determine at what points the tangent line is horizontal; these are the points where the derivative is zero. This procedure was applied in Illustration 1 of Sec. 2.4, where we found the maximum value of a profit function by locating the point on its graph where the derivative was zero. Before applying the derivative to draw sketches of graphs and to solve maximum and minimum problems, we need some definitions and theorems.

Definition of relative maximum value

> The function f is said to have a **relative maximum value** at c if there exists an open interval containing c, on which f is defined, such that $f(c) \geq f(x)$ for all x in this interval.

Figures 4.1.1 and 4.1.2 each show a sketch of a portion of the graph of a function having a relative maximum value at c.

Definition of relative minimum value

> The function f is said to have a **relative minimum value** at c if there exists an open interval containing c, on which f is defined, such that $f(c) \leq f(x)$ for all x in this interval.

Figures 4.1.3 and 4.1.4 each show a sketch of a portion of the graph of a function having a relative minimum value at c.

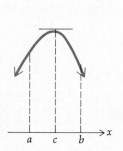

Figure 4.1.1

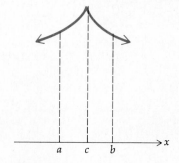

Figure 4.1.2

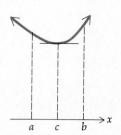

Figure 4.1.3

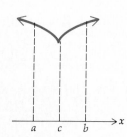

Figure 4.1.4

If the function f has either a relative maximum or a relative minimum value at c, then f is said to have a **relative extremum** at c.

The following theorem is used to locate the possible values of c for which there is a relative extremum.

> **Theorem 4.1.1** If $f(x)$ exists for all values of x in the open interval (a, b), and if f has a relative extremum at c, where $a < c < b$, then if $f'(c)$ exists, $f'(c) = 0$.

The geometrical interpretation of Theorem 4.1.1 is that if f has a relative extremum at c, and if $f'(c)$ exists, then the graph of $y = f(x)$ must have a horizontal tangent line at the point where $x = c$. A proof of the theorem is omitted.

If f is a differentiable function, then the only possible values of x for which f can have a relative extremum are those for which $f'(x) = 0$. However, $f'(x)$ can be equal to zero for a specific value of x, and yet f may not have a relative extremum there, as shown in the following illustration.

● ILLUSTRATION 1

Consider the function f defined by

$$f(x) = (x - 1)^3$$

A sketch of the graph of this function is shown in Fig. 4.1.5. $f'(x) = 3(x - 1)^2$, and so $f'(1) = 0$. However, $f(x) < 0$ if $x < 1$, and $f(x) > 0$ if $x > 1$. So f does not have a relative extremum at 1.　　　　　　　　　●

A function f may have a relative extremum at a number and f' may fail to exist there. This situation is shown in Illustration 2.

● ILLUSTRATION 2

Let the function f be defined as follows:

$$f(x) = \begin{cases} 2x - 1 & \text{if } x \leq 3 \\ 8 - x & \text{if } 3 < x \end{cases}$$

A sketch of the graph of this function is shown in Fig. 4.1.6. The function f has a relative maximum value at 3. Observe that f is not differentiable at 3; that is, $f'(3)$ does not exist.　　　　　　　　　●

Illustration 2 demonstrates why the condition "$f'(c)$ exists" must be included in the hypothesis of Theorem 4.1.1.

In summary, then, if a function f is defined at a number c, a necessary condition for f to have a relative extremum there is that either $f'(c) = 0$ or $f'(c)$ does not exist. But we have noted that this condition is not sufficient.

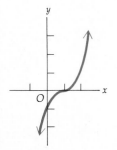

Figure 4.1.5

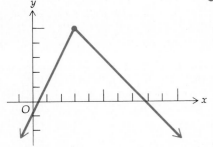

Figure 4.1.6

Definition of a critical number of a function

> If c is a number in the domain of the function f, and if either $f'(c) = 0$ or $f'(c)$ does not exist, then c is called a **critical number** of f.

Because of this definition and the previous discussion, a necessary condition for a function to have a relative extremum at a number c is for c to be a critical number.

EXAMPLE 1 Find the critical numbers of the function f defined by $f(x) = x^{4/3} + 4x^{1/3}$.

Solution

$$f(x) = x^{4/3} + 4x^{1/3}$$

$$f'(x) = \tfrac{4}{3}x^{1/3} + \tfrac{4}{3}x^{-2/3} = \tfrac{4}{3}x^{-2/3}(x + 1) = \frac{4(x + 1)}{3x^{2/3}}$$

$f'(x) = 0$ when $x = -1$, and $f'(x)$ does not exist when $x = 0$. Both -1 and 0 are in the domain of f; therefore the critical numbers of f are -1 and 0.

We are frequently concerned with a function defined on a given interval, and we wish to find the largest or smallest function value on the interval. These intervals can be either closed, open, or closed at one end and open at the other. The greatest function value on an interval is called the *absolute maximum value*, and the smallest function value on an interval is called the *absolute minimum value*. Following are the precise definitions.

Definition of absolute maximum value on an interval	The function f is said to have an **absolute maximum value on an interval** if there is some number c in the interval such that $f(c) \geq f(x)$ for all x in the interval. In such a case, $f(c)$ is the absolute maximum value of f on the interval.

Definition of absolute minimum value on an interval	The function f is said to have an **absolute minimum value on an interval** if there is some number c in the interval such that $f(c) \leq f(x)$ for all x in the interval. In such a case, $f(c)$ is the absolute minimum value of f on the interval.

An **absolute extremum** of a function on an interval is either an absolute maximum value or an absolute minimum value of the function on the interval. A function may or may not have an absolute extremum on a given interval. In each of the following illustrations, a function and an interval are given, and we determine the absolute extrema of the function on the interval if there are any.

● ILLUSTRATION 3

Suppose f is the function defined by

$$f(x) = 2x$$

A sketch of the graph of f on $[1, 4)$ is shown in Fig. 4.1.7. The function f has an absolute minimum value of 2 on $[1, 4)$. There is no absolute maximum value of f on $[1, 4)$ because $\lim\limits_{x \to 4^-} f(x) = 8$, but $f(x)$ is always less than 8 on the given interval.   ●

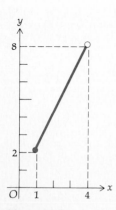

Figure 4.1.7

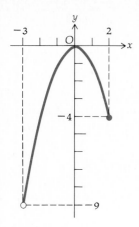

Figure 4.1.8

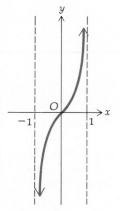

Figure 4.1.9

● ILLUSTRATION 4

Consider the function f defined by

$$f(x) = -x^2$$

A sketch of the graph of f on $(-3, 2]$ is shown in Fig. 4.1.8. The function f has an absolute maximum value of 0 on $(-3, 2]$. There is no absolute minimum value of f on $(-3, 2]$ because $\lim\limits_{x \to -3^+} f(x) = -9$, but $f(x)$ is always greater than -9 on the given interval. ●

● ILLUSTRATION 5

The function f defined by

$$f(x) = \frac{x}{1 - x^2}$$

has neither an absolute maximum value nor an absolute minimum value on $(-1, 1)$. A sketch of the graph of f on $(-1, 1)$ is shown in Fig. 4.1.9. Observe that

$$\lim_{x \to -1^+} f(x) = -\infty \quad \text{and} \quad \lim_{x \to 1^-} f(x) = +\infty \qquad ●$$

● ILLUSTRATION 6

Let f be the function defined by

$$f(x) = \begin{cases} x + 1 & \text{if } x < 1 \\ x^2 - 6x + 7 & \text{if } 1 \leq x \end{cases}$$

A sketch of the graph of f on $[-5, 4]$ is shown in Fig. 4.1.10. The absolute maximum value of f on $[-5, 4]$ occurs at 1, and $f(1) = 2$; the absolute minimum value of f on $[-5, 4]$ occurs at -5, and $f(-5) = -4$. Note that f has a relative maximum value at 1 and a relative minimum value at 3. Also note that 1 is a critical number of f because f' does not exist at 1, and 3 is a critical number of f because $f'(3) = 0$. ●

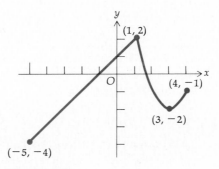

Figure 4.1.10

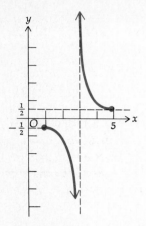

Figure 4.1.11

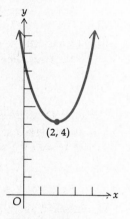

Figure 4.1.12

● ILLUSTRATION 7

The function f defined by

$$f(x) = \frac{1}{x - 3}$$

has neither an absolute maximum value nor an absolute minimum value on [1, 5]. See Fig. 4.1.11 for a sketch of the graph of f. $\lim\limits_{x \to 3^-} f(x) = -\infty$; so $f(x)$ can be made less than any negative number by taking $(3 - x) > 0$ and sufficiently small. Also, $\lim\limits_{x \to 3^+} f(x) = +\infty$; so $f(x)$ can be made greater than any positive number by taking $(x - 3) > 0$ and sufficiently small. ●

We may speak of an absolute extremum of a function when no interval is specified. In such a case we are referring to an absolute extremum of the function on the entire domain of the function.

● ILLUSTRATION 8

The graph of the function f defined by

$$f(x) = x^2 - 4x + 8$$

is a parabola, and a sketch is shown in Fig. 4.1.12. The lowest point of the parabola is at (2, 4), and the parabola opens upward. The function has an absolute minimum value of 4 at 2. There is no absolute maximum value of f. ●

Referring back to Illustrations 3–8, we see that the only case in which there is both an absolute maximum function value and an absolute minimum function value is in Illustration 6, where the function is continuous on the closed interval $[-5, 4]$. In the other illustrations, either we do not have a closed interval or we do not have a continuous function. If a function is continuous on a closed interval, there is a theorem, called the *extreme-value theorem*, that assures that the function has both an absolute maximum value and an absolute minimum value on the interval. The proof of the theorem is omitted.

The extreme-value theorem

> **Theorem 4.1.2** If the function f is continuous on the closed interval $[a, b]$, then f has an absolute maximum value and an absolute minimum value on $[a, b]$.

The extreme-value theorem states that continuity of a function on a closed interval is a sufficient condition to guarantee that the function has both an absolute maximum value and an absolute minimum value on the interval. However, it is not a necessary condition. For example, the function

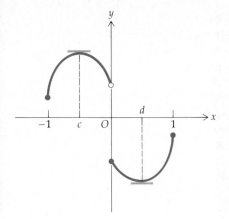

Figure 4.1.13

Table 4.1.1

x	-2	-1	$\frac{1}{3}$	$\frac{1}{2}$
$f(x)$	-1	2	$\frac{22}{27}$	$\frac{7}{8}$

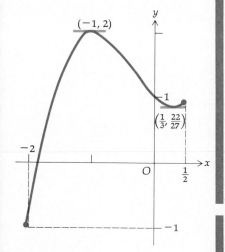

Figure 4.1.14

Table 4.1.2

x	1	2	5
$f(x)$	1	0	$\sqrt[3]{9}$

whose graph appears in Fig. 4.1.13 has an absolute maximum value at $x = c$ and an absolute minimum value at $x = d$, even though the function is discontinuous on the closed interval $[-1, 1]$.

An absolute extremum of a function continuous on a closed interval must be either a relative extremum or a function value at an endpoint of the interval. Because a necessary condition for a function to have a relative extremum at a number c is for c to be a critical number, the absolute maximum value and the absolute minimum value of a continuous function f on a closed interval $[a, b]$ can be determined by the following procedure:

1. Find the function values at the critical numbers of f on $[a, b]$.
2. Find the values of $f(a)$ and $f(b)$.
3. The largest of the values from steps 1 and 2 is the absolute maximum value, and the smallest of the values from steps 1 and 2 is the absolute minimum value.

EXAMPLE 2   Given $f(x) = x^3 + x^2 - x + 1$, find the absolute extrema of f on $[-2, \frac{1}{2}]$.

Solution   Because f is continuous on $[-2, \frac{1}{2}]$, the extreme-value theorem applies. To find the critical numbers of f, first find f':

$$f(x) = x^3 + x^2 - x + 1 \qquad f'(x) = 3x^2 + 2x - 1$$

$f'(x)$ exists for all real numbers, and so the only critical numbers of f will be the values of x for which $f'(x) = 0$. Set $f'(x) = 0$.

$$(3x - 1)(x + 1) = 0$$

$$x = \tfrac{1}{3} \quad x = -1$$

The critical numbers of f are -1 and $\frac{1}{3}$, and each of these numbers is in the given closed interval $[-2, \frac{1}{2}]$. The function values at the critical numbers and at the endpoints of the interval are given in Table 4.1.1.

The absolute maximum value of f on $[-2, \frac{1}{2}]$ is therefore 2, which occurs at -1, and the absolute minimum value of f on $[-2, \frac{1}{2}]$ is -1, which occurs at the left endpoint -2. A sketch of the graph of this function on $[-2, \frac{1}{2}]$ is shown in Fig. 4.1.14.

EXAMPLE 3   Given $f(x) = (x - 2)^{2/3}$, find the absolute extrema of f on $[1, 5]$.

Solution   Because f is continuous on $[1, 5]$, the extreme-value theorem applies.

$$f(x) = (x - 2)^{2/3} \qquad f'(x) = \frac{2}{3(x - 2)^{1/3}}$$

There is no value of x for which $f'(x) = 0$. However, because $f'(x)$ does not exist at 2, we conclude that 2 is a critical number of f; so the absolute extrema occur either at 2 or at one of the endpoints of the interval. The function values at these numbers are given in Table 4.1.2.

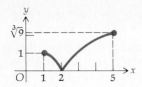

From the table we conclude that the absolute minimum value of f on $[1, 5]$ is 0, occurring at 2, and the absolute maximum value of f on $[1, 5]$ is $\sqrt[3]{9}$, occurring at 5. A sketch of the graph of this function on $[1, 5]$ is shown in Fig. 4.1.15.

Figure 4.1.15

Exercises 4.1

In Exercises 1 through 12, find the critical numbers of the given function.

1. $f(x) = x^3 + 7x^2 - 5x$

2. $g(x) = 2x^3 - 2x^2 - 16x + 1$

3. $f(x) = x^4 + 4x^3 - 2x^2 - 12x$

4. $f(x) = x^{7/3} + x^{4/3} - 3x^{1/3}$

5. $g(x) = x^{6/5} - 12x^{1/5}$

6. $f(x) = x^4 + 11x^3 + 34x^2 + 15x - 2$

7. $f(t) = (t^2 - 4)^{2/3}$

8. $f(x) = (x^2 - 3x - 4)^{1/3}$

9. $h(x) = \dfrac{x - 3}{x + 7}$

10. $f(t) = t^{5/3} - 3t^{2/3}$

11. $f(x) = \dfrac{x}{x^2 - 9}$

12. $f(x) = \dfrac{x + 1}{x^2 - 5x + 4}$

In Exercises 13 through 24, find the absolute extrema of the given function on the indicated interval, if there are any, and find the values of x at which the absolute extrema occur. Draw a sketch of the graph of the function on the interval.

13. $f(x) = 4 - 3x; (-1, 2]$

14. $f(x) = x^2 - 2x + 4; (-\infty, +\infty)$

15. $f(x) = \dfrac{1}{x}; [-2, 3]$

16. $f(x) = \dfrac{1}{x}; [2, 3)$

17. $f(x) = \sqrt{3 + x}; [-3, +\infty)$

18. $f(x) = \dfrac{3x}{9 - x^2}; (-3, 2)$

19. $f(x) = \dfrac{4}{(x - 3)^2}; [2, 5]$

20. $f(x) = \sqrt{4 - x^2}; (-2, 2)$

21. $f(x) = |x - 4| + 1; (0, 6)$

22. $f(x) = |3 - x|; (-\infty, +\infty)$

23. $f(x) = \begin{cases} \dfrac{2}{x - 5} & \text{if } x \neq 5 \\ 2 & \text{if } x = 5 \end{cases}; [3, 5]$

24. $f(x) = \begin{cases} |x + 1| & \text{if } x \neq -1 \\ 3 & \text{if } x = -1 \end{cases}; [-2, 1]$

In Exercises 25 through 34, find the absolute maximum value and the absolute minimum value of the given function on the indicated interval by the method used in Examples 2 and 3 of this section. Draw a sketch of the graph of the function on the interval.

25. $g(x) = x^3 + 5x - 4; [-3, -1]$

26. $f(x) = x^3 + 3x^2 - 9x; [-4, 4]$

27. $f(x) = x^4 - 8x^2 + 16; [-4, 0]$

28. $f(x) = x^4 - 8x^2 + 16; [-3, 2]$

29. $f(x) = x^4 - 8x^2 + 16; [0, 3]$

30. $g(x) = x^4 - 8x^2 + 16; [-1, 4]$

31. $f(x) = \dfrac{x}{x + 2}; [-1, 2]$

32. $f(x) = \dfrac{x + 5}{x - 3}; [-5, 2]$

33. $f(x) = (x + 1)^{2/3}; [-2, 1]$

34. $f(x) = 1 - (x - 3)^{2/3}; [-5, 4]$

4.2 Applications involving an absolute extremum on a closed interval

We consider some problems in which the solution is an absolute extremum of a function on a closed interval. Use is made of the extreme-value theorem, which assures that both an absolute maximum value and an absolute minimum value of a function exist on a closed interval if the function is continuous on that closed interval. The procedure is illustrated by some examples.

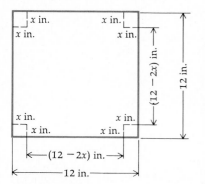

Figure 4.2.1

EXAMPLE 1   Example 2 of Sec. 1.5 and Illustration 10 of Sec. 2.3 pertained to the following situation: A cardboard box manufacturer wishes to make open boxes from square pieces of cardboard of side 12 in. by cutting equal squares from the four corners and turning up the sides. Find the length of the side of the square to be cut out to obtain a box of the largest possible volume.

Solution   Let x be the number of inches in the length of the side of the square to be cut out and $V(x)$ in.^{3} be the volume of the box. Figure 4.2.1 shows a given piece of cardboard, and Fig. 4.2.2 shows the box.

In Example 2 of Sec. 1.5 we obtained the following equation that defines $V(x)$:

$$V(x) = 144x - 48x^2 + 4x^3 \qquad x \in [0, 6]$$

Because V is continuous on the closed interval $[0, 6]$, it follows from the extreme-value theorem that V has an absolute maximum value on this interval. We also know that this absolute maximum value of V must occur either at a critical number or at an endpoint of the interval. To find the critical numbers of V we find $V'(x)$ and then find the values of x for which either $V'(x) = 0$ or $V'(x)$ does not exist.

$$V'(x) = 144 - 96x + 12x^2$$

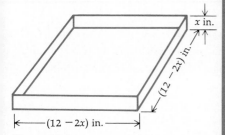

Figure 4.2.2

$V'(x)$ exists for all values of x. Setting $V'(x) = 0$ we have

$$12(x^2 - 8x + 12) = 0$$

$$12(x - 6)(x - 2) = 0$$

$$x = 6 \quad x = 2$$

The critical numbers of V are 2 and 6, both of which are in the closed interval $[0, 6]$. Because $V(0) = 0$ and $V(6) = 0$ while $V(2) = 128$, we conclude that the absolute maximum value of V on $[0, 6]$ is 128, occurring at 2.

Therefore the largest possible volume is 128 in.^{3}, and this is obtained when the length of the side of the square cut out is 2 in.

EXAMPLE 2   Points A and B are opposite each other on shores of a straight river that is 3 km wide. Point C is on the same shore as B but 2 km down the river from B. A telephone company wishes to lay a cable from A to C. If the cost per kilometer of the cable is 25 percent more under the water than it is on land, what line of cable would be least expensive for the company?

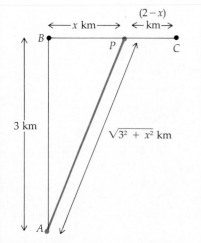

Figure 4.2.3

Solution   Refer to Fig. 4.2.3. Let P be a point on the same shore as B and C and between B and C so that the cable will run from A to P to C. Let x km be the distance from B to P. Then $(2 - x)$ km is the distance from P to C, and

$x \in [0, 2]$. Let k dollars be the cost per kilometer on land and $\frac{5}{4}k$ dollars be the cost per kilometer under the water (k is a constant). If $C(x)$ dollars is the total cost of running the cable from A to P and from P to C, then

$$C(x) = \tfrac{5}{4}k\sqrt{3^2 + x^2} + k(2 - x)$$

$$C'(x) = \frac{5kx}{4\sqrt{9 + x^2}} - k$$

$C'(x)$ exists for all values of x. Setting $C'(x) = 0$ and solving for x we have

$$\frac{5kx}{4\sqrt{9 + x^2}} - k = 0$$

$$5x = 4\sqrt{9 + x^2} \qquad (1)$$

$$25x^2 = 16(9 + x^2)$$

$$9x^2 = 16 \cdot 9$$

$$x^2 = 16$$

$$x = \pm 4$$

The number -4 is an extraneous root of (1), and 4 is not in the interval $[0, 2]$. Therefore there are no critical numbers of C in $[0, 2]$. The absolute minimum value of C on $[0, 2]$ must therefore occur at an endpoint of the interval. Computing $C(0)$ and $C(2)$ we get

$$C(0) = \tfrac{23}{4}k \quad \text{and} \quad C(2) = \tfrac{5}{4}k\sqrt{13}$$

Because $\frac{5}{4}k\sqrt{13} < \frac{23}{4}k$, the absolute minimum value of C on $[0, 2]$ is $\frac{5}{4}k\sqrt{13}$, occurring when $x = 2$. Therefore, for the cost of the cable to be the least, the cable should go directly from A to C under the water.

EXAMPLE 3   In Example 3 of Sec. 1.5 we had the following situation: A rectangular field is to be fenced off along the bank of a river, and no fence is required along the river. The material for the fence costs $8 per running foot for the two ends and $12 per running foot for the side parallel to the river; $3600 worth of fence is to be used. Find the dimensions of the field of largest possible area that can be enclosed with the $3600 worth of fence.

Solution   Let x ft be the length of an end of the field. See Fig. 4.2.4. In Example 3 of Sec. 1.5 we showed that the number of feet in the length of the side parallel to the river is $300 - \frac{4}{3}x$, and if $A(x)$ ft^2 is the area of the field,

$$A(x) = 300x - \tfrac{4}{3}x^2 \qquad x \in [0, 225] \qquad (2)$$

Because A is continuous on the closed interval $[0, 225]$, from the extreme-value theorem A has an absolute maximum value on this interval. From (2),

$$A'(x) = 300 - \tfrac{8}{3}x$$

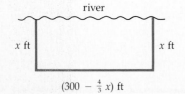

Figure 4.2.4

Because $A'(x)$ exists for all x, the critical numbers of A are found by setting $A'(x) = 0$, which gives

$$900 - 8x = 0$$

$$x = 112\tfrac{1}{2}$$

The only critical number of A is $112\tfrac{1}{2}$, which is in the closed interval $[0, 225]$. Thus the absolute maximum value of A must occur at 0, $112\tfrac{1}{2}$, or 225. Because $A(0) = 0$ and $A(225) = 0$, while $A(112\tfrac{1}{2}) = 16{,}875$, the absolute maximum value of A on $[0, 225]$ is 16,875, occurring when $x = 112\tfrac{1}{2}$ and $300 - \tfrac{4}{3}x = 150$. Therefore the largest possible area that can be enclosed for \$3600 is 16,875 ft^2, and this is obtained when the side parallel to the river is 150 ft long and the ends are each $112\tfrac{1}{2}$ ft long.

EXAMPLE 4   Example 5 of Sec. 1.5 and Illustration 11 of Sec. 2.3 pertained to the following situation: In the planning of a coffee shop it is estimated that if there are places for 40 to 80 people, the daily gross profit will be \$8 per place. However, if the seating capacity is above 80 places, the daily gross profit on each place will be decreased by 4 cents times the number of places over 80. What should be the seating capacity to yield the greatest daily gross profit?

Solution   Let x be the number of places in the seating capacity and $P(x)$ dollars be the daily gross profit. In Example 5 of Sec. 1.5 we obtained the following formula that defines $P(x)$.

$$P(x) = \begin{cases} 8x & \text{if } 40 \le x \le 80 \\ 11.20x - 0.04x^2 & \text{if } 80 < x \le 280 \end{cases}$$

Even though x, by definition, is an integer, to have a continuous function we let x be any real number in the interval $[40, 280]$. In Illustration 11 of Sec. 2.3 we showed that P is continuous on the closed interval $[40, 280]$. Therefore, by the extreme-value theorem there is an absolute maximum value of P on this interval.

$$P'(x) = 8 \qquad \text{when } 40 < x < 80$$

and

$$P'(x) = 11.20 - 0.08x \qquad \text{when } 80 < x < 280 \tag{3}$$

To find when $P'(x) = 0$ we equate the right-hand side of (3) to zero.

$$11.20 - 0.08x = 0$$

$$x = 140$$

Thus 140 is a critical number of P, and the graph has a horizontal tangent line at $x = 140$. Because P is not differentiable at 80, then 80 is also a critical

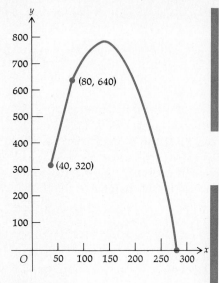

Figure 4.2.5

number of P. We evaluate $P(x)$ at the endpoints of the interval $[40, 280]$ and at the critical numbers.

$$P(40) = 320 \qquad P(80) = 640 \qquad P(140) = 784 \qquad P(280) = 0$$

The absolute maximum value of P, then, is 784, occurring when $x = 140$.

The seating capacity should be 140 places, which gives a daily gross profit of $784.

A sketch of the graph of P is shown in Fig. 4.2.5.

The following example shows an interesting application of absolute extrema in biology.

EXAMPLE 5 During a cough there is a decrease in the radius of a person's trachea (or windpipe). Suppose that the normal radius of the trachea is R cm and the radius of the trachea during a cough is r cm, where R is a constant and r is a variable. The velocity of air through the trachea can be shown to be a function of r, and if $V(r)$ cm/sec is this velocity, then

$$V(r) = kr^2(R - r) \qquad r \in \left[\tfrac{1}{2}R, R\right] \tag{4}$$

where k is a positive constant. Determine the radius of the trachea during a cough for which the velocity of air through the trachea is greatest.

Solution We wish to find the value of r that makes V an absolute maximum. Because V is continuous on the closed interval $\left[\tfrac{1}{2}R, R\right]$, from the extreme-value theorem V has an absolute maximum value on this interval. From (4),

$$V(r) = kRr^2 - kr^3$$

$$V'(r) = 2kRr - 3kr^2$$

$V'(r)$ exists for all r. Thus the only critical numbers of V are found by setting $V'(r) = 0$. We have

$$2kRr - 3kr^2 = 0$$

$$kr(2R - 3r) = 0$$

$$r = 0 \qquad r = \tfrac{2}{3}R$$

Because 0 is not in $\left[\tfrac{1}{2}R, R\right]$, the only critical number is $\tfrac{2}{3}R$. The absolute maximum value of V must occur at $\tfrac{1}{2}R$, $\tfrac{2}{3}R$, or R. Because $V(\tfrac{1}{2}R) = \tfrac{1}{8}kR^3$, $V(\tfrac{2}{3}R) = \tfrac{4}{27}kR^3$, and $V(R) = 0$, the absolute maximum value of V occurs when $r = \tfrac{2}{3}R$. Therefore, during a cough, the velocity of air through the trachea is greatest when the radius of the trachea is two-thirds of its normal radius.

Exercises 4.2

1. Find the number in the interval $[0, 1]$ such that the difference between the number and its square is a maximum.
2. Find the number in the interval $[\tfrac{1}{3}, 2]$ such that the sum of the number and its reciprocal is a maximum.
3. Find two numbers in the interval $[-20, 20]$ whose difference is 20 and whose product is a minimum.
4. Find two numbers in the interval $[0, 50]$ whose sum is 50 and whose product is a maximum.

5. Refer to Exercise 37 in Exercises 2.3. A manufacturer of open tin boxes wishes to make use of pieces of tin with dimensions 8 in. by 15 in. by cutting equal squares from the four corners and turning up the sides. Find the length of the side of the square to be cut out if a box having the largest possible volume is to be obtained from each piece of tin.

6. Refer to Exercise 38 in Exercises 2.3. The manufacturer of Exercise 5 makes the open boxes from square pieces of tin that measure k cm on a side. Find the length of the side of the square to be cut out if a box having the largest possible volume is to be obtained from each piece of tin.

7. Refer to Exercise 39 in Exercises 2.3. Find the dimensions of the largest rectangular field that can be enclosed with 240 m of fence.

8. Refer to Exercise 6 in Exercises 1.5. Find the dimensions of the largest rectangular garden that can be fenced off with 100 ft of fencing material.

9. Refer to Exercise 7 in Exercises 1.5. If one side of a rectangular field is to have a river as a natural boundary, find the dimensions of the largest rectangular field that can be enclosed by using 240 m of fence for the other three sides.

10. Refer to Exercise 40 in Exercises 2.3. Find the dimensions of the largest rectangular garden that can be placed so that a side of a house serves as a boundary and 100 ft of fencing material is to be used for the other three sides.

11. An island is at point A, 6 km offshore from the nearest point B on a straight beach. A woman on the island wishes to go to a point C 9 km down the beach from B. The woman can rent a boat for $2.50 per kilometer and travel by water to a point P between B and C, and then she can take a taxi at a cost of $2 per kilometer and travel a straight road from P to C. Find the least expensive route from point A to point C.

12. Solve Exercise 11 if point C is only 7 km down the beach from B.

13. Solve Example 2 of this section if point C is 6 km down the river from B.

14. Two towns A and B are to get their water supply from the same pumping station to be located on the bank of a straight river that is 15 km from town A and 10 km from town B. If the points on the river nearest to A and B are 20 km apart and A and B are on the same side of the river, where should the pumping station be located so that the least amount of piping is required?

15. Refer to Exercise 9 in Exercises 1.5. A rectangular plot of ground is to be enclosed by a fence and then divided down the middle by another fence. If the fence down the middle costs $2 per running foot and the other fence costs $5 per running foot, find the dimensions of the plot of largest possible area that can be enclosed with $960 worth of fence.

16. Refer to Exercise 10 in Exercises 1.5. For a package to be accepted by a particular mailing service, the sum of the length and girth (the perimeter of a cross section) must not be greater than 100 in. If a package is to be in the shape of a rectangular box with a square cross section, find the dimensions of the package having the greatest possible volume that can be mailed by the service.

17. Refer to Exercise 41 in Exercises 2.3. Orange trees grown in California produce 600 oranges per year if not more than 20 trees are planted per acre. For each additional tree planted per acre, the yield decreases by 15 oranges. How many trees per acre should be planted to obtain the greatest number of oranges?

18. Refer to Exercise 42 in Exercises 2.3. A manufacturer can make a profit of $20 on each item if not more than 800 items are produced each week. If the profit on each item decreases 2 cents for every item over 800, how many items should the manufacturer produce each week to have the greatest profit?

19. Refer to Exercise 19 in Exercises 1.5. A private club charges each member annual membership dues of $100 less 50 cents for each member over 600 and plus 50 cents for each member under 600. How many members would give the club the most revenue from annual dues?

20. Refer to Exercise 20 in Exercises 1.5. A charity theatrical performance will cost each person $15 if not more than 150 tickets are sold. However, the cost per ticket will be reduced 7 cents for each ticket in excess of 150. How many tickets should be sold to maximize the charity's receipts?

21. Refer to Exercise 43 in Exercises 2.3. The maximum number of bacteria supportable by a particular environment is 900,000, and the rate of bacterial growth is jointly proportional to the number present and the difference between 900,000 and the number present. Determine the number of bacteria for which the rate of growth is a maximum.

22. Refer to Exercise 44 in Exercises 2.3. A particular lake can support a maximum of 14,000 fish, and the rate of growth of the fish population is jointly proportional to the number of fish present and the difference between 14,000 and the number present. What should be the size of the fish population for the growth rate to be a maximum?

23. Two products A and B are manufactured at a particular factory. If C dollars is the total cost of production for an 8-hour day, then $C = 3x^2 + 42y$, where x machines are used to produce product A and y machines are used to produce product B. If during an 8-hour day there are 15 machines working, determine how many of these machines should be used to produce A and how many should be used to produce B for the total cost to be least.

24. A piece of wire 20 cm long is cut into two pieces, and each piece is bent into the shape of a square. How should the wire be cut so that the total area of the two squares is as small as possible?

25. Assume the decrease in a person's blood pressure depends on the amount of a particular drug taken by the person. Thus if x mg of the drug is taken, the decrease in blood pressure is a function of x. Suppose that $f(x)$ defines this function, and

$$f(x) = \tfrac{1}{2}x^2(k - x) \qquad x \in [0, k]$$

where k is a positive constant. Determine the value of x that causes the greatest decrease in blood pressure.

26. The strength of a rectangular beam is jointly proportional to its breadth and the square of its depth. Find the dimensions of the strongest beam that can be cut from a log in the shape of a right-circular cylinder of radius 72 cm.

27. In a particular small town the rate at which a rumor spreads is jointly proportional to the number of people who have heard the rumor and the number of people who have not heard it. Show that the rumor is being spread at the greatest rate when half the population of the town knows the rumor.

4.3 Increasing and decreasing functions and the first-derivative test

Suppose that Fig. 4.3.1 represents a sketch of the graph of a function f for all x in the closed interval $[x_1, x_7]$. In drawing this sketch we have assumed that f is continuous on $[x_1, x_7]$.

Figure 4.3.1 shows that as a point moves along the curve from A to B, the function values increase as the abscissa increases, and that as a point moves along the curve from B to C, the function values decrease as the abscissa increases. We say, then, that f is *increasing* on the closed interval $[x_1, x_2]$ and that f is *decreasing* on the closed interval $[x_2, x_3]$. Following are the precise definitions of a function increasing or decreasing on an interval.

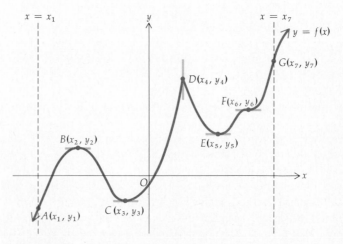

Figure 4.3.1

Definition of a function increasing on an interval	A function f defined on an interval is said to be **increasing** on that interval if and only if $$f(x_1) < f(x_2) \qquad \text{whenever } x_1 < x_2$$ where x_1 and x_2 are any numbers in the interval.

The function of Fig. 4.3.1 is increasing on the following closed intervals: $[x_1, x_2]$; $[x_3, x_4]$; $[x_5, x_6]$; $[x_6, x_7]$; $[x_5, x_7]$.

Definition of a function decreasing on an interval	A function f defined on an interval is said to be **decreasing** on that interval if and only if $$f(x_1) > f(x_2) \qquad \text{whenever } x_1 < x_2$$ where x_1 and x_2 are any numbers in the interval.

The function of Fig. 4.3.1 is decreasing on the following closed intervals: $[x_2, x_3]$; $[x_4, x_5]$.

If a function f is either increasing on an interval or decreasing on an interval, then f is said to be **monotonic** on the interval.

Before stating Theorem 4.3.1, which gives a test for determining if a given function is monotonic on an interval, let us see what is happening geometrically. Refer to Fig. 4.3.1, and observe that when the slope of the tangent line is positive, the function is increasing, and when the slope of the tangent line is negative, the function is decreasing. Because $f'(x)$ is the slope of the tangent line to the curve $y = f(x)$, f is increasing when $f'(x) > 0$, and f is decreasing when $f'(x) < 0$. Also, because $f'(x)$ is the rate of change of the function values $f(x)$ with respect to x, when $f'(x) > 0$, the function values are increasing as x increases; and when $f'(x) < 0$, the function values are decreasing as x increases. We have the following theorem.

Theorem 4.3.1 Let the function f be continuous on the closed interval $[a, b]$ and differentiable on the open interval (a, b):   (i) if $f'(x) > 0$ for all x in (a, b), then f is increasing on $[a, b]$;   (ii) if $f'(x) < 0$ for all x in (a, b), then f is decreasing on $[a, b]$.

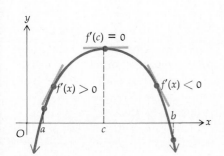

Figure 4.3.2

Refer to Fig. 4.3.2, which shows a sketch of the graph of a function f that we assume is continuous on the open interval (a, b) containing the number c. If $a < x < c$, $f'(x) > 0$ and f is increasing; the slope of the tangent line is positive for these values of x. If $c < x < b$, $f'(x) < 0$ and f is decreasing;

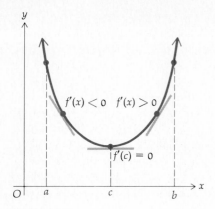

Figure 4.3.3

the slope of the tangent line is negative for these values of x. Because $f'(c) = 0$, the graph has a horizontal tangent line at the point where $x = c$; further-more, f has a relative maximum value at $x = c$. This is an illustration of part (i) of the first-derivative test given by Theorem 4.3.2. It states that if f is continuous at c and $f'(x)$ changes algebraic sign from positive to negative as x increases through the number c, then f has a relative maximum value at c.

In Fig. 4.3.3 there is a sketch of the graph of another function f that we assume is continuous on the open interval (a, b) containing the number c. For this function, when $a < x < c$, $f'(x) < 0$ and f is decreasing; the slope of the tangent line is negative for these values of x. Also for this function, when $c < x < b$, $f'(x) > 0$ and f is increasing; the slope of the tangent line is positive for these values of x. In Fig. 4.3.3 we observe that the graph has a horizontal tangent line at the point where $x = c$, which follows from the fact that $f'(c) = 0$. For this function there is a relative minimum value at $x = c$. We have here an illustration of part (ii) of the first-derivative test. This part states that if f is continuous at c and $f'(x)$ changes algebraic sign from negative to positive as x increases through the number c, then f has a relative minimum value at c. Following is the formal statement of the first-derivative test for relative extrema of a function.

First-derivative test for
relative extrema

Theorem 4.3.2   Let the function f be continuous at all points of the open interval (a, b) containing the number c, and suppose that f' exists at all points of (a, b) except possibly at c:

(i) if $f'(x) > 0$ for all values of x in some open interval having c as its right endpoint, and if $f'(x) < 0$ for all values of x in some open interval having c as its left endpoint, then f has a relative maximum value at c;

(ii) if $f'(x) < 0$ for all values of x in some open interval having c as its right endpoint, and if $f'(x) > 0$ for all values of x in some open interval having c as its left endpoint, then f has a relative minimum value at c.

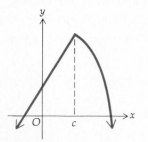

Figure 4.3.4

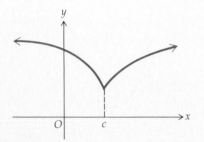

Figure 4.3.5

Figures 4.3.2 and 4.3.3 illustrate parts (i) and (ii), respectively, of the first-derivative test when $f'(c)$ exists. The first-derivative test can be applied even when $f'(c)$ does not exist, provided that $f'(x)$ exists for all other values of x in the open interval (a, b). Figure 4.3.4 shows a sketch of the graph of a function f that has a relative maximum value at a number c, but $f'(c)$ does not exist; however, $f'(x) > 0$ when $x < c$, and $f'(x) < 0$ when $x > c$.

The function of Fig. 4.3.5 has a relative minimum value at c where $f'(c)$ does not exist. For this function, $f'(x) < 0$ when $x < c$ and $f'(x) > 0$ when $x > c$, and the first-derivative test can be applied.

If c is a critical number of the function f but $f'(x)$ does not change algebraic sign as x increases through the number c, then the first-derivative

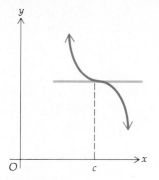

Figure 4.3.6

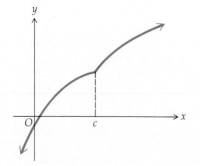

Figure 4.3.7

test cannot be applied. For instance, in Fig. 4.3.6 there is a sketch of the graph of a function for which c is a critical number where $f'(c) = 0$, and $f'(x) < 0$ when $x < c$, and $f'(x) < 0$ when $x > c$; f does not have a relative extremum at c. Figure 4.3.7 shows a sketch of the graph of a function f for which $f'(c)$ does not exist, and $f'(x) > 0$ when $x < c$, and $f'(x) > 0$ when $x > c$; f does not have a relative extremum at c.

Further illustrations of the first-derivative test occur in Fig. 4.3.1. At the critical numbers x_2 and x_4, $f'(x)$ changes algebraic sign from positive to negative, and f has a relative maximum value. At the critical numbers x_3 and x_5, $f'(x)$ changes algebraic sign from negative to positive, and f has a relative minimum value. Observe in the figure that even though x_6 is a critical number of f, there is no relative extremum at x_6 because $f'(x) > 0$ if $x_5 < x < x_6$ and $f'(x) > 0$ if $x_6 < x < x_7$.

In summary, to determine the relative extrema of a function f:

1. Find $f'(x)$.
2. Find the critical numbers of f, that is, the values of x for which $f'(x) = 0$ or for which $f'(x)$ does not exist.
3. Apply the first-derivative test (Theorem 4.3.2).

The following examples illustrate this procedure.

EXAMPLE 1 Given

$$f(x) = x^3 - 6x^2 + 9x + 1$$

find the relative extrema of f by applying the first-derivative test. Determine the values of x at which the relative extrema occur, as well as the intervals on which f is increasing and the intervals on which f is decreasing. Draw a sketch of the graph.

Solution $f'(x) = 3x^2 - 12x + 9$. $f'(x)$ exists for all values of x. Setting $f'(x) = 0$ we have

$$3x^2 - 12x + 9 = 0$$

$$3(x - 3)(x - 1) = 0$$

$$x = 3 \quad x = 1$$

Thus the critical numbers of f are 1 and 3. To determine whether f has a relative extremum at either of these numbers, we apply the first-derivative test. The results are summarized in Table 4.3.1.

Table 4.3.1

	$f(x)$	$f'(x)$	Conclusion
$x < 1$		$+$	f is increasing
$x = 1$	5	0	f has a relative maximum value
$1 < x < 3$		$-$	f is decreasing
$x = 3$	1	0	f has a relative minimum value
$3 < x$		$+$	f is increasing

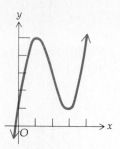

Figure 4.3.8

From the table, 5 is a relative maximum value of f occurring at 1, and 1 is a relative minimum value of f occurring at 3. A sketch of the graph is shown in Fig. 4.3.8.

EXAMPLE 2   Given $f(x) = x^{4/3} + 4x^{1/3}$, find the relative extrema of f, determine the values of x at which the relative extrema occur, and determine the intervals on which f is increasing and the intervals on which f is decreasing. Draw a sketch of the graph.

Solution

$$f(x) = x^{4/3} + 4x^{1/3}$$

$$f'(x) = \tfrac{4}{3}x^{1/3} + \tfrac{4}{3}x^{-2/3}$$

$$f'(x) = \frac{4}{3}\left(x^{1/3} + \frac{1}{x^{2/3}}\right)$$

$$f'(x) = \frac{4(x+1)}{3x^{2/3}}$$

Because $f'(x)$ does not exist when $x = 0$ and $f'(x) = 0$ when $x = -1$, the critical numbers of f are -1 and 0. We apply the first-derivative test and summarize the results in Table 4.3.2. A sketch of the graph appears in Fig. 4.3.9.

Table 4.3.2

	$f(x)$	$f'(x)$	Conclusion
$x < -1$		$-$	f is decreasing
$x = -1$	-3	0	f has a relative minimum value
$-1 < x < 0$		$+$	f is increasing
$x = 0$	0	does not exist	f does not have a relative extremum at $x = 0$
$0 < x$		$+$	f is increasing

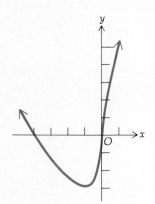

Figure 4.3.9

Exercises 4.3

In Exercises 1 through 22, do each of the following: (a) find the relative extrema of f by applying the first-derivative test; (b) determine the value of x at which the relative extrema occur; (c) determine the intervals on which f is increasing; (d) determine the intervals on which f is decreasing; (e) draw a sketch of the graph.

1. $f(x) = x^2 - 4x - 1$
2. $f(x) = 3x^2 - 3x + 2$
3. $f(x) = x^3 - x^2 - x$
4. $f(x) = x^3 - 9x^2 + 15x - 5$
5. $f(x) = 2x^3 - 9x^2 + 2$
6. $f(x) = x^3 - 3x^2 - 9x$
7. $f(x) = \tfrac{1}{4}x^4 - x^3 + x^2$
8. $f(x) = x^4 + 4x$
9. $f(x) = x^5 - 5x^3 - 20x - 2$

10. $f(x) = \tfrac{1}{5}x^5 - \tfrac{5}{3}x^3 + 4x + 1$
11. $f(x) = \sqrt{x} - \dfrac{1}{\sqrt{x}}$
12. $f(x) = \dfrac{x-2}{x+2}$

13. $f(x) = 2x\sqrt{3 - x}$

14. $f(x) = x\sqrt{5 - x^2}$

15. $f(x) = (1 - x)^2(1 + x)^3$

16. $f(x) = (x + 2)^2(x - 1)^2$

17. $f(x) = 2 - 3(x - 4)^{2/3}$

18. $f(x) = 2 - (x - 1)^{1/3}$

19. $f(x) = x^{2/3} - x^{1/3}$

20. $f(x) = x^{2/5} + 1$

21. $f(x) = (x + 1)^{2/3}(x - 2)^{1/3}$

22. $f(x) = x^{2/3}(x - 1)^2$

23. Find a and b such that the function defined by $f(x) = x^3 + ax^2 + b$ will have a relative extremum at $(2, 3)$.

24. Find a, b, and c such that the function defined by $f(x) = ax^2 + bx + c$ will have a relative maximum value of 7 at 1 and the graph of $y = f(x)$ will go through the point $(2, -2)$.

25. Find a, b, c, and d such that the function defined by $f(x) = ax^3 + bx^2 + cx + d$ will have relative extrema at $(1, 2)$ and $(2, 3)$.

4.4 Derivatives of higher order and the second-derivative test

If f' is the derivative of the function f, then f' is also a function, and it is the **first derivative** of f. It is sometimes referred to as the **first derived function**. If the derivative of f' exists, it is called the **second derivative** of f, or the second derived function, and can be denoted by f'' (read "f double prime"). Similarly, we define the **third derivative** of f, or the third derived function, as the first derivative of f'' if it exists. We denote the third derivative of f by f''' (read "f triple prime").

The **nth derivative** of the function f, where n is a positive integer greater than 1, is the first derivative of the $(n - 1)$st derivative of f. We denote the nth derivative of f by $f^{(n)}$. Thus, if $f^{(n)}$ denotes the nth derived function, we can denote the function f itself by $f^{(0)}$. Another symbol for the nth derivative of f is $D_x^n f(x)$.

EXAMPLE 1 Find all the derivatives of the function f defined by

$$f(x) = 8x^4 + 5x^3 - x^2 + 7$$

Solution

$$f'(x) = 32x^3 + 15x^2 - 2x$$

$$f''(x) = 96x^2 + 30x - 2$$

$$f'''(x) = 192x + 30$$

$$f^{(4)}(x) = 192$$

$$f^{(5)}(x) = 0$$

$$f^{(n)}(x) = 0 \qquad n \geq 5$$

If the function f is defined by the equation $y = f(x)$, we can denote the first derivative of f by the Leibniz notation $\dfrac{dy}{dx}$. Corresponding to this notation we have the symbol $\dfrac{d^2 y}{dx^2}$ for the second derivative of y with respect to x. However, $\dfrac{d^2 y}{dx^2}$ must not be thought of as a quotient. The symbol $\dfrac{d^n y}{dx^n}$ is a notation for the nth derivative of y with respect to x.

Because $f'(x)$ gives the instantaneous rate of change of $f(x)$ with respect to x, $f''(x)$, being the derivative of $f'(x)$, gives the instantaneous rate of change of $f'(x)$ with respect to x. Furthermore, if (x, y) is any point on the graph of $y = f(x)$, then $\dfrac{dy}{dx}$ gives the slope of the tangent line to the graph at the point (x, y). Thus $\dfrac{d^2y}{dx^2}$ is the instantaneous rate of change of the slope of the tangent line with respect to x at the point (x, y).

EXAMPLE 2   Let $m(x)$ be the slope of the tangent line to the curve

$$y = x^3 - 2x^2 + x$$

at the point (x, y). Find the instantaneous rate of change of $m(x)$ with respect to x at the point $(2, 2)$.

Solution

$$m(x) = \frac{dy}{dx} = 3x^2 - 4x + 1$$

The instantaneous rate of change of $m(x)$ with respect to x is given by $m'(x)$ or, equivalently, $\dfrac{d^2y}{dx^2}$.

$$m'(x) = \frac{d^2y}{dx^2} = 6x - 4$$

At the point $(2, 2)$, $\dfrac{d^2y}{dx^2} = 8$. Hence, at the point $(2, 2)$ the change in $m(x)$ is 8 times the change in x.

In the following illustration there is an economic situation in which the second derivative indicates a rate of inflation.

● ILLUSTRATION 1

It is estimated that t months after January 1 until July 1, the price of a certain commodity will be $P(t)$ cents, where

$$P(t) = 40 + 3t^2 - \tfrac{1}{3}t^3 \qquad 0 \le t \le 6$$

In Table 4.4.1 there are values of $P(t)$ for $t = 0, 1, 2, 3, 4, 5, 6$, and these values give the price of the commodity on the first of the month for each of the months from January through July.

Computing the first and second derivatives of P we obtain

$$P'(t) = 6t - t^2$$

$$P''(t) = 6 - 2t$$

Table 4.4.1

t	$P(t)$	Conclusion
0	40	price on January 1 is 40 cents
1	42.7	price on February 1 is 43 cents
2	49.3	price on March 1 is 49 cents
3	58.0	price on April 1 is 58 cents
4	66.7	price on May 1 is 67 cents
5	73.3	price on June 1 is 73 cents
6	76.0	price on July 1 is 76 cents

Observe that when $0 < t < 6$, $P'(t) > 0$. Therefore the price is increasing on $[0, 6]$. Also observe that when $0 < t < 3$, $P''(t) > 0$. Hence, when $0 < t < 3$, $P'(t)$ is increasing; that is, the rate of the price increase is increasing. Thus we can say that when $0 < t < 3$, the price is increasing and the rate of inflation of this commodity is increasing. When $3 < t < 6$, $P''(t) < 0$. Therefore, when $3 < t < 6$, $P'(t)$ is decreasing; that is, the rate of the price increase is decreasing. Hence, when $3 < t < 6$, the price is increasing but the rate of inflation of the commodity is decreasing.

For instance, when $t = 2$ (on March 1), we have $P'(2) = 8$ and $P''(2) = 2$. Hence, on March 1 the price is increasing at the rate of 8 cents per month, and the price increase is increasing at the rate of 2 cents a month per month. When $t = 5$ (on June 1), we obtain $P'(5) = 5$ and $P''(5) = -4$. Thus, on June 1 the price is increasing at the rate of 5 cents per month, and the price increase is decreasing at the rate of 4 cents a month per month.

In Illustration 5 of Sec. 5.1 we return to this function and discuss its behavior at $t = 3$, where the second derivative changes sign from positive to negative. ●

Further applications of the second derivative are its uses in the second-derivative test for relative extrema (Theorem 4.4.1) and the sketching of the graph of a function (Sec. 5.2). An important application of other higher-ordered derivatives is to determine infinite series.

The following example illustrates how the second derivative is found for functions defined implicitly.

EXAMPLE 3 Given

$$4x^2 + 9y^2 = 36$$

find $\dfrac{d^2y}{dx^2}$ by implicit differentiation.

Solution   Differentiating implicitly with respect to x, we have

$$8x + 18y \frac{dy}{dx} = 0$$

$$\frac{dy}{dx} = \frac{-4x}{9y} \tag{1}$$

To find $\dfrac{d^2 y}{dx^2}$ we compute the derivative of a quotient and keep in mind that y is a function of x. Thus

$$\frac{d^2 y}{dx^2} = \frac{9y(-4) - (-4x)\left(9 \cdot \dfrac{dy}{dx}\right)}{81y^2} \tag{2}$$

Substituting the value of $\dfrac{dy}{dx}$ from (1) into (2) we get

$$\frac{d^2 y}{dx^2} = \frac{-36y + (36x)\dfrac{-4x}{9y}}{81y^2}$$

$$= \frac{-36y^2 - 16x^2}{81y^3}$$

$$= \frac{-4(9y^2 + 4x^2)}{81y^3} \tag{3}$$

Because any values of x and y satisfying (3) must also satisfy the original equation, we can replace $(9y^2 + 4x^2)$ by 36 and obtain

$$\frac{d^2 y}{dx^2} = \frac{-4(36)}{81y^3} = -\frac{16}{9y^3}$$

In Sec. 4.3 we learned how to determine whether a function has a relative maximum value or a relative minimum value at a critical number c by checking the algebraic sign of f' at numbers in intervals to the left and right of c. Another test for relative extrema is one that involves only the critical number c. Before stating the test in the form of a theorem, we give an informal geometric discussion that should appeal to your intuition.

Suppose that f is a function such that f' and f'' exist on some open interval (a, b) containing c and that $f'(c) = 0$; furthermore suppose that f'' is negative on (a, b). From Theorem 4.3.1(ii), because $f''(x) < 0$ on (a, b), then f' is decreasing on $[a, b]$. Since the value of f' at a point on the graph of f gives the slope of the tangent line at the point, it follows that the slope of the tangent line is decreasing on $[a, b]$. In Fig. 4.4.1 there is a sketch of the graph

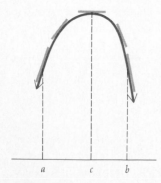

Figure 4.4.1

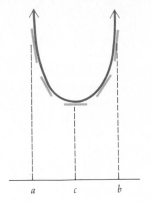

a   c   b

Figure 4.4.2

of a function f having these properties. Segments of a few tangent lines are shown in the figure. Observe that the slope of the tangent line is decreasing on $[a, b]$. Note that f has a relative maximum value at c where $f'(c) = 0$ and $f''(c) < 0$.

Now suppose that f is a function having the properties of the function in the previous paragraph except that f'' is positive on (a, b). Then from Theorem 4.3.1(i), because $f''(x) > 0$ on (a, b), it follows that f' is increasing on $[a, b]$. Thus the slope of the tangent line is increasing on $[a, b]$. Figure 4.4.2 shows a sketch of the graph of a function f having these properties. In the figure there are segments of a few tangent lines whose slopes are increasing on $[a, b]$. The function f has a relative minimum value at c where $f'(c) = 0$ and $f''(c) > 0$.

The facts in the preceding two paragraphs are given in the **second-derivative test for relative extrema**, which is now stated.

Second-derivative test for relative extrema

> **Theorem 4.4.1**   Let c be a critical number of a function f at which $f'(c) = 0$, and let f' exist for all values of x in some open interval containing c. If $f''(c)$ exists and
>
>   (i) if $f''(c) < 0$, then f has a relative maximum value at c;
>   (ii) if $f''(c) > 0$, then f has a relative minimum value at c.

EXAMPLE 4   Given

$$f(x) = x^4 + \tfrac{4}{3}x^3 - 4x^2$$

find the relative maxima and the relative minima of f by applying the second-derivative test.

Solution

$$f'(x) = 4x^3 + 4x^2 - 8x$$
$$f''(x) = 12x^2 + 8x - 8$$

Set $f'(x) = 0$.

$$4x(x + 2)(x - 1) = 0$$
$$x = 0 \qquad x = -2 \qquad x = 1$$

Thus the critical numbers of f are -2, 0, and 1. We determine whether or not there is a relative extremum at any of these critical numbers by finding the sign of the second derivative there. The results are summarized in Table 4.4.2.

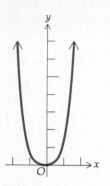

Figure 4.4.3

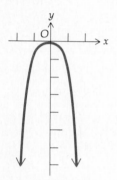

Figure 4.4.4

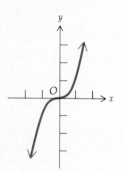

Figure 4.4.5

Table 4.4.2

	$f(x)$	$f'(x)$	$f''(x)$	Conclusion
$x = -2$	$-\frac{32}{3}$	0	+	f has a relative minimum value
$x = 0$	0	0	−	f has a relative maximum value
$x = 1$	$-\frac{5}{3}$	0	+	f has a relative minimum value

If $f''(c) = 0$ as well as $f'(c) = 0$, nothing can be concluded regarding a relative extremum of f at c. The following three illustrations justify this statement.

● ILLUSTRATION 2

If $f(x) = x^4$, then $f'(x) = 4x^3$ and $f''(x) = 12x^2$. Thus $f(0)$, $f'(0)$, and $f''(0)$ all have the value zero. By applying the first-derivative test, we see that f has a relative minimum value at 0. A sketch of the graph of f is shown in Fig. 4.4.3.

●

● ILLUSTRATION 3

If $g(x) = -x^4$, then $g'(x) = -4x^3$ and $g''(x) = -12x^2$. Hence $g(0) = g'(0) = g''(0) = 0$. In this case g has a relative maximum value at 0, as can be seen by applying the first-derivative test. A sketch of the graph of g is shown in Fig. 4.4.4.

●

● ILLUSTRATION 4

If $h(x) = x^3$, then $h'(x) = 3x^2$ and $h''(x) = 6x$; so $h(0) = h'(0) = h''(0) = 0$. The function h does not have a relative extremum at 0 because if $x < 0, h(x) < h(0)$; and if $x > 0, h(x) > h(0)$. A sketch of the graph of h is shown in Fig. 4.4.5. ●

In Illustrations 2, 3, and 4 we have examples of three functions, each of which has zero for its second derivative at a number for which its first derivative is zero; yet one function has a relative minimum value at that number, another function has a relative maximum value at that number, and the third function has neither a relative maximum value nor a relative minimum value at that number.

Exercises 4.4

In Exercises 1 through 10, find the first and second derivative of the function defined by the given equation.

1. $f(x) = x^5 - 2x^3 + x$

2. $F(x) = 7x^3 - 8x^2$

3. $g(s) = 2s^4 - 4s^3 + 7s - 1$

4. $G(t) = t^3 - t^2 + 1$

5. $F(x) = x^2 \sqrt{x} - 5x$

6. $g(r) = \sqrt{r} + \dfrac{1}{\sqrt{r}}$

7. $f(x) = \sqrt{x^2 + 1}$

8. $h(y) = \sqrt[3]{2y^3 + 5}$

9. $g(x) = \dfrac{x^2}{x^2 + 4}$

10. $f(x) = \dfrac{2 - \sqrt{x}}{2 + \sqrt{x}}$

11. Find $\dfrac{d^3y}{dx^3}$ if $y = x^4 - 2x^2 + x - 5$.

12. Find $\dfrac{d^4y}{dx^4}$ if $y = 3x^5 - 2x^4 + 4x^2 - 2$.

13. Find $\dfrac{d^4y}{dx^4}$ if $y = x^{7/2} - 2x^{5/2} + x^{1/2}$.

14. Find $\dfrac{d^3s}{dt^3}$ if $s = \sqrt{4t + 1}$.

15. Find $f^{(4)}(x)$ if $f(x) = \dfrac{3}{2x - 1}$.

16. Find $D_x^4 f(x)$ if $f(x) = \dfrac{2}{x - 1}$.

17. Find $D_x^3 f(x)$ if $f(x) = \dfrac{x}{(1 - x)^2}$.

18. Find $\dfrac{d^3u}{dv^3}$ if $u = v\sqrt{v - 2}$.

19. Given $x^2 + y^2 = 1$, show that $\dfrac{d^2y}{dx^2} = -\dfrac{1}{y^3}$.

20. Given $x^2 + 25y^2 = 100$, show that $\dfrac{d^2y}{dx^2} = -\dfrac{4}{25y^3}$.

21. Given $x^3 + y^3 = 1$, show that $\dfrac{d^2y}{dx^2} = -\dfrac{2x}{y^5}$.

22. Given $x^{1/2} + y^{1/2} = 2$, show that $\dfrac{d^2y}{dx^2} = \dfrac{1}{x^{3/2}}$.

23. Given $x^4 + y^4 = a^4$, where a is a constant, find $\dfrac{d^2y}{dx^2}$ in simplest form.

24. Given $b^2x^2 - a^2y^2 = a^2b^2$, where a and b are constants, find $\dfrac{d^2y}{dx^2}$ in simplest form.

25. Find the instantaneous rate of change of the slope of the tangent line to the graph of $y = 2x^3 - 6x^2 - x + 1$ at the point $(3, -2)$.

26. Find the slope of the tangent line at each point of the graph of $y = x^4 + x^3 - 3x^2$ where the rate of change of the slope is zero.

In Exercises 27 through 44, find the relative extrema of the given function by using the second-derivative test if it can be applied. If the second-derivative test cannot be applied, use the first-derivative test.

27. $f(x) = 3x^2 - 2x + 1$

28. $g(x) = 7 - 6x - 3x^2$

29. $f(x) = -4x^3 + 3x^2 + 18x$

30. $h(x) = 2x^3 - 9x^2 + 27$

31. $g(x) = \frac{1}{3}x^3 - x^2 + 3$

32. $f(y) = y^3 - 5y + 6$

33. $f(z) = (4 - z)^4$

34. $G(x) = (x + 2)^3$

35. $h(x) = x^4 - \frac{1}{3}x^3 - \frac{3}{2}x^2$

36. $f(x) = \frac{1}{5}x^5 - \frac{2}{3}x^3$

37. $f(x) = 4x^{1/2} + 4x^{-1/2}$

38. $g(t) = (t - 2)^{7/3}$

39. $f(x) = x(x + 2)^3$

40. $f(x) = x\sqrt{8 - x^2}$

41. $h(x) = x\sqrt{x + 3}$

42. $f(x) = x(x - 1)^3$

43. $F(x) = 6x^{1/3} - x^{2/3}$

44. $g(x) = \dfrac{9}{x} + \dfrac{x^2}{9}$

45. For the commodity of Illustration 1, suppose that for the t months after July 1 until November 1 it is estimated that the price will be $P(t)$ cents, where

$$P(t) = 76 + t^2 - \tfrac{1}{6}t^3 \qquad 0 \le t \le 4$$

(a) Determine the price on July 1, August 1, September 1, October 1, and November 1; (b) show that the price is increasing on $[0, 4]$; (c) show that the rate of inflation is increasing when $0 < t < 2$ and decreasing when $2 < t < 4$.

4.5 Additional problems involving absolute extrema

The extreme-value theorem (Theorem 4.1.2) guarantees an absolute maximum value and an absolute minimum value for a function that is continuous on a closed interval. We now consider some functions defined on intervals for which the extreme-value theorem does not apply and which may or may not have absolute extrema.

EXAMPLE 1   Given

$$f(x) = \frac{x^2 - 27}{x - 6}$$

find the absolute extrema of f on the interval $[0, 6)$ if there are any.

Solution f is continuous on the interval $[0, 6)$ because the only discontinuity of f is at 6, which is not in the interval.

$$f'(x) = \frac{2x(x - 6) - (x^2 - 27)}{(x - 6)^2} = \frac{x^2 - 12x + 27}{(x - 6)^2} = \frac{(x - 3)(x - 9)}{(x - 6)^2}$$

$f'(x)$ exists for all values of x in $[0, 6)$, and $f'(x) = 0$ when $x = 3$ or 9; so the only critical number of f in the interval $[0, 6)$ is 3. The first-derivative test is applied to determine if f has a relative extremum at 3. The results are summarized in Table 4.5.1.

Table 4.5.1

	$f(x)$	$f'(x)$	Conclusion
$0 \le x < 3$		$+$	f is increasing
$x = 3$	6	0	f has a relative maximum value
$3 < x < 6$		$-$	f is decreasing

Because f has a relative maximum value at 3, and f is increasing on the interval $[0, 3)$ and decreasing on the interval $(3, 6)$, then on $[0, 6)$ f has an absolute maximum value at 3, and it is $f(3)$, which is 6. Note that $\lim\limits_{x \to 6^-} f(x) = -\infty$, and conclude that there is no absolute minimum value of f on $[0, 6)$.

EXAMPLE 2   Given

$$f(x) = \frac{-x}{(x^2 + 6)^2}$$

find the absolute extrema of f on $(0, +\infty)$ if there are any.

Solution f is continuous for all values of x.

$$f'(x) = \frac{-1(x^2 + 6)^2 + 4x^2(x^2 + 6)}{(x^2 + 6)^4} = \frac{-(x^2 + 6) + 4x^2}{(x^2 + 6)^3} = \frac{3x^2 - 6}{(x^2 + 6)^3}$$

$f'(x)$ exists for all values of x. Setting $f'(x) = 0$ we obtain $x = \pm\sqrt{2}$; so $\sqrt{2}$ is the only critical number of f in $(0, +\infty)$. The first-derivative test is applied at $\sqrt{2}$, and the results are summarized in Table 4.5.2.

Table 4.5.2

	$f(x)$	$f'(x)$	Conclusion
$0 < x < \sqrt{2}$			f is decreasing
$x = \sqrt{2}$	$-\frac{1}{64}\sqrt{2}$	0	f has a relative minimum value
$\sqrt{2} < x$		$+$	f is increasing

Because f has a relative minimum value at $\sqrt{2}$, and because f is decreasing on $(0, \sqrt{2})$ and increasing on $(\sqrt{2}, +\infty)$, f has an absolute minimum value at $\sqrt{2}$ on $(0, +\infty)$. The absolute minimum value is $-\frac{1}{64}\sqrt{2}$. There is no absolute maximum value of f on $(0, +\infty)$.

For certain functions the following theorem is useful to determine if a relative extremum is an absolute extremum.

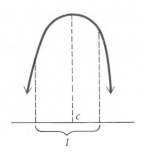

Figure 4.5.1

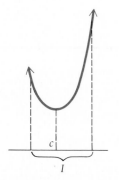

Figure 4.5.2

> **Theorem 4.5.1** Let the function f be continuous on the interval I containing the number c. If $f(c)$ is a relative extremum of f on I and c is the only number in I for which f has a relative extremum, then
>
> (i) if $f(c)$ is a relative maximum value of f on I, $f(c)$ is an absolute maximum value of f on I;
> (ii) if $f(c)$ is a relative minimum value of f on I, $f(c)$ is an absolute minimum value of f on I.

The proof of Theorem 4.5.1 is omitted. However, we give a geometric interpretation of the statement of the theorem. In each of Figs. 4.5.1 and 4.5.2 there is a sketch of the graph of a continuous function having only one relative extremum on an interval. The relative extremum is an absolute extremum. In Fig. 4.5.3 there is a sketch of the graph of a continuous function having two relative extrema on an interval. Neither is an absolute extremum.

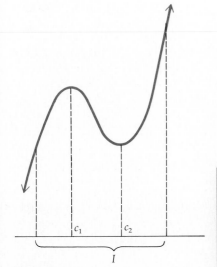

Figure 4.5.3

EXAMPLE 3 Example 4 of Sec. 1.5 pertained to the following situation: A closed box with a square base is to have a volume of 2000 in.^{3}; the material for the top and bottom of the box is to cost \$3 per square inch, and the material for the sides is to cost \$1.50 per square inch. Find the dimensions of the box for which the cost of the material is least.

Solution Let x in. be the length of a side of the square base and $C(x)$ dollars be the cost of the material. In Example 4 of Sec. 1.5 we obtained the following equation that defines $C(x)$:

$$C(x) = 6x^2 + \frac{12{,}000}{x} \tag{1}$$

The domain of C is $(0, +\infty)$, and C is continuous on its domain. From (1),

$$C'(x) = 12x - \frac{12{,}000}{x^2} \tag{2}$$

$C'(x)$ does not exist when $x = 0$, but 0 is not in the domain of C. Therefore the only critical numbers are those obtained by setting $C'(x) = 0$. Doing this we have

$$\frac{12x^3 - 12{,}000}{x^2} = 0$$

$$x^3 = 1000$$

$$x = 10$$

To determine if $x = 10$ makes C a relative minimum, we apply the second-derivative test. From (2) it follows that

$$C''(x) = 12 + \frac{24{,}000}{x^3}$$

The results of the second-derivative test are summarized in Table 4.5.3.

Table 4.5.3

	$C'(x)$	$C''(x)$	Conclusion
$x = 10$	0	+	C has a relative minimum value

Because C is continuous on its domain $(0, +\infty)$ and the one and only relative extremum of C on $(0, +\infty)$ is at $x = 10$, it follows from Theorem 4.5.1(ii) that this relative minimum value of C is the absolute minimum value of C. Thus the total cost of the material will be least when the side of the square base is 10 in. and the depth is 20 in.

EXAMPLE 4   If a closed tin can of volume 16π in.^{3} is to be in the form of a right-circular cylinder, find the height and radius if the least amount of material is to be used in its manufacture.

Solution   Let r in. be the base radius of the cylinder, h in. be the height of the cylinder, and S in.^{2} be the total surface area of the cylinder. See Fig. 4.5.4. The lateral surface area is $2\pi rh$ in.^{2}, the area of the top is πr^2 in.^{2}, and the area of the bottom is πr^2 in.^{2} Therefore

$$S = 2\pi rh + 2\pi r^2 \tag{3}$$

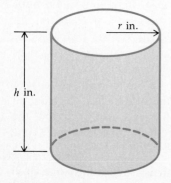

r in.

h in.

Figure 4.5.4

The formula for the volume of a right-circular cylinder is $V = \pi r^2 h$. Thus

$$16\pi = \pi r^2 h \tag{4}$$

Solving (4) for h and substituting into (3) we obtain S as a function of r:

$$S(r) = 2\pi r\left(\frac{16}{r^2}\right) + 2\pi r^2$$

$$S(r) = \frac{32\pi}{r} + 2\pi r^2 \tag{5}$$

The domain of S is $(0, +\infty)$, and S is continuous on its domain. From (5),

$$S'(r) = -\frac{32\pi}{r^2} + 4\pi r \tag{6}$$

$S'(r)$ does not exist when $r = 0$, but 0 is not in the domain of S. The only critical numbers are those obtained by setting $S'(r) = 0$, from which we have

$$4\pi r^3 = 32\pi$$

$$r^3 = 8$$

$$r = 2$$

The only critical number of S is 2. From (6),

$$S''(r) = \frac{64\pi}{r^3} + 4\pi$$

The results of applying the second-derivative test are summarized in Table 4.5.4.

Table 4.5.4

	$S'(r)$	$S''(r)$	Conclusion
$r = 2$	0	+	S has a relative minimum value

Because S is continuous on its domain $(0, +\infty)$ and the one and only relative extremum of S on $(0, +\infty)$ is at $r = 2$, it follows from Theorem 4.5.1(ii) that this relative minimum value of S is the absolute minimum value of S. When $r = 2$, we have, from (4), $h = 4$. Therefore the least amount of material will be used in the manufacture of the can when the radius is 2 in. and the height is 4 in.

An important business application of finding absolute extrema is **inventory control**, where a company is concerned with minimizing the costs of obtaining and storing inventory.

In addition to the costs of producing each item, a manufacturing company has two other costs, which are called the **setup cost** and the **carrying**

cost, which is often a cost of storage. The setup cost is the expense of preparing a production run, and the carrying cost is the amount incurred when the produced items are stored until they are sold. We make two assumptions:

1. Production is instantaneous, in which case the inventory will become zero before there is a new production run.
2. No shortages are permitted, which prohibits a negative inventory.

We assume that if there is a uniform demand when a production run consists of x units, the average inventory is $\frac{1}{2}x$ units. This assumption should seem reasonable, and it can be proved using the definite integral, as shown in Sec. 7.4 (see Exercise 19 in Exercises 7.4). Suppose, for example, that there is a uniform monthly demand of 600 units of a particular commodity. There could be one production run of 600 units at the beginning of the month, and these items would be stored and sold during the month. Because of uniform demand, the average inventory during the month would be 300 units. If there were two production runs—say, 300 units produced at the beginning of the month and another 300 units produced after 15 days—then the average monthly inventory would be reduced to 150 units, and if 200 units were produced each 10 days, this would be further reduced to 100 units. As the number of production runs increases, the average monthly inventory decreases, and so as the total setup cost increases, the total carrying cost decreases. We wish to minimize the sum of these two costs.

EXAMPLE 5   A company has a uniform demand of 6000 items per month. The setup cost for each production run is $60, and the monthly storage cost for each item is 50 cents. If the production is instantaneous and shortages are not permitted, how many items should be produced in each run to minimize the total monthly cost of obtaining and storing inventory?

Solution   Let x be the number of items produced in each run; $0 < x \leq 6000$. Let $C(x)$ be the number of dollars in the total monthly cost of obtaining and storing inventory.

Because 6000 items are to be produced each month, the number of runs per month is $6000/x$. The cost of preparing each run is $60, and hence the number of dollars in the total setup cost is

$$60\left(\frac{6000}{x}\right)$$

Because we are assuming a uniform demand, and since production is instantaneous, the average number of items on inventory is $x/2$. The monthly storage cost is 50 cents per item, and therefore the number of dollars in the total carrying cost is

$$\frac{1}{2} \cdot \frac{x}{2}$$

Hence

$$C(x) = 60 \left(\frac{6000}{x} \right) + \frac{1}{2} \cdot \frac{x}{2}$$

By the definition of x, it is a positive integer. However, we consider the function C to have as its domain the set of all real numbers in the interval $(0, 6000]$, and if necessary we shall adjust the solution to the nearest positive integer. Hence C is continuous on $(0, 6000]$.

$$C'(x) = -\frac{360,000}{x^2} + \frac{1}{4}$$

$$C''(x) = \frac{720,000}{x^3}$$

$C'(x)$ exists for all x in the interval $(0, 6000]$. Hence the only critical numbers of C are obtained by setting $C'(x) = 0$. We get

$$x^2 = 1,440,000$$

from which we obtain the critical number 1200 (-1200 is rejected since it is not in the interval $(0, 6000]$). The results of applying the second-derivative test at 1200 are summarized in Table 4.5.5.

Table 4.5.5

	$C'(x)$	$C''(x)$	Conclusion
$x = 1200$	0	+	C has a relative minimum value

The only extremum of C on the interval $(0, 6000]$ occurs at 1200, and therefore the relative minimum value of C is an absolute minimum value. We conclude, then, that to minimize the total monthly cost of obtaining and storing inventory, 1200 items should be produced in each run and the number of runs per month should be $6000 \div 1200 = 5$.

Exercises 4.5

In Exercises 1 through 16, find the absolute extrema of the given function on the indicated interval if there are any.

1. $f(x) = x^2; (-3, 2]$

2. $g(x) = (x - 2)^2; [-1, 6)$

3. $F(x) = x^3 - 3x + 5; (-\infty, 0)$

4. $f(x) = -x^3 + 12x - 6; (0, +\infty)$

5. $g(x) = \dfrac{x + 2}{x - 2}; [-4, 4]$

6. $F(x) = \dfrac{x^2}{x + 3}; [-4, -1]$

7. $f(x) = 4x^2 - 2x + 1; (-\infty, +\infty)$

8. $G(x) = (x - 5)^{2/3}; (-\infty, +\infty)$

9. $g(x) = (x + 1)^{2/3}; (-\infty, 0]$

10. $f(x) = (x + 3)^{1/3}; (-\infty, +\infty)$

11. $f(x) = \dfrac{x}{(x^2 + 4)^{3/2}}; [0, +\infty)$

12. $f(x) = \dfrac{x^2 - 12}{x + 4}; (-\infty, -4)$

13. $f(x) = 3x^4 - 4x^3 + 6x^2 - 12x + 1; (-\infty, +\infty)$ **14.** $g(x) = 3x^4 - 2x^3 + 6x^2 - 6x - 1; (-\infty, +\infty)$

15. $F(x) = 3 + 12x - 2x^2 + 8x^3 - 2x^4; (-\infty, +\infty)$ **16.** $f(x) = x^4 + x^3 - 8x^2 - 12x; (-\infty, +\infty)$

17. Refer to Exercise 11 in Exercises 1.5. A rectangular field having an area of 2700 m² is to be enclosed by a fence, and an additional fence is to be used to divide the field down the middle. The cost of the fence down the middle is $12 per running meter, and the fence along the sides costs $18 per running meter. Find the dimensions of the field such that the cost of the fencing will be the least.

18. Refer to Exercise 12 in Exercises 1.5. A rectangular open tank is to have a square base, and its volume is to be 125 m³. The cost per square meter for the bottom is $24 and for the sides is $12. Find the dimensions of the tank for the cost of the material to be the least.

19. Refer to Exercise 13 in Exercises 1.5. A box manufacturer is to produce a closed box of volume 288 in.³, where the base is a rectangle having a length three times its width. Find the dimensions of the box constructed from the least amount of material.

20. Refer to Exercise 14 in Exercises 1.5. Solve Exercise 19 if the box is to have an open top.

21. Refer to Exercise 15 in Exercises 1.5. A page of print is to contain 24 in.² of printed region, a margin of $1\frac{1}{2}$ in. at the top and bottom, and a margin of 1 in. at the sides. What are the dimensions of the smallest page that will fill these requirements?

22. Refer to Exercise 16 in Exercises 1.5. A one-story building having a rectangular floor space of 13,200 ft² is to be constructed where a walkway 22 ft wide is required in the front and back and a walkway 15 ft wide is required on each side. Find the dimensions of the lot having the least area on which this building can be located.

23. For the tin can of Example 4, suppose that the cost of material for the top and bottom is twice as much as it is for the side. Find the height and radius for the cost of the material to be the least.

24. A cardboard poster containing 32 in.² of printed region is to have a margin of 2 in. at the top and bottom and $1\frac{1}{3}$ in. at the sides. Determine the dimensions of the smallest piece of cardboard that can be used to make the poster.

25. In a particular community, a certain epidemic spreads in such a way that x months after the start of the epidemic, P percent of the population is infected, where

$$P = \frac{30x^2}{(1 + x^2)^2}$$

In how many months will the most people be infected, and what percent of the population is this?

26. A direct current generator has an electromotive force of E volts and an internal resistance of r ohms, where E and r are constants. If R ohms is the external resistance, the total resistance is $(r + R)$ ohms, and if P watts is the power, then

$$P = \frac{E^2 R}{(r + R)^2}$$

Show that the most power is consumed when the external resistance is equal to the internal resistance.

27. A Norman window consists of a rectangle surmounted by a semicircle. If the perimeter of a Norman window is to be 32 ft, determine what should be the radius of the semicircle and the height of the rectangle such that the window will admit the most light.

28. Solve Exercise 27 if the window is to be such that the semicircle transmits only half as much light per square foot of area as the rectangle.

29. It is determined that if salaries are excluded, the number of dollars in the cost per kilometer for operating a truck is $8 + \frac{1}{300}x$, where x km/hr is the speed of the truck. If the combined salary of the driver and the driver's assistant is $27 per hour, what should be the average speed of the truck for the cost per kilometer to be the least?

30. The number of dollars in the cost per hour of fuel for a cargo ship is $\frac{1}{50}v^3$, where v knots (nautical miles per hour) is the speed of the ship. If there are additional costs of $400 per hour, what should be the average speed of the ship for the cost per nautical mile to be the least?

31. A table manufacturer has a contract to supply 3000 tables per year at a uniform rate. The cost of starting a production run is $96, and the annual storage cost is $40 per table. If the excess tables are stored until delivery, how many tables should be produced in each run if the total cost of obtaining and storing inventory is to be minimized? Assume that production is instantaneous and shortages are not permitted.

32. A firm has a contract to supply 5 units of a certain commodity each day. The cost of starting a production run is $250, and the storage cost of each unit is $1 per day. Assuming that production is instantaneous and shortages are not permitted, determine the number of units that should be produced in each run to minimize the total cost of obtaining and storing inventory.

33. A distributor of a certain nonperishable food product has a monthly demand of 12,000 cases, and they are being bought at a constant rate. The cost of placing an order, which includes such expenses as clerical costs, delivery charges, and handling costs, is $25 regardless of the size of the order. The monthly storage cost is 60 cents per case. Assuming that deliveries are instantaneous and that shortages are not allowed, determine how many orders should be placed each month and the size of each order such that the cost of maintaining an inventory is minimized.

34. An appliance store sells 2500 television sets per year, and they are being sold at a constant rate. The cost of placing an order is $20 regardless of the size of the order, and the annual storage cost is $10 per set. Assuming that deliveries are instantaneous and that shortages are not allowed, determine how many orders should be placed each year and the size of each order such that the cost of maintaining an inventory is minimized.

35. Solve Exercise 33 if the monthly demand is R cases, the cost of placing an order is K dollars, and the monthly storage cost is S dollars per case.

Review Exercises for Chapter 4

In Exercises 1 through 4, find the first and second derivative of the function defined by the given equation.

1. $f(x) = 3x^4 - 2x^3 + 7x^2 - 5x + 1$

2. $f(t) = \dfrac{t}{t-1}$

3. $g(x) = \dfrac{2x-1}{x+2}$

4. $F(x) = \sqrt[3]{2-3x}$

5. Find $\dfrac{d^3 y}{dx^3}$ if $y = \sqrt{3-2x}$.

6. Given $4x^2 + 9y^2 = 36$, show that $\dfrac{d^2 y}{dx^2} = -\dfrac{16}{9y^3}$

In Exercises 7 through 14, find the absolute extrema of the given function on the indicated interval, if there are any, and find the values of x at which the absolute extrema occur. Draw a sketch of the graph of the function on the interval.

7. $f(x) = \sqrt{x-5}; [5, +\infty)$

8. $f(x) = \sqrt{9-x^2}; [-3, 3]$

9. $f(x) = \frac{5}{2}x^6 - 3x^5; [-1, 2]$

10. $f(x) = \dfrac{3}{x-2}; [0, 4]$

11. $f(x) = x^4 - 2x^2 + 1; [-1, 2]$

12. $f(x) = x^3 - 12x; [-1, 3]$

13. $f(x) = (x-1)^{1/3}; (-\infty, +\infty)$

14. $f(x) = (x-5)^{2/3}; (-\infty, +\infty)$

In Exercises 15 through 20, do each of the following: (a) find the relative extrema of f by applying the first-derivative test; (b) determine the values of x at which the relative extrema occur; (c) determine the intervals on which f is increasing; (d) determine the intervals on which f is decreasing; (e) draw a sketch of the graph.

15. $f(x) = x^3 - 9x^2 + 24x - 10$

16. $f(x) = 2x^3 + 3x^2 - 18x - 18$

17. $f(x) = (x-4)^2(x+2)^3$

18. $f(x) = (x-2)^{4/3}$

19. $f(x) = x + \dfrac{1}{x^2}$

20. $f(x) = 2x + \dfrac{1}{2x}$

In Exercises 21 through 28, find the relative extrema of the given function by using the second-derivative test if it can be applied. If the second-derivative test cannot be applied, use the first-derivative test.

21. $f(x) = x^3 + 3x^2 + 2$

22. $f(x) = x^3 - 3x + 2$

23. $g(x) = (x-2)^4$

24. $F(x) = (x-3)^3$

25. $f(t) = \dfrac{4}{t} + \dfrac{t^2}{4}$ **26.** $g(y) = (y+2)(y-1)^3$ **27.** $G(x) = x(x-2)^3$ **28.** $g(w) = 5w^3 - 3w^5$

29. Find two positive numbers whose sum is 12 such that the sum of their squares is an absolute minimum.

30. Find two positive numbers whose sum is 12 such that their product is an absolute maximum.

31. Show that among all the rectangles having a perimeter of 36 in., the square of side 9 in. has the greatest area.

32. Show that among all the rectangles having an area of 81 in.^{2}, the square of side 9 in. has the least perimeter.

33. Refer to Exercise 35 in Review Exercises for Chapter 2. Square pieces of metal of side 20 in. are used to construct open boxes by cutting equal squares from the four corners and turning up the sides. Find the length of the side of the square to be cut out to obtain a box of the largest possible volume.

34. Refer to Exercise 36 in Review Exercises for Chapter 2. A wholesaler offers to deliver to a dealer 300 chairs at $90 per chair and to reduce the price per chair on the entire order by 25 cents for each additional chair over 300. Find the dollar total involved in the largest possible transaction between the wholesaler and the dealer under these circumstances.

35. Refer to Exercise 37 in Review Exercises for Chapter 2. A school-sponsored trip that can accommodate up to 250 students will cost each student $15 if not more than 150 students make the trip; however, the cost per student will be reduced 5 cents for each student in excess of 150 until the cost reaches $10 per student. How many students should make the trip for the school to receive the largest gross income?

36. Refer to Exercise 38 in Review Exercises for Chapter 2. In a town of population 11,000 the rate of growth of an epidemic is jointly proportional to the number of people infected and the number of people not infected. Determine the number of people infected when the epidemic is growing at a maximum rate.

37. Refer to Exercise 41 in Review Exercises for Chapter 1. An open box having a square base is to have a volume of 32 ft^3. Find the dimensions of the box that can be constructed with the least amount of material.

38. Refer to Exercise 42 in Review Exercises for Chapter 1. Solve Exercise 37 if the box is to be closed.

39. Refer to Exercise 43 in Review Exercises for Chapter 1. A sign is to contain 50 m^2 of printed material. If margins of 4 m at the top and bottom and 2 m on the side are required, find the dimensions of the smallest sign that will meet these requirements.

40. A manufacturer has to supply 10,000 desks per year at a uniform rate. The cost of starting a production run is $180, the excess desks are stored until delivery, and the annual storage cost is $40 per desk. Assuming that deliveries are instantaneous and that shortages are not allowed, how many desks should be produced in each run if the total cost of obtaining and storing inventory is to be minimized?

41. A company that manufactures lamps has a contract to supply 3000 lamps per year at a uniform rate. The cost of starting a production run is $120. The excess lamps are stored until delivery, and the annual storage cost is $50 per lamp. How many lamps should be produced in each run if the total cost of obtaining and storing inventory is to be a minimum? Assume that production is instantaneous and shortages are not permitted.

42. The stiffness of a rectangular beam is jointly proportional to its breadth and the cube of its depth. Find the dimensions of the stiffest beam that can be cut from a log in the shape of a right-circular cylinder of radius 72 cm.

43. Because of various restrictions, the size of a particular community is limited to 3000 inhabitants, and the rate of increase of the population is jointly proportional to its size and the difference between 3000 and its size. Determine the size of the population for which the rate of growth of the population is a maximum.

44. A closed tin can having a volume of 27 in.^{3} is to be in the form of a right-circular cylinder. If the circular top and bottom are cut from square pieces of tin, find the radius and height of the can if the least amount of tin is to be used in its manufacture. Include the tin that is wasted when obtaining the top and bottom.

CHAPTER 5

MORE APPLICATIONS
OF THE DERIVATIVE

5.1 Concavity and points of inflection

Figure 5.1.1 shows a sketch of the graph of a function f whose first and second derivatives exist on the closed interval $[x_1, x_7]$. Because both f and f' are differentiable there, f and f' are continuous on $[x_1, x_7]$.

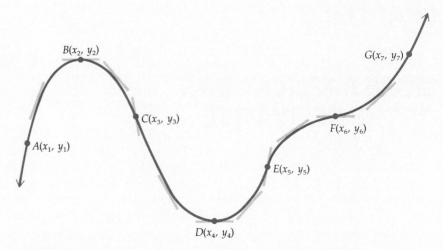

Figure 5.1.1

If we consider a point P moving along the graph of Fig. 5.1.1 from A to G, then the position of P varies as we increase x from x_1 to x_7. As P moves along the graph from A to B, the slope of the tangent line to the graph is positive and is decreasing; that is, the tangent line is turning clockwise, and the graph lies below the tangent line. When the point P is at B, the slope of the tangent line is zero and is still decreasing. As P moves along the graph from B to C, the slope of the tangent line is negative and is still decreasing; the tangent line is still turning clockwise, and the graph is below its tangent line. We say that the graph is *concave downward* from A to C. As P moves along the graph from C to D, the slope of the tangent line is negative and is increasing; that is, the tangent line is turning counterclockwise, and the graph is above its tangent line. At D, the slope of the tangent line is zero and is still increasing. From D to E, the slope of the tangent line is positive and increasing; the tangent line is still turning counterclockwise, and the graph is above its tangent line. We say that the graph is *concave upward* from C to E. We have the following definitions.

Definition of concave upward

> The graph of a function f is said to be **concave upward** at the point $(c, f(c))$ if $f'(c)$ exists and if there is an open interval I containing c such that for all values of $x \neq c$ in I the point $(x, f(x))$ on the graph is above the tangent line to the graph at $(c, f(c))$.

Definition of concave downward

The graph of a function f is said to be **concave downward** at the point $(c, f(c))$ if $f'(c)$ exists and if there is an open interval I containing c such that for all values of $x \neq c$ in I the point $(x, f(x))$ on the graph is below the tangent line to the graph at $(c, f(c))$.

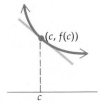

Figure 5.1.2

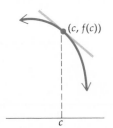

Figure 5.1.3

- ILLUSTRATION 1

Figure 5.1.2 shows a sketch of a portion of the graph of a function f that is concave upward at the point $(c, f(c))$, and Fig. 5.1.3 shows a sketch of a portion of the graph of a function f that is concave downward at the point $(c, f(c))$.   •

The graph of the function f of Fig. 5.1.1 is concave downward at all points $(x, f(x))$ for which x is in either of the open intervals (x_1, x_3) or (x_5, x_6). Similarly, the graph of the function f in Fig. 5.1.1 is concave upward at all points $(x, f(x))$ for which x is in either (x_3, x_5) or (x_6, x_7).

- ILLUSTRATION 2

If f is the function defined by $f(x) = x^2$, then $f'(x) = 2x$ and $f''(x) = 2$. Thus $f''(x) > 0$ for all x. Furthermore, the graph of f, shown in Fig. 5.1.4, is above all of its tangent lines. Thus the graph of f is concave upward at all of its points.

If g is the function defined by $g(x) = -x^2$, then $g'(x) = -2x$ and $g''(x) = -2$. Hence $g''(x) < 0$ for all x. Also, the graph of g, shown in Fig. 5.1.5, is below all of its tangent lines. Therefore the graph of g is concave downward at all of its points.   •

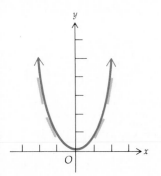

Figure 5.1.4

The function f of Illustration 2 is such that $f''(x) > 0$ for all x, and the graph of f is concave upward everywhere. For function g of Illustration 2, $g''(x) < 0$ for all x, and the graph of g is concave downward everywhere. These two functions are special cases of the following theorem.

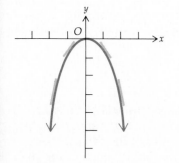

Figure 5.1.5

Theorem 5.1.1 Let f be a function which is differentiable on some open interval containing c. Then

   (i) if $f''(c) > 0$, the graph of f is concave upward at $(c, f(c))$;
   (ii) if $f''(c) < 0$, the graph of f is concave downward at $(c, f(c))$.

The proof of Theorem 5.1.1 is omitted; however, the following geometric argument should appeal to your intuition.

See Fig. 5.1.6(a–c), where the graphs are concave upward. In Fig. 5.1.6(a), $f''(x) > 0$, where $f'(x) < 0$ and f' is increasing. In Fig. 5.1.6(b), $f''(x) > 0$, where $f'(x) > 0$ and f' is increasing. In Fig. 5.1.6(c), $f''(x) > 0$, where f' is increasing from negative values to positive values.

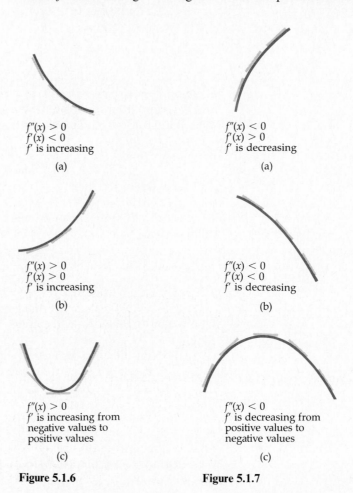

$f''(x) > 0$
$f'(x) < 0$
f' is increasing

(a)

$f''(x) < 0$
$f'(x) > 0$
f' is decreasing

(a)

$f''(x) > 0$
$f'(x) > 0$
f' is increasing

(b)

$f''(x) < 0$
$f'(x) < 0$
f' is decreasing

(b)

$f''(x) > 0$
f' is increasing from
negative values to
positive values

(c)

$f''(x) < 0$
f' is decreasing from
positive values to
negative values

(c)

Figure 5.1.6   **Figure 5.1.7**

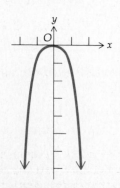

Figure 5.1.8

Figure 5.1.9

Now refer to Fig. 5.1.7(a–c), where the graphs are concave downward. In Fig. 5.1.7(a), $f''(x) < 0$, where $f'(x) > 0$ and f' is decreasing. In Fig. 5.1.7(b), $f''(x) < 0$, where $f'(x) < 0$ and f' is decreasing. In Fig. 5.1.7(c), $f''(x) < 0$, where f' is decreasing from positive values to negative values.

Theorem 5.1.1 does not apply when $f''(c) = 0$. Actually, if $f''(c) = 0$, it's possible for the graph to be concave upward at the point $(c, f(c))$, or it may be concave downward at $(c, f(c))$. For example, in Fig. 5.1.8 there is a sketch of the graph of the function defined by $f(x) = x^4$; $f''(0) = 0$, and the graph is concave upward at $(0, 0)$. Figure 5.1.9 shows a sketch of the graph of the function defined by $g(x) = -x^4$; $g''(0) = 0$, and the graph is concave downward at $(0, 0)$.

Table 5.1.1

t	1	2	3	4	5
$f(t)$	29	70	117	164	205

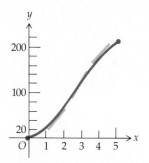

Figure 5.1.10

Figure 5.1.11

● ILLUSTRATION 3

Suppose it is estimated that t hours after starting work at 7 A.M. a factory worker on an assembly line has performed a particular task on $f(t)$ units, and

$$f(t) = 21t + 9t^2 - t^3 \qquad 0 \le t \le 5 \tag{1}$$

In Table 5.1.1 we have function values for integer values of t from 1 through 5, and a sketch of the graph of f on $[0, 5]$ is shown in Fig. 5.1.10. From Eq. (1) we obtain

$$f'(t) = 21 + 18t - 3t^2$$

and

$$\begin{aligned} f''(t) &= 18 - 6t \\ &= 6(3 - t) \end{aligned}$$

Observe that $f''(t) > 0$ if $0 < t < 3$ and $f''(t) < 0$ if $3 < t < 5$. Because $f''(t) > 0$ when $0 < t < 3$, $f'(t)$ is increasing on $[0, 3]$; the graph is concave upward when $0 < t < 3$. Because $f''(t) < 0$ when $3 < t < 5$, $f'(t)$ is decreasing on $[3, 5]$; the graph is concave downward when $3 < t < 5$. Since $f'(t)$ is the rate of change of $f(t)$ with respect to t, we conclude that in the first 3 hours (from 7 A.M. until 10 A.M.) the worker is performing the task at an increasing rate, and during the remaining 2 hours (from 10 A.M. until noon) the worker is performing the task at a decreasing rate. At $t = 3$ (at 10 A.M.) the worker is producing most efficiently, and when $3 < t < 5$ (after 10 A.M.) there is a reduction in the worker's production rate. The point at which the worker is producing most efficiently is called the *point of diminishing returns.* ●

In Illustration 3, at the point of diminishing returns there is a change in the sense of concavity of the graph. Such a point is called a *point of inflection.*

Definition of a point
of inflection

The point $(c, f(c))$ is a **point of inflection** of the graph of the function f if the graph has a tangent line there, and if there exists an open interval I containing c such that if x is in I, then either

(i) $f''(x) < 0$ if $x < c$ and $f''(x) > 0$ if $x > c$; or
(ii) $f''(x) > 0$ if $x < c$ and $f''(x) < 0$ if $x > c$

● ILLUSTRATION 4

Figure 5.1.11 illustrates a point of inflection where condition (i) of the above definition holds; in this case the graph is concave downward at points immediately to the left of the point of inflection, and the graph is concave

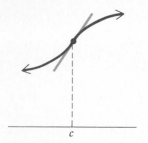

Figure 5.1.12

Figure 5.1.13

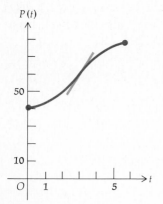

Figure 5.1.14

upward at points immediately to the right of the point of inflection. Condition (ii) is shown in Fig. 5.1.12, where the sense of concavity changes from upward to downward at the point of inflection. Figure 5.1.13 is another illustration of condition (i), where the sense of concavity changes from downward to upward at the point of inflection. Note that in Fig. 5.1.13 there is a horizontal tangent line at the point of inflection. ●

For the graph in Fig. 5.1.1 there are points of inflection at C, E, and F.

● ILLUSTRATION 5

In Illustration 1 of Sec. 4.4 we had the following situation: It is estimated that t months after January 1 until July 1, the price of a certain commodity will be $P(t)$ cents, where

$$P(t) = 40 + 3t^2 - \tfrac{1}{3}t^3 \qquad 0 \le t \le 6$$

Furthermore, $P'(t) = 6t - t^2$, and $P''(t) = 6 - 2t$. When $0 < t < 3$, $P''(t) > 0$, and when $3 < t < 6$, $P''(t) < 0$. Thus, from part (ii) of the definition, the graph of P has a point of inflection at $t = 3$. See Fig. 5.1.14 for a sketch of the graph of P.

In Illustration 1 of Sec. 4.4 we concluded that when $0 < t < 3$, the price is increasing and the rate of inflation of the commodity is increasing; when $3 < t < 6$, the price is increasing but the rate of inflation of the commodity is decreasing. Thus, at the point of inflection (at $t = 3$) the price is increasing at the greatest rate, and the rate of inflation changes from increasing to decreasing. ●

The definition of a point of inflection indicates nothing about the value of the second derivative there. The following theorem states that if the second derivative exists at a point of inflection, it must be zero there.

Theorem 5.1.2   If the function f is differentiable on some open interval containing c, and if $(c, f(c))$ is a point of inflection of the graph of f, then if $f''(c)$ exists, $f''(c) = 0$.

Proof   Let g be the function such that $g(x) = f'(x)$; then $g'(x) = f''(x)$. Because $(c, f(c))$ is a point of inflection of the graph of f, then $f''(x)$ changes sign at c and so $g'(x)$ changes sign at c. Therefore, by the first-derivative test (Theorem 4.3.2), g has a relative extremum at c, and c is a critical number of g. Because $g'(c) = f''(c)$, and because by hypothesis $f''(c)$ exists, it follows that $g'(c)$ exists. Therefore, by Theorem 4.1.1, $g'(c) = 0$ and $f''(c) = 0$, which is what we wanted to prove. ■

The converse of Theorem 5.1.2 is not true. That is, if the second derivative of a function is zero at a number c, it is not necessarily true that the graph of the function has a point of inflection at the point where $x = c$. This fact is shown in the following illustration.

● ILLUSTRATION 6

Consider the function f defined by $f(x) = x^4$. $f'(x) = 4x^3$ and $f''(x) = 12x^2$. Further, $f''(0) = 0$; but because $f''(x) > 0$ if $x < 0$ and $f''(x) > 0$ if $x > 0$, the graph is concave upward at points on the graph immediately to the left of $(0, 0)$ and at points immediately to the right of $(0, 0)$. Consequently, $(0, 0)$ is not a point of inflection. In Illustration 2 of Section 4.4 we showed that this function f has a relative minimum value at zero. Furthermore, the graph is concave upward at the point $(0, 0)$ (see Fig. 5.1.8). ●

The graph of a function may have a point of inflection at a point, and the second derivative may fail to exist there, as shown in the next illustration.

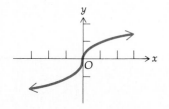

Figure 5.1.15

● ILLUSTRATION 7

If f is the function defined by $f(x) = x^{1/3}$, then

$$f'(x) = \tfrac{1}{3}x^{-2/3} \quad \text{and} \quad f''(x) = -\tfrac{2}{9}x^{-5/3}$$

$f''(0)$ does not exist; but if $x < 0$, $f''(x) > 0$, and if $x > 0$, $f''(x) < 0$. Hence f has a point of inflection at $(0, 0)$. A sketch of the graph of this function is shown in Fig. 5.1.15. Note that for this function $f'(0)$ also fails to exist. The tangent line to the graph at $(0, 0)$ is the y axis. ●

In drawing a sketch of a graph having points of inflection it is helpful to draw a segment of the tangent line at a point of inflection. Such a tangent line is called an **inflectional tangent**.

EXAMPLE 1 For the function in Example 1 of Sec. 4.3 find the points of inflection of the graph of the function, and determine where the graph is concave upward and where it is concave downward.

Solution

$$f(x) = x^3 - 6x^2 + 9x + 1$$

$$f'(x) = 3x^2 - 12x + 9$$

$$f''(x) = 6x - 12$$

$f''(x)$ exists for all values of x; so the only possible point of inflection is where $f''(x) = 0$, which occurs at $x = 2$. To determine whether there is a point of inflection at $x = 2$ we must check to see if $f''(x)$ changes sign; at the same time we determine the concavity of the graph for the respective intervals. The results are summarized in Table 5.1.2.

Table 5.1.2

	$f(x)$	$f'(x)$	$f''(x)$	Conclusion
$x < 2$				graph is concave downward
$x = 2$	3	-3	0	graph has a point of inflection
$2 < x$			$+$	graph is concave upward

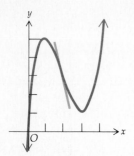

Figure 5.1.16

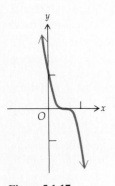

Figure 5.1.17

In Example 1 of Sec. 4.3 we showed that f has a relative maximum value at 1 and a relative minimum value at 3. A sketch of the graph showing a segment of the inflectional tangent appears in Fig. 5.1.16.

EXAMPLE 2 If $f(x) = (1 - 2x)^3$, find the points of inflection of the graph of f, and determine where the graph is concave upward and where it is concave downward. Draw a sketch of the graph of f.

Solution

$$f(x) = (1 - 2x)^3$$
$$f'(x) = -6(1 - 2x)^2$$
$$f''(x) = 24(1 - 2x)$$

Because $f''(x)$ exists for all values of x, the only possible point of inflection is where $f''(x) = 0$, that is, at $x = \frac{1}{2}$. By using the results summarized in Table 5.1.3 we see that $f''(x)$ changes sign from "$+$" to "$-$" at $x = \frac{1}{2}$, and so the graph has a point of inflection there. Note also that because $f'(\frac{1}{2}) = 0$, the graph has a horizontal tangent line at the point of inflection. A sketch of the graph is shown in Fig. 5.1.17.

Table 5.1.3

	$f(x)$	$f'(x)$	$f''(x)$	Conclusion
$x < \frac{1}{2}$			$+$	graph is concave upward
$x = \frac{1}{2}$	0	0	0	graph has a point of inflection
$\frac{1}{2} < x$				graph is concave downward

Exercises 5.1

In Exercises 1 through 12, determine where the graph of the given function is concave upward and where it is concave downward, and find the points of inflection, if there are any.

1. $f(x) = x^3 + 9x$
2. $g(x) = x^3 + 3x^2 - 3x - 3$
3. $g(x) = 2x^3 + 3x^2 - 7x + 1$
4. $f(x) = \frac{1}{12}x^4 + \frac{1}{6}x^3 - x^2$
5. $F(x) = x^4 - 8x^3 + 24x^2$
6. $f(x) = 16x^4 + 32x^3 + 24x^2 - 5x - 20$

7. $g(x) = \dfrac{x}{x^2 - 1}$
8. $G(x) = \dfrac{2x}{(x^2 + 4)^{3/2}}$
9. $f(x) = (x - 2)^{1/5}$

10. $F(x) = (2x - 6)^{3/2} + 1$
11. $g(x) = \dfrac{x - 2}{x + 4}$
12. $f(x) = \dfrac{x + 5}{x - 3}$

In Exercises 13 through 24, draw a portion of the graph of a function f through the point where $x = c$ if the given conditions are satisfied. It is assumed that f is continuous on some open interval containing c.

13. $f'(x) > 0$ if $x < c$; $f'(x) < 0$ if $x > c$; $f''(x) < 0$ if $x < c$; $f''(x) < 0$ if $x > c$.
14. $f'(x) > 0$ if $x < c$; $f'(x) > 0$ if $x > c$; $f''(x) > 0$ if $x < c$; $f''(x) < 0$ if $x > c$.

15. $f'(x) > 0$ if $x < c$; $f'(x) < 0$ if $x > c$; $f''(x) > 0$ if $x < c$; $f''(x) > 0$ if $x > c$.
16. $f'(x) < 0$ if $x < c$; $f'(x) > 0$ if $x > c$; $f''(x) > 0$ if $x < c$; $f''(x) < 0$ if $x > c$.
17. $f''(c) = 0$; $f'(c) = 0$; $f''(x) > 0$ if $x < c$; $f''(x) < 0$ if $x > c$.
18. $f'(c) = 0$; $f'(x) > 0$ if $x < c$; $f''(x) > 0$ if $x > c$.
19. $f''(c) = 0$; $f'(c) = 0$; $f''(x) > 0$ if $x < c$; $f''(x) > 0$ if $x > c$.
20. $f'(c) = 0$; $f'(x) < 0$ if $x < c$; $f''(x) > 0$ if $x > c$.
21. $f''(c) = 0$; $f'(c) = -1$; $f''(x) < 0$ if $x < c$; $f''(x) > 0$ if $x > c$.
22. $f''(c) = 0$; $f'(c) = \frac{1}{2}$; $f''(x) > 0$ if $x < c$; $f''(x) < 0$ if $x > c$.
23. $f'(c)$ does not exist; $f''(x) > 0$ if $x < c$; $f''(x) > 0$ if $x > c$.
24. $f'(c)$ does not exist; $f''(c)$ does not exist; $f''(x) < 0$ if $x < c$; $f''(x) > 0$ if $x > c$.

25. Draw a sketch of the graph of a function f for which $f(x)$, $f'(x)$, and $f''(x)$ exist and are positive for all x.
26. Draw a sketch of the graph of a function f for which $f(x)$, $f'(x)$, and $f''(x)$ exist and are negative for all x.
27. A construction worker starts a job at 8 A.M., and x hours later the worker has done a particular operation on $f(x)$ units, where

$$f(x) = 4x + 9x^2 - x^3 \qquad 0 \leq x \leq 5$$

Find at what time the worker is performing the operation most efficiently; that is, at what time does the worker reach the point of diminishing returns?
28. Refer to Exercise 12 in Exercises 3.2. It is estimated that a worker in a shop that makes picture frames can paint y frames x hours after starting work at 8 A.M., and

$$y = 3x + 8x^2 - x^3 \qquad 0 \leq x \leq 4$$

Find at what time the worker is painting most efficiently; that is, at what time does the worker reach the point of diminishing returns?
29. Refer to Exercise 45 in Exercises 4.4. For the commodity of Illustration 5 of this section, suppose that for the t months after July 1 until November 1 it is estimated that the price will be $P(t)$ cents, where

$$P(t) = 76 + t^2 - \tfrac{1}{6}t^3 \qquad 0 \leq t \leq 4$$

(a) Draw a sketch of the graph of P on $[0, 4]$. (b) Find the point of inflection of the graph of P, and show that at this point the price is increasing at the greatest rate and the rate of inflation changes from increasing to decreasing.
30. If $f(x) = ax^3 + bx^2$, determine a and b so that the graph of f will have a point of inflection at $(1, 2)$.
31. If $f(x) = ax^3 + bx^2 + cx$, determine a, b, and c so that the graph of f will have a point of inflection at $(1, 2)$ and so that the slope of the inflectional tangent there will be -2.
32. If $f(x) = ax^3 + bx^2 + cx + d$, determine a, b, c, and d so that f will have a relative extremum at $(0, 3)$ and so that the graph of f will have a point of inflection at $(1, -1)$.

5.2 Applications to drawing a sketch of the graph of a function

We now apply the discussions in Secs. 4.3, 4.4, and 5.1 to drawing a sketch of the graph of a function. If you are given $f(x)$ and wish to draw a sketch of the graph of f, proceed as follows. First find $f'(x)$ and $f''(x)$. Then the critical numbers of f are the values of x in the domain of f for which either $f'(x)$ does not exist or $f'(x) = 0$. Next apply the first-derivative test (Theorem 4.3.2) or the second-derivative test (Theorem 4.4.1) to determine whether at a critical number there is a relative maximum value, a relative minimum value, or neither. To determine the intervals on which f is increasing, find the values of x for which $f'(x)$ is positive; to determine the intervals on which

which f is decreasing, find the values of x for which $f'(x)$ is negative. In determining the intervals on which f is monotonic, also check the critical numbers at which f does not have a relative extremum. The values of x for which $f''(x) = 0$ or $f''(x)$ does not exist give the possible points of inflection; check to see if $f''(x)$ changes sign at each of these values of x to determine whether there actually is a point of inflection. The values of x for which $f''(x)$ is positive and those for which $f''(x)$ is negative will give points at which the graph is concave upward and points at which the graph is concave downward. It is also helpful to find the slope of each inflectional tangent. It is suggested that all the information so obtained be incorporated into a table, as in the following examples.

EXAMPLE 1 Given $f(x) = x^3 - 3x^2 + 3$, find the relative extrema of f; the points of inflection of the graph of f; the intervals on which f is increasing; the intervals on which f is decreasing; where the graph is concave upward; where the graph is concave downward; and the slope of any inflectional tangent. Draw a sketch of the graph.

Solution $f(x) = x^3 - 3x^2 + 3$ and $f'(x) = 3x^2 - 6x$; $f''(x) = 6x - 6$. Set $f'(x) = 0$ to obtain $x = 0$ and $x = 2$. From $f''(x) = 0$ we obtain $x = 1$. In making the table, consider the points at which $x = 0$, $x = 1$, and $x = 2$, and the intervals excluding these values of x:

$$x < 0 \qquad 0 < x < 1 \qquad 1 < x < 2 \qquad 2 < x$$

Table 5.2.1

	$f(x)$	$f'(x)$	$f''(x)$	Conclusion
$x < 0$		$+$	$-$	f is increasing; graph is concave downward
$x = 0$	3	0	$-$	f has a relative maximum value; graph is concave downward
$0 < x < 1$		$-$	$-$	f is decreasing; graph is concave downward
$x = 1$	1	-3	0	f is decreasing; graph has a point of inflection
$1 < x < 2$		$-$	$+$	f is decreasing; graph is concave upward
$x = 2$	-1	0	$+$	f has a relative minimum value; graph is concave upward
$2 < x$		$+$	$+$	f is increasing; graph is concave upward

From the information in Table 5.2.1 and by plotting a few points, we obtain the sketch of the graph shown in Fig. 5.2.1.

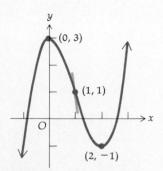

Figure 5.2.1

EXAMPLE 2 Given

$$f(x) = \frac{x^2 + 4}{x} \tag{1}$$

find the relative extrema of f; the points of inflection of the graph of f; the intervals on which f is increasing; the intervals on which f is decreasing; where the graph is concave upward; where the graph is concave downward; and the slope of any inflectional tangent. Draw a sketch of the graph.

Solution

$$f(x) = \frac{x^2 + 4}{x}$$

$$f'(x) = \frac{2x(x) - 1(x^2 + 4)}{x^2} = \frac{x^2 - 4}{x^2}$$

$$f''(x) = \frac{2x(x^2) - 2x(x^2 - 4)}{x^4} = \frac{8}{x^3}$$

Set $f'(x) = 0$ to obtain $x = \pm 2$; $f''(x)$ is never zero. For Table 5.2.2 consider the points at which $x = \pm 2$ and observe that 0 is not in the domain of f. Also for the table consider the intervals excluding ± 2 and 0:

$$x < -2 \qquad -2 < x < 0 \qquad 0 < x < 2 \qquad 2 < x$$

Table 5.2.2

	$f(x)$	$f'(x)$	$f''(x)$	Conclusion
$x < -2$		+	−	f is increasing; graph is concave downward
$x = -2$	−4	0	−	f has a relative maximum value; graph is concave downward
$-2 < x < 0$		−	−	f is decreasing; graph is concave downward
$x = 0$	does not exist	does not exist	does not exist	
$0 < x < 2$		−	+	f is decreasing; graph is concave upward
$x = 2$	4	0	+	f has a relative minimum value; graph is concave upward
$2 < x$		+	+	f is increasing; graph is concave upward

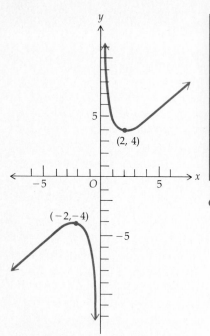

Figure 5.2.2

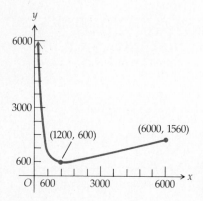

Figure 5.2.3

Observe that

$$\lim_{x \to 0^-} f(x) = \lim_{x \to 0^-} \frac{x^2 + 4}{x} = -\infty$$

and

$$\lim_{x \to 0^+} f(x) = \lim_{x \to 0^+} \frac{x^2 + 4}{x} = +\infty$$

With these facts and those from Table 5.2.2, and the plotting of a few points, we obtain the sketch of the graph of f shown in Fig. 5.2.2.

● ILLUSTRATION 1

Equation (1), defining the function of Example 2, can be written as

$$f(x) = \frac{4}{x} + x \tag{2}$$

A function defined by an equation similar to (2) arises in problems involving inventory control discussed in Sec. 4.5. In Example 5 of that section we had the function C defined by

$$C(x) = \frac{360,000}{x} + \frac{1}{4}x \qquad x \in (0, 6000] \tag{3}$$

where $C(x)$ dollars is a company's total monthly cost of obtaining and storing inventory when x items are produced in each run. Observe the resemblance of Eqs. (2) and (3). In the solution of Example 5 of Sec. 4.5 we determined that C has a relative minimum value at the point where $x = 1200$. A sketch of the graph of C appears in Fig. 5.2.3. Notice the similarity of the graph in Fig. 5.2.3 and the first-quadrant portion of the graph in Fig. 5.2.2.

●

EXAMPLE 3   Given $f(x) = 5x^{2/3} - x^{5/3}$, find the relative extrema of f; the points of inflection of the graph of f; the intervals on which f is increasing; the intervals on which f is decreasing; where the graph is concave upward; where the graph is concave downward; and the slope of any inflectional tangent. Draw a sketch of the graph.

Solution

$$f(x) = 5x^{2/3} - x^{5/3}$$

$$f'(x) = \tfrac{10}{3}x^{-1/3} - \tfrac{5}{3}x^{2/3}$$

$$f''(x) = -\tfrac{10}{9}x^{-4/3} - \tfrac{10}{9}x^{-1/3}$$

$f'(x)$ does not exist when $x = 0$. Set $f'(x) = 0$ to obtain $x = 2$. Therefore the critical numbers of f are 0 and 2. $f''(x)$ does not exist when $x = 0$.

From $f''(x) = 0$ we obtain $x = -1$. In making the table, consider the points at which $x = -1$, $x = 0$, and $x = 2$, and the following intervals:

$$x < -1 \qquad -1 < x < 0 \qquad 0 < x < 2 \qquad 2 < x$$

A sketch of the graph, drawn from the information in Table 5.2.3 and by plotting a few points, is shown in Fig. 5.2.4.

Table 5.2.3

	$f(x)$	$f'(x)$	$f''(x)$	Conclusion
$x < -1$		$-$	$+$	f is decreasing; graph is concave upward
$x = -1$	6	-5	0	f is decreasing; graph has a point of inflection
$-1 < x < 0$		$-$	$-$	f is decreasing; graph is concave downward
$x = 0$	0	does not exist	does not exist	f has a relative minimum value
$0 < x < 2$		$+$	$-$	f is increasing; graph is concave downward
$x = 2$	$3\sqrt[3]{4} = 4.8$	0	$-$	f has a relative maximum value; graph is concave downward
$2 < x$		$-$	$-$	f is decreasing; graph is concave downward

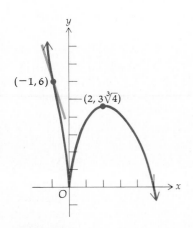

$(-1, 6)$

$(2, 3\sqrt[3]{4})$

Figure 5.2.4

Exercises 5.2

For each of the functions in Exercises 1 through 24, find the relative extrema of f; the points of inflection of the graph of f; the intervals on which f is increasing; the intervals on which f is decreasing; where the graph is concave upward; where the graph is concave downward; and the slope of any inflectional tangent. Draw a sketch of the graph.

1. $f(x) = 2x^3 - 6x + 1$
2. $f(x) = x^3 + x^2 - 5x$
3. $f(x) = x^4 - 2x^3$
4. $f(x) = 3x^4 + 2x^3$
5. $f(x) = x^3 + 5x^2 + 3x - 4$
6. $f(x) = 2x^3 - \frac{1}{2}x^2 - 12x + 1$
7. $f(x) = x^4 - 3x^3 + 3x^2 + 1$
8. $f(x) = x^4 - 4x^3 + 16x$
9. $f(x) = \frac{1}{4}x^4 - \frac{1}{3}x^3 - x^2 + 1$
10. $f(x) = \frac{1}{4}x^4 - x^3$
11. $f(x) = \frac{1}{2}x^4 - 2x^3 + 3x^2 + 2$
12. $f(x) = 3x^4 + 4x^3 + 6x^2 - 4$
13. $f(x) = (x + 1)^3(x - 2)^2$
14. $f(x) = x^2(x + 4)^3$
15. $f(x) = 3x^5 + 5x^4$

16. $f(x) = 3x^5 + 5x^3$
17. $f(x) = \dfrac{x^2 + 1}{x}$
18. $f(x) = x + \dfrac{9}{x}$

19. $f(x) = \dfrac{x^2}{x - 1}$
20. $f(x) = \dfrac{2x}{x^2 + 1}$
21. $f(x) = 3x^{2/3} - 2x$

22. $f(x) = x^{1/3} + 2x^{4/3}$
23. $f(x) = 3x^{4/3} - 4x$
24. $f(x) = 3x^{1/3} - x$

25. Refer to Exercise 31 in Exercises 4.5. A table manufacturer has a contract to supply 3000 tables per year at a uniform rate. The cost of starting a production run is $96, and the annual storage cost is $40 per table. The excess tables are stored until delivery, production is instantaneous, and shortages are not permitted. If $C(x)$ dollars is the total annual cost of obtaining and storing inventory when x tables are produced in each run, draw a sketch of the graph of C.

26. Refer to Exercise 32 in Exercises 4.5. A firm has a contract to supply 5 units of a certain commodity each day. The cost of starting a production run is $250, and the storage cost of each unit is $1 per day. Assume that production is instantaneous and shortages are not permitted. If $C(x)$ dollars is the total daily cost of obtaining and storing inventory when x units are produced in each run, draw a sketch of the graph of C.

5.3 Graphs of functions in economics

In Sec. 3.1 we discussed the total cost function, the marginal cost function, and the average cost function, but the treatment of the graphs of these functions was postponed until now so that the methods of sketching graphs learned in this chapter could be applied. It is suggested that you review the definitions of these functions in Sec. 3.1.

Let C be a total cost function so that $C(x)$ dollars is the total cost of producing x units of a commodity. In normal situations x and $C(x)$ are nonnegative. Furthermore, we assume that x is a nonnegative real number for C to be a continuous function. In addition, certain economic restrictions must be imposed on C in normal situations, and they are as follows:

1. $C(0) \geq 0$. That is, when nothing is produced, the total cost must be positive or zero. $C(0)$ is called the **overhead cost** of production.
2. $C'(x) > 0$ for all x. That is, the total cost must increase as the number of units produced increases.
3. $C''(x) \geq 0$ for x greater than some positive number N. When the number of units of the commodity produced is large, the marginal cost will eventually be increasing or zero. So unless $C''(x) = 0$, the graph of the total cost function is concave upward for $x > N$. However, the marginal cost may decrease for some values of x; hence, for these values of x, $C''(x) < 0$, and therefore the graph of the total cost function will be concave downward for these values of x (see Example 2).

● ILLUSTRATION 1

Consider a linear total cost function.

$$C(x) = mx + b$$

Observe that b represents the overhead cost. The marginal cost is given by $C'(x) = m$. If Q is the average cost function,

$$Q(x) = m + \frac{b}{x}$$

and

$$Q'(x) = -\frac{b}{x^2}$$

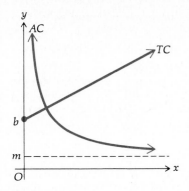

Figure 5.3.1

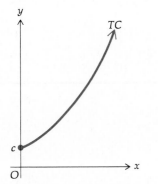

Figure 5.3.2

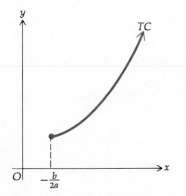

Figure 5.3.3

Refer to Fig. 5.3.1 for sketches of the total cost curve (labeled TC) and the average cost curve (labeled AC). The total cost curve is a segment of a straight line in the first quadrant having slope m and y intercept b. The average cost curve is a branch of an equilateral hyperbola in the first quadrant that has the line $y = m$ as horizontal asymptote. Because $Q'(x)$ is always negative, the average cost function is always decreasing, and as x increases, the value of $Q(x)$ gets closer and closer to m. The concept of the average cost approaching a constant frequently occurs in large manufacturing situations where the number of units produced is very great. ●

● ILLUSTRATION 2

Suppose C is a quadratic total cost function and

$$C(x) = ax^2 + bx + c$$

where a and c are positive. Here c is the number of dollars in the overhead cost. The total cost curve is a parabola opening upward. Because $C'(x) = 2ax + b$, a critical number of C is $-b/2a$, which gives the vertex of the parabola. We distinguish two cases: $b \geq 0$ and $b < 0$.

Case 1: $b \geq 0$. $-b/2a$ is then either negative or zero, and the vertex of the parabola is either to the left of the y axis or on the y axis. Hence the domain of C is the set of all nonnegative numbers. A sketch of TC for which $b > 0$ is shown in Fig. 5.3.2.

Case 2: $b < 0$. $-b/2a$ is positive; so the vertex of the parabola is to the right of the y axis, and the domain of C is restricted to numbers in the interval $[-b/2a, +\infty)$. A sketch of TC for which $b < 0$ is shown in Fig. 5.3.3. ●

EXAMPLE 1 Suppose that $100C(x)$ dollars is the total cost of producing $100x$ units of a commodity, and $C(x) = \frac{1}{2}x^2 - 2x + 5$. Find the function giving (a) the average cost and (b) the marginal cost. (c) Find the absolute minimum average unit cost. (d) Draw sketches of the total cost curve, average cost curve, and marginal cost curve on the same set of axes.

Solution $C(x) = \frac{1}{2}x^2 - 2x + 5$.

(a) If Q is the average cost function, $Q(x) = C(x)/x$. Thus

$$Q(x) = \frac{1}{2}x - 2 + \frac{5}{x}$$

(b) The marginal cost function is C', and

$$C'(x) = x - 2$$

(c) We compute $Q'(x)$ and $Q''(x)$.

$$Q'(x) = \frac{1}{2} - \frac{5}{x^2} \qquad Q''(x) = \frac{10}{x^3}$$

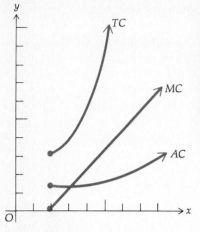

Figure 5.3.4

Setting $Q'(x) = 0$ we obtain $\sqrt{10}$ as a critical number of Q, and $Q(\sqrt{10}) = \sqrt{10} - 2 = 1.16$. Because $Q''(\sqrt{10}) > 0$, Q has a relative minimum value of 1.16 at $x = \sqrt{10}$. From the equation defining $Q(x)$ we see that Q is continuous on $(0, +\infty)$. Because the only relative extremum of Q on $(0, +\infty)$ is at $x = \sqrt{10}$, it follows from Theorem 4.5.1(ii) that Q has an absolute minimum value there. When $x = \sqrt{10} = 3.16$, $100x = 316$; so we conclude that the absolute minimum average unit cost is \$1.16 when 316 units are produced.

The sketches of the curves TC, AC, and MC (the marginal cost curve) are shown in Fig. 5.3.4.

In Fig. 5.3.4, observe that the lowest point on curve AC is at the point of intersection of curves AC and MC, which is where the average cost and marginal cost are equal. This situation occurs because $Q'(x) = 0$ here. Since $Q(x) = C(x)/x$, then

$$Q'(x) = \frac{xC'(x) - C(x)}{x^2}$$

Thus $Q'(x) = 0$ when $xC'(x) - C(x) = 0$, or equivalently, when

$$C'(x) = \frac{C(x)}{x}$$

You should note the economic significance that when the marginal cost and the average cost are equal, the commodity is being produced at the very lowest average unit cost.

The following example involving a cubic cost function illustrates the case in which the concavity of the graph of the total cost function changes.

EXAMPLE 2   Draw a sketch of the graph of the total cost function C for which $C(x) = x^3 - 6x^2 + 13x + 1$. Determine where the graph is concave upward and where it is concave downward. Find any points of inflection and an equation of any inflectional tangent. Draw a segment of the inflectional tangent.

Solution

$$C(x) = x^3 - 6x^2 + 13x + 1$$
$$C'(x) = 3x^2 - 12x + 13$$
$$C''(x) = 6x - 12$$

$C'(x)$ can be written as $3(x - 2)^2 + 1$. Hence $C'(x) > 0$ for all x. $C''(x) = 0$ when $x = 2$. To determine the concavity of the graph for the intervals $(0, 2)$ and $(2, +\infty)$ and if the graph has a point of inflection at $x = 2$, we use the results summarized in Table 5.3.1.

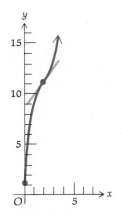

Figure 5.3.5

Table 5.3.1

	$C(x)$	$C'(x)$	$C''(x)$	Conclusion
$0 < x < 2$		+	−	C is increasing; graph is concave downward
$x = 2$	11	1	0	graph has a point of inflection
$2 < x$		+	+	C is increasing; graph is concave upward

An equation of the inflectional tangent is

$$y - 11 = 1(x - 2)$$

$$x - y + 9 = 0$$

A sketch of the graph of the total cost function together with a segment of the inflectional tangent is shown in Fig. 5.3.5.

Recall from Secs. 1.6 and 3.1 that a demand equation gives the relationship between p and x, where x units of a certain commodity are demanded when p dollars is the price of one unit. If the demand equation is solved for p, we get

$$p = f(x)$$

where f is the price function, and if it is solved for x, we have

$$x = g(p)$$

where g is the demand function. Both x and p are nonnegative, and the functions f and g are assumed to be continuous.

Unless the demand is constant, both the price function and the demand function are decreasing, because if p_1 and p_2 are the number of dollars in the prices of x_1 and x_2 units, respectively, of a commodity, then $x_2 > x_1$ if and only if $p_2 < p_1$.

● ILLUSTRATION 3

If the demand equation is linear (and the demand is not constant), then

$$p = mx + b \qquad 0 \le x \le -\frac{b}{m} \tag{1}$$

where $m < 0$, because the price function is decreasing, and $b > 0$. The total revenue function R is given by $R(x) = px$; thus

$$R(x) = mx^2 + bx \tag{2}$$

The marginal revenue function is R', and

$$R'(x) = 2mx + b \tag{3}$$

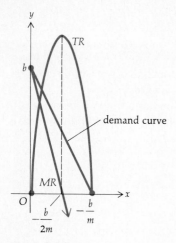

Figure 5.3.6

Furthermore,

$$R''(x) = 2m \qquad (4)$$

The demand curve, given from (1), and the marginal revenue curve, given from (3), have the same y intercept b, and for values of $x \neq 0$ the marginal revenue curve lies below the demand curve. From (3) we see that if $R'(x) = 0$, $x = -b/2m$, which is a positive number. From (4) it follows that $R''(x)$ is always negative, and hence R has a relative maximum value when $x = -b/2m$. This relative maximum value of R is an absolute maximum value on the interval $[0, -b/m]$. Hence we see that the marginal revenue curve intersects the x axis at the point whose abscissa is the value of x for which the total revenue is greatest, and the demand curve intersects the x axis at the point whose abscissa is twice that. Refer to Fig. 5.3.6. ●

Example 3 of Sec. 3.1 is a particular case of the situation described in Illustration 3. A case where the demand equation is nonlinear is shown in the following example.

EXAMPLE 3   The demand equation for a particular commodity is

$$p^2 + x - 12 = 0$$

Find the total revenue function and the marginal revenue function. Draw sketches of the demand curve, the total revenue curve, and the marginal revenue curve on the same set of axes.

Solution   If the demand equation is solved for p, we find $p = \pm \sqrt{12 - x}$. Because $R(x) = px$ and $p \geq 0$, we have

$$R(x) = x\sqrt{12 - x}$$

and

$$R'(x) = \sqrt{12 - x} - \frac{x}{2\sqrt{12 - x}}$$

$$R'(x) = \frac{24 - 3x}{2\sqrt{12 - x}}$$

Setting $R'(x) = 0$ we obtain $x = 8$. Using the information in Table 5.3.2 we see that the required sketches are drawn as shown in Fig. 5.3.7.

Table 5.3.2

x	p	$R(x)$	$R'(x)$
0	$\sqrt{12}$	0	$\sqrt{12}$
3	3	9	$\frac{5}{2}$
8	2	16	0
11	1	11	$-\frac{9}{2}$
12	0	0	does not exist

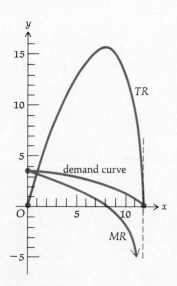

Figure 5.3.7

Exercises 5.3

1. The number of dollars in the total cost of producing x units of a commodity is given by $C(x) = x^2 + 4x + 8$. Find (a) the average cost function and (b) the marginal cost function. (c) Find the absolute minimum average unit cost. (d) Draw sketches of the total cost, average cost, and marginal cost curves on the same set of axes. Observe that the average cost and marginal cost are equal when the average cost has its least value.

2. The total cost of producing x units of a commodity is $C(x)$ dollars, and $C(x) = 3x^2 + x + 3$. Find (a) the average cost function and (b) the marginal cost function. (c) Find the absolute minimum average unit cost. (d) Draw sketches of the total cost, average cost, and marginal cost curves on the same set of axes. Observe that the average cost and marginal cost are equal when the average cost has its least value.

3. If $C(x)$ dollars is the total cost of producing x units of a commodity, and $C(x) = 3x^2 - 6x + 4$, find (a) the average cost function and (b) the marginal cost function. (c) What is the range of C? (d) Find the absolute minimum average unit cost. (e) Draw sketches of the total cost, average cost, and marginal cost curves on the same set of axes. Observe that the average cost and marginal cost are equal when the average cost has its least value.

4. If $C(x)$ dollars is the total cost of producing x units of a commodity, and $C(x) = 2x^2 - 8x + 18$, find (a) the domain and range of C, (b) the average cost function, (c) the absolute minimum average unit cost, and (d) the marginal cost function. (e) Draw sketches of the total cost, average cost, and marginal cost curves on the same set of axes.

5. The total cost function C is given by $C(x) = \frac{1}{3}x^3 - 2x^2 + 5x + 2$. (a) Determine the range of C. (b) Find the marginal cost function. (c) Find the interval on which the marginal cost function is decreasing and the interval on which it is increasing. (d) Draw a sketch of the graph of the total cost function; determine where the graph is concave upward and where it is concave downward, and find the points of inflection and an equation of any inflectional tangent.

6. The number of dollars in the total cost of producing x bowls per day in a certain factory is given by $C(x) = 4x + 500$. Find (a) the average cost function and (b) the marginal cost function. (c) Show that there is no absolute minimum average unit cost. (d) What is the smallest number of bowls that the factory must produce in a day so that the average cost per bowl is less than $7? (e) Draw sketches of the total cost, average cost, and marginal cost curves on the same set of axes.

7. The overhead cost of a manufacturer of children's toys is $400 per week, and other costs amount to $3 for each toy produced. Find (a) the total cost function, (b) the average cost function, and (c) the marginal cost function. (d) Show that there is no absolute minimum average unit cost. (e) What is the smallest number of toys that must be produced so that the average cost per toy is less than $3.42? (f) Draw sketches of the graphs of the functions in (a), (b), and (c) on the same set of axes.

8. The demand equation for a particular commodity is $px^2 + 9p = 18$, where $100x$ units are demanded when p dollars is the price per unit. Find (a) the total revenue function and (b) the marginal revenue function. (c) Find the absolute maximum total revenue.

9. Follow the instructions of Exercise 8 if the demand equation is $x^2 + p^2 = 36$.

10. Follow the instructions of Exercise 8 if the demand equation is $(p + 4)(x + 3) = 48$.

11. The total revenue received from the sale of x desks is $R(x)$ dollars, and $R(x) = 200x - \frac{1}{3}x^2$. Find (a) the demand equation and (b) the marginal revenue function. (c) Find the absolute maximum total revenue. (d) Draw sketches of the demand, total revenue, and marginal revenue curves on the same set of axes.

12. If $R(x)$ dollars is the total revenue received from the sale of x television sets, and $R(x) = 600x - \frac{1}{20}x^3$, find (a) the demand equation and (b) the marginal revenue function. (c) Find the absolute maximum total revenue. (d) Draw sketches of the demand, total revenue, and marginal revenue curves on the same set of axes.

13. $R(x)$ dollars is the total revenue obtained when x units of a commodity are demanded, and $R(x) = 30 + 50\sqrt{x + 1}$, where x is in the closed interval $[3, 24]$. Find (a) the demand equation and (b) the marginal revenue function. (c) Find the absolute maximum total revenue. (d) Draw sketches of the demand, total revenue, and marginal revenue curves on the same set of axes.

14. The total revenue received from the sale of x units of a commodity is $R(x)$ dollars, and $R(x) = 20 + 30\sqrt{x - 1}$, where x is in the closed interval $[2, 17]$. Find (a) the demand equation and (b) the marginal revenue function. (c) Find the absolute maximum total revenue. (d) Draw sketches of the demand, total revenue, and marginal revenue curves on the same set of axes.

5.4 Price elasticity of demand

Consider a demand equation involving p and x, where p dollars is the unit price of a certain commodity for which x units are demanded at that price. If the demand equation is solved for x, we obtain the demand function g given by

$$x = g(p)$$

We assume that p is a nonnegative real number and that the demand function is continuous.

If p changes by an amount Δp, then x changes by an amount Δx. The relative change in p is then $\Delta p/p$, and the relative change in x is $\Delta x/x$. The average relative change in x (the quantity demanded) per unit relative change in p (the price) is given by

$$\frac{\Delta x}{x} \div \frac{\Delta p}{p} \tag{1}$$

or, equivalently,

$$\frac{p}{x} \cdot \frac{\Delta x}{\Delta p} \tag{2}$$

Because $\Delta x = g(p + \Delta p) - g(p)$, (2) can be written as

$$\frac{p}{x} \cdot \frac{g(p + \Delta p) - g(p)}{\Delta p}$$

Taking the limit of the above as Δp approaches zero we have, if $g'(p)$ exists,

$$\lim_{\Delta p \to 0} \frac{p}{x} \cdot \frac{g(p + \Delta p) - g(p)}{\Delta p} = \frac{p}{x} \cdot g'(p)$$

We have shown that the limit of quotient (1) can be expressed as

$$\frac{p}{x} \frac{dx}{dp}$$

This limit gives the approximate percent change in the demand that corresponds to a change of 1 percent in the price. It is called the *price elasticity of demand* and is denoted by the Greek letter eta, η. Following is the formal definition.

Definition of price elasticity of demand

> The **price elasticity of demand** gives the approximate percent change in the demand that corresponds to a change of 1 percent in the price. If the demand equation is $x = g(p)$ and η is the price elasticity of demand, then
>
> $$\eta = \frac{p}{x} \frac{dx}{dp} \tag{3}$$

Because the demand function is decreasing, $\dfrac{dx}{dp} < 0$, and if $p \neq 0$, η is negative. Of course if $x = 0$, then η is not defined.

EXAMPLE 1 The demand equation for a certain commodity is

$$x = 18 - 2p^2$$

where x units are demanded when p dollars is the price per unit. (a) Find the relative decrease in the demand when the price of one unit is increased from \$2 to \$2.06. (b) Use the result in part (a) to obtain an approximation to the price elasticity of demand at $p = 2$. (c) Find the exact price elasticity of demand at $p = 2$, and interpret the result.

Solution   (a) When $p = 2$, $x = 10$. When $p = 2.06$,

$$x = 18 - 2(2.06)^2$$
$$= 9.51$$

The decrease in x, then, is 0.49, and the relative decrease in x is $0.49/10 = 0.049 = 4.9$ percent.

(b) In part (a), p is increased from \$2 to \$2.06, which is an increase of 3 percent. Thus an increase in p of 3 percent causes a decrease in x of 4.9 percent. Hence an approximation to the price elasticity of demand at $p = 2$ is

$$\frac{-4.9}{3} = -1.63$$

(c) The price elasticity of demand is, from formula (3),

$$\eta = \frac{p}{x}\frac{dx}{dp} = \frac{p}{x}(-4p) = -\frac{4p^2}{x}$$

So when $p = 2$ and $x = 10$,

$$\eta = -\frac{4(2)^2}{10} = -1.60$$

We interpret the result of $\eta = -1.60$ at $p = 2$ as meaning that when the unit price is \$2, an increase of 1 percent in the unit price will cause an approximate decrease of 1.60 percent in the demand (or a decrease of 1 percent in the unit price will cause an approximate increase of 1.60 percent in the demand).

EXAMPLE 2 The demand equation for a certain candy bar is

$$x = 960 - 31p + \tfrac{1}{4}p^2$$

where p cents is the unit price and $1000x$ candy bars are demanded weekly at that price. (a) Find the price elasticity of demand when $p = 50$. (b) If the price of 50 cents is increased by 2 percent, what approximate change is there in the weekly demand?

Solution   (a) When $p = 50$, $x = 35$. Because

$$x = 960 - 31p + \tfrac{1}{4}p^2$$

then

$$\frac{dx}{dp} = -31 + \tfrac{1}{2}p$$

Therefore

$$\frac{dx}{dp}\bigg]_{p=50} = -31 + 25 = -6$$

From formula (3), when $p = 50$, we get for the price elasticity of demand

$$\eta = \tfrac{50}{35}(-6) = -8.57$$

We interpret this answer as meaning that when the price per candy bar is 50 cents, an increase of 1 percent in the price will cause an approximate decrease of 8.57 percent in the weekly demand.

(b) If the price of 50 cents is increased by 2 percent, there is an approximate decrease of 17.14 percent in the weekly demand.

EXAMPLE 3   Suppose that $1000x$ cans of coffee are demanded when p dollars is the price per can, and

$$x = 25 - p^2$$

(a) Find the price elasticity of demand when the price of a can of coffee is \$3.50. (b) What percent decrease in the price would yield an approximate 5 percent increase in the demand?

Solution   (a) Because $x = 25 - p^2$, then

$$\frac{dx}{dp} = -2p$$

We compute the price elasticity of demand from formula (3), and we have

$$\eta = \frac{p}{x}\frac{dx}{dp} = \frac{p}{x}(-2p) = -\frac{2p^2}{x}$$

When $p = 3.50$, $x = 25 - (3.50)^2 = 12.75$. Therefore, when $p = 3.50$,

$$\eta = -\frac{2(3.50)^2}{12.75} = -1.92$$

(b) From the result of part (a), a decrease of 1 percent in the unit price would cause an approximate increase of 1.92 percent in the demand. Because $5/1.92 = 2.60$, it follows that a 2.6 percent decrease in the price of a can of coffee would cause an approximate 5 percent increase in the demand.

Some interesting properties of the total revenue function can be obtained from the price elasticity of demand. If R is the total revenue function, $R(x)$

dollars is the total revenue for a demand of x units at p dollars per unit, and

$$R(x) = xp \tag{4}$$

where x is a function of p (recall that the demand equation is $x = g(p)$). We now differentiate implicitly with respect to p on both sides of Eq. (4), and we obtain

$$D_x R(x) \cdot \frac{dx}{dp} = x + p\frac{dx}{dp} \tag{5}$$

We divide on both sides of (5) by $\dfrac{dx}{dp}$ and obtain

$$D_x R(x) = \frac{x}{\dfrac{dx}{dp}} + p$$

or, equivalently,

$$D_x R(x) = p\left(1 + \frac{x}{p} \cdot \frac{1}{\dfrac{dx}{dp}}\right) \tag{6}$$

Because

$$\eta = \frac{p}{x}\frac{dx}{dp}$$

then

$$\frac{1}{\eta} = \frac{x}{p} \cdot \frac{1}{\dfrac{dx}{dp}} \tag{7}$$

Substituting from (7) into (6), and replacing $D_x R(x)$ by $R'(x)$, we have

$$R'(x) = p\left(1 + \frac{1}{\eta}\right) \tag{8}$$

From Eq. (8) we can make the following conclusions (bear in mind that η is negative).

(i) If $|\eta| > 1$, then for a given demand and price, a decrease in price results in a greater relative increase in the quantity demanded. The demand is said to be **elastic**. Furthermore, the marginal revenue is positive, and so the total revenue increases as the price decreases and hence the demand increases.

(ii) If $|\eta| = 1$, then for a given demand and price, a decrease in price results in the same relative increase in the quantity demanded, and the demand is said to be **unitary**. Furthermore, the marginal revenue is zero, and so the total revenue may have an extremum (usually a maximum).

(iii) If $|\eta| < 1$, then for a given demand and price, a decrease in price results in a smaller relative increase in the quantity demanded. The demand is then said to be **inelastic**. Furthermore, the marginal revenue is negative, and so the total revenue decreases as the price decreases and hence the demand increases.

Observe from conclusion (i) that if $|\eta| > 1$ (that is, if the demand is elastic), then the total revenue can be increased by decreasing the price.

As the demand increases, with a decrease in price, the absolute value of the price elasticity of demand decreases continuously from values greater than 1, when the demand is small, to values less than 1, when the demand is large. As the demand increases, it becomes more inelastic. From conclusions (i)–(iii) we see that the total revenue increases or decreases as the demand increases, in response to a decrease in price, according to whether the demand is elastic or inelastic. At first the total revenue increases as the price decreases and the demand increases ($|\eta| > 1$), then attains an absolute maximum value for a particular demand (where $|\eta| = 1$), and finally decreases as the demand increases further ($|\eta| < 1$). See Fig. 5.4.1.

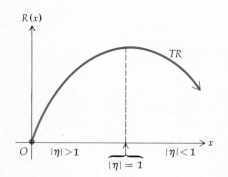

Figure 5.4.1

EXAMPLE 4 For the demand equation of Example 1, determine the values of x for which the demand is (a) elastic, (b) unitary, and (c) inelastic. (d) Find the total revenue function and the marginal revenue function, and show that (8) holds. (e) Draw a sketch of the graph of the total revenue function and indicate where $|\eta| > 1$, $|\eta| = 1$, and $|\eta| < 1$.

Solution In Example 1 the demand equation is

$$x = 18 - 2p^2$$

and

$$\eta = -\frac{4p^2}{x} = \frac{2x - 36}{x}$$

Because x and p must be nonnegative, it follows that x is in the interval $[0, 18]$ and p is in the interval $[0, 3]$ (of course η does not exist when $x = 0$). Therefore

$$|\eta| = \frac{36 - 2x}{x}$$

(a) The demand is elastic when $|\eta| > 1$, that is, when

$$\frac{36 - 2x}{x} > 1 \quad \text{and} \quad x \in (0, 18]$$

or, equivalently, when

$$36 - 2x > x \quad \text{and} \quad x \in (0, 18]$$

or, equivalently, when

$$0 < x < 12$$

(b) The demand is unitary when $|\eta| = 1$, that is, when

$$\frac{36 - 2x}{x} = 1$$

or, equivalently,

$$x = 12$$

(c) The demand is inelastic when $|\eta| < 1$, that is, when

$$\frac{36 - 2x}{x} < 1 \quad \text{and} \quad x \in (0, 18]$$

or, equivalently, when

$$12 < x \leq 18$$

(d) We solve the demand equation for p and obtain, because $p \geq 0$,

$$p = \sqrt{9 - \tfrac{1}{2}x} \tag{9}$$

If R is the total revenue function,

$$R(x) = x\sqrt{9 - \tfrac{1}{2}x} \tag{10}$$

R' is the marginal revenue function and

$$R'(x) = \sqrt{9 - \tfrac{1}{2}x} - \frac{x}{4\sqrt{9 - \tfrac{1}{2}x}} \tag{11}$$

To verify that (8) holds, start with the right-hand side of (8) and take $\eta = -4p^2/x$, and get

$$p\left(1 + \frac{1}{\eta}\right) = p\left(1 - \frac{x}{4p^2}\right)$$

$$= p - \frac{x}{4p}$$

Because $p = \sqrt{9 - \tfrac{1}{2}x}$,

$$p\left(1 + \frac{1}{\eta}\right) = \sqrt{9 - \tfrac{1}{2}x} - \frac{x}{4\sqrt{9 - \tfrac{1}{2}x}} \tag{12}$$

Comparing (11) and (12) we have

$$R'(x) = p\left(1 + \frac{1}{\eta}\right)$$

which is (8).

(e) The total revenue function is given by (10). To draw a sketch of the graph of R we first find when $R'(x) = 0$. From (11) we can determine that $R'(x) = 0$ when $x = 12$. We can prove that R has a relative maximum value when $x = 12$. From this fact and by plotting a few points we have the required sketch shown in Fig. 5.4.2.

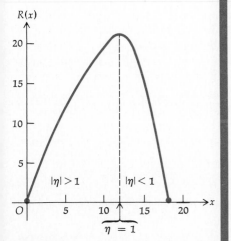

Figure 5.4.2

Exercises 5.4

1. When p dollars is the price per unit, x units of a certain commodity are demanded, and $x = 2100 - 100p^2 - 400p$. (a) If the price per unit is increased from $2 to $2.20, find the relative decrease in the demand. (b) Use the result in part (a) to obtain an approximation to the price elasticity of demand at $p = 2$. (c) Find the exact price elasticity of demand at $p = 2$, and interpret the result.

2. The demand equation for a certain commodity is $p^2 + p + \frac{1}{50}x = 40$, where x units are demanded when p dollars is the price of one unit. (a) If the price of one unit is increased from $4 to $4.24, find the relative decrease in the demand. (b) Use the result in part (a) to obtain an approximation to the price elasticity of demand at $p = 4$. (c) Find the exact price elasticity of demand at $p = 1$, and interpret the result.

3. For a particular commodity the demand equation is $x = 20 - 4p^2$, where x units are demanded when p dollars is the price per unit. (a) Find the relative increase in the demand when the price per unit is decreased from $1.00 to 95 cents. (b) Use the result in part (a) to obtain an approximation to the price elasticity of demand at $p = 1$. (c) Find the exact price elasticity of demand at $p = 1$, and interpret the result.

4. The demand equation for a particular article of merchandise is $xp^3 = 24{,}000$ where x units are demanded when the price of one unit is p dollars. (a) Find the relative increase in the demand when the price of one unit is decreased from $2 to $1.90. (b) Use the result in part (a) to obtain an approximation to the price elasticity of demand at $p = 2$. (c) Find the exact price elasticity of demand at $p = 2$, and interpret the result.

5. For a particular kind of bread, $100x$ loaves are demanded weekly when p cents is the price of one loaf, and the demand equation is $p^2 + 400x = 18{,}000$. (a) Find the price elasticity of demand when the price is 60 cents per loaf. (b) If the price of 60 cents is decreased by 6 percent, what approximate change is there in the weekly demand?

6. The demand equation for a certain children's toy is $x = 60 - 3p^2$, where $1000x$ toys are demanded when p dollars is the price per toy. (a) Find the price elasticity of demand when the price is $4 per toy. (b) Find the approximate change in the demand if the price of $4 is decreased by 4 percent.

7. Suppose that $100x$ units of a certain commodity are demanded when p dollars is the price per unit and $p(x + 1) = 16$, where $p \in [1, 8]$. (a) Find the price elasticity of demand when the price is $2 per unit. (b) Find the approximate change in the demand if the price of $2 is increased by 3 percent.

8. A particular commodity has the demand equation $x = \sqrt{10 - p^2}$, where $100x$ units are demanded when p dollars is the price per unit. (a) Find the price elasticity of demand when the price is $3. (b) Find the approximate change in the demand if the price of $3 is increased by 6 percent.

9. The demand equation for a particular book is $p = \sqrt{100 - x} + 8$, where $100x$ books are demanded when the price is p dollars per book. (a) Find the price elasticity of demand when the price is $14. (b) What percent decrease in the price would yield an approximate 4 percent increase in the demand?

10. The demand for alarm clocks is $100x$ clocks when the price per clock is p dollars, and the demand equation is $p = 8\sqrt{25 - x^2}$. (a) Find the price elasticity of demand when the price is $32. (b) What percent decrease in the price would yield an approximate 10 percent increase in the demand?

11. For the demand equation of Exercise 3, determine the values of x for which the demand is (a) elastic, (b) unitary, and (c) inelastic. (d) Find the total revenue function and the marginal revenue function, and show that (8) holds. (e) Draw a sketch of the graph of the total revenue function, and indicate where $|\eta| > 1$, $|\eta| = 1$, and $|\eta| < 1$.

12. Follow the instructions of Exercise 11 if the demand equation is $x = 25 - p^2$ and x units are demanded when p dollars is the price per unit.

13. Follow the instructions of Exercise 11 for the demand equation of Exercise 5.

14. Follow the instructions of Exercise 11 for the demand equation of Exercise 6.

15. If the demand equation is $xp^n = a$, where n is a positive integer and a is a real number other than zero, prove that the price elasticity of demand is a constant, $-n$. Give the economic interpretation of this result.

16. For each of the following demand equations, find the price elasticity of demand, and determine the value of x for which the demand is unitary: (a) $p = \sqrt{5 - 2x}$; (b) $p = (5 - 2x)^2$; (c) $p = 5 - 2x^2$.

5.5 Profit

The **total profit** earned by a business is defined to be the difference between the total revenue and the total cost. That is, if $P(x)$ dollars is the total profit obtained by producing and selling x units of a commodity, then

$$P(x) = R(x) - C(x) \qquad (1)$$

where $R(x)$ dollars is the total revenue and $C(x)$ dollars is the total cost. The function P is called the **total profit function**. From Eq. (1) we obtain

$$P'(x) = R'(x) - C'(x) \qquad (2)$$

The function P' is called the **marginal profit function**, and $P'(k)$ is the approximate profit realized from the $(k + 1)$st unit after k units have been produced and sold. From Eq. (2) we see that the marginal profit is the marginal revenue minus the marginal cost.

In Fig. 5.5.1(a) we have sketches of the graphs of a total cost function and a total revenue function of a particular company, and in Fig. 5.5.1(b) we have the sketch of the graph of the company's corresponding total profit function (denoted by TP). Observe in Fig. 5.5.1(a) that when curve TC is above curve TR (when the cost is greater than the revenue), then in Fig. 5.5.1(b) curve TP is below the x axis (the profit is negative; that is, the company shows a loss); this occurs when $0 \le x < x_2$ and when $x > x_3$. When curve TC is below curve TR (when the cost is less than the revenue), curve TP is above the x axis (the profit is positive; that is, the company shows a profit); this occurs when $x_2 < x < x_3$. At the points of intersection of curves TC and TR (when the cost equals the revenue), curve TP intersects the x axis (the profit is zero; that is, the company is breaking even); this occurs at x_2 and x_3.

The company is showing a profit when its output is between x_2 and x_3. Let us determine what level of production is necessary to obtain the maximum profit. In Fig. 5.5.1(a) the vertical distance between curves TR and TC for a particular value of x is $P(x)$, which gives the total profit corresponding to that value of x. When this vertical distance is largest, $P(x)$ has its maximum value. Observe from Fig. 5.5.1(a) that the distance AB is the largest vertical distance between the two curves in the interval $[x_2, x_3]$, and this occurs at the critical number x_1 of the function P. From Eq. (2) we see that $P'(x) = 0$ if and only if $R'(x) = C'(x)$. Thus, *for maximum profit the marginal revenue must equal the marginal cost.* Note in Fig. 5.5.1(a) that at points B and A on the two curves the tangent lines are parallel, and hence $R'(x_1) = C'(x_1)$.

From (2) we obtain

$$P''(x) = R''(x) - C''(x) \qquad (3)$$

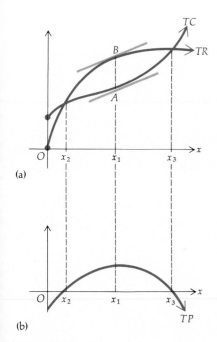

(a)

(b)

Figure 5.5.1

Because P will have a relative maximum function value at a number x for which $P'(x) = 0$ and $P''(x) < 0$, we can conclude from Eqs. (2) and (3) that this will occur for a value of x for which $R'(x) = C'(x)$ and $R''(x) < C''(x)$. Therefore the profit function will have a relative maximum function value when the marginal revenue equals the marginal cost and the slope of the graph of the marginal revenue function is less than the slope of the graph of the marginal cost function. These concepts are illustrated in the following example, in which we consider *perfect competition*. When a company is operating under **perfect competition**, there is a great number of small firms, and so any one firm cannot affect price by increasing production. Therefore, under perfect competition, the price of a commodity is constant, and the company can sell as much as it wishes to sell at this constant price. The demand curve in such a case is horizontal; it is a line parallel to the x axis.

EXAMPLE 1 Under perfect competition a firm can sell at a price of $100 per unit all of a particular commodity it produces. If x units are produced each day, the number of dollars in the total cost of each day's production is $x^2 + 20x + 700$. Find the number of units that should be produced each day for the firm to have the greatest daily total profit.

Solution If $R(x)$ dollars is the total revenue from the sale of x units, then

$$R(x) = 100x \tag{4}$$

If C is the total cost function, then

$$C(x) = x^2 + 20x + 700 \tag{5}$$

If P is the total profit function, then

$$\begin{aligned} P(x) &= R(x) - C(x) \\ &= 100x - (x^2 + 20x + 700) \\ &= -x^2 + 80x - 700 \end{aligned} \tag{6}$$

Observe that x is in $[0, +\infty)$.

From (4) and (5) we obtain

$$R'(x) = 100 \quad \text{and} \quad R''(x) = 0$$

$$C'(x) = 2x + 20 \quad \text{and} \quad C''(x) = 2$$

Equating $C'(x)$ and $R'(x)$ we get

$$2x + 20 = 100$$

$$x = 40$$

Because $R''(x) < C''(x)$ for all x, $P''(x) = R''(x) - C''(x) < 0$. Therefore $P(40)$ is a relative maximum value. It is also an absolute maximum value, because P is continuous on $[0, +\infty)$ and $P(40)$ is the only extremum of P for x in $[0, +\infty)$.

Therefore the daily total profit is greatest when 40 units are produced for the day.

Note that an alternate method of solution is to find $P'(x)$ and $P''(x)$ from (6) and proceed as customary in extremum problems.

For a given demand equation, the amount demanded by the consumer depends only on the price of the commodity. Under a **monopoly**, which means that there is only one producer of a certain commodity, price and, hence, demand can be controlled by regulating the quantity of the commodity produced. Generally the price decreases as the producer increases the output. The producer under a monopoly is called a **monopolist**. The monopolist wishes to control the quantity produced and, hence, the price per unit (determined from the demand equation) so that the profit will be as large as possible.

EXAMPLE 2 Suppose that under a monopoly the demand equation for a certain commodity is

$$p = 4 - 0.0002x$$

where x units are produced each day and p dollars is the price of each unit. The number of dollars in the total cost of producing x units is $600 + 3x$. If the daily profit is to be as large as possible, find the number of units that the monopolist should produce each day, the price of each unit, and the daily profit.

Solution Because the demand equation is $p = 4 - 0.0002x$ and x and p must be nonnegative, it follows that x is in the closed interval $[0, 20,000]$. Because $R(x) = xp$, we have

$$R(x) = 4x - 0.0002x^2 \qquad x \in [0, 20,000] \tag{7}$$

The total cost function is given by

$$C(x) = 600 + 3x \tag{8}$$

If $P(x)$ dollars is the total profit, $P(x) = R(x) - C(x)$, and so

$$P(x) = x - 0.0002x^2 - 600 \qquad x \in [0, 20,000]$$

From (7) and (8) we obtain

$$R'(x) = 4 - 0.0004x \quad \text{and} \quad R''(x) = -0.0004$$

$$C'(x) = 3 \quad \text{and} \quad C''(x) = 0$$

Equating $R'(x)$ and $C'(x)$ we get

$$4 - 0.0004x = 3$$

$$x = 2500$$

Because $R''(x)$ is always less than $C''(x)$, we can conclude that $P(2500)$ is an absolute maximum value (observe that $P(x)$ is negative for $x = 0$ and $x = 20,000$). $P(2500) = 650$, and when $x = 2500$, $p = 3.50$.

Therefore, to have the greatest daily profit, 2500 units should be produced each day to be sold at $3.50 each, for a total profit of $650.

EXAMPLE 3 Solve Example 2 if a tax of 20 cents is levied by the government on the monopolist for each unit produced.

Solution With the added tax, the total cost function is now given by

$$C(x) = (600 + 3x) + 0.20x$$

and so the marginal cost function is given by

$$C'(x) = 3.20$$

Equating $R'(x)$ and $C'(x)$ we get

$$4 - 0.0004x = 3.20$$

$$x = 2000$$

As in Example 2, it follows that when $x = 2000$, P has an absolute maximum value on $[0, 20,000]$. From $P(x) = 0.8x - 0.0002x^2 - 600$ we have $P(2000) = 200$. From the demand equation $p = 4 - 0.0002x$ we obtain $p = 3.60$ when $x = 2000$.

Therefore, if the tax of 20 cents per unit is levied, the monopolist should produce only 2000 units each day, and they should be sold at $3.60 per unit to attain a maximum daily total profit of $200.

It is interesting to note that in comparing the results of Examples 2 and 3, the entire 20 cent increase should not be passed on to the consumer to achieve the greatest daily profit. That is, it is most profitable to raise the unit price by only 10 cents. The economic significance of this result is that consumers are sensitive to price changes, which prohibits the monopolist from passing on the tax completely to the consumer. This fact is demonstrated in the following illustration, where we compute the price elasticity of demand for the demand equation of the two examples.

● ILLUSTRATION 1

The demand equation of Examples 2 and 3 is $p = 4 - 0.0002x$. If we solve this equation for x, we obtain

$$x = 20,000 - 5000p$$

If η is the price elasticity of demand, then

$$\eta = \frac{p}{x} \frac{dx}{dp}$$

Because $\dfrac{dx}{dp} = -5000$, it follows that

$$\eta = -\frac{5000p}{x}$$

In Example 2 we determined that for maximum profit, $x = 2500$ and $p = 3.50$. We compute η for these values of x and p, and we obtain

$$\eta = -\frac{5000(3.50)}{2500} = -7$$

In Example 3 a tax of 20 cents was levied by the government on the monopolist for each unit produced. If the entire 20 cents is passed on to the consumer, the price will increase from $3.50 by 5.7 percent (0.20 is 5.7 percent of $3.50). However, if the price is increased by only 10 cents, then the price will increase from $3.50 by 2.9 percent. Because $\eta = -7$ when $p = 3.50$, it follows that a 5.7 percent increase in p will cause a decrease in the demand of about 40 percent, while a 2.9 percent increase in p will cause a decrease in the demand of only about 20 percent.   ●

It should be pointed out that the relationship of the increase in price by the monopolist to the tax levy imposed upon the monopolist depends on the given demand equation, because the demand equation determines the price elasticity of demand. In Exercise 15(b) at the end of this section we have a situation where only about one-fourth of the tax imposed on the monopolist should be passed on to the consumer. However, in Exercise 17(b) the increase in price after the tax levy on the monopolist is greater than the tax.

A further note of interest in Examples 2 and 3 is how the overhead cost of a company does not affect the determination of the number of units to be produced or the unit price such that maximum profit is obtained. Regardless of whether anything is produced, the overhead cost must be met. In Examples 2 and 3, because $C(x) = 600 + 3x$ and $600 + 3.2x$, respectively, the overhead cost is \$600. If the 600 in the expression for $C(x)$ is replaced by any constant k, $C'(x)$ is not affected; hence the value of x for which the marginal cost equals the marginal revenue is not affected by any such change. Of course, a change in overhead cost affects the unit cost and hence the actual profit; however, if a company is to have the greatest profit possible, a change in its overhead cost will not affect the number of units to be produced, nor the price per unit.

EXAMPLE 4 For the monopolist in Example 2, how much tax should be levied by the government for each unit produced in order for the tax revenue received by the government to be maximized?

Solution Let t be the number of dollars in the tax levied by the government on the monopolist for each unit produced. If x units are produced daily, the total daily tax is tx dollars. The total cost function is then given by

$$C(x) = (600 + 3x) + tx$$

Thus

$$C'(x) = 3 + t$$

Equating $R'(x)$ and $C'(x)$ we obtain

$$4 - 0.0004x = 3 + t$$
$$x = 2500 - 2500t$$

This value of x maximizes the producer's daily profit. If we let T be the number of dollars in the daily tax revenue received by the government, we have $T = tx$. Hence

$$T(t) = t(2500 - 2500t) \qquad 0 \le t \le 1$$

$$T'(t) = 2500 - 5000t$$

$$T''(t) = -5000$$

Setting $T'(t) = 0$ we get $t = 0.50$, and since $T''(t) < 0$, T has a relative maximum value at $t = 0.50$. Because $T = 0$ for both $t = 0$ and $t = 1$, we conclude that T has an absolute maximum value when $t = 0.50$.

Therefore the government should levy a tax of 50 cents for each unit produced in order to maximize its tax revenue.

Exercises 5.5

1. Under a monopoly the demand equation for a certain commodity is $x + p = 140$, where x units are demanded daily when p dollars is the price per unit. The number of dollars in the total cost of producing x units is given by $C(x) = x^2 + 20x + 300$, and x is in the closed interval $[0, 140]$. (a) Find the total profit function, and draw a sketch of its graph. (b) On a set of axes different from that in (a), draw sketches of the total revenue and total cost curves, and show the geometrical interpretation of the total profit function. (c) Find the marginal revenue and marginal cost functions. (d) Find the maximum daily profit. (e) Draw sketches of the graphs of the marginal revenue function and the marginal cost function on the same set of axes, and show that they intersect at the point for which the value of x makes the total profit a maximum.

2. Follow the instructions of Exercise 1 if the demand equation is $x^2 + p = 320$, $C(x) = 20x$, and x is in the closed interval $[0, 8\sqrt{5}]$.

3. A company that builds and sells desks is operating under perfect competition and can sell at a price of $200 per desk all the desks it produces. If x desks are produced and sold each week and $C(x)$ dollars is the total cost of the week's production, then $C(x) = x^2 + 40x + 3000$. Determine how many desks should be built each week in order for the manufacturer to have the greatest weekly total profit. What is the greatest weekly total profit?

4. A firm operating under perfect competition manufactures and sells portable radios. The firm can sell at a price of $75 per radio all the radios it produces. If x radios are manufactured each day and $C(x)$ dollars is the daily total cost of production, then $C(x) = x^2 + 25x + 100$. How many radios should be produced each day in order for the firm to have the greatest daily total profit? What is the greatest daily total profit?

5. Suppose that under a monopoly the demand equation of a particular article of merchandise is $p = 6 - \frac{1}{5}\sqrt{x - 100}$, where x articles are demanded when p dollars is the price per article, and $x \in [100, 1000]$. If $C(x)$ dollars is the total cost of producing x articles, then $C(x) = 2x + 100$. (a) Find the marginal revenue and marginal cost functions. (b) Find the value of x that yields the maximum profit.

6. Under a monopoly the demand equation for a certain commodity is $p = (8 - \frac{1}{100}x)^2$, where x units are demanded when p dollars is the price per unit, and $x \in [0, 800]$. The total cost function is given by $C(x) = 18x - \frac{1}{100}x^2$, where $C(x)$ dollars is the total cost of producing x units. (a) Find the marginal revenue and marginal cost functions. (b) Find the value of x that yields the maximum total profit.

7. A monopolist determines that if $C(x)$ cents is the total cost of producing x units of a certain commodity, then $C(x) = 20,000 + 25x$. The demand equation is $p = 100 - 0.02x$, where x units are demanded when p cents is the price per unit. If the total profit is to be maximized, find (a) the number of units that should be produced, (b) the price of each unit, and (c) the total profit.

8. Under a monopoly the demand equation for a particular article of merchandise is $x + 2500p = 20,000$ where x units are demanded when p dollars is the price per unit. The number of dollars in the total cost of producing x units is $300 + 4x$. If the total profit is to be as large as possible, find (a) the number of units that should be produced, (b) the price of each unit, and (c) the total profit.

9. Solve Exercise 7 if the government levies a tax on the monopolist of 10 cents per unit produced.

10. Solve Exercise 8 if the government levies a tax on the monopolist of 30 cents per unit produced.

11. For the monopolist of Exercise 7, determine the amount of tax that should be levied by the government on each unit produced in order for the tax revenue received by the government to be maximized.

12. Follow the instructions of Exercise 11 for the monopolist of Exercise 8.

13. Under a monopoly the demand equation for a certain commodity is $p + 2\sqrt{x - 1} = 6$, and the cost function is given by $C(x) = 2x + 1$. (a) Determine the permissible values of x. (b) Find the marginal revenue and marginal cost functions. (c) Find the value of x that yields the maximum profit.

14. Find the maximum tax revenue that can be received by the government if an additive tax for each unit produced is levied on a monopolist for which the demand equation is $x + 3p = 75$, where x units are demanded when p dollars is the price per unit, and $C(x) = 3x + 100$, where $C(x)$ dollars is the total cost of producing x units.

15. Under a monopoly the demand equation for a certain commodity is $x + 2p = 24$, where x units are demanded when p dollars is the price per unit. If $C(x)$ dollars is the total cost of producing x units, then $C(x) = \frac{1}{100}x^3 + \frac{3}{10}x^2$. (a) Find the price per

unit if the total profit is to be maximized. (b) Solve part (a) if the government levies on the monopolist a tax of $2 for each unit produced.

16. The demand equation for a certain commodity produced by a monopolist is $p = a - bx$, where x units are demanded when p dollars is the price of one unit, and the total cost, $C(x)$ dollars, of producing x units is determined by $C(x) = c + dx$. The numbers a, b, c, and d are positive constants. If the government levies a tax on the monopolist of t dollars per unit produced, show that to maximize profits the monopolist should pass on to the consumer only one-half of the tax; that is, the monopolist should increase the unit price by $\frac{1}{2}t$ dollars. *Note:* The demand equation and the total cost function of Example 3 and Exercises 9 and 10 are special cases of this more general situation.

17. The demand equation for a monopolist is $100p = (100 - x)^2$, where x units are demanded when p dollars is the price per unit. The total cost function is given by $C(x) = 55x - \frac{4}{5}x^2$, where $C(x)$ dollars is the total cost of producing x units. (a) Find the price the monopolist should charge to make the greatest total profit. (b) If the government imposes on the monopolist a tax of $9 for each unit produced, determine the price that should be charged for the greatest total profit. (c) Show that the increase in price from part (a) to part (b) is greater than the tax levied by the government.

18. Under a monopoly the demand equation for a certain commodity is

$$10^6 px = 10^9 - 2 \cdot 10^6 x + 18 \cdot 10^3 x^2 - 6x^3$$

where x units are produced per week when p dollars is the price per unit, and $x \geq 100$. The number of dollars in the average cost of producing each unit is given by $Q(x) = \frac{1}{50}x - 24 + 11 \cdot 10^3 x^{-1}$. Find the number of units that should be produced each week and the price of each unit for the weekly profit to be maximized.

Review Exercises for Chapter 5

In Exercises 1 through 4, determine where the graph of the given function is concave upward and where it is concave downward, and find the points of inflection, if there are any.

1. $f(x) = x^3 + 3x^2 + 12x + 10$
2. $g(x) = x^3 - 6x + 2$
3. $g(x) = (x - 1)^{1/3}$
4. $f(x) = \dfrac{x - 1}{x - 2}$

In Exercises 5 through 10, find the relative extrema of f; the points of inflection of the graph of f; the intervals on which f is increasing; the intervals on which f is decreasing; where the graph is concave upward; where the graph is concave downward; and the slope of any inflectional tangent. Draw a sketch of the graph.

5. $f(x) = x^3 + 3x^2 - 4$
6. $f(x) = (x + 2)^{4/3}$
7. $f(x) = (x - 3)^{5/3} + 1$

8. $f(x) = (x - 4)^2(x - 1)$
9. $f(x) = \dfrac{x^2}{x - 3}$
10. $f(x) = (x - 2)^4$

11. The total cost of producing x units of a commodity is $C(x)$ dollars, and $C(x) = 2x^2 + 4x + 32$. Find (a) the average cost function and (b) the marginal cost function. (c) Find the absolute minimum average unit cost. (d) Draw sketches of the total cost, average cost, and marginal cost curves on the same set of axes.

12. The number of dollars in the total cost of producing x radios per day at a certain factory is given by $C(x) = 20x + 1000$. Find (a) the average cost function and (b) the marginal cost function. (c) Show that there is no absolute minimum average unit cost. (d) What is the smallest number of radios that the factory must produce in a day so that the average cost per radio is at most $30? (e) Draw sketches of the total cost, average cost, and marginal cost curves on the same set of axes.

13. The demand equation for a certain article of merchandise is $x = 16 - p^2$, where x units are demanded when p dollars is the price per unit. Find (a) the total revenue function and (b) the marginal revenue function. (c) Draw sketches of the demand, total revenue, and marginal revenue curves on the same set of axes.

14. The total revenue received from the sale of x units of a particular commodity is $R(x)$ dollars, and $R(x) = x\sqrt{72 - x^2}$. Find (a) the demand equation and (b) the marginal revenue function. (c) Draw sketches of the demand, total revenue, and marginal revenue curves on the same set of axes.

15. If $R(x)$ dollars is the total revenue received from the sale of x units of a particular commodity, and $R(x) = 1350x - \frac{1}{2}x^3$, find (a) the demand equation and (b) the marginal revenue function. (c) Find the absolute maximum total revenue. (d) Draw sketches of the demand, total revenue, and marginal revenue curves on the same set of axes.

16. When p dollars is the price per unit, x units of a particular commodity are demanded, and $x = 1000 - 50p^2 - 150p$. (a) If the price per unit is increased from $3 to $3.15, find the relative decrease in the demand. (b) Use the result in part (a) to obtain an approximation to the price elasticity of demand at $p = 3$. (c) Find the exact price elasticity of demand at $p = 3$, and interpret the result.

17. The demand equation for a certain kind of pastry is $p^2 + 50x = 10,000$, where x pastries are demanded when p cents is the price per pastry. (a) Find the price elasticity of demand when the price is 50 cents per pastry. (b) Find the approximate change in the demand if the price of 50 cents is increased by 2 percent.

18. Suppose that x units of a certain article of merchandise are demanded when p dollars is the price per unit, and $p(x + 2) = 200$, where $p \in [1, 10]$. (a) Find the price elasticity of demand when the price is $4 per unit. (b) Find the approximate change in the demand if the price of $4 is decreased by 10 percent.

19. For the demand equation of Exercise 17, determine the values of x for which the demand is (a) elastic, (b) unitary, and (c) inelastic, and show that Eq. (8) of Sec. 5.4 holds. (e) Draw a sketch of the graph of the total revenue function, and indicate where $|\eta| > 1$, $|\eta| = 1$, and $|\eta| < 1$.

20. Follow the instructions of Exercise 19 for the demand equation of Exercise 18.

21. A furniture company operating under perfect competition can sell at a price of $600 per table all the tables it manufactures. If x tables are produced and sold per week, and $C(x)$ dollars is the weekly total cost of production, then $C(x) = 4000 + 60x + 2x^2$. How many tables should be manufactured each week for the company to have the greatest weekly total profit?

22. Under a monopoly the demand equation for a particular decorative item is $x^2 - 375x + 60,000 - 300p = 0$, where x items are demanded when the price is p dollars per item. If $C(x)$ dollars is the total cost of producing x items, then $C(x) = 50x + 100$. (a) Find the marginal revenue and marginal cost functions. (b) Find the value of x that yields the maximum total profit.

23. Under a monopoly the demand equation for a certain commodity is $p = 190 - 0.03x$, where x units are demanded when p cents is the price per unit. If $C(x)$ cents is the total cost of producing x units, then $C(x) = 50,000 + 40x$. If the total profit is to be maximized, find (a) the number of units that should be produced, (b) the price of each unit, and (c) the total profit.

24. Solve Exercise 23 if the government levies a tax on the monopolist of 12 cents per unit produced.

25. For the monopolist of Exercise 23, determine the amount of tax that should be levied by the government on each unit produced for the tax revenue received by the government to be maximized.

26. Under a monopoly the demand equation for a particular article of merchandise is $2x + 5p = 100$, where x units are demanded when p dollars is the price per unit. The total cost of producing x units is $C(x)$ dollars, where $C(x) = \frac{3}{20}x^2 + \frac{1}{100}x^3$. Find the price per unit if the total profit is to be maximized.

27. If the government levies on the monopolist of Exercise 26 a tax of $1 for each unit produced, find the price per unit if the total profit is to be maximized.

28. Refer to Exercise 40 in Review Exercises for Chapter 4. A manufacturer has to supply 10,000 desks per year at a uniform rate. The cost of starting a production run is $180, the excess desks are stored until delivery, and the annual storage cost is $40 per desk. The deliveries are instantaneous, and shortages are not allowed. If $C(x)$ dollars is the total annual cost of obtaining and storing inventory when x desks are produced in each run, draw a sketch of the graph of C.

29. If $f(x) = ax^3 + bx^2$, determine a and b so that the graph of f will have a point of inflection at $(2, 16)$.

30. If $f(x) = ax^3 + bx^2 + cx$, determine a, b, and c so that the graph of f will have a point of inflection at $(1, -1)$ and so that the slope of the inflectional tangent there will be -3.

CHAPTER 6

THE DIFFERENTIAL AND ANTIDIFFERENTIATION

6.1 The differential

In Fig. 6.1.1 an equation of the curve is $y = f(x)$. The line PT is tangent to the curve at $P(x, y)$, Q is the point $(x + \Delta x, y + \Delta y)$, and the directed distance $\overline{MQ}$ is Δy. In the figure Δx and Δy are both positive; however, they could be negative. For a small value of $|\Delta x|$, the slope of the secant line PQ and the slope of the tangent line at P are approximately equal; that is,

$$\frac{\Delta y}{\Delta x} \approx f'(x)$$

$$\Delta y \approx f'(x)\,\Delta x \tag{1}$$

The right side of (1) is defined to be the *differential* of y.

Definition of the differential

> If the function f is defined by $y = f(x)$, then the **differential of y**, denoted by dy, is given by
>
> $$dy = f'(x)\,\Delta x \tag{2}$$
>
> where x is in the domain of f' and Δx is an arbitrary increment of x.

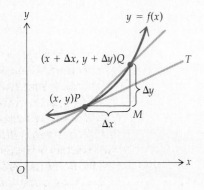

Figure 6.1.1

● ILLUSTRATION 1

If $y = 3x^2 - x$, then $f(x) = 3x^2 - x$; so $f'(x) = 6x - 1$. Therefore, from the definition of the differential,

$$dy = (6x - 1)\,\Delta x$$

In particular, if $x = 2$, then $dy = 11\,\Delta x$.   ●

When $y = f(x)$, the above definition indicates what is meant by dy, the differential of the dependent variable. We also wish to define the differential of the independent variable, or dx. To arrive at a suitable definition for dx that is consistent with the definition of dy, we consider the identity function, which is the function f defined by $f(x) = x$. Then $f'(x) = 1$ and $y = x$; so $dy = 1 \cdot \Delta x = \Delta x$. Because $y = x$, we want dx to be equal to dy for this particular function; that is, for this function we want $dx = \Delta x$. It is this reasoning that leads to the following definition.

Definition of the differential of the independent variable

> If the function f is defined by $y = f(x)$, then the **differential of x**, denoted by dx, is given by
>
> $$dx = \Delta x \tag{3}$$
>
> where Δx is an arbitrary increment of x, and x is any number in the domain of f'.

From (2) and (3),

$$dy = f'(x)\, dx \tag{4}$$

Dividing on both sides of the above by dx we have

$$\frac{dy}{dx} = f'(x) \qquad \text{if } dx \neq 0 \tag{5}$$

Equation (5) expresses the derivative as the quotient of two differentials. Recall that the notation $\dfrac{dy}{dx}$ for a derivative was introduced in Sec. 2.4, at which time dy and dx had not been given independent meaning.

EXAMPLE 1 Given $y = 4x^2 - 3x + 1$, find Δy, dy, and $\Delta y - dy$ for (a) any x and Δx; (b) $x = 2, \Delta x = 0.1$; (c) $x = 2, \Delta x = 0.01$; (d) $x = 2, \Delta x = 0.001$.

Solution   (a) Because $y = 4x^2 - 3x + 1$, then

$$y + \Delta y = 4(x + \Delta x)^2 - 3(x + \Delta x) + 1$$

$$(4x^2 - 3x + 1) + \Delta y = 4x^2 + 8x\,\Delta x + 4(\Delta x)^2 - 3x - 3\,\Delta x + 1$$

$$\Delta y = (8x - 3)\,\Delta x + 4(\Delta x)^2$$

Also, if $y = f(x)$,

$$dy = f'(x)\, dx$$

Thus

$$dy = (8x - 3)\, dx$$
$$= (8x - 3)\,\Delta x$$

$$\Delta y - dy = 4(\Delta x)^2$$

The results for parts (b), (c), and (d) are given in Table 6.1.1, where $\Delta y = (8x - 3)\,\Delta x + 4(\Delta x)^2$ and $dy = (8x - 3)\,\Delta x$.

Table 6.1.1

x	Δx	Δy	dy	$\Delta y - dy$
2	0.1	1.34	1.3	0.04
2	0.01	0.1304	0.13	0.0004
2	0.001	0.013004	0.013	0.000004

Note from Table 6.1.1 that the closer Δx is to zero, the smaller is the difference between Δy and dy. Furthermore, observe that for each value of Δx, the corresponding value of $\Delta y - dy$ is smaller than the value of Δx. More generally, dy is an approximation of Δy when Δx is small, and the approximation is of better accuracy than the size of Δx.

For a fixed value of x, say x_0,

$$dy = f'(x_0)\, dx$$

That is, dy is a linear function of dx; consequently, dy is usually easier to compute than Δy (this was seen in Example 1). Because $f(x_0 + \Delta x) - f(x_0) = \Delta y$,

$$f(x_0 + \Delta x) = f(x_0) + \Delta y$$

Thus

$$f(x_0 + \Delta x) \approx f(x_0) + dy \tag{6}$$

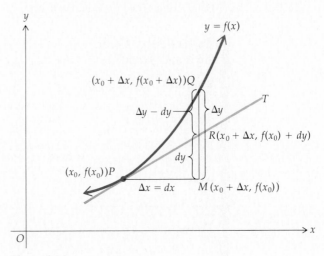

Figure 6.1.2

Our results are illustrated in Fig. 6.1.2. The equation of the curve in the figure is $y = f(x)$, and the graph is concave upward. The line PT is tangent to the curve at $P(x_0, f(x_0))$; Δx and dx are equal and are represented by the directed distance $\overline{PM}$, where M is the point $(x_0 + \Delta x, f(x_0))$. We let Q be the point $(x_0 + \Delta x, f(x_0 + \Delta x))$, and the directed distance $\overline{MQ}$ is Δy or, equivalently, $f(x_0 + \Delta x) - f(x_0)$. The slope of PT is $f'(x) = dy/dx$. Also, the slope of PT is $\overline{MR}/\overline{PM}$, and because $\overline{PM} = dx$, we have $dy = \overline{MR}$ and $\overline{RQ} = \Delta y - dy$. Note that the smaller the value of dx (i.e., the closer the point Q is to the point P), then the smaller will be the value of $\Delta y - dy$ (i.e., the smaller will be the length of the line segment RQ). An equation of the tangent line PT is

$$y = f(x_0) + f'(x_0)(x - x_0)$$

Thus, if $\bar{y}$ is the ordinate of R, then

$$\bar{y} = f(x_0) + dy \tag{7}$$

By comparing (6) and (7), observe that when using $f(x_0) + dy$ to approximate the value of $f(x_0 + \Delta x)$, we are approximating the ordinate of the point $Q(x_0 + \Delta x, f(x_0 + \Delta x))$ on the curve by the ordinate of the point $R(x_0 + \Delta x, f(x_0) + dy)$ on the line that is tangent to the curve at $P(x_0, f(x_0))$.

In Fig. 6.1.2 the graph is concave upward and the function is increasing. In Exercises 1 and 2 you are asked to draw similar figures. In Exercise 1 the graph is to be concave downward and the function increasing, while in Exercise 2 the graph is to be concave upward and the function decreasing.

The calculations in Examples 2 and 3 are intended to illustrate the concept of the differential.

EXAMPLE 2 Use differentials to find an approximate value for $\sqrt[3]{28}$.

Solution Consider the function f defined by $f(x) = \sqrt[3]{x}$, and let $y = f(x)$. Hence

$$y = \sqrt[3]{x}$$

and

$$dy = f'(x)\,dx$$

$$= \frac{1}{3x^{2/3}}\,dx$$

The nearest perfect cube to 28 is 27. Thus we compute dy with $x = 27$ and $dx = \Delta x = 1$.

$$dy = \frac{1}{3(27)^{2/3}}\,(1) = \frac{1}{27}$$

By applying formula (6) with $x_0 = 27$, $\Delta x = 1$, and $dy = \frac{1}{27}$,

$$f(27 + 1) \approx f(27) + \tfrac{1}{27}$$

$$\sqrt[3]{27 + 1} \approx \sqrt[3]{27} + \tfrac{1}{27}$$

$$\sqrt[3]{28} \approx 3 + \tfrac{1}{27}$$

Therefore $\sqrt[3]{28} \approx 3.037$.

From Table 1 in the back of the book, $\sqrt[3]{28} = 3.037$. Thus the approximation is accurate to three decimal places.

EXAMPLE 3 A paperweight is in the shape of a hollow spherical ball having an inner radius of 6 cm and a thickness of $\frac{1}{3}$ cm. If it is made of metal costing 25 cents per cubic centimeter, use differentials to find the approximate cost of the metal to be used in the manufacture of the paperweight.

Solution We consider the volume of the spherical shell as an increment of the volume of a sphere. Let r be the number of centimeters in the radius of a sphere; V be the number of cubic centimeters in the volume of a sphere; and ΔV be the number of cubic centimeters in the volume of a spherical shell.

$$V = \tfrac{4}{3}\pi r^3$$

$$dV = 4\pi r^2\,dr$$

Substituting $r = 6$ and $dr = \Delta r = \frac{1}{3}$ into the above, we obtain

$$dV = 4\pi(6)^2\tfrac{1}{3}$$
$$= 48\pi$$

Thus

$$\Delta V \approx 48\pi$$

If C dollars is the total cost of the metal used in the manufacture of the paperweight, then

$$C \approx (0.25)(48\pi) = 12\pi = 12(3.1416) = 37.70$$

Therefore the total cost of the metal is approximately \$37.70.

EXAMPLE 4   For the company of Example 5, Sec. 4.5 use differentials to find the approximate change in the total monthly cost of obtaining and storing inventory if a production run of 1000 items is increased to 1010 items.

Solution   If C dollars is the total monthly cost when x items are produced in a run, then from the solution of Example 5, Sec. 4.5, we have

$$C = \frac{360,000}{x} + \frac{x}{4}$$

Thus

$$dC = \frac{dC}{dx} \cdot dx = \left(-\frac{360,000}{x^2} + \frac{1}{4} \right) dx$$

When $x = 1000$ and $dx = \Delta x = 10$, we obtain from the above

$$dC = \left(-\frac{360,000}{1,000,000} + \frac{1}{4} \right) 10$$
$$= (-\tfrac{9}{25} + \tfrac{1}{4})10$$
$$= -1.1$$

Therefore

$$\Delta C \approx -1.1$$

We conclude that when a production run of 1000 items is increased to 1010 items, the total monthly cost of obtaining and storing inventory is decreased by approximately \$1.10.

● ILLUSTRATION 2

In Illustration 1 of Sec. 3.1, $C(x)$ is the number of dollars in the total cost of manufacturing x toys, and

$$C(x) = 110 + 4x + 0.02x^2$$

We showed in Sec. 3.1 that an approximation to $C(51) - C(50)$, which is the number of dollars in the cost of producing the fifty-first toy, is given by $C'(50)$, the value of the marginal cost function at 50. This fact can be proved by using the concept of the differential. Let

$$y = C(x)$$

Then

$$\Delta y = C(x + \Delta x) - C(x) \tag{8}$$

and

$$dy = C'(x)\, dx \tag{9}$$

If $x = 50$, and $dx = \Delta x = 1$, then from (8) we have

$$\Delta y = C(51) - C(50)$$

and from (9) we have

$$dy = [C'(50)](1) = C'(50)$$

Because dy is an approximation of Δy, we can conclude that $C'(50)$ is an approximation of $C(51) - C(50)$. ●

    Formulas for computing derivatives were obtained in Chapter 2. These formulas are now stated with Leibniz notation along with a corresponding formula for the differential. In these formulas, u and v are functions of x, and it is understood that the formulas hold if $\dfrac{du}{dx}$ and $\dfrac{dv}{dx}$ exist. When c appears, it is a constant.

$$\text{I} \quad \frac{d(c)}{dx} = 0 \qquad\qquad\qquad \text{I}' \ \ d(c) = 0$$

$$\text{II} \quad \frac{d(x^n)}{dx} = nx^{n-1} \qquad\qquad \text{II}' \ \ d(x^n) = nx^{n-1}\, dx$$

$$\text{III} \quad \frac{d(cu)}{dx} = c\,\frac{du}{dx} \qquad\qquad \text{III}' \ \ d(cu) = c\, du$$

$$\text{IV} \quad \frac{d(u+v)}{dx} = \frac{du}{dx} + \frac{dv}{dx} \qquad \text{IV}' \ \ d(u+v) = du + dv$$

$$\text{V} \quad \frac{d(uv)}{dx} = u\,\frac{dv}{dx} + v\,\frac{du}{dx} \qquad \text{V}' \ \ d(uv) = u\, dv + v\, du$$

$$\text{VI} \quad \frac{d\left(\dfrac{u}{v}\right)}{dx} = \frac{v\,\dfrac{du}{dx} - u\,\dfrac{dv}{dx}}{v^2} \qquad \text{VI}' \ \ d\left(\frac{u}{v}\right) = \frac{v\, du - u\, dv}{v^2}$$

$$\text{VII} \quad \frac{d(u^n)}{dx} = nu^{n-1}\,\frac{du}{dx} \qquad \text{VII}' \ \ d(u^n) = nu^{n-1}\, du$$

    The operation of differentiation is extended to include the process of finding the differential as well as finding the derivative. If $y = f(x)$, dy can be found either by applying formulas I′–VII′ or by finding $f'(x)$ and multiplying it by dx.

EXAMPLE 5   Given

$$y = \frac{\sqrt{x^2 + 1}}{2x + 1}$$

find dy.

Solution   Applying formula VI′ we obtain

$$dy = \frac{(2x + 1)\, d(\sqrt{x^2 + 1}) - \sqrt{x^2 + 1}\, d(2x + 1)}{(2x + 1)^2} \tag{10}$$

From formula VII′,

$$d(\sqrt{x^2 + 1}) = \tfrac{1}{2}(x^2 + 1)^{-1/2}\, 2x\, dx = x(x^2 + 1)^{-1/2}\, dx \tag{11}$$

and

$$d(2x + 1) = 2\, dx \tag{12}$$

Substituting values from (11) and (12) into (10) we get

$$dy = \frac{x(2x + 1)(x^2 + 1)^{-1/2}\, dx - 2(x^2 + 1)^{1/2}\, dx}{(2x + 1)^2}$$

$$= \frac{(2x^2 + x)\, dx - 2(x^2 + 1)\, dx}{(2x + 1)^2 (x^2 + 1)^{1/2}}$$

$$= \frac{x - 2}{(2x + 1)^2 \sqrt{x^2 + 1}}\, dx$$

Exercises 6.1

1. Draw a figure similar to Fig. 6.1.2 if the graph is concave downward and the function is increasing. Indicate the line segments whose lengths represent the following quantities: Δx, Δy, dx, and dy.

2. Draw a figure similar to Fig. 6.1.2 if the graph is concave upward, the function is decreasing, and $\Delta x < 0$. Indicate the line segments whose lengths represent the following quantities: $|\Delta x|$, Δy, $|dx|$, and dy.

In Exercises 3 through 10, find (a) Δy; (b) dy; (c) $\Delta y - dy$.

3. $y = x^2$
4. $y = 3x^2 - x$
5. $y = 6 - 3x - 2x^2$
6. $y = x^3 - x^2$

7. $y = \sqrt{x}$
8. $y = \dfrac{1}{x^2 + 1}$
9. $y = \dfrac{2}{x - 1}$
10. $y = \dfrac{1}{\sqrt{x}}$

In Exercises 11 through 16, find, for the given values, (a) Δy; (b) dy; (c) $\Delta y - dy$.

11. $y = x^2 - 3x$; $x = 2$; $\Delta x = 0.03$
12. $y = x^2 - 3x$; $x = -1$; $\Delta x = 0.02$
13. $y = \dfrac{1}{x}$; $x = -2$; $\Delta x = -0.1$

14. $y = \dfrac{1}{x}$; $x = 3$; $\Delta x = -0.2$
15. $y = x^3 + 1$; $x = 1$; $\Delta x = -0.5$
16. $y = x^3 + 1$; $x = -1$; $\Delta x = 0.1$

In Exercises 17 through 22, find dy.

17. $y = (3x^2 - 2x + 1)^3$ **18.** $y = \sqrt{4 - x^2}$ **19.** $y = x^2 \sqrt[3]{2x + 3}$

20. $y = \dfrac{3x}{x^2 + 2}$ **21.** $y = \sqrt{\dfrac{x - 1}{x + 1}}$ **22.** $y = \sqrt{3x + 4}\,\sqrt[3]{x^2 - 1}$

In Exercises 23 through 26, use differentials to find an approximate value for the given quantity. Express each answer to three significant digits.

23. $\sqrt{82}$ **24.** $\sqrt[4]{82}$ **25.** $\sqrt[3]{7.5}$ **26.** $\sqrt[3]{0.0098}$

27. The measurement of an edge of a cube is found to be 15 cm with a possible error of 0.01 cm. Using differentials find the approximate error in computing from this measurement (a) the volume; (b) the area of one of the faces.

28. A metal box in the form of a cube is to have an interior volume of 1000 cm^3. The six sides are to be made of metal $\frac{1}{2}$ cm thick. If the cost of the metal to be used is 20 cents per cubic centimeter, use differentials to find the approximate total cost of the metal to be used in the manufacture of the box.

29. An open cylindrical tank is to have an outside coating of thickness 2 cm. If the inner radius is 6 m and the altitude is 10 m, find by differentials the approximate amount of coating material to be used.

30. The stem of a particular mushroom is cylindrical in shape, and a stem of height 2 cm and radius r cm has a volume of V cm^3, where $V = 2\pi r^2$. Use the differential to find the approximate increase in the volume of the stem when the radius increases from 0.4 cm to 0.5 cm.

31. A burn on a person's skin is in the shape of a circle such that if r cm is the radius and A cm^2 is the area of the burn, then $A = \pi r^2$. Use the differential to find the approximate decrease in the area of the burn when the radius decreases from 1 cm to 0.8 cm.

32. A certain bacterial cell is spherical in shape such that if r micrometers (μm) is its radius and V cubic micrometers is its volume, then $V = \frac{4}{3}\pi r^3$. Use the differential to find the approximate increase in the volume of the cell when the radius increases from 2.2 μm to 2.3 μm.

33. A tumor in a person's body is spherical in shape such that if r cm is the radius and V cm^3 is the volume of the tumor, then $V = \frac{4}{3}\pi r^3$. Use the differential to find the approximate increase in the volume of the tumor when the radius increases from 1.5 cm to 1.6 cm.

34. A contractor agrees to paint on both sides of 1000 circular signs each of radius 3 m. Upon receiving the signs it is discovered that the radius is 1 cm too large. Use differentials to find the approximate percent increase of paint that will be needed.

35. If the possible error in the measurement of the volume of a gas is 0.1 ft^3 and the allowable error in the pressure is $0.001C$ lb/ft^2, find the size of the smallest container for which Boyle's law (Exercise 29 in Exercises 3.5) holds.

36. For the adiabatic law for the expansion of air (Exercise 30 in Exercises 3.5), prove that $dP/P = -1.4\, dV/V$.

37. Solve Example 4 of this section if the number of items in a production run is increased from 1400 to 1410.

38. For the manufacturer of Exercise 31 in Exercises 4.5, use differentials to find the approximate change in the total cost of obtaining and storing inventory if the number of tables in a production run is increased from 100 to 105.

39. For the distributor of Exercise 33 in Exercises 4.5, use differentials to find the approximate change in the total cost of maintaining an inventory if the number of cases in an order is increased from 800 to 810.

40. For the appliance store of Exercise 34 in Exercises 4.5, use differentials to find the approximate change in the total cost of maintaining an inventory if the number of television sets in an order is increased from 80 to 84.

6.2 Antidifferentiation

You are already familiar with *inverse operations*. Addition and subtraction are inverse operations; multiplication and division are inverse operations, as are raising to powers and extracting roots. The inverse operation of differentiation is called **antidifferentiation**. Suppose, for example, that a total cost function C is known; then the marginal cost function is obtained from

C by finding C', the derivative of C. If, however, the marginal cost function C' is given and it is desired to find the total cost function C, the procedure involves antidifferentiation, and the function found from this operation is called an *antiderivative*. In this section we develop some techniques for computing antiderivatives, and in Secs. 6.3 and 6.4 we apply these techniques to solving some practical problems.

<table>
<tr>
<td>Definition of an antiderivative
of a function</td>
<td>A function F is called an **antiderivative** of a function f on an interval I if $F'(x) = f(x)$ for every value of x in I.</td>
</tr>
</table>

● ILLUSTRATION 1

If F is defined by $F(x) = 4x^3 + x^2 + 5$, then $F'(x) = 12x^2 + 2x$. Thus if f is the function defined by $f(x) = 12x^2 + 2x$, we state that f is the derivative of F and that F is an antiderivative of f. If G is the function defined by $G(x) = 4x^3 + x^2 - 17$, then G is also an antiderivative of f because $G'(x) = 12x^2 + 2x$. Actually, any function whose function value is given by $4x^3 + x^2 + C$, where C is any constant, is an antiderivative of f.   ●

In general, if a function F is an antiderivative of a function f on an interval I, and if G is defined by

$$G(x) = F(x) + C$$

where C is an arbitrary constant, then

$$G'(x) = F'(x) = f(x)$$

and G is also an antiderivative of f on the interval I.

We now proceed to show that if F is any particular antiderivative of f on an interval I, then the set of all antiderivatives of f on I is defined by $F(x) + C$, where C is an arbitrary constant. First, two preliminary theorems are needed.

<table>
<tr>
<td>**Theorem 6.2.1**   If f is a function, I is an interval, and

$\quad f'(x) = 0 \qquad$ for all x in I

then f is constant on I; that is, there is a constant K such that

$\quad f(x) = K \qquad$ for all x in I</td>
</tr>
</table>

A formal proof of Theorem 6.2.1 is omitted. However, the following informal geometric argument should appeal to your intuition.

Because $f'(x) = 0$ for all x in the interval I, then for the graph of the equation $y = f(x)$, the slope of the tangent line is zero everywhere in I. Thus the tangent line is horizontal everywhere in I, and for this situation to occur, the graph of $y = f(x)$ in I must be a horizontal line. Therefore $f(x) = K$ in I, where K is a constant.

Suppose now that g and h are two functions having the same derivative on an interval I; that is,

$$g'(x) = h'(x) \qquad \text{for all } x \text{ in } I \tag{1}$$

Let f be the function defined on I by

$$f(x) = g(x) - h(x)$$

so that for all values of x in I

$$f'(x) = g'(x) - h'(x)$$

But from (1), $g'(x) = h'(x)$ for all values of x in I. Therefore

$$f'(x) = 0 \qquad \text{for all } x \text{ in } I$$

Thus Theorem 6.2.1 applies to the function f, and there is a constant K such that

$$f(x) = K \qquad \text{for all } x \text{ in } I$$

Replacing $f(x)$ by $g(x) - h(x)$ we have

$$g(x) = h(x) + K \qquad \text{for all } x \text{ in } I$$

We have proved the following theorem.

Theorem 6.2.2 If g and h are two functions such that $g'(x) = h'(x)$ for all values of x in an interval I, then there is a constant K such that

$$g(x) = h(x) + K \qquad \text{for all } x \text{ in } I$$

We are now in a position to obtain the significant theorem of this section.

Let F be a particular antiderivative of f on an interval I. Then

$$F'(x) = f(x) \qquad \text{for all } x \text{ in } I \tag{2}$$

If G represents any antiderivative of f on I, then

$$G'(x) = f(x) \qquad \text{for all } x \text{ in } I \tag{3}$$

From (2) and (3) it follows that

$$G'(x) = F'(x) \qquad \text{for all } x \text{ in } I$$

Therefore, from Theorem 6.2.2, there is a constant K such that

$$G(x) = F(x) + K \qquad \text{for all } x \text{ in } I$$

Because G represents any antiderivative of f on I, it follows that all antiderivatives of f can be obtained from $F(x) + C$, where C is an arbitrary constant. We have proved the following theorem.

> **Theorem 6.2.3**   If F is a particular antiderivative of f on an interval I, then the set of all antiderivatives of f on I is given by
>
> $$F(x) + C \qquad\qquad\qquad (4)$$
>
> where C is an arbitrary constant, and all antiderivatives of f on I can be obtained from (4) by assigning particular values to C.

If F is an antiderivative of f, then $F'(x) = f(x)$, and so

$$d(F(x)) = f(x)\,dx$$

Antidifferentiation is the process of finding the set of all antiderivatives of a given function. The symbol

$$\int$$

denotes the operation of antidifferentiation, and we write

$$\int f(x)\,dx = F(x) + C \qquad\qquad\qquad (5)$$

where

$$F'(x) = f(x)$$

or, equivalently,

$$d(F(x)) = f(x)\,dx \qquad\qquad\qquad (6)$$

From (5) and (6) we can write

$$\int d(F(x)) = F(x) + C \qquad\qquad\qquad (7)$$

Equation (7) states that when we antidifferentiate the differential of a function, we obtain that function plus an arbitrary constant. So we can think of the $\int$ symbol for antidifferentiation as meaning that operation which is the inverse of the operation denoted by d for finding the differential.

Because antidifferentiation is the inverse operation of differentiation, antidifferentiation theorems can be obtained from those on differentiation. Thus the following theorems can be proved from the corresponding ones for differentiation.

> **Theorem 6.2.4**
>
> $$\int dx = x + C$$

Theorem 6.2.5

$$\int af(x)\,dx = a \int f(x)\,dx$$

where a is a constant.

Theorem 6.2.5 states that to find an antiderivative of a constant times a function, find an antiderivative of the function and multiply it by the constant.

Theorem 6.2.6 If f_1 and f_2 are defined on the same interval, then

$$\int [f_1(x) + f_2(x)]\,dx = \int f_1(x)\,dx + \int f_2(x)\,dx$$

Theorem 6.2.6 states that to find an antiderivative of the sum of two functions, find an antiderivative of each of the functions separately and then add the results, it being understood that both functions are defined on the same interval. Theorem 6.2.6 can be extended to any finite number of functions. Combining Theorem 6.2.6 with Theorem 6.2.5, we have the following theorem.

Theorem 6.2.7 If $f_1, f_2, \ldots, f_n$ are defined on the same interval,

$$\int [c_1 f_1(x) + c_2 f_2(x) + \ldots + c_n f_n(x)]\,dx$$

$$= c_1 \int f_1(x)\,dx + c_2 \int f_2(x)\,dx + \ldots + c_n \int f_n(x)\,dx$$

where $c_1, c_2, \ldots, c_n$ are constants.

Theorem 6.2.8

$$\int x^n\,dx = \frac{x^{n+1}}{n+1} + C \qquad \text{if } n \neq -1$$

As stated above, these theorems follow from the corresponding theorems for finding the differential. Following is the proof of Theorem 6.2.8.

$$d\left(\frac{x^{n+1}}{n+1} + C\right) = \frac{(n+1)x^n}{n+1}\,dx$$

$$= x^n\,dx$$

Applications of the above theorems are illustrated in the following examples.

EXAMPLE 1   Evaluate $\displaystyle\int (3x + 5)\, dx$

Solution

$$\int (3x + 5)\, dx = 3 \int x\, dx + 5 \int dx \quad \text{(by Theorem 6.2.7)}$$

$$= 3\left(\frac{x^2}{2} + C_1\right) + 5(x + C_2) \quad \text{(by Theorems 6.2.8 and 6.2.4)}$$

$$= \tfrac{3}{2}x^2 + 5x + 3C_1 + 5C_2$$

Because $3C_1 + 5C_2$ is an arbitrary constant, it may be denoted by C, and so our answer is

$$\tfrac{3}{2}x^2 + 5x + C$$

This answer may be checked by finding the derivative. Doing this we have

$$D_x(\tfrac{3}{2}x^2 + 5x + C) = 3x + 5$$

EXAMPLE 2   Evaluate $\displaystyle\int \sqrt[3]{x^2}\, dx$

Solution

$$\int \sqrt[3]{x^2}\, dx = \int x^{2/3}\, dx$$

$$= \frac{x^{2/3 + 1}}{\frac{2}{3} + 1} + C \quad \text{(from Theorem 6.2.8)}$$

$$= \tfrac{3}{5}x^{5/3} + C$$

EXAMPLE 3   Evaluate $\displaystyle\int \left(\frac{1}{x^4} + \frac{1}{\sqrt[4]{x}}\right) dx$

Solution

$$\int \left(\frac{1}{x^4} + \frac{1}{\sqrt[4]{x}}\right) dx = \int (x^{-4} + x^{-1/4})\, dx$$

$$= \frac{x^{-4 + 1}}{-4 + 1} + \frac{x^{-1/4 + 1}}{-\frac{1}{4} + 1} + C$$

$$= \frac{x^{-3}}{-3} + \frac{x^{3/4}}{\frac{3}{4}} + C$$

$$= -\frac{1}{3x^3} + \frac{4}{3}x^{3/4} + C$$

Many antiderivatives cannot be found directly by applying formulas. However, sometimes it is possible to find an antiderivative by the formulas after changing the variable.

● ILLUSTRATION 2

Suppose that we wish to find

$$\int 2x\sqrt{1 + x^2}\, dx \tag{8}$$

If we make the substitution $u = 1 + x^2$, then $du = 2x\, dx$, and (8) becomes

$$\int u^{1/2}\, du$$

which by Theorem 6.2.8 gives

$$\tfrac{2}{3}u^{3/2} + C$$

Then, replacing u by $(1 + x^2)$, we have as our result

$$\tfrac{2}{3}(1 + x^2)^{3/2} + C$$ ●

Justification of the procedure used in Illustration 2 is provided by the following theorem, which is analogous to the chain rule for differentiation, and hence may be called the *chain rule for antidifferentiation*.

The chain rule for antidifferentiation

Theorem 6.2.9 Let g be a differentiable function of x, and let the range of g be an interval I. Suppose that f is a function defined on I and that F is an antiderivative of f on I. Then if $u = g(x)$,

$$\int f(g(x))g'(x)\, dx = \int f(u)\, du = F(u) + C = F(g(x)) + C$$

The proof of Theorem 6.2.9 is omitted. As a particular case of Theorem 6.2.9, from Theorem 6.2.8 we have the *generalized power formula for antiderivatives*, which we now state.

The generalized power formula for antiderivatives

Theorem 6.2.10 If g is a differentiable function, then if $u = g(x)$,

$$\int [g(x)]^n g'(x)\, dx = \int u^n\, du = \frac{u^{n+1}}{n + 1} + C = \frac{[g(x)]^{n+1}}{n + 1} + C$$

where $n \neq -1$.

Examples 4, 5, and 6 illustrate the application of Theorem 6.2.10.

EXAMPLE 4 Evaluate $\displaystyle\int \sqrt{3x + 4}\, dx$

Solution To apply Theorem 6.2.10 we make the substitution $u = 3x + 4$; then $du = 3\, dx$, or $\frac{1}{3}\, du = dx$. Hence

$$\int \sqrt{3x + 4}\, dx = \int u^{1/2} \cdot \frac{du}{3}$$

$$= \frac{1}{3} \int u^{1/2}\, du$$

$$= \frac{1}{3} \cdot \frac{u^{3/2}}{\frac{3}{2}} + C$$

$$= \tfrac{2}{9} u^{3/2} + C$$

$$= \tfrac{2}{9}(3x + 4)^{3/2} + C$$

EXAMPLE 5 Evaluate $\displaystyle\int t(5 + 3t^2)^8\, dt$

Solution Observe that $d(5 + 3t^2) = 6t\, dt$. Thus we make the substitution

$$u = 5 + 3t^2$$

and then

$$du = 6t\, dt$$

$$\tfrac{1}{6}\, du = t\, dt$$

Therefore

$$\int t(5 + 3t^2)^8\, dt = \int (5 + 3t^2)^8 (t\, dt)$$

$$= \int u^8 (\tfrac{1}{6}\, du)$$

$$= \frac{1}{6} \int u^8\, du$$

$$= \frac{1}{6} \cdot \frac{u^9}{9} + C$$

$$= \tfrac{1}{54}(5 + 3t^2)^9 + C$$

EXAMPLE 6 Evaluate $\displaystyle\int x^2 \sqrt[5]{7 - 4x^3}\, dx$

Solution Because $d(7 - 4x^3) = -12x^2\, dx$, we make the substitution

$$u = 7 - 4x^3$$

Then

$$du = -12x^2 \, dx$$

$$-\tfrac{1}{12} \, du = x^2 \, dx$$

We have, therefore,

$$\int x^2 \sqrt[5]{7 - 4x^3} \, dx = \int (7 - 4x^3)^{1/5} (x^2 \, dx)$$

$$= \int u^{1/5} (-\tfrac{1}{12} \, du)$$

$$= -\frac{1}{12} \int u^{1/5} \, du$$

$$= -\frac{1}{12} \cdot \frac{u^{6/5}}{\frac{6}{5}} + C$$

$$= -\tfrac{5}{72} (7 - 4x^3)^{6/5} + C$$

The details of the solutions of Examples 4, 5, and 6 can be shortened by not explicitly stating the substitution of u. The solution of Example 4 takes the following form:

$$\int \sqrt{3x + 4} \, dx = \frac{1}{3} \int (3x + 4)^{1/2} (3 \, dx)$$

$$= \frac{1}{3} \cdot \frac{(3x + 4)^{3/2}}{\frac{3}{2}} + C$$

$$= \tfrac{2}{9} (3x + 4)^{3/2} + C$$

The solution of Example 5 can be written as

$$\int t(5 + 3t^2)^8 \, dt = \frac{1}{6} \int (5 + 3t^2)^8 (6t \, dt)$$

$$= \frac{1}{6} \cdot \frac{(5 + 3t^2)^9}{9} + C$$

$$= \tfrac{1}{54} (5 + 3t^2)^9 + C$$

and the solution of Example 6 can be shortened as follows:

$$\int x^2 \sqrt[5]{7 - 4x^3} \, dx = -\frac{1}{12} \int (7 - 4x^3)^{1/5} (-12x^2 \, dx)$$

$$= -\frac{1}{12} \cdot \frac{(7 - 4x^3)^{6/5}}{\frac{6}{5}} + C$$

$$= -\tfrac{5}{72} (7 - 4x^3)^{6/5} + C$$

EXAMPLE 7 Evaluate $\displaystyle\int \frac{4x^2\,dx}{(1-8x^3)^4}$

Solution Because $d(1-8x^3) = -24x^2\,dx$, we write

$$\int \frac{4x^2\,dx}{(1-8x^3)^4} = 4\int (1-8x^3)^{-4}(x^2\,dx)$$

$$= 4\left(-\frac{1}{24}\right)\int (1-8x^3)^{-4}(-24x^2\,dx)$$

$$= -\frac{1}{6}\cdot\frac{(1-8x^3)^{-3}}{-3} + C$$

$$= \frac{1}{18(1-8x^3)^3} + C$$

EXAMPLE 8 Evaluate $\displaystyle\int x^2\sqrt{1+x}\,dx$

Solution Let $u = 1 + x$. Then $x = u - 1$ and $dx = du$. Making these substitutions we have

$$\int x^2\sqrt{1+x}\,dx = \int (u-1)^2 u^{1/2}\,du$$

$$= \int (u^2 - 2u + 1)u^{1/2}\,du$$

$$= \int u^{5/2}\,du - 2\int u^{3/2}\,du + \int u^{1/2}\,du$$

$$= \frac{u^{7/2}}{\frac{7}{2}} - 2\cdot\frac{u^{5/2}}{\frac{5}{2}} + \frac{u^{3/2}}{\frac{3}{2}} + C$$

$$= \tfrac{2}{7}(1+x)^{7/2} - \tfrac{4}{5}(1+x)^{5/2} + \tfrac{2}{3}(1+x)^{3/2} + C$$

● ILLUSTRATION 3

An alternate method for the solution of Example 8 is to make the substitution $v = \sqrt{1+x}$. Then $v^2 = 1 + x$, and therefore $x = v^2 - 1$ and $dx = 2v\,dv$. With these substitutions the computation takes the following form:

$$\int x^2\sqrt{1+x}\,dx = \int (v^2-1)^2 \cdot v \cdot (2v\,dv)$$

$$= 2\int v^6\,dv - 4\int v^4\,dv + 2\int v^2\,dv$$

$$= \tfrac{2}{7}v^7 - \tfrac{4}{5}v^5 + \tfrac{2}{3}v^3 + C$$

$$= \tfrac{2}{7}(1+x)^{7/2} - \tfrac{4}{5}(1+x)^{5/2} + \tfrac{2}{3}(1+x)^{3/2} + C$$ ●

The results for each of the above examples can be checked by finding the derivative (or the differential) of the answer.

● ILLUSTRATION 4

In Example 5 we have

$$\int t(5 + 3t^2)^8 \, dt = \tfrac{1}{54}(5 + 3t^2)^9 + C$$

Checking by differentiation gives

$$D_t[\tfrac{1}{54}(5 + 3t^2)^9] = \tfrac{1}{54} \cdot 9(5 + 3t^2)^8 \cdot 6t$$
$$= t(5 + 3t^2)^8$$

● ILLUSTRATION 5

In Example 8 we have

$$\int x^2 \sqrt{1 + x} \, dx = \tfrac{2}{7}(1 + x)^{7/2} - \tfrac{4}{5}(1 + x)^{5/2} + \tfrac{2}{3}(1 + x)^{3/2} + C$$

Checking by differentiation gives

$$D_x[\tfrac{2}{7}(1 + x)^{7/2} - \tfrac{4}{5}(1 + x)^{5/2} + \tfrac{2}{3}(1 + x)^{3/2}]$$
$$= (1 + x)^{5/2} - 2(1 + x)^{3/2} + (1 + x)^{1/2}$$
$$= (1 + x)^{1/2}[(1 + x)^2 - 2(1 + x) + 1]$$
$$= (1 + x)^{1/2}[1 + 2x + x^2 - 2 - 2x + 1]$$
$$= x^2 \sqrt{1 + x}$$

Exercises 6.2

In Exercises 1 through 46, perform the antidifferentiation. In Exercises 1 through 4, 11 through 14, and 21 through 24, check by finding the derivative of your answer.

1. $\displaystyle\int 3x^4 \, dx$

2. $\displaystyle\int (3x^5 - 2x^3) \, dx$

3. $\displaystyle\int (3 - 2t + t^2) \, dt$

4. $\displaystyle\int (4x^3 - 3x^2 + 6x - 1) \, dx$

5. $\displaystyle\int (8x^4 + 4x^3 - 6x^2 - 4x + 5) \, dx$

6. $\displaystyle\int (2 + 3x^2 - 8x^3) \, dx$

7. $\displaystyle\int \sqrt{x}(x + 1) \, dx$

8. $\displaystyle\int (ax^2 + bx + c) \, dx$

9. $\displaystyle\int (x^{3/2} - x) \, dx$

10. $\displaystyle\int \left(\sqrt{x} - \dfrac{1}{\sqrt{x}}\right) dx$

11. $\displaystyle\int \left(\dfrac{2}{x^3} + \dfrac{3}{x^2} + 5\right) dx$

12. $\displaystyle\int \left(3 - \dfrac{1}{x^4} + \dfrac{1}{x^2}\right) dx$

13. $\displaystyle\int \dfrac{x^2 + 4x - 4}{\sqrt{x}} \, dx$

14. $\displaystyle\int \dfrac{y^4 + 2y^2 - 1}{\sqrt{y}} \, dy$

15. $\displaystyle\int \left(\sqrt{2x} - \dfrac{1}{\sqrt{2x}}\right) dx$

16. $\displaystyle\int \dfrac{27t^3 - 1}{\sqrt[3]{t}} \, dt$

17. $\displaystyle\int \sqrt{1 - 4y} \, dy$

18. $\displaystyle\int \sqrt[3]{3x - 4} \, dx$

19. $\displaystyle\int \sqrt[3]{6-2x}\,dx$

20. $\displaystyle\int \sqrt{5r+1}\,dr$

21. $\displaystyle\int x\sqrt{x^2-9}\,dx$

22. $\displaystyle\int 3x\sqrt{4-x^2}\,dx$

23. $\displaystyle\int x^2\sqrt{x^3-1}\,dx$

24. $\displaystyle\int x(2x^2+1)^6\,dx$

25. $\displaystyle\int 5x\sqrt[3]{(9-4x^2)^2}\,dx$

26. $\displaystyle\int \frac{x\,dx}{\sqrt[3]{x^2+1}}$

27. $\displaystyle\int \frac{y^3}{(1-2y^4)^5}\,dy$

28. $\displaystyle\int (x^2-4x+4)^{4/3}\,dx$

29. $\displaystyle\int \frac{s\,ds}{\sqrt{3s^2+1}}$

30. $\displaystyle\int \frac{2r\,dr}{(1-r)^{2/3}}$

31. $\displaystyle\int \frac{t\,dt}{\sqrt{t+3}}$

32. $\displaystyle\int \sqrt{\frac{1}{t}-1}\,\frac{dt}{t^2}$

33. $\displaystyle\int \sqrt{1+\frac{1}{3x}}\,\frac{dx}{x^2}$

34. $\displaystyle\int x^4\sqrt{3x^5-5}\,dx$

35. $\displaystyle\int \sqrt{3-x}\,x^2\,dx$

36. $\displaystyle\int (x^3+3)^{1/4}x^5\,dx$

37. $\displaystyle\int \frac{(x^2+2x)\,dx}{\sqrt{x^3+3x^2+1}}$

38. $\displaystyle\int x(x^2+1)\sqrt{4-2x^2-x^4}\,dx$

39. $\displaystyle\int \frac{x(3x^2+1)\,dx}{(3x^4+2x^2+1)^2}$

40. $\displaystyle\int \sqrt{3+s}(s+1)^2\,ds$

41. $\displaystyle\int \frac{y+3}{(3-y)^{2/3}}\,dy$

42. $\displaystyle\int (2t^2+1)^{1/3}t^3\,dt$

43. $\displaystyle\int \frac{(r^{1/3}+2)^4}{\sqrt[3]{r^2}}\,dr$

44. $\displaystyle\int \left(t+\frac{1}{t}\right)^{3/2}\left(\frac{t^2-1}{t^2}\right)dt$

45. $\displaystyle\int \frac{x^3}{(x^2+4)^{3/2}}\,dx$

46. $\displaystyle\int \frac{x^3}{\sqrt{1-2x^2}}\,dx$

47. Evaluate $\displaystyle\int (2x+1)^3\,dx$ by two methods: (a) expand $(2x+1)^3$ by the binomial theorem, and apply Theorems 6.2.4, 6.2.7, and 6.2.8; (b) make the substitution $u=2x+1$. Explain the difference in appearance of the answers obtained in (a) and (b).

48. Evaluate $\displaystyle\int \sqrt{x-1}\,x^2\,dx$ by two methods: (a) make the substitution $u=x-1$; (b) make the substitution $v=\sqrt{x-1}$.

6.3 Differential equations with variables separable

An equation containing derivatives is called a **differential equation**. Some simple differential equations are

$$\frac{dy}{dx}=2x \tag{1}$$

$$\frac{dy}{dx}=\frac{2x^2}{3y^3} \tag{2}$$

$$\frac{d^2y}{dx^2}=4x+3 \tag{3}$$

The **order** of a differential equation is the order of the derivative of highest order that appears in the equation. Therefore (1) and (2) are first-order differential equations and (3) is of the second order. The simplest type of differential equation is a first-order equation of the form

$$\frac{dy}{dx} = f(x) \tag{4}$$

for which (1) is a particular example. Writing (4) with differentials we have

$$dy = f(x)\,dx \tag{5}$$

Another type of differential equation of the first order is one of the form

$$\frac{dy}{dx} = \frac{g(x)}{h(y)} \tag{6}$$

Equation (2) is a particular example of an equation of this type. If (6) is written with differentials, we have

$$h(y)\,dy = g(x)\,dx \tag{7}$$

In both (5) and (7) the left side involves only the variable y and the right side involves only the variable x. Thus the variables are separated, and we say that these are differential equations with variables separable.

Consider (5). To solve this equation we must find all functions G for which $y = g(x)$ such that the equation is satisfied. So if F is an antiderivative of f, all functions G are defined by $G(x) = F(x) + C$, where C is an arbitrary constant. That is, if $d(G(x)) = d(F(x) + C) = f(x)\,dx$, then what is called the **complete solution** of (5) is given by

$$y = F(x) + C \tag{8}$$

Equation (8) represents a family of functions depending on an arbitrary constant C. This is called a **one-parameter family**. The graphs of these functions form a one-parameter family of curves in the plane, and through any particular point (x_1, y_1) there passes just one curve of the family.

● ILLUSTRATION 1

Suppose that we wish to find the complete solution of the differential equation

$$dy = 2x\,dx \tag{9}$$

The set of all antiderivatives of the left side of (9) is $(y + C_1)$ and the set of all antiderivatives of $2x$ is $(x^2 + C_2)$. Thus

$$y + C_1 = x^2 + C_2$$

Because $(C_2 - C_1)$ is an arbitrary constant if C_2 and C_1 are arbitrary, we can replace $(C_2 - C_1)$ by C, thereby obtaining

$$y = x^2 + C \tag{10}$$

which is the complete solution of the given differential equation.

Equation (10) represents a one-parameter family of functions. Figure 6.3.1 shows sketches of the graphs of the functions corresponding to $C = -4$, $C = -1$, $C = 0$, $C = 1$, and $C = 2$.

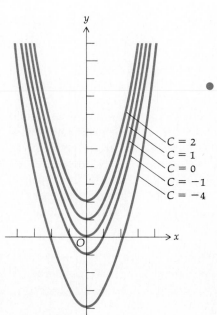

Figure 6.3.1

Now consider (7), which is

$$h(y)\,dy = g(x)\,dx$$

If we antidifferentiate on both sides of the equation, we have

$$\int h(y)\,dy = \int g(x)\,dx$$

If H is an antiderivative of h and G is an antiderivative of g, the complete solution of (7) is given by

$$H(y) = G(x) + C$$

EXAMPLE 1 Find the complete solution of the differential equation

$$\frac{dy}{dx} = \frac{2x^2}{3y^3}$$

Solution   If the given equation is written with differentials, we have

$$3y^3\,dy = 2x^2\,dx$$

and the variables are separated. We antidifferentiate on both sides of the equation and obtain

$$\int 3y^3\,dy = \int 2x^2\,dx$$

$$\frac{3y^4}{4} = \frac{2x^3}{3} + \frac{C}{12} \tag{11}$$

$$9y^4 = 8x^3 + C$$

which is the complete solution.

In (11) the arbitrary constant was written as $C/12$ so that when both sides of the equation are multiplied by 12, the arbitrary constant becomes C.

Often in problems involving differential equations it is desired to find particular solutions that satisfy certain conditions called **boundary conditions** or **initial conditions**. For example, if a first-order differential equation is given as well as the boundary condition that $y = y_1$ when $x = x_1$, then after the complete solution is found, if x and y in the complete solution are replaced by x_1 and y_1, a particular value of C is determined. When this value of C is substituted back into the complete solution, a particular solution is obtained.

● ILLUSTRATION 2

To find the particular solution of differential equation (9) satisfying the boundary condition that $y = 6$ when $x = 2$, we substitute these values in (10) and solve for C, giving $6 = 4 + C$, or $C = 2$. Substituting this value of C in (10) we obtain

$$y = x^2 + 2$$

which is the particular solution desired.   ●

Equation (3) is an example of a particular type of differential equation of the second order

$$\frac{d^2y}{dx^2} = f(x) \tag{12}$$

Two successive antidifferentiations are necessary to solve (12), and two arbitrary constants occur in the complete solution. The complete solution of (12) therefore represents a **two-parameter family** of functions, and the graphs of these functions form a two-parameter family of curves in the plane. The following example shows the method of obtaining the complete solution of an equation of this kind.

EXAMPLE 2 Find the complete solution of the differential equation

$$\frac{d^2y}{dx^2} = 4x + 3$$

Solution Because $\dfrac{d^2y}{dx^2} = \dfrac{dy'}{dx}$, the given equation can be written as

$$\frac{dy'}{dx} = 4x + 3$$

Writing this in differential form we have

$$dy' = (4x + 3)\, dx$$

Antidifferentiating we have

$$\int dy' = \int (4x + 3)\, dx$$

from which we get

$$y' = 2x^2 + 3x + C_1$$

Because $y' = \dfrac{dy}{dx}$, we make this substitution in the above equation and get

$$\frac{dy}{dx} = 2x^2 + 3x + C_1$$

Using differentials we have

$$dy = (2x^2 + 3x + C_1)\, dx$$

Antidifferentiating we obtain

$$\int dy = \int (2x^2 + 3x + C_1)\, dx$$

from which we get

$$y = \tfrac{2}{3}x^3 + \tfrac{3}{2}x^2 + C_1x + C_2$$

which is the complete solution.

EXAMPLE 3 Find the particular solution of the differential equation in Example 2 for which $y = 2$ and $y' = -3$ when $x = 1$.

Solution Because $y' = 2x^2 + 3x + C_1$, we substitute -3 for y' and 1 for x, giving $-3 = 2 + 3 + C_1$, or $C_1 = -8$. Substituting this value of C_1 into the complete solution gives

$$y = \tfrac{2}{3}x^3 + \tfrac{3}{2}x^2 - 8x + C_2$$

Because $y = 2$ when $x = 1$, we substitute these values in the above equation and get $2 = \tfrac{2}{3} + \tfrac{3}{2} - 8 + C_2$, from which we obtain $C_2 = \tfrac{47}{6}$. The particular solution desired, then, is

$$y = \tfrac{2}{3}x^3 + \tfrac{3}{2}x^2 - 8x + \tfrac{47}{6}$$

EXAMPLE 4 A business firm has made an analysis of its production facilities and its personnel. With the present equipment and number of workers, the firm can produce 3000 units per day. It was estimated that without any change in equipment, the rate of change of the number of units produced per day with respect to a change in the number of additional workers is $80 - 6x^{1/2}$, where x is the number of additional workers. Find the daily production if 25 workers are added to the labor force.

Solution Let y be the number of units produced per day. Then

$$\frac{dy}{dx} = 80 - 6x^{1/2}$$

from which we get

$$dy = (80 - 6x^{1/2})\, dx$$

Antidifferentiating gives

$$y = 80x - 4x^{3/2} + C$$

Because $y = 3000$ when $x = 0$, we get $C = 3000$. Hence

$$y = 80x - 4x^{3/2} + 3000$$

Letting y_{25} be the value of y when $x = 25$ we have

$$y_{25} = 2000 - 500 + 3000$$
$$= 4500$$

Therefore 4500 units are produced per day if the labor force is increased by 25 workers.

Exercises 6.3

In Exercises 1 through 16, find the complete solution of the differential equation.

1. $\dfrac{dy}{dx} = 4x - 5$ **2.** $\dfrac{dy}{dx} = 6 - 3x^2$ **3.** $\dfrac{dy}{dx} = 3x^2 + 2x - 7$ **4.** $\dfrac{dy}{dx} = x^4 + 2x^2 - \dfrac{1}{x^2}$

5. $\dfrac{du}{dv} = \dfrac{2v^2 - 3}{v^2}$

6. $\dfrac{dy}{dx} = (3x + 1)^3$

7. $\dfrac{dy}{dx} = 3\sqrt{2x - 1}$

8. $\dfrac{ds}{dt} = 5\sqrt{s}$

9. $\dfrac{dy}{dx} = 3xy^2$

10. $\dfrac{dy}{dx} = \dfrac{\sqrt{x} + x}{\sqrt{y} - y}$

11. $\dfrac{dy}{dx} = \dfrac{3x\sqrt{1 + y^2}}{y}$

12. $\dfrac{dy}{dx} = \dfrac{x^2\sqrt{x^3 - 3}}{y^2}$

13. $\dfrac{d^2y}{dx^2} = 5x^2 + 1$

14. $\dfrac{d^2y}{dx^2} = \sqrt{2x - 3}$

15. $\dfrac{d^2u}{dv^2} = \sqrt{3v + 1}$

16. $\dfrac{d^2s}{dt^2} = \sqrt[3]{4t + 5}$

In Exercises 17 through 24, find the particular solution of the given differential equation determined by the boundary conditions.

17. $\dfrac{dy}{dx} = x^2 - 2x - 4$; $y = -6$ when $x = 3$

18. $\dfrac{dy}{dx} = (x + 1)(x + 2)$; $y = -\frac{3}{2}$ when $x = -3$

19. $\dfrac{dx}{y} = \dfrac{4\,dy}{x}$; $y = -2$ when $x = 4$

20. $\dfrac{dy}{dx} = \dfrac{x}{4\sqrt{(1 + x^2)^3}}$; $y = 0$ when $x = 1$

21. $\dfrac{d^2y}{dx^2} = x^2 + 3x$; $y = 2$ and $y' = 1$ when $x = 1$

22. $\dfrac{d^2y}{dx^2} = -\dfrac{3}{x^4}$; $y = \frac{1}{2}$ and $y' = -1$ when $x = 1$

23. $\dfrac{d^2u}{dv^2} = 4(1 + 3v)^2$; $u = -1$ and $\dfrac{du}{dv} = -2$ when $v = -1$

24. $\dfrac{d^2y}{dx^2} = \sqrt[3]{3x - 1}$; $y = 2$ and $y' = 5$ when $x = 3$

25. The point $(3, 2)$ is on a curve, and at any point (x, y) on the curve the tangent line has a slope equal to $2x - 3$. Find an equation of the curve.

26. An equation of the tangent line to a curve at the point $(1, 3)$ is $y = x + 2$. If at any point (x, y) on the curve $\dfrac{d^2y}{dx^2} = 6x$, find an equation of the curve.

27. The points $(-1, 3)$ and $(0, 2)$ are on a curve, and at any point (x, y) on the curve, $\dfrac{d^2y}{dx^2} = 2 - 4x$. Find an equation of the curve.

28. At any point (x, y) on a curve, $\dfrac{d^3y}{dx^3} = 2$, and $(1, 3)$ is a point of inflection at which the slope of the inflectional tangent is -2. Find an equation of the curve.

29. The cost of a certain piece of machinery is \$700, and its value depreciates according to the formula $\dfrac{dV}{dt} = -500(t + 1)^{-2}$, where V dollars is its value t years after its purchase. What is its value 3 years after its purchase?

30. An art collector purchased a painting for \$1000 by an artist, the value of whose works are currently increasing with respect to time according to the formula $\dfrac{dV}{dt} = 5(t + 4)^{3/2}$, where V dollars is the expected value of a painting t years after its purchase. If this formula were valid for the next 6 years, what would be the expected value of the painting 5 years from now?

31. A firm estimates that its growth in income from sales for the next 10 years is given by the formula $\dfrac{dS}{dt} = \frac{1}{5}(t + 1)^{1/3}$, where S millions of dollars is the gross income from sales t years hence, and $0 \le t \le 10$. If this year's gross income from sales is \$1 million, what is the expected gross income from sales 7 years hence?

32. The efficiency of a factory worker is expressed as a percent. For instance, if the worker's efficiency at a particular time is given as 70 percent, then the worker is performing at 70 percent of his or her full potential. Suppose that E percent is a factory worker's efficiency t hours after beginning work, and the rate at which E is changing is $(35 - 8t)$ percent per hour. If the worker's efficiency is 81 percent after working 3 hours, find his or her efficiency after working (a) 4 hours and (b) 8 hours.

33. A wound is healing in such a way that t days since Monday the area of the wound has been decreasing at a rate of $-3(t + 2)^{-2}$ square centimeters per day. If on Tuesday the area of the wound was 2 cm², (a) what was the area of the wound on Monday, and (b) what is the anticipated area of the wound on Friday if it continues to heal at the same rate?

34. The population of a particular town has been increasing at a rate of $400(t + 1)^{-1/2}$ people per year t years after 1981. If the population was 6000 in 1984, (a) what was the population in 1981 and (b) what is the expected population in 1989 if it continues to increase at the same rate?

35. For the first 10 days in December a plant cell grew in such a way that t days after December 1 the volume of the cell was increasing at a rate of $(12 - t)^{-2}$ cubic micrometers(μm) per day. If on December 3 the volume of the cell was 3 μm³, what was the volume on December 8?

36. The volume of water in a tank is V cubic meters when the depth of the water is h meters. If the rate of change of V with respect to h is given by $\dfrac{dV}{dh} = \pi(2h + 3)^2$, find the volume of water in the tank when the depth is 3 m.

6.4 Applications of antidifferentiation in economics

Because the marginal cost function C' and the marginal revenue function R' are the first derivatives of the total cost function C and the total revenue function R, respectively, C and R can be obtained from C' and R' by antidifferentiation. When finding the function C from C', the arbitrary constant can be evaluated if we know either the overhead cost (that is, the cost when no units are produced) or the cost of production of a specific number of units of the commodity. Because it is generally true that the total revenue is zero when the number of units produced is zero, this fact may be used to evaluate the arbitrary constant when finding the function R from R'.

EXAMPLE 1   The marginal cost function C' is given by $C'(x) = 4x - 8$, where $C(x)$ dollars is the total cost of producing x units. If the cost of producing 5 units is $20, find the total cost function. Draw sketches of the total cost curve and the marginal cost curve on the same set of axes.

Solution   The marginal cost must be nonnegative. Hence $4x - 8 \geq 0$, and therefore the permissible values of x are $x \geq 2$. Because

$$C'(x) = 4x - 8$$

$$C(x) = \int (4x - 8)\, dx$$

$$= 2x^2 - 8x + k$$

Because $C(5) = 20$, we obtain $k = 10$. Hence

$$C(x) = 2x^2 - 8x + 10 \qquad x \geq 2$$

The required sketches are shown in Fig. 6.4.1.

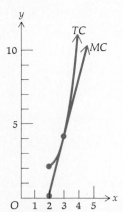

Figure 6.4.1

EXAMPLE 2   If the marginal revenue is given by $R'(x) = 27 - 12x + x^2$, find the total revenue function and the demand equation. Draw sketches of the demand curve, the total revenue curve, and the marginal revenue curve on the same set of axes.

Solution   If R is the total revenue function, and

$$R'(x) = 27 - 12x + x^2$$

then

$$R(x) = \int (27 - 12x + x^2) \, dx$$

$$= 27x - 6x^2 + \tfrac{1}{3}x^3 + C$$

Because $R(0) = 0$, we get $C = 0$. Hence

$$R(x) = 27x - 6x^2 + \tfrac{1}{3}x^3$$

If f is the price function, $R(x) = xf(x)$; so

$$f(x) = 27 - 6x + \tfrac{1}{3}x^2$$

If p dollars is the price of one unit of the commodity when x units are demanded, then because $p = f(x)$, the demand equation is

$$3p = 81 - 18x + x^2$$

To determine the permissible values of x, we use the facts that $x \geq 0$, $p \geq 0$, and f is a decreasing function. Because

$$f'(x) = -6 + \tfrac{2}{3}x$$

f is decreasing when $x < 9$ (that is, when $f'(x) < 0$). Also, when $x = 9$, $p = 0$; so the permissible values of x are the numbers in the closed interval $[0, 9]$. The required sketches are shown in Fig. 6.4.2.

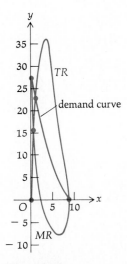

Figure 6.4.2

EXAMPLE 3 After experimentation, a certain manufacturer determined that if x units of a certain article of merchandise are produced per week, the marginal cost is given by $C'(x) = 0.3x - 11$, where $C(x)$ dollars is the total cost of producing x units. If the selling price of the article is fixed at \$19 per unit and the overhead cost is \$200 per week, find the maximum weekly total profit that can be obtained.

Solution   Let $R(x)$ dollars be the total revenue obtained by selling x units, and $P(x)$ dollars be the total profit obtained by selling x units. Because x units are sold at \$19 per unit,

$$R(x) = 19x$$

Thus the marginal revenue is given by

$$R'(x) = 19$$

We are given that

$$C'(x) = 0.3x - 11$$

The total profit will be a maximum when the marginal revenue equals the marginal cost (see Sec. 5.5). Equating $R'(x)$ and $C'(x)$ we get $x = 100$. So 100 units should be produced each week for maximum total profit.

$$C(x) = \int (0.3x - 11) \, dx$$

$$= 0.15x^2 - 11x + k$$

Because the overhead cost is \$200, $C(0) = 200$, and hence $k = 200$. Therefore

$$C(x) = 0.15x^2 - 11x + 200$$

Because $P(x) = R(x) - C(x)$,

$$P(x) = 19x - (0.15x^2 - 11x + 200)$$
$$= -0.15x^2 + 30x - 200$$

Hence $P(100) = 1300$. Therefore the maximum weekly total profit is \$1300, which is obtained if 100 units are produced weekly.

Exercises 6.4

1. The marginal revenue function for a certain commodity is given by $R'(x) = 12 - 3x$. If x units are demanded when p dollars is the price per unit, find (a) the total revenue function and (b) the demand equation. (c) Draw sketches of the total revenue curve and the demand curve on the same set of axes.

2. For a particular article of merchandise the marginal revenue function is given by $R'(x) = 15 - 4x$. Find (a) the total revenue function and (b) the demand equation. (c) Draw sketches of the total revenue curve and the demand curve on the same set of axes.

3. The marginal cost function for a particular article of merchandise is given by $C'(x) = 3(5x + 4)^{-1/2}$. If the overhead cost is \$10, find the total cost function.

4. For a certain commodity the marginal cost function is given by $C'(x) = 3\sqrt{2x + 4}$. If the overhead cost is zero, find the total cost function.

5. Find the demand equation of a commodity for which the marginal revenue function is given by $R'(x) = 4 + 10(x + 5)^{-2}$.

6. Find the price function for a commodity for which the marginal revenue function is given by $R'(x) = 14 - 9x + x^2$.

7. The marginal cost function for a particular commodity is given by $C'(x) = 6x - 17$. If the cost of producing 2 units is \$25, find the total cost function.

8. If the marginal revenue function is given by $R'(x) = 3x^2 - 12x + 10$, find (a) the total revenue function and (b) the price function.

9. The marginal revenue function is given by $R'(x) = 16 - 3x^2$. Find (a) the total revenue function and (b) the demand equation. (c) Draw sketches of the demand curve, the total revenue curve, and the marginal revenue curve on the same set of axes.

10. Find (a) the total revenue function and (b) the demand equation if the marginal revenue function is given by $R'(x) = \frac{3}{4}x^2 - 10x + 12$. (c) Draw sketches of the demand curve, the total revenue curve, and the marginal revenue curve on the same set of axes.

11. The marginal cost function is given by $C'(x) = 3x^2 + 8x + 4$, and the overhead cost is \$6. If $C(x)$ dollars is the total cost of x units, (a) find the total cost function, and (b) draw sketches of the total cost curve and the marginal cost curve on the same set of axes.

12. The marginal cost function is defined by $C'(x) = 6x$, where $C(x)$ is the number of hundreds of dollars in the total cost of x hundred units of a certain commodity. If the cost of 200 units is \$2000, find (a) the total cost function and (b) the overhead cost. (c) Draw sketches of the total cost curve and the marginal cost curve on the same set of axes.

13. A company has determined that the marginal cost function for the production of a particular commodity is given by $C'(x) = 125 + 10x + \frac{1}{9}x^2$, where $C(x)$ dollars is the total cost of producing x units of the commodity. If the overhead cost is \$250, what is the cost of producing 15 units?

14. The marginal revenue function is given by $R'(x) = ab(x + b)^{-2} - c$. Find (a) the total revenue function and (b) the demand equation.

15. The marginal cost function for a manufacturer of vases is $C'(x) = 4 - 9\sqrt{3x}/2x$, where $C(x)$ dollars is the total cost of producing x vases. The overhead cost is \$54. If 27 vases are produced, find (a) the marginal cost and (b) the average unit cost.

16. The marginal cost function for a particular article of merchandise is $C'(x) = 30x^{1/2}$, where $C(x)$ dollars is the total cost of producing x articles. The overhead cost is \$100. If 25 units are produced, find (a) the marginal cost and (b) the average unit cost.

17. The rate of change of the slope of the total cost curve of a particular company is the constant 2, and the total cost curve contains the points (2, 12) and (3, 18). Find the total cost function.

18. For a particular commodity the rate of change of the total cost function per unit change in x is $3x$ and the total cost curve contains the point (5, 45). Find the total cost function.

19. A manufacturer of children's toys has a new toy coming on the market and wishes to determine a selling price for the toy such that the total profit will be a maximum. From analyzing the price and demand of another similar toy, it is anticipated that if x toys are demanded when p dollars is the price per toy, then $\dfrac{dp}{dx} = -\dfrac{p^2}{30,000}$, and the demand should be 1800 when the price is \$10. If $C(x)$ dollars is the total cost of producing x toys, then $C(x) = x + 7500$. Find the price that should be charged for the manufacturer's total profit to be a maximum.

20. If the marginal revenue function is given by $R'(x) = ax^{-1/3}$, where a is a constant, show that the price elasticity of demand is constant.

Review Exercises for Chapter 6

In Exercises 1 through 14, perform the antidifferentiation.

1. $\displaystyle\int (2x^3 - x^2 + 3)\,dx$

2. $\displaystyle\int (5x^4 + 3x - 1)\,dx$

3. $\displaystyle\int (4y + 6\sqrt{y})\,dy$

4. $\displaystyle\int \sqrt{x}(1 + x^2)\,dx$

5. $\displaystyle\int \left(\frac{2}{x^4} - \frac{5}{x^2}\right) dx$

6. $\displaystyle\int \left(\sqrt[3]{t} - \frac{1}{\sqrt[3]{t}}\right) dt$

7. $\displaystyle\int 5x\sqrt{2 + 3x^2}\,dx$

8. $\displaystyle\int x^4\sqrt{x^5 - 1}\,dx$

9. $\displaystyle\int \left(\sqrt{3x} + \frac{1}{\sqrt{5x}}\right) dx$

10. $\displaystyle\int \sqrt[3]{7w + 3}\,dw$

11. $\displaystyle\int \frac{x^3 + x}{\sqrt{x^4 + 2x^2}}\,dx$

12. $\displaystyle\int (x^3 + x)\sqrt{x^2 + 3}\,dx$

13. $\displaystyle\int \sqrt{4x + 3}(x^2 + 1)\,dx$

14. $\displaystyle\int \frac{s}{\sqrt{2s + 3}}\,ds$

In Exercises 15 through 20, find the complete solution of the given differential equation.

15. $x^2 y \dfrac{dy}{dx} = (y^2 - 1)^2$

16. $\dfrac{dy}{dx} = \dfrac{x + 1}{\sqrt{x}}$

17. $y^2\,dx + y^2\,dy = dy$

18. $\dfrac{d^2y}{dx^2} = 12x^2 - 30x$

19. $\dfrac{d^2y}{dx^2} = \sqrt{2x - 1}$

20. $y\sqrt{2x^2 + 1}\,dy = x\sqrt{1 - y^2}\,dx$

21. If $y = 80x - 16x^2$, find the difference $\Delta y - dy$ if (a) $x = 2$ and $\Delta x = 0.1$; (b) $x = 4$ and $\Delta x = -0.2$.

22. Use differentials to find an approximate value to three decimal places of $\sqrt[3]{126}$.

23. Use differentials to approximate the volume of material needed to make a rubber ball if the radius of the hollow inner core is 2 in. and the thickness of the rubber is $\frac{1}{8}$ in.

24. The slope of the tangent line at any point (x, y) on a curve is $10 - 4x$, and the point $(1, -1)$ is on the curve. Find an equation of the curve.

25. At any point (x, y) on a curve, $\dfrac{d^2y}{dx^2} = 4 - x^2$, and an equation of the tangent line to the curve at the point $(1, -1)$ is $2x - 3y = 3$. Find an equation of the curve.

26. Find the particular solution of the differential equation $x^2\,dy = y^3\,dx$ for which $y = 1$ when $x = 4$.

27. Find the particular solution of the differential equation $\dfrac{d^2y}{dx^2} = \sqrt{x + 4}$ for which $y = 3$ and $y' = 2$ when $x = 4$.

28. If $x^3 + y^3 - 3xy^2 + 1 = 0$, find dy at the point $(1, 1)$ if $dx = 0.1$.

29. The marginal cost for a particular commodity is given by $C'(x) = 4(3x + 1)^{-1/2}$, where $C(x)$ dollars is the total cost of producing x units. If the cost of producing 5 units is \$20, find the total cost function.

30. Find the demand equation of a commodity for which the marginal revenue function is given by $R'(x) = 20(x + 4)^{-2} + 10$.

31. The marginal revenue for a particular commodity is given by $R'(x) = x^2 - 16x + 48$, where $R(x)$ dollars is the total revenue received from the sale of x units of the commodity. Find (a) the total revenue function and (b) the demand equation. (c) Draw sketches of the demand curve, the total revenue curve, and the marginal revenue curve on the same set of axes.

32. The marginal cost function is given by $C'(x) = 3x^2 + 12x + 9$ and the overhead cost is \$12. If $C(x)$ dollars is the total cost of x units, (a) find the total cost function, and (b) draw sketches of the total cost curve and the marginal cost curve on the same set of axes.

33. If x units of a certain article of merchandise are produced per day, the marginal cost is given by $C'(x) = 0.5x - 10$, where $C(x)$ dollars is the total cost of producing x units and the overhead cost is \$50 per day. If the selling price of the article is \$20 per unit, find the maximum daily profit that can be obtained.

34. A container in the form of a cube having a volume of 1000 cm^3 is to be made using six identical squares of material costing 12 cents per square centimeter. Use differentials to determine how accurately the side of each square must be made so that the total cost of the material shall be correct to within \$3.

35. A company estimated that the rate of change of the number of units produced per day with respect to a change in the number of additional workers is $100 - 9x^{1/2}$, where x is the number of additional workers. Find the daily production when 9 workers are added to the labor force if 4000 units are produced when there are no additional workers.

36. The cost of a certain piece of machinery is \$900, and its value depreciates with time according to the formula $\dfrac{dV}{dt} = -300(t + 1)^{-2}$, where V dollars is its value t years after its purchase. What is its value 2 years after its purchase?

37. A company purchased some equipment on April 1, 1982, for \$1200. If the value depreciates according to the formula $\dfrac{dV}{dt} = -300(2t + 1)^{-1/2}$, where V dollars is the value t years after the purchase, determine the value of the equipment on April 1, 1986.

38. Suppose that a particular company estimates its growth income from sales by the formula $\dfrac{dS}{dt} = 2(t - 1)^{2/3}$, where S millions of dollars is the gross income from sales t years hence. If the gross income from the current year's sales is \$8 million, what should be the expected gross income from sales 2 years from now?

39. It is July 31 and a tumor has been growing inside a person's body in such a way that t days since July 1 the volume of the tumor has been increasing at a rate of $\frac{1}{100}(t + 6)^{1/2} \text{ cm}^3$ per day. If the volume of the tumor on July 4 was 0.20 cm^3, what is the volume today?

40. The enrollment at a certain college has been increasing at the rate of $1000(t + 1)^{-1/2}$ students per year since 1981. If the enrollment in 1984 was 10,000, (a) what was the enrollment in 1981, and (b) what is the anticipated enrollment in 1989 if it is expected to be increasing at the same rate?

41. The volume of a balloon is increasing according to the formula $\dfrac{dV}{dt} = \sqrt{t + 1} + \frac{2}{3}t$, where $V \text{ cm}^3$ is the volume of the balloon at t sec. If $V = 33$ when $t = 3$, find (a) a formula for V in terms of t; (b) the volume of the balloon at 8 sec.

CHAPTER 7

THE DEFINITE INTEGRAL

7.1 The sigma notation and limits at infinity

In this chapter we are concerned with sums of many terms, and so a notation called the **sigma notation** is introduced to facilitate writing these sums. This notation involves the use of the symbol $\sum$, the capital sigma of the Greek alphabet, which corresponds to the letter S. Some examples of the sigma notation are given in the following illustration.

● ILLUSTRATION 1

$$\sum_{i=1}^{5} i^2 = 1^2 + 2^2 + 3^2 + 4^2 + 5^2$$

$$\sum_{i=-2}^{2} (3i + 2) = [3(-2) + 2] + [3(-1) + 2] + [3 \cdot 0 + 2] \\ + [3 \cdot 1 + 2] + [3 \cdot 2 + 2] \\ = (-4) + (-1) + 2 + 5 + 8$$

$$\sum_{j=1}^{n} j^3 = 1^3 + 2^3 + 3^3 + \ldots + n^3$$

$$\sum_{k=3}^{8} \frac{1}{k} = \frac{1}{3} + \frac{1}{4} + \frac{1}{5} + \frac{1}{6} + \frac{1}{7} + \frac{1}{8}$$

●

We have the formal definition of the sigma notation.

Definition of the sigma notation

$$\sum_{i=m}^{n} F(i) = F(m) + F(m + 1) + F(m + 2) + \ldots + F(n - 1) + F(n) \qquad (1)$$

where m and n are integers, and $m \leq n$.

The right side of (1) consists of the sum of $(n - m + 1)$ terms, the first of which is obtained by replacing i by m in $F(i)$, the second by replacing i by $(m + 1)$ in $F(i)$, and so on, until the last term is obtained by replacing i by n in $F(i)$.

The number m is called the **lower limit** of the sum, and n is called the **upper limit**. The symbol i is called the **index of summation**. It is a "dummy" symbol because any other letter can be used for this purpose. For example,

$$\sum_{k=3}^{5} k^2 = 3^2 + 4^2 + 5^2$$

is equivalent to

$$\sum_{i=3}^{5} i^2 = 3^2 + 4^2 + 5^2$$

● ILLUSTRATION 2

From the definition of the sigma notation,

$$\sum_{i=3}^{6} \frac{i^2}{i+1} = \frac{3^2}{3+1} + \frac{4^2}{4+1} + \frac{5^2}{5+1} + \frac{6^2}{6+1} \qquad ●$$

Sometimes the terms of a sum involve subscripts, as shown in the next illustration.

● ILLUSTRATION 3

$$\sum_{i=1}^{n} A_i = A_1 + A_2 + \ldots + A_n$$

$$\sum_{k=4}^{9} kb_k = 4b_4 + 5b_5 + 6b_6 + 7b_7 + 8b_8 + 9b_9$$

$$\sum_{i=1}^{5} f(x_i)\,\Delta x = f(x_1)\,\Delta x + f(x_2)\,\Delta x + f(x_3)\,\Delta x + f(x_4)\,\Delta x + f(x_5)\,\Delta x \qquad ●$$

The following theorems involving the sigma notation are useful for computation and are easily proved.

Theorem 7.1.1

$$\sum_{i=1}^{n} c = cn, \text{ where } c \text{ is any constant}$$

Proof

$$\sum_{i=1}^{n} c = c + c + \ldots + c \qquad (n \text{ terms})$$

$$= cn \qquad \blacksquare$$

Theorem 7.1.2

$$\sum_{i=1}^{n} c \cdot F(i) = c \sum_{i=1}^{n} F(i), \text{ where } c \text{ is any constant}$$

Proof

$$\sum_{i=1}^{n} c \cdot F(i) = c \cdot F(1) + c \cdot F(2) + c \cdot F(3) + \ldots + c \cdot F(n)$$

$$= c[F(1) + F(2) + F(3) + \ldots + F(n)]$$

$$= c \sum_{i=1}^{n} F(i) \qquad \blacksquare$$

Theorem 7.1.3

$$\sum_{i=1}^{n} [F(i) + G(i)] = \sum_{i=1}^{n} F(i) + \sum_{i=1}^{n} G(i)$$

Proof

$$\sum_{i=1}^{n} [F(i) + G(i)]$$

$$= [F(1) + G(1)] + [F(2) + G(2)] + \ldots + [F(n) + G(n)]$$
$$= [F(1) + F(2) + \ldots + F(n)] + [G(1) + G(2) + \ldots + G(n)]$$

$$= \sum_{i=1}^{n} F(i) + \sum_{i=1}^{n} G(i) \qquad \blacksquare$$

Theorem 7.1.3 can be extended to the sum of any number of functions. In the following theorem there are four formulas that are useful for computation with sigma notation. They are numbered for future reference.

Theorem 7.1.4 If n is a positive integer, then

$$\sum_{i=1}^{n} i = \frac{n(n+1)}{2} \qquad \text{(Formula 1)}$$

$$\sum_{i=1}^{n} i^2 = \frac{n(n+1)(2n+1)}{6} \qquad \text{(Formula 2)}$$

$$\sum_{i=1}^{n} i^3 = \frac{n^2(n+1)^2}{4} \qquad \text{(Formula 3)}$$

$$\sum_{i=1}^{n} i^4 = \frac{n(n+1)(6n^3 + 9n^2 + n - 1)}{30} \qquad \text{(Formula 4)}$$

Proof of Formula 1

$$\sum_{i=1}^{n} i = 1 + 2 + 3 + \ldots + (n-1) + n$$

and

$$\sum_{i=1}^{n} i = n + (n-1) + (n-2) + \ldots + 2 + 1$$

If these two equations are added term by term, the left side is

$$2 \sum_{i=1}^{n} i$$

and on the right side are n terms, each having the value $(n + 1)$. Hence

$$2 \sum_{i=1}^{n} i = (n + 1) + (n + 1) + (n + 1) + \ldots + (n + 1) \qquad n \text{ terms}$$

$$= n(n + 1)$$

Therefore

$$\sum_{i=1}^{n} i = \frac{n(n + 1)}{2}$$

∎

The proofs of Formulas 2, 3, and 4 are omitted.

EXAMPLE 1   Use Formulas 1–4 to evaluate the following summations:

(a) $\displaystyle\sum_{i=1}^{4} i$;   (b) $\displaystyle\sum_{i=1}^{4} i^2$;   (c) $\displaystyle\sum_{i=1}^{4} i^3$;   (d) $\displaystyle\sum_{i=1}^{4} i^4$

Solution

(a) $\displaystyle\sum_{i=1}^{4} i = 1 + 2 + 3 + 4$

From Formula 1, with $n = 4$,

$$\sum_{i=1}^{4} i = \frac{4(4 + 1)}{2} = \frac{4 \cdot 5}{2} = 10$$

(b) $\displaystyle\sum_{i=1}^{4} i^2 = 1^2 + 2^2 + 3^2 + 4^2$

From Formula 2, with $n = 4$,

$$\sum_{i=1}^{4} i^2 = \frac{4(4 + 1)(8 + 1)}{6} = \frac{4 \cdot 5 \cdot 9}{6} = 30$$

(c) $\displaystyle\sum_{i=1}^{4} i^3 = 1^3 + 2^3 + 3^3 + 4^3$

From Formula 3, with $n = 4$,

$$\sum_{i=1}^{4} i^3 = \frac{4^2(4+1)^2}{4} = \frac{16 \cdot 25}{4} = 100$$

(d) $\displaystyle\sum_{i=1}^{4} i^4 = 1^4 + 2^4 + 3^4 + 4^4$

From Formula 4, with $n = 4$,

$$\sum_{i=1}^{4} i^4 = \frac{4(4+1)(6 \cdot 64 + 9 \cdot 16 + 4 - 1)}{30} = \frac{4(5)(531)}{30} = 354$$

EXAMPLE 2 Evaluate $\displaystyle\sum_{i=1}^{n} (12i^2 - 2i + 5)$

Solution

$$\sum_{i=1}^{n} (12i^2 - 2i + 5) = \sum_{i=1}^{n} 12i^2 + \sum_{i=1}^{n} (-2i) + \sum_{i=1}^{n} 5$$

$$= 12 \sum_{i=1}^{n} i^2 - 2 \sum_{i=1}^{n} i + \sum_{i=1}^{n} 5$$

We use Formulas 2 and 1 and Theorem 7.1.1 to obtain

$$12 \sum_{i=1}^{n} i^2 - 2 \sum_{i=1}^{n} i + \sum_{i=1}^{n} 5 = \frac{12n(n+1)(2n+1)}{6} - \frac{2n(n+1)}{2} + 5n$$

$$= 4n^3 + 6n^2 + 2n - n^2 - n + 5n$$
$$= 4n^3 + 5n^2 + 6n$$

EXAMPLE 3 In Sec. 7.2 we have the following expression:

$$\sum_{i=1}^{n} \frac{1}{n} \left[60 + 288 \left(1 + \frac{i-1}{n} \right)^2 \right]$$

Simplify the summation.

Solution

$$\sum_{i=1}^{n} \frac{1}{n}\left[60 + 288\left(1 + \frac{i-1}{n}\right)^2\right]$$

$$= \frac{1}{n}\sum_{i=1}^{n} 60 + \frac{288}{n}\sum_{i=1}^{n}\frac{(n+i-1)^2}{n^2}$$

$$= \frac{1}{n}(60n) + \frac{288}{n^3}\sum_{i=1}^{n}(n^2 + i^2 + 1 + 2ni - 2n - 2i)$$

$$= 60 + \frac{288}{n^3}\left[\sum_{i=1}^{n} i^2 + \sum_{i=1}^{n}(2n-2)i + \sum_{i=1}^{n}(n^2 - 2n + 1)\right]$$

$$= 60 + \frac{288}{n^3}\left[\sum_{i=1}^{n} i^2 + 2(n-1)\sum_{i=1}^{n} i + (n^2 - 2n + 1)\sum_{i=1}^{n} 1\right]$$

$$= 60 + \frac{288}{n^3}\left[\frac{n(n+1)(2n+1)}{6} + \frac{2(n-1)n(n+1)}{2} + (n^2 - 2n + 1)n\right]$$

$$= 60 + \frac{288}{n^3}\left[\frac{2n^3 + 3n^2 + n}{6} + n^3 - n + n^3 - 2n^2 + n\right]$$

$$= 60 + \frac{288}{n^3}\left[\frac{7}{3}n^3 - \frac{3}{2}n^2 + \frac{1}{6}n\right]$$

$$= 60 + 672 - \frac{432}{n} + \frac{48}{n^2}$$

$$= 732 - \frac{432}{n} + \frac{48}{n^2}$$

In this chapter we need to consider limits of functions when the independent variable either increases without bound or decreases without bound. Let the function f be defined by the equation

$$f(x) = \frac{2x^2}{x^2 + 1}$$

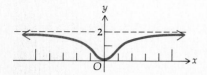

Figure 7.1.1

A sketch of the graph of this function is shown in Fig. 7.1.1. Let x take on the values 0, 1, 2, 3, 4, 5, 10, 100, 1000, and so on, allowing x to increase without bound. The corresponding function values are given in Table 7.1.1.

Observe from Table 7.1.1 that as x increases through positive values, the function values $f(x)$ get closer and closer to 2. In particular, when $x = 4$,

$$2 - \frac{2x^2}{x^2 + 1} = 2 - \frac{32}{17} = \frac{2}{17}$$

Table 7.1.1

x	0	1	2	3	4	5	10	100	1000
$f(x) = \dfrac{2x^2}{x^2 + 1}$	0	1	$\dfrac{8}{5}$	$\dfrac{18}{10}$	$\dfrac{32}{17}$	$\dfrac{50}{26}$	$\dfrac{200}{101}$	$\dfrac{20{,}000}{10{,}001}$	$\dfrac{2{,}000{,}000}{1{,}000{,}001}$

Therefore the difference between 2 and $f(x)$ is $\frac{2}{17}$ when $x = 4$. When $x = 100$,

$$2 - \frac{2x^2}{x^2 + 1} = 2 - \frac{20{,}000}{10{,}001} = \frac{2}{10{,}001}$$

Hence the difference between 2 and $f(x)$ is $2/10{,}001$ when $x = 100$.

Continuing on, we see intuitively that we can make the value of $f(x)$ as close to 2 as we please by taking x large enough. In other words, we can make the difference between 2 and $f(x)$ as small as we please by taking x greater than some sufficiently large positive number.

When an independent variable is increasing without bound through positive values, we write "$x \to +\infty$." From the illustrative example above, then, we can say that

$$\lim_{x \to +\infty} \frac{2x^2}{x^2 + 1} = 2$$

In general, we have the following definition.

Definition of $\lim_{x \to +\infty} f(x) = L$

> Let f be a function which is defined at every number in some interval $(a, +\infty)$. The **limit of $f(x)$, as x increases without bound, is L,** written
>
> $$\lim_{x \to +\infty} f(x) = L \tag{1}$$
>
> if $|f(x) - L|$ can be made as small as we please by taking x greater than some sufficiently large positive number.

Note: When $x \to +\infty$ is written, it does not have the same meaning as, for instance, $x \to 1000$. The symbol $x \to +\infty$ indicates the behavior of the variable x. However, (1) can be read as "the limit of $f(x)$ as x approaches positive infinity is L," bearing in mind this note.

Now consider the same function and let x take on the values 0, -1, -2, -3, -4, -5, -10, -100, -1000, and so on, allowing x to decrease through negative values without bound. Table 7.1.2 gives the corresponding function values of $f(x)$.

Table 7.1.2

x	0	-1	-2	-3	-4	-5	-10	-100	-1000
$f(x) = \dfrac{2x^2}{x^2 + 1}$	0	1	$\dfrac{8}{5}$	$\dfrac{18}{10}$	$\dfrac{32}{17}$	$\dfrac{50}{26}$	$\dfrac{200}{101}$	$\dfrac{20{,}000}{10{,}001}$	$\dfrac{2{,}000{,}000}{1{,}000{,}001}$

Observe that the function values are the same for the negative numbers as for the corresponding positive numbers. So we see intuitively that as x decreases without bound, $f(x)$ approaches 2, and more formally we say that we can make $|f(x) - 2|$ as small as we please by taking x less than some negative number having a sufficiently large absolute value. Using the symbolism "$x \to -\infty$" to denote that the variable x is decreasing without bound, we write

$$\lim_{x \to -\infty} \frac{2x^2}{x^2 + 1} = 2$$

In general, we have the following definition.

Definition of
$\lim\limits_{x \to -\infty} f(x) = L$

> Let f be a function which is defined at every number in some interval $(-\infty, a)$. The **limit of $f(x)$, as x decreases without bound, is L**, written
>
> $$\lim_{x \to -\infty} f(x) = L \qquad (2)$$
>
> if $|f(x) - L|$ can be made as small as we please by taking x less than some negative number having a sufficiently large absolute value.

Note: As in the note following the previous definition, the symbol $x \to -\infty$ indicates only the behavior of the variable x, but (2) can be read as "the limit of $f(x)$ as x approaches negative infinity is L."

Limit theorems 2, 4, 5, 6, 7, 8, 9, and 10 in Sec. 2.1 and Limit theorems 11 and 12 in Sec. 2.2 remain unchanged when "$x \to a$" is replaced by "$x \to +\infty$" or "$x \to -\infty$." We have the following additional limit theorem.

> **Limit Theorem 13** If r is any positive integer, then
>
> (i) $\displaystyle\lim_{x \to +\infty} \frac{1}{x^r} = 0$
>
> (ii) $\displaystyle\lim_{x \to -\infty} \frac{1}{x^r} = 0$

● ILLUSTRATION 4

From Example 3,

$$\sum_{i=1}^{n} \frac{1}{n}\left[60 + 288\left(1 + \frac{i-1}{n}\right)^2\right] = 732 - \frac{432}{n} + \frac{48}{n^2}$$

Thus

$$\lim_{n \to +\infty} \sum_{i=1}^{n} \frac{1}{n} \left[60 + 288 \left(1 + \frac{i-1}{n} \right)^2 \right] = \lim_{n \to +\infty} \left(732 - \frac{432}{n} + \frac{48}{n^2} \right)$$

$$= \lim_{n \to +\infty} 732 - 432 \lim_{n \to +\infty} \frac{1}{n} + 48 \lim_{n \to +\infty} \frac{1}{n^2}$$

$$= 732 - 432(0) + 48(0)$$

$$= 732 \qquad \bullet$$

EXAMPLE 4 Find each of the following limits:

(a) $\displaystyle \lim_{x \to +\infty} \frac{4x - 3}{2x + 5}$ (b) $\displaystyle \lim_{x \to -\infty} \frac{2x^2 - x + 5}{4x^3 - 1}$

Solution   (a) To use Limit theorem 13 we divide the numerator and denominator by x, thus giving

$$\lim_{x \to +\infty} \frac{4x - 3}{2x + 5} = \lim_{x \to +\infty} \frac{4 - \dfrac{3}{x}}{2 + \dfrac{5}{x}}$$

$$= \frac{\displaystyle \lim_{x \to +\infty} 4 - 3 \lim_{x \to +\infty} \frac{1}{x}}{\displaystyle \lim_{x \to +\infty} 2 + 5 \lim_{x \to +\infty} \frac{1}{x}}$$

$$= \frac{4 - 3 \cdot 0}{2 + 5 \cdot 0}$$

$$= 2$$

(b) To use Limit theorem 13 we divide the numerator and denominator by the highest power of x occurring in either the numerator or denominator, which in this case is x^3.

$$\lim_{x \to -\infty} \frac{2x^2 - x + 5}{4x^3 - 1} = \lim_{x \to -\infty} \frac{\dfrac{2}{x} - \dfrac{1}{x^2} + \dfrac{5}{x^3}}{4 - \dfrac{1}{x^3}}$$

$$= \frac{\displaystyle 2 \lim_{x \to -\infty} \frac{1}{x} - \lim_{x \to -\infty} \frac{1}{x^2} + 5 \lim_{x \to -\infty} \frac{1}{x^3}}{\displaystyle \lim_{x \to -\infty} 4 - \lim_{x \to -\infty} \frac{1}{x^3}}$$

$$= \frac{2 \cdot 0 - 0 + 5 \cdot 0}{4 - 0}$$

$$= 0$$

"Infinite" limits at infinity can be considered. There are formal definitions for each of the following.

$$\lim_{x \to +\infty} f(x) = +\infty \qquad \lim_{x \to -\infty} f(x) = +\infty$$

$$\lim_{x \to +\infty} f(x) = -\infty \qquad \lim_{x \to -\infty} f(x) = -\infty$$

For example, $\lim_{x \to +\infty} f(x) = +\infty$ if the function f is defined on some interval $(a, +\infty)$ and if $f(x)$ can be made larger than any preassigned positive number by taking x greater than some sufficiently large positive number.

EXAMPLE 5 Find each of the following limits:

(a) $\lim_{x \to +\infty} \dfrac{x^2}{x + 1}$ (b) $\lim_{x \to +\infty} \dfrac{2x - x^2}{3x + 5}$

Solution

(a) $\lim_{x \to +\infty} \dfrac{x^2}{x + 1} = \lim_{x \to +\infty} \dfrac{1}{\dfrac{1}{x} + \dfrac{1}{x^2}}$

Evaluating the limit of the denominator we have

$$\lim_{x \to +\infty} \left(\frac{1}{x} + \frac{1}{x^2} \right) = \lim_{x \to +\infty} \frac{1}{x} + \lim_{x \to +\infty} \frac{1}{x^2} = 0 + 0 = 0$$

Therefore the limit of the denominator is 0, and the denominator is approaching 0 through positive values.

The limit of the numerator is 1, and so by Limit theorem 12 it follows that

$$\lim_{x \to +\infty} \frac{x^2}{x + 1} = +\infty$$

(b) $\lim_{x \to +\infty} \dfrac{2x - x^2}{3x + 5} = \lim_{x \to +\infty} \dfrac{\dfrac{2}{x} - 1}{\dfrac{3}{x} + \dfrac{5}{x^2}}$

The limits of the numerator and the denominator are considered separately.

$$\lim_{x \to +\infty} \left(\frac{2}{x} - 1 \right) = \lim_{x \to +\infty} \frac{2}{x} - \lim_{x \to +\infty} 1 = 0 - 1 = -1$$

$$\lim_{x \to +\infty} \left(\frac{3}{x} + \frac{5}{x^2} \right) = \lim_{x \to +\infty} \frac{3}{x} + \lim_{x \to +\infty} \frac{5}{x^2} = 0 + 0 = 0$$

Therefore we have the limit of a quotient in which the limit of the numerator is -1 and the limit of the denominator is 0, where the denominator is approaching 0 through positive values. By Limit theorem 12 it follows that

$$\lim_{x \to +\infty} \frac{2x - x^2}{3x + 5} = -\infty$$

Exercises 7.1

In Exercises 1 through 8, write the terms of the sumation, and find the sum.

1. $\displaystyle\sum_{i=1}^{6} (3i - 2)$

2. $\displaystyle\sum_{i=1}^{7} (i + 1)^2$

3. $\displaystyle\sum_{i=2}^{5} \frac{i}{i - 1}$

4. $\displaystyle\sum_{j=3}^{6} \frac{2}{j(j - 2)}$

5. $\displaystyle\sum_{i=-2}^{3} 2^i$

6. $\displaystyle\sum_{i=0}^{3} \frac{1}{1 + i^2}$

7. $\displaystyle\sum_{k=1}^{4} \frac{(-1)^{k+1}}{k}$

8. $\displaystyle\sum_{k=-2}^{3} \frac{k}{k + 3}$

In Exercises 9 through 16, evaluate the indicated summation by using Theorems 7.1.1 through 7.1.4.

9. $\displaystyle\sum_{i=1}^{30} (i^2 + 3i + 1)$

10. $\displaystyle\sum_{i=1}^{40} (2i^2 - 4i + 1)$

11. $\displaystyle\sum_{i=1}^{25} 2i(i - 1)$

12. $\displaystyle\sum_{i=1}^{20} 3i(i^2 + 2)$

13. $\displaystyle\sum_{i=1}^{n} (i^3 + i + 5)$

14. $\displaystyle\sum_{i=1}^{n} (4i^2 - 3i - 5)$

15. $\displaystyle\sum_{i=1}^{n} 4i^2(i - 2)$

16. $\displaystyle\sum_{i=1}^{n} 2i(1 + i^2)$

In Exercises 17 through 32, evaluate the limit.

17. $\displaystyle\lim_{x \to +\infty} \frac{1}{x^3}$

18. $\displaystyle\lim_{x \to -\infty} \frac{3}{x^4}$

19. $\displaystyle\lim_{t \to +\infty} \frac{2t + 1}{5t - 2}$

20. $\displaystyle\lim_{x \to -\infty} \frac{6x - 4}{3x + 1}$

21. $\displaystyle\lim_{x \to -\infty} \frac{2x + 7}{4 - 5x}$

22. $\displaystyle\lim_{x \to +\infty} \frac{1 + 5x}{2 - 3x}$

23. $\displaystyle\lim_{x \to +\infty} \frac{7x^2 - 2x + 1}{3x^2 + 8x + 5}$

24. $\displaystyle\lim_{s \to -\infty} \frac{4s^2 + 3}{2s^2 - 1}$

25. $\displaystyle\lim_{x \to +\infty} \frac{x + 4}{3x^2 - 5}$

26. $\displaystyle\lim_{x \to +\infty} \frac{x^2 + 5}{x^3}$

27. $\displaystyle\lim_{y \to +\infty} \frac{2y^2 - 3y}{y + 1}$

28. $\displaystyle\lim_{x \to +\infty} \frac{x^2 - 2x + 5}{7x^3 + x + 1}$

29. $\displaystyle\lim_{x \to -\infty} \frac{4x^3 + 2x^2 - 5}{8x^3 + x + 2}$

30. $\displaystyle\lim_{x \to +\infty} \frac{3x^4 - 7x^2 + 2}{2x^4 + 1}$

31. $\displaystyle\lim_{y \to +\infty} \frac{2y^3 - 4}{5y + 3}$

32. $\displaystyle\lim_{x \to -\infty} \frac{5x^3 - 12x + 7}{4x^2 - 1}$

In Exercises 33 through 36, find the limit.

33. $\displaystyle\lim_{n \to +\infty} \sum_{i=1}^{n} (i - 1)^2 \cdot \frac{8}{n^3}$

34. $\displaystyle\lim_{n \to +\infty} \sum_{i=1}^{n} i^3 \cdot \frac{81}{n^4}$

35. $\displaystyle\lim_{n \to +\infty} \sum_{i=1}^{n} \left(1 + \frac{3i}{n} + \frac{3i^2}{n^2} + \frac{i^3}{n^3}\right) \frac{1}{n}$

36. $\displaystyle\lim_{n \to +\infty} \sum_{i=1}^{n} \left(4 - \frac{i^2}{n^2}\right) \frac{1}{n}$

7.2 Area

Suppose that a producer of a modern electronic calculator determines that during the first 3 years of production, if x years have elapsed since the calculator was first introduced, then $f(x)$ units per year should be produced, where

$$f(x) = 60 + 288x^2 \qquad 0 \le x \le 3$$

How should this equation be interpreted? Because

$$f(1) = 60 + 288(1)^2 = 348$$

one might conclude that 348 calculators are produced 1 year after the commodity first came on the market. However, this interpretation is valid only if the rate of production is constant on a yearly basis. That is, 348 units would be produced during the second year only if the annual rate of production during the second year is a constant 348 units, which is the level at which it was at the end of the first year. This situation does not occur. For instance,

$$f(1\tfrac{1}{2}) = 60 + 288(\tfrac{3}{2})^2 = 708$$

and

$$f(2) = 60 + 288(2)^2 = 1212$$

Since when $x = 1$, the production is 348 units, and when $x = 2$, the production is 1212 units, it follows that the number of calculators produced during the second year is between 348 and 1212. A better approximation arises from the argument that during the first half of the second year the number of units produced should be at least $\tfrac{1}{2}(348) = 174$ and during the second half it should be at least $\tfrac{1}{2}(708) = 354$; thus during the second year the number of calculators produced should be at least $174 + 354 = 528$. A geometric interpretation of this reasoning is shown in Fig. 7.2.1. The figure shows a sketch of the graph of

$$f(x) = 60 + 288x^2 \qquad 0 \le x \le 3$$

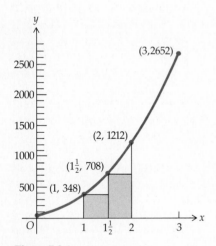

Figure 7.2.1

and two shaded rectangles. The number of square units in the area of the first rectangle is $\tfrac{1}{2}(348) = 174$, which would be the amount of calculators produced during the first half of the second year if production during that time is at a constant yearly rate of $f(1) = 348$. The number of square units in the area of the second rectangle is $\tfrac{1}{2}(708) = 354$, which is the amount of calculators that would be produced during the second half of the year if production during that time is at a constant yearly rate of $f(1\tfrac{1}{2}) = 708$.

For an even better approximation of the second-year production, we assume that during each month the production rate is the constant amount at which it is at the first of the month. With this assumption, as well as that each month is $\tfrac{1}{12}$ of a year, the number of units produced during the first

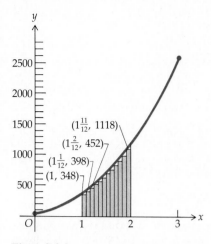

Figure 7.2.2

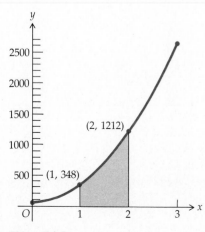

Figure 7.2.3

month of the second year is $\frac{1}{12}f(1) = \frac{1}{12}(348) = 29$; during the second month the number of units produced is $\frac{1}{12}f(1\frac{1}{12}) = \frac{1}{12}[60 + 288(\frac{13}{12})^2] = 33\frac{1}{6}$; during the third month, the number of units produced is $\frac{1}{12}f(1\frac{2}{12}) = \frac{1}{12}[60 + 288(\frac{7}{6})^2] = 37\frac{2}{3}$; and so on; during the twelfth month, the number of units produced is $\frac{1}{12}f(1\frac{11}{12}) = \frac{1}{12}[60 + 288(\frac{23}{12})^2] = 93\frac{1}{6}$. In a similar way we compute the number of units produced during the other eight months; and the total number of units produced during the second year is then

$$\frac{1}{12}[f(1) + f(1\tfrac{1}{12}) + f(1\tfrac{2}{12}) + f(1\tfrac{3}{12}) + f(1\tfrac{4}{12}) + f(1\tfrac{5}{12}) + f(1\tfrac{6}{12}) + f(1\tfrac{7}{12})$$
$$+ f(1\tfrac{8}{12}) + f(1\tfrac{9}{12}) + f(1\tfrac{10}{12}) + f(1\tfrac{11}{12})]$$

$$= 29 + 33\tfrac{1}{6} + 37\tfrac{2}{3} + 42\tfrac{1}{2} + 47\tfrac{2}{3} + 53\tfrac{1}{6} + 59 + 65\tfrac{1}{6} + 71\tfrac{2}{3} + 78\tfrac{1}{2}$$
$$+ 85\tfrac{2}{3} + 93\tfrac{1}{6}$$

$$= 696\tfrac{1}{3} \tag{1}$$

We can conclude from this approximation that the number of calculators produced during the second year is at least 696.

Summation (1) represents the sum of the areas of the 12 shaded rectangles in Fig. 7.2.2. The first rectangle has a height of $f(1)$ units and a width of $\frac{1}{12}$ unit; thus the number of square units in its area is $\frac{1}{12}f(1) = \frac{1}{12}(348) = 29$. For the second rectangle the height is $f(1\frac{1}{12})$ units and its width is $\frac{1}{12}$ unit, and so the number of square units in its area is $\frac{1}{12}f(1\frac{1}{12}) = \frac{1}{12}(398) = 33\frac{1}{6}$. And so on. The twelfth rectangle has a height of $f(1\frac{11}{12})$ units and a width of $\frac{1}{12}$ unit; therefore the number of square units in its area is $\frac{1}{12}f(1\frac{11}{12}) = \frac{1}{12}(1118) = 93\frac{1}{6}$.

We can improve the approximation of the second-year production by keeping the production rate constant during a length of time shorter than a month. For instance, suppose that we assume that during each week the production rate is the constant amount at which it is at the beginning of the week. Then the number of calculators produced during the second year would be given by the number of square units in the sum of the areas of 52 rectangles. We can improve upon the approximation even further by assuming that during each day the production rate is the constant amount at which it is at the start of the day. Then the number of square units in the sum of the areas of 365 rectangles will be the approximation of the number of calculators produced during the second year. If we take smaller and smaller lengths of time over which the production rate is assumed to be constant, the number of rectangles increases without bound and the total number of calculators produced will get closer and closer to the number of square units in the area of the shaded region shown in Fig. 7.2.3.

We proceed now to obtain a formula for computing the sum of the areas of n rectangles inscribed in the shaded region of Fig. 7.2.3. Recall that when there are two inscribed rectangles, as in Fig. 7.2.1, the sum of the areas of the rectangles is

$$\tfrac{1}{2}[f(1) + f(1\tfrac{1}{2})] = 528 \tag{2}$$

When there are twelve inscribed rectangles, as in Fig. 7.2.2, the sum of the areas of the rectangles is

$$\tfrac{1}{12}[f(1) + f(1\tfrac{1}{12}) + f(1\tfrac{2}{12}) + \cdots + f(1\tfrac{11}{12})] = 696\tfrac{1}{3} \tag{3}$$

The formula we want will be a generalization of the expressions on the left-hand sides of (2) and (3).

Figure 7.2.4 shows an enlargement of the shaded region in Fig. 7.2.3. The closed interval $[1, 2]$ along the x axis is divided into n equal subintervals. Each of these subintervals has a length of $\dfrac{1}{n}$ units. The left endpoints of the subintervals are $1,\ 1 + \dfrac{1}{n},\ 1 + \dfrac{2}{n},\ 1 + \dfrac{3}{n}, \ldots, 1 + \dfrac{i-1}{n}, \ldots, 1 + \dfrac{n-1}{n}$, where $1 + \dfrac{i-1}{n}$ denotes the left endpoint of the ith subinterval. Consider the n rectangles having as bases these subintervals and as altitudes the function values at the left endpoints of the subintervals. In Fig. 7.2.4 you'll see the first three rectangles, the ith rectangle, and the nth rectangle. Let S_n be

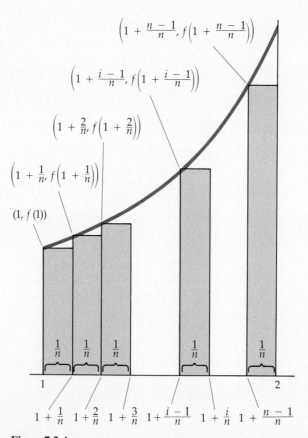

Figure 7.2.4

the sum of the areas of the n rectangles. Then

$$S_n = \frac{1}{n} f(1) + \frac{1}{n} f\left(1 + \frac{1}{n}\right) + \frac{1}{n} f\left(1 + \frac{2}{n}\right) + \ldots + \frac{1}{n} f\left(1 + \frac{i-1}{n}\right) +$$

$$\ldots + \frac{1}{n} f\left(1 + \frac{n-1}{n}\right)$$

$$= \frac{1}{n}\left[f(1) + f\left(1 + \frac{1}{n}\right) + f\left(1 + \frac{2}{n}\right) + \ldots + f\left(1 + \frac{i-1}{n}\right) + \right.$$

$$\left. \ldots + f\left(1 + \frac{n-1}{n}\right)\right]$$

$$= \sum_{i=1}^{n} \frac{1}{n} f\left(1 + \frac{i-1}{n}\right)$$

Because $f(x) = 60 + 288x^2$, then

$$f\left(1 + \frac{i-1}{n}\right) = 60 + 288\left(1 + \frac{i-1}{n}\right)^2$$

Therefore

$$S_n = \sum_{i=1}^{n} \frac{1}{n}\left[60 + 288\left(1 + \frac{i-1}{n}\right)^2 \right]$$

From Example 3 in Sec. 7.1,

$$\sum_{i=1}^{n} \frac{1}{n}\left[60 + 288\left(1 + \frac{i-1}{n}\right)^2 \right] = 732 - \frac{432}{n} + \frac{48}{n^2}$$

Thus

$$S_n = 732 - \frac{432}{n} + \frac{48}{n^2} \tag{4}$$

If in (4), $n = 2$, then we have

$$S_2 = 732 - \frac{432}{2} + \frac{48}{2^2}$$

$$= 732 - 216 + 12$$

$$= 528$$

which agrees with (2). If $n = 12$ in (4), we have

$$S_{12} = 732 - \frac{432}{12} + \frac{48}{12^2}$$

$$= 732 - 36 + \tfrac{1}{3}$$

$$= 696\tfrac{1}{3}$$

which is consistent with (3).

By using formula (4) we can compute the sum of the areas of any number of rectangles inscribed in the shaded region of Fig. 7.2.3. If the production rate is held constant weekly at the amount at which it is at the beginning of the week, then we have 52 rectangles, and from (4),

$$S_{52} = 732 - \frac{432}{52} + \frac{48}{(52)^2}$$

$$= 732 - 8\frac{4}{13} + \frac{3}{169}$$

$$= 723.71$$

If the production rate is held constant each day, for 365 days per year, at the amount at which it is at the start of the day, there are 365 rectangles, and from (4),

$$S_{365} = 732 - \frac{432}{365} + \frac{48}{(365)^2}$$

$$= 732 - 1.18356 + 0.00036$$

$$= 730.82$$

If the number of rectangles increases without bound, that is, $n \to +\infty$, we have, as in Illustration 4 of Sec. 7.1,

$$\lim_{n \to +\infty} S_n = \lim_{n \to +\infty} \left(732 - \frac{432}{n} + \frac{48}{n^2} \right)$$

$$= \lim_{n \to +\infty} 732 - 432 \lim_{n \to +\infty} \frac{1}{n} + 48 \lim_{n \to +\infty} \frac{1}{n^2}$$

$$= 732 - 432(0) + 48(0)$$

$$= 732$$

This result of 732 represents the actual number of calculators produced during the second year. From a geometric standpoint, 732 is the number of square units in the area of the region bounded above by the graph of $f(x) = 60 + 288x^2$, below by the x axis, and on the sides by the lines $x = 1$ and $x = 2$; that is, the area of the shaded region of Fig. 7.2.3 is 732 square units. We now apply the method used to find this area to obtain a formula for the area of a plane region bounded by a curve.

The word *measure* is used in the development. A measure refers to a number (no units are included). For example, if the area of a triangle is 10 in.^{2}, we say that the measure of the area of the triangle is 10. You have an intuitive idea of what is meant by the measure of the area of certain geometrical figures; it is a number that in some way gives the size of the region enclosed by the figure. The area of a rectangle is the product of its length and width, and the area of a triangle is half the product of the lengths of a base and its corresponding altitude.

The area of a polygon can be defined as the sum of the areas of triangles into which it is decomposed, and it can be proved that the area thus obtained is independent of how the polygon is decomposed into triangles (see Fig.

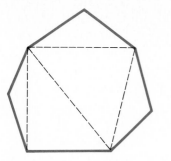

Figure 7.2.5

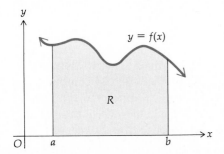

Figure 7.2.6

7.2.5). We proceed to define the measure of the area of a region in a plane if the region is bounded by a curve.

Consider a region R in the plane, as shown in Fig. 7.2.6. The region R is bounded by the x axis, the lines $x = a$ and $x = b$, and the curve having the equation $y = f(x)$, where f is a function continuous on the closed interval $[a, b]$. For simplicity, take $f(x) \geq 0$ for all x in $[a, b]$. We wish to assign a number A to be the measure of the area of R, and we use a limiting process similar to the one used in defining the area of a circle: The area of a circle is defined as the limit of the areas of inscribed regular polygons as the number of sides increases without bound. We realize intuitively that, whatever number is chosen to represent A, that number must be at least as great as the measure of the area of any polygonal region contained in R, and it must be no greater than the measure of the area of any polygonal region containing R.

We first define a polygonal region contained in R. Divide the closed interval $[a, b]$ into n subintervals. For convenience we take each of these subintervals as being of equal length, for instance, Δx. Therefore $\Delta x = (b - a)/n$. Denote the endpoints of these subintervals by $x_0, x_1, x_2, \ldots, x_{n-1}, x_n$, where $x_0 = a$, $x_1 = a + \Delta x, \ldots, x_i = a + i \Delta x, \ldots, x_{n-1} = a + (n - 1) \Delta x$, $x_n = b$. Let the ith subinterval be denoted by $[x_{i-1}, x_i]$. Because f is continuous on the closed interval $[a, b]$, it is continuous on each closed subinterval. By the extreme-value theorem (4.1.2) there is a number in each subinterval for which f has an absolute minimum value. In the ith subinterval let this number be c_i, so that $f(c_i)$ is the absolute minimum value of f on the subinterval $[x_{i-1}, x_i]$. Consider n rectangles, each having a width Δx units and an altitude $f(c)$ units (see Fig. 7.2.7). Let the sum of the areas

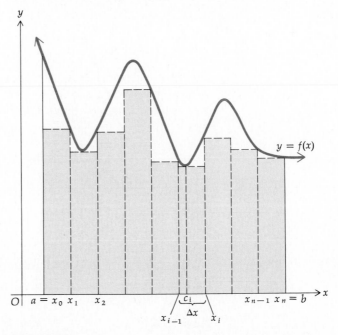

Figure 7.2.7

of these n rectangles be given by S_n square units; then

$$S_n = f(c_1)\,\Delta x + f(c_2)\,\Delta x + \ldots + f(c_i)\,\Delta x + \ldots + f(c_n)\Delta x$$

or, with the sigma notation,

$$S_n = \sum_{i=1}^{n} f(c_i)\,\Delta x \tag{5}$$

The summation on the right side of (5) gives the sum of the measures of the areas of n inscribed rectangles. Thus, however we define A, it must be such that

$$A \geq S_n$$

In Fig. 7.2.7 the shaded region has an area of S_n square units. Now, let n increase. Specifically, multiply n by 2; then the number of rectangles is doubled, and the width of each rectangle is halved. This is illustrated in Fig. 7.2.8, showing twice as many rectangles as Fig. 7.2.7. By comparing the two figures, notice that the shaded region in Fig. 7.2.8 appears to approximate the region R more nearly than that of Fig. 7.2.7. So the sum of the measures of the areas of the rectangles in Fig. 7.2.8 is closer to the number we wish to represent the measure of the area of R.

As n increases, the values of S_n found from (5) increase, and successive values of S_n differ from each other by amounts that become arbitrarily small. This is proved in advanced calculus by a theorem that states that if f

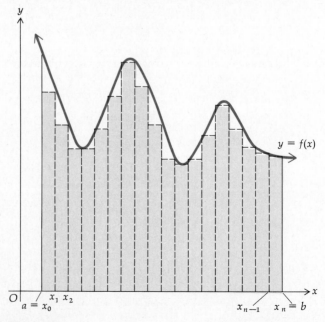

Figure 7.2.8

is continuous on $[a, b]$, then as n increases without bound, the value of S_n given by (5) approaches a limit. It is this limit that we take as the measure of the area of region R.

Definition of the area of a plane region bounded by a curve

> Suppose that the function f is continuous on the closed interval $[a, b]$, with $f(x) \geq 0$ for all x in $[a, b]$, and that R is the region bounded by the curve $y = f(x)$, the x axis, and the lines $x = a$ and $x = b$. Divide the interval $[a, b]$ into n subintervals, each of length $\Delta x = (b - a)/n$, and denote the ith subinterval by $[x_{i-1}, x_i]$. Then if $f(c_i)$ is the absolute minimum function value on the ith subinterval, the measure of the area of region R is given by
>
> $$A = \lim_{n \to +\infty} \sum_{i=1}^{n} f(c_i) \, \Delta x \qquad (6)$$

Equation (6) means that

$$\left| \sum_{i=1}^{n} f(c_i) \, \Delta x - A \right|$$

can be made as small as we please by taking n as a positive integer greater than some sufficiently large positive number.

We could take circumscribed rectangles instead of inscribed rectangles. In this case we take as the measures of the altitudes of the rectangles the absolute maximum value of f on each subinterval. The existence of an absolute maximum value of f on each subinterval is guaranteed by the extreme-value theorem. The corresponding sums of the measures of the areas of the circumscribed rectangles are at least as great as the measure of the area of the region R, and it can be shown that the limit of these sums as n increases without bound is exactly the same as the limit of the sum of the measures of the areas of the inscribed rectangles. This is also proved in advanced calculus. Thus we could define the measure of the area of the region R by

$$A = \lim_{n \to +\infty} \sum_{i=1}^{n} f(d_i) \, \Delta x \qquad (7)$$

where $f(d_i)$ is the absolute maximum value of f on $[x_{i-1}, x_i]$.

The measure of the altitude of the rectangle in the ith subinterval actually can be taken as the function value of any number in that subinterval, and the limit of the sum of the measures of the areas of the rectangles is the same no matter what numbers are selected. This is also proved in advanced calculus,

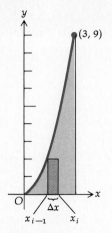

Figure 7.2.9

and in Sec. 7.3 we extend the definition of the measure of the area of a region to be the limit of such a sum.

EXAMPLE 1 Find the area of the region bounded by the curve $y = x^2$, the x axis, and the line $x = 3$ by taking inscribed rectangles.

Solution Figure 7.2.9 shows the region and the ith inscribed rectangle. Divide the closed interval $[0, 3]$ into n subintervals, each of length Δx: $x_0 = 0$, $x_1 = \Delta x$, $x_2 = 2 \Delta x, \ldots, x_i = i \Delta x, \ldots, x_{n-1} = (n-1) \Delta x$, $x_n = 3$.

$$\Delta x = \frac{3 - 0}{n} = \frac{3}{n} \qquad f(x) = x^2$$

Because f is increasing on $[0, 3]$, the absolute minimum value of f on the ith subinterval $[x_{i-1}, x_i]$ is $f(x_{i-1})$. Therefore, from (6),

$$A = \lim_{n \to +\infty} \sum_{i=1}^{n} f(x_{i-1}) \Delta x \tag{8}$$

Because $x_{i-1} = (i-1) \Delta x$ and $f(x) = x^2$,

$$f(x_{i-1}) = [(i-1) \Delta x]^2$$

Therefore

$$\sum_{i=1}^{n} f(x_{i-1}) \Delta x = \sum_{i=1}^{n} (i-1)^2 (\Delta x)^3$$

But $\Delta x = 3/n$; so

$$\sum_{i=1}^{n} f(x_{i-1}) \Delta x = \sum_{i=1}^{n} (i-1)^2 \frac{27}{n^3} = \frac{27}{n^3} \sum_{i=1}^{n} (i-1)^2$$

$$= \frac{27}{n^3} \left[\sum_{i=1}^{n} i^2 - 2 \sum_{i=1}^{n} i + \sum_{i=1}^{n} 1 \right]$$

and using Formulas 2 and 1 from Sec. 7.1 and Theorem 7.1.1 we get

$$\sum_{i=1}^{n} f(x_{i-1}) \Delta x = \frac{27}{n^3} \left[\frac{n(n+1)(2n+1)}{6} - 2 \cdot \frac{n(n+1)}{2} + n \right]$$

$$= \frac{27}{n^3} \cdot \frac{2n^3 + 3n^2 + n - 6n^2 - 6n + 6n}{6}$$

$$= \frac{9}{2} \cdot \frac{2n^2 - 3n + 1}{n^2}$$

Then, from (8),

$$A = \lim_{n \to +\infty} \left[\frac{9}{2} \cdot \frac{2n^2 - 3n + 1}{n^2} \right]$$

$$= \frac{9}{2} \cdot \lim_{n \to +\infty} \left(2 - \frac{3}{n} + \frac{1}{n^2} \right)$$

$$= \tfrac{9}{2}(2 - 0 + 0)$$

$$= 9$$

Therefore the area of the region is 9 square units.

EXAMPLE 2 Find the area of the region in Example 1 by taking circumscribed rectangles.

Solution With circumscribed rectangles the measure of the altitude of the ith rectangle is the absolute maximum value of f on the ith subinterval $[x_{i-1}, x_i]$, which is $f(x_i)$. From (7),

$$A = \lim_{n \to +\infty} \sum_{i=1}^{n} f(x_i) \, \Delta x \tag{9}$$

Because $x_i = i \, \Delta x$, then $f(x_i) = (i \, \Delta x)^2$, and so

$$\sum_{i=1}^{n} f(x_i) \, \Delta x = \sum_{i=1}^{n} i^2 (\Delta x)^3 = \frac{27}{n^3} \sum_{i=1}^{n} i^2$$

$$= \frac{27}{n^3} \left[\frac{n(n+1)(2n+1)}{6} \right]$$

$$= \frac{9}{2} \cdot \frac{2n^2 + 3n + 1}{n^2}$$

Therefore, from (9),

$$A = \lim_{n \to +\infty} \frac{9}{2} \left(2 + \frac{3}{n} + \frac{1}{n^2} \right)$$

$$= 9$$

as in Example 1.

EXAMPLE 3 Find the area of the trapezoid that is the region bounded by the line $2x + y = 8$, the x axis, and the lines $x = 1$ and $x = 3$. Take inscribed rectangles.

Solution The region and the ith inscribed rectangle are shown in Fig. 7.2.10. The closed interval $[1, 3]$ is divided into n subintervals, each of length

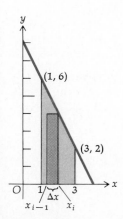

Figure 7.2.10

Δx: $x_0 = 1$, $x_1 = 1 + \Delta x$, $x_2 = 1 + 2 \Delta x$, ..., $x_i = 1 + i \Delta x$, ..., $x_{n-1} = 1 + (n-1) \Delta x$, $x_n = 3$.

$$\Delta x = \frac{3-1}{n} = \frac{2}{n}$$

Solving the equation of the line for y we obtain $y = -2x + 8$. Therefore $f(x) = -2x + 8$, and because f is decreasing on $[1, 3]$, the absolute minimum value of f on the ith subinterval $[x_{i-1}, x_i]$ is $f(x_i)$. Because $x_i = 1 + i \Delta x$ and $f(x) = -2x + 8$, then $f(x_i) = -2(1 + i \Delta x) + 8 = 6 - 2i \Delta x$. From (6),

$$A = \lim_{n \to +\infty} \sum_{i=1}^{n} f(x_i) \Delta x$$

$$= \lim_{n \to +\infty} \sum_{i=1}^{n} (6 - 2i \Delta x) \Delta x$$

$$= \lim_{n \to +\infty} \sum_{i=1}^{n} [6 \Delta x - 2i(\Delta x)^2]$$

$$= \lim_{n \to +\infty} \sum_{i=1}^{n} \left[6\left(\frac{2}{n}\right) - 2i\left(\frac{2}{n}\right)^2 \right]$$

$$= \lim_{n \to +\infty} \left[\frac{12}{n} \sum_{i=1}^{n} 1 - \frac{8}{n^2} \sum_{i=1}^{n} i \right]$$

From Theorem 7.1.1 and Formula 1 in Sec. 7.1,

$$A = \lim_{n \to +\infty} \left[\frac{12}{n} \cdot n - \frac{8}{n^2} \cdot \frac{n(n+1)}{2} \right]$$

$$= \lim_{n \to +\infty} \left(8 - \frac{4}{n} \right)$$

$$= 8$$

Therefore the area is 8 square units. Using the formula from plane geometry for the area of a trapezoid, $A = \frac{1}{2}h(b_1 + b_2)$, where h, b_1, and b_2 are, respectively, the number of units in the lengths of the altitude and the two bases, we get $A = \frac{1}{2}(2)(6 + 2) = 8$, which agrees with the result.

Exercises 7.2

In Exercises 1 through 16, use the method of this section to find the area of the given region; use inscribed or circumscribed rectangles as indicated. For each exercise draw a figure showing the region and the ith rectangle.

1. The region bounded by $y = x^2$, the x axis, and the line $x = 2$; inscribed rectangles.
2. The region of Exercise 1; circumscribed rectangles.
3. The region bounded by $y = 2x$, the x axis, and the lines $x = 1$ and $x = 4$; circumscribed rectangles.

4. The region of Exercise 3; inscribed rectangles.
5. The region above the x axis and to the right of the line $x = 1$ bounded by the x axis, the line $x = 1$, and the curve $y = 4 - x^2$; inscribed rectangles.
6. The region of Exercise 5; circumscribed rectangles.
7. The region lying to the left of the line $x = 1$ bounded by the curve and lines of Exercise 5; circumscribed rectangles.
8. The region of Exercise 7; inscribed rectangles.
9. The region bounded by $y = 3x^4$, the x axis, and the line $x = 1$; inscribed rectangles.
10. The region of Exercise 9; circumscribed rectangles.
11. The region bounded by $y = x^3$, the x axis, and the lines $x = -1$ and $x = 2$; inscribed rectangles.
12. The region of Exercise 11; circumscribed rectangles.
13. The region bounded by $y = x^3 + x$, the x axis, and the lines $x = -2$ and $x = 1$; circumscribed rectangles.
14. The region of Exercise 13; inscribed rectangles.
15. The region bounded by $y = mx$, with $m > 0$, the x axis, and the lines $x = a$ and $x = b$, with $b > a > 0$; circumscribed rectangles.
16. The region of Exercise 15; inscribed rectangles.
17. Use the method of this section to find the area of an isosceles trapezoid whose bases have measures b_1 and b_2 and whose altitude has measure h.
18. The graph of $y = 4 - |x|$ and the x axis from $x = -4$ to $x = 4$ form a triangle. Use the method of this section to find the area of this triangle.

In Exercises 19 through 24, find the area of the region by taking as the measure of the altitude of the ith rectangle $f(m_i)$, where m_i is the midpoint of the ith subinterval. ($Hint$: $m_i = \frac{1}{2}(x_i + x_{i-1})$.)

19. The region of Example 1.   20. The region of Exercise 1.   21. The region of Exercise 3.
22. The region of Exercise 5.   23. The region of Exercise 7.   24. The region of Exercise 9.
25. Suppose that the function discussed at the beginning of this section is linear, so that

$$f(x) = 60 + 288x \qquad 0 \leq x \leq 3$$

where $f(x)$ units per year should be produced when x years have elapsed since the calculator was first introduced. (a) Interpret the number of calculators produced during the second year as the measure of the area of a region enclosed by a trapezoid, and draw a figure showing the trapezoid. (b) Use the method of this section to compute the number of calculators produced during the second year. (c) Verify the result of part (b) by using the formula for the area of a region enclosed by a trapezoid: $A = \frac{1}{2}(b_1 + b_2)h$, where b_1 units and b_2 units are the lengths of the parallel sides and h units is the altitude.

26. During the first 4 months, the sales of a new product is expected to be $f(x)$ units per month x months after the product was first put on the market, where

$$f(x) = 100 + 150x^2 \qquad 0 \leq x \leq 4$$

(a) Interpret the expected second-month sales as the measure of the area of a plane region, and draw a figure showing the plane region. (b) Use the method of this section to compute the expected second-month sales by finding the area of the plane region in part (a).

27. For the product of Exercise 26, do the following: (a) Interpret the expected first 3 months' sales as the measure of the area of a plane region, and draw a figure showing the plane region. (b) Use the method of this section to compute the expected first 3 months' sales by finding the area of the region in part (a).

7.3 The definite integral

In Sec. 7.2 the measure of the area of a region was defined as the following limit:

$$\lim_{n \to +\infty} \sum_{i=1}^{n} f(c_i) \, \Delta x \tag{1}$$

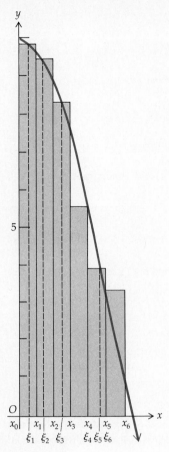

Figure 7.3.1

To lead up to this definition we divided the closed interval $[a, b]$ into subintervals of equal length and then took c_i as the point in the ith subinterval for which f has an absolute minimum value. We also restricted the function values $f(x)$ to be nonnegative on $[a, b]$ and further required f to be continuous on $[a, b]$.

The summation in (1) is a special case of the summation we use to define the *definite integral*. Let f be a function defined on the closed interval $[a, b]$. Divide this interval into n equal subintervals each of length Δx. Therefore $\Delta x = (b - a)/n$. Denote the endpoints of these subintervals by $x_0, x_1, x_2, \ldots, x_{n-1}, x_n$, where $x_0 = a$, $x_1 = a + \Delta x, \ldots, x_i = a + i \Delta x$, $\ldots, x_{n-1} = a + (n - 1) \Delta x$, $x_n = b$. Let the ith subinterval be denoted by $[x_{i-1}, x_i]$.

Choose a point in each subinterval: Let ξ_1 be the point chosen in $[x_0, x_1]$ such that $x_0 \leq \xi_1 \leq x_1$. Let ξ_2 be the point chosen in $[x_1, x_2]$ such that $x_1 \leq \xi_2 \leq x_2$, and so forth, such that ξ_i is the point chosen in $[x_{i-1}, x_i]$, and $x_{i-1} \leq \xi_i \leq x_i$. Form the sum

$$f(\xi_1) \Delta x + f(\xi_2) \Delta x + \ldots + f(\xi_i) \Delta x + \ldots + f(\xi_n) \Delta x$$

or

$$\sum_{i=1}^{n} f(\xi_i) \Delta x \tag{2}$$

Such a sum is called a **Riemann sum**, named for the mathematician Georg Friedrich Bernhard Riemann (1826–1866).

● ILLUSTRATION 1

Suppose $f(x) = 10 - x^2$ with $0 \leq x \leq 3$. We will find the Riemann sum (2) for the function f on $[0, 3]$ if there are six subintervals with $x_0 = 0$, $x_1 = \frac{1}{2}$, $x_2 = 1$, $x_3 = \frac{3}{2}$, $x_4 = 2$, $x_5 = \frac{5}{2}$, $x_6 = 3$, and $\xi_1 = \frac{1}{4}$, $\xi_2 = \frac{2}{3}$, $\xi_3 = \frac{5}{4}$, $\xi_4 = 2$, $\xi_5 = \frac{7}{3}$, $\xi_6 = \frac{5}{2}$.

Figure 7.3.1 shows a sketch of the graph of f on $[0, 3]$ and the six rectangles, the measures of whose areas are the terms of the Riemann sum.

$$\Delta x = \frac{3 - 0}{6} = \frac{1}{2}$$

$$\sum_{i=1}^{6} f(\xi_i) \Delta x = f(\xi_1) \Delta x + f(\xi_2) \Delta x + f(\xi_3) \Delta x + f(\xi_4) \Delta x + f(\xi_5) \Delta x + f(\xi_6) \Delta x$$

$$= f(\tfrac{1}{4})\tfrac{1}{2} + f(\tfrac{2}{3})\tfrac{1}{2} + f(\tfrac{5}{4})\tfrac{1}{2} + f(2)\tfrac{1}{2} + f(\tfrac{7}{3})\tfrac{1}{2} + f(\tfrac{5}{2})\tfrac{1}{2}$$

$$= \tfrac{1}{2}(9\tfrac{15}{16} + 9\tfrac{5}{9} + 8\tfrac{7}{16} + 6 + 4\tfrac{5}{9} + 3\tfrac{3}{4})$$

$$= 21\tfrac{17}{144}$$

●

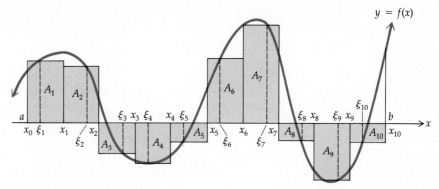

Figure 7.3.2

Because the function values $f(x)$ are not restricted to nonnegative values, some of the $f(\xi_i)$ could be negative. In such a case, the geometric interpretation of the Riemann sum would be the sum of the measures of the areas of the rectangles lying above the x axis plus the negatives of the measures of the areas of the rectangles lying below the x axis. This situation is illustrated in Fig. 7.3.2. Here

$$\sum_{i=1}^{10} f(\xi_i)\,\Delta x = A_1 + A_2 - A_3 - A_4 - A_5 + A_6 + A_7 - A_8 - A_9 - A_{10}$$

because $f(\xi_3)$, $f(\xi_4)$, $f(\xi_5)$, $f(\xi_8)$, $f(\xi_9)$, and $f(\xi_{10})$ are negative numbers.

We now define the *definite integral*.

Definition of the definite integral

If f is a function defined on the closed interval $[a, b]$, then the **definite integral** of f from a to b, denoted by $\int_a^b f(x)\,dx$, is given by

$$\int_a^b f(x)\,dx = \lim_{n \to +\infty} \sum_{i=1}^{n} f(\xi_i)\,\Delta x \qquad (3)$$

if the limit exists, where ξ_i is any number in the closed interval $[x_{i-1}, x_i]$, $i = 1, 2, \ldots, n$.

The definite integral can be defined more generally by allowing the n subintervals to be of different widths and requiring the largest width to approach zero as n increases without bound. However, the more restricted definition given is sufficient for our purposes.

In the notation for the definite integral $\int_a^b f(x)\,dx$, $f(x)$ is called the

integrand, a is called the **lower limit**, and b is called the **upper limit**. The symbol

$$\int$$

is called an **integral sign**. The integral sign resembles a capital S, which is appropriate because the definite integral is the limit of a sum. It is the same symbol we used in Chapter 6 to indicate the operation of antidifferentiation. The reason for the common symbol is that a theorem (7.3.2), called the fundamental theorem of the calculus, enables us to evaluate a definite integral by finding an antiderivative (also called an **indefinite integral**).

The statement "the function f is **integrable** on the closed interval $[a, b]$" is equivalent to the statement "the definite integral of f from a to b exists." The following theorem gives a condition for which a function is integrable.

Theorem 7.3.1 If a function f is continuous on the closed interval $[a, b]$, then f is integrable on $[a, b]$.

The definition of the area of a plane region bounded by a curve given in Sec. 7.2 stated that

$$A = \lim_{n \to +\infty} \sum_{i=1}^{n} f(c_i)\, \Delta x$$

where $f(c_i)$ is the absolute minimum function value on the ith subinterval. We now give a more general definition that allows $f(\xi_i)$ to be any function value on the ith subinterval.

Definition of the area of a plane region bounded by a curve

Let the function f be continuous on $[a, b]$, and $f(x) \geq 0$ for all x in $[a, b]$. Let R be the region bounded by the curve $y = f(x)$, the x axis, and the lines $x = a$ and $x = b$. Then the measure of the area of region R is given by

$$A = \lim_{n \to +\infty} \sum_{i=1}^{n} f(\xi_i)\, \Delta x = \int_a^b f(x)\, dx$$

The above definition states that if $f(x) \geq 0$ for all x in $[a, b]$, the definite integral $\int_a^b f(x)\, dx$ can be interpreted geometrically as the measure of the area of the region R shown in Fig. 7.3.3.

Equation (3) can be used to find the exact value of a definite integral, as illustrated in the following example.

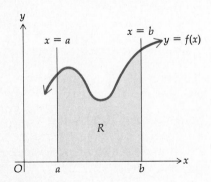

Figure 7.3.3

EXAMPLE 1 Find the exact value of the definite integral $\int_1^3 x^2\,dx$. Interpret the result geometrically.

Solution Divide the interval $[1, 3]$ into n subintervals of equal length. Then $\Delta x = 2/n$.

If we choose ξ_i as the right endpoint of each subinterval, we have

$$\xi_1 = 1 + \frac{2}{n}, \xi_2 = 1 + 2\left(\frac{2}{n}\right), \xi_3 = 1 + 3\left(\frac{2}{n}\right), \ldots, \xi_i = 1 + i\left(\frac{2}{n}\right),$$

$$\ldots, \xi_n = 1 + n\left(\frac{2}{n}\right)$$

Because $f(x) = x^2$,

$$f(\xi_i) = \left(1 + \frac{2i}{n}\right)^2 = \left(\frac{n + 2i}{n}\right)^2$$

Therefore, by using (3) and applying theorems from Sec. 7.1, we get

$$\int_1^3 x^2\,dx = \lim_{n \to +\infty} \sum_{i=1}^n \left(\frac{n + 2i}{n}\right)^2 \frac{2}{n}$$

$$= \lim_{n \to +\infty} \frac{2}{n^3} \sum_{i=1}^n (n^2 + 4ni + 4i^2)$$

$$= \lim_{n \to +\infty} \frac{2}{n^3} \left[n^2 \sum_{i=1}^n 1 + 4n \sum_{i=1}^n i + 4 \sum_{i=1}^n i^2 \right]$$

$$= \lim_{n \to +\infty} \frac{2}{n^3} \left[n^2 n + 4n \cdot \frac{n(n + 1)}{2} + \frac{4n(n + 1)(2n + 1)}{6} \right]$$

$$= \lim_{n \to +\infty} \frac{2}{n^3} \left[n^3 + 2n^3 + 2n^2 + \frac{2n(2n^2 + 3n + 1)}{3} \right]$$

$$= \lim_{n \to +\infty} \left[6 + \frac{4}{n} + \frac{8n^2 + 12n + 4}{3n^2} \right]$$

$$= \lim_{n \to +\infty} \left[6 + \frac{4}{n} + \frac{8}{3} + \frac{4}{n} + \frac{4}{3n^2} \right]$$

$$= 6 + 0 + \tfrac{8}{3} + 0 + 0$$

$$= 8\tfrac{2}{3}$$

We interpret the result geometrically. Because $x^2 \geq 0$ for all x in $[1, 3]$, the region bounded by the curve $y = x^2$, the x axis, and the lines $x = 1$ and $x = 3$ has an area of $8\tfrac{2}{3}$ square units. The region is shown in Fig. 7.3.4.

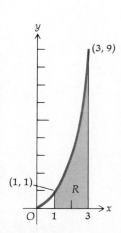

Figure 7.3.4

Historically, the basic concepts of the definite integral were used by the ancient Greeks, principally Archimedes (287–212 B.C.), more than 2000 years ago, which was many years before the differential calculus was invented. In the seventeenth century, almost simultaneously but working independently, Newton and Leibniz showed how the calculus could be used to find the area of a region bounded by a curve or a set of curves by evaluating a definite integral by antidifferentiation. The procedure involves what is known as the *fundamental theorem of the calculus*. We now state this important theorem.

Fundamental theorem
of the calculus

Theorem 7.3.2   Let the function f be continuous on the closed interval $[a, b]$, and let g be any antiderivative of f on $[a, b]$; that is,

$$g'(x) = f(x)$$

for all x in $[a, b]$. Then

$$\int_a^b f(x)\, dx = g(b) - g(a)$$

A complete proof of the fundamental theorem is beyond the scope of this book. However, in Sec. 7.7 a discussion of the proof is given. We now give some illustrations and examples showing how Theorem 7.3.2 is applied. When using the theorem we denote

$$[g(b) - g(a)] \qquad \text{by} \qquad g(x) \Big]_a^b$$

● ILLUSTRATION 2

We apply the fundamental theorem of the calculus to evaluate the definite integral of Example 1:

$$\int_1^3 x^2\, dx$$

Here $f(x) = x^2$. Because g can be any antiderivative of f, we may choose

$$g(x) = \tfrac{1}{3}x^3$$

Therefore, from Theorem 7.3.2 we get

$$\int_1^3 x^2\, dx = \tfrac{1}{3}x^3 \Big]_1^3 = 9 - \tfrac{1}{3} = 8\tfrac{2}{3}$$

This result agrees with that of Example 1.   ●

In Illustration 2, any antiderivative of f is of the form $\tfrac{1}{3}x^3 + C$, where C is an arbitrary constant, and if this form is selected for the antiderivative of f, the solution is written as follows:

$$\int_1^3 x^2 \, dx = \frac{1}{3}x^3 + C \Big]_1^3$$
$$= (9 + C) - (\tfrac{1}{3} + C)$$
$$= 9 + C - \tfrac{1}{3} - C$$
$$= 8\tfrac{2}{3}$$

Observe that the constant C is eliminated by subtraction.

Because of the connection between definite integrals and antiderivatives we used the integral sign $\int$ for the notation $\int f(x) \, dx$ for an antiderivative. We now dispense with the terminology of antiderivatives and antidifferentiation and begin to call $\int f(x) \, dx$ the **indefinite integral** of "f of x, dx." The process of evaluating an indefinite integral or a definite integral is called **integration**.

The difference between an indefinite integral and a definite integral should be emphasized. The indefinite integral $\int f(x) \, dx$ is defined as a set of functions $g(x) + C$ such that $D_x[g(x)] = f(x)$. However, the definite integral $\int_a^b f(x) \, dx$ is a number whose value depends on the function f and the numbers a and b, and it is defined as the limit of a Riemann sum. The definition of the definite integral makes no reference to differentiation.

The indefinite integral involves an arbitrary constant; for instance,

$$\int x^2 \, dx = \frac{x^3}{3} + C$$

This arbitrary constant C is called a **constant of integration**. In applying the fundamental theorem to evaluate a definite integral, we do not need to include the arbitrary constant C in the expression for $g(x)$ because the fundamental theorem permits us to select *any* antiderivative, including the one for which $C = 0$.

EXAMPLE 2   Evaluate $\displaystyle\int_2^4 (x^3 - 6x^2 + 9x + 1) \, dx$

Solution

$$\int_2^4 (x^3 - 6x^2 + 9x + 1) \, dx = \frac{x^4}{4} - 6 \cdot \frac{x^3}{3} + 9 \cdot \frac{x^2}{2} + x \Big]_2^4$$
$$= (64 - 128 + 72 + 4) - (4 - 16 + 18 + 2)$$
$$= 4$$

EXAMPLE 3   Evaluate $\int_{-1}^{1} (x^{4/3} + 4x^{1/3}) \, dx$

Solution

$$\int_{-1}^{1} (x^{4/3} + 4x^{1/3}) \, dx = \tfrac{3}{7}x^{7/3} + 4 \cdot \tfrac{3}{4}x^{4/3} \Big]_{-1}^{1}$$

$$= \tfrac{3}{7} + 3 - (-\tfrac{3}{7} + 3)$$

$$= \tfrac{6}{7}$$

EXAMPLE 4   Evaluate $\int_{0}^{2} 2x^2 \sqrt{x^3 + 1} \, dx$

Solution

$$\int_{0}^{2} 2x^2 \sqrt{x^3 + 1} \, dx = \frac{2}{3} \int_{0}^{2} \sqrt{x^3 + 1}(3x^2 \, dx)$$

$$= \frac{2}{3} \cdot \frac{(x^3 + 1)^{3/2}}{\frac{3}{2}} \Big]_{0}^{2}$$

$$= \tfrac{4}{9}(8 + 1)^{3/2} - \tfrac{4}{9}(0 + 1)^{3/2}$$

$$= \tfrac{4}{9}(27 - 1)$$

$$= \tfrac{104}{9}$$

EXAMPLE 5   Evaluate $\int_{0}^{3} x\sqrt{1 + x} \, dx$

Solution   To evaluate the indefinite integral $\int x\sqrt{1 + x} \, dx$ we let

$$u = \sqrt{1 + x} \qquad u^2 = 1 + x \qquad x = u^2 - 1 \qquad dx = 2u \, du$$

Substituting, we have

$$\int x\sqrt{1 + x} \, dx = \int (u^2 - 1)u(2u \, du)$$

$$= 2 \int (u^4 - u^2) \, du$$

$$= \tfrac{2}{5}u^5 - \tfrac{2}{3}u^3 + C$$
$$= \tfrac{2}{5}(1 + x)^{5/2} - \tfrac{2}{3}(1 + x)^{3/2} + C$$

Therefore the definite integral

$$\int_{0}^{3} x\sqrt{1 + x} \, dx = \tfrac{2}{5}(1 + x)^{5/2} - \tfrac{2}{3}(1 + x)^{3/2} \Big]_{0}^{3}$$

$$= \tfrac{2}{5}(4)^{5/2} - \tfrac{2}{3}(4)^{3/2} - \tfrac{2}{5}(1)^{5/2} + \tfrac{2}{3}(1)^{3/2}$$

$$= \tfrac{64}{5} - \tfrac{16}{3} - \tfrac{2}{5} + \tfrac{2}{3}$$

$$= \tfrac{116}{15}$$

Another method for evaluating the definite integral in Example 5 involves changing the limits of the definite integral to values of u. The procedure is shown in the following illustration. Often this second method is shorter and its justification follows immediately from Theorems 6.2.9 and 7.3.2.

● ILLUSTRATION 3

Because $u = \sqrt{1 + x}$, we see that when $x = 0$, $u = 1$; and when $x = 3$, $u = 2$. Thus we have

$$\int_0^3 x\sqrt{1 + x}\, dx = 2 \int_1^2 (u^4 - u^2)\, du$$

$$= \tfrac{2}{5}u^5 - \tfrac{2}{3}u^3 \Big]_1^2$$

$$= \tfrac{64}{5} - \tfrac{16}{3} - \tfrac{2}{5} + \tfrac{2}{3}$$

$$= \tfrac{116}{15}$$ ●

EXAMPLE 6 Find the area of the region in the first quadrant bounded by the curve whose equation is $y = 10 - x^2$, the x axis, the y axis, and the line $x = 3$. Make a sketch.

Solution See Fig. 7.3.5. The region is shown together with one of the rectangular elements of area. We divide the interval $[0, 3]$ into n subintervals of equal length. The width of each rectangle is Δx units, and the altitude of the ith rectangle is $10 - \xi_i^2$ units, where ξ_i is any number in the ith subinterval. Therefore the measure of the area of the rectangular element is $(10 - \xi_i^2)\, \Delta x$. The sum of the measures of the areas of n such rectangles is

$$\sum_{i=1}^{n} (10 - \xi_i^2)\, \Delta x$$

which is a Riemann sum. The limit of this sum as $n \to +\infty$ gives the measure of the desired area. The limit of the Riemann sum is a definite integral, which we evaluate by the fundamental theorem of the calculus.

Let A square units be the area of the region. Then

$$A = \lim_{n \to +\infty} \sum_{i=1}^{n} (10 - \xi_i^2)\, \Delta x$$

$$= \int_0^3 (10 - x^2)\, dx$$

$$= 10x - \frac{x^3}{3} \bigg]_0^3$$

$$= 30 - 9$$

$$= 21$$

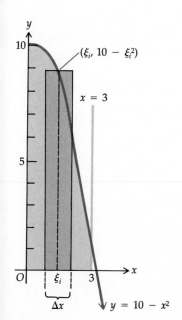

Figure 7.3.5

Compare the solution of Example 6 with the Riemann sum in Illustration 1. The result of $21\frac{17}{144}$ in Illustration 1 is an approximation to the measure of the area in Example 6.

Exercises 7.3

In Exercises 1 through 6, find the Riemann sum for the function on the interval, for the given number of subintervals and the given values of ξ_i. Draw a sketch of the graph of the function on the interval, and show the rectangles the measures of whose areas are the terms of the Riemann sum. (See Illustration 1 and Fig. 7.3.1.)

1. $f(x) = x^2$, $0 \le x \le 3$; four subintervals: $x_0 = 0$, $x_1 = \frac{3}{4}$, $x_2 = 1\frac{1}{2}$, $x_3 = 2\frac{1}{4}$, $x_4 = 3$; $\xi_1 = \frac{1}{4}$, $\xi_2 = 1$, $\xi_3 = 1\frac{1}{2}$, $\xi_4 = 2\frac{1}{2}$

2. $f(x) = x^2$, $0 \le x \le 3$; six subintervals: $x_0 = 0$, $x_1 = \frac{1}{2}$, $x_2 = 1$, $x_3 = 1\frac{1}{2}$, $x_4 = 2$, $x_5 = 2\frac{1}{2}$, $x_6 = 3$; $\xi_1 = \frac{1}{4}$, $\xi_2 = \frac{2}{3}$, $\xi_3 = 1\frac{1}{3}$, $\xi_4 = 2$, $\xi_5 = 2\frac{1}{4}$, $\xi_6 = 2\frac{2}{3}$

3. $f(x) = \dfrac{1}{x}$, $1 \le x \le 3$; six subintervals: $x_0 = 1$, $x_1 = 1\frac{1}{3}$, $x_2 = 1\frac{2}{3}$, $x_3 = 2$, $x_4 = 2\frac{1}{3}$, $x_5 = 2\frac{2}{3}$, $x_6 = 3$; $\xi_1 = 1\frac{1}{4}$, $\xi_2 = 1\frac{1}{2}$, $\xi_3 = 2$, $\xi_4 = 2$, $\xi_5 = 2\frac{1}{2}$, $\xi_6 = 2\frac{3}{4}$

4. $f(x) = \dfrac{1}{x}$, $1 \le x \le 3$; four subintervals: $x_0 = 1$, $x_1 = 1\frac{1}{2}$, $x_2 = 2$, $x_3 = 2\frac{1}{2}$, $x_4 = 3$; $\xi_1 = 1\frac{1}{4}$, $\xi_2 = 2$, $\xi_3 = 2\frac{1}{2}$, $\xi_4 = 2\frac{3}{4}$

5. $f(x) = x^2 - x + 1$, $0 \le x \le 1$; four subintervals: $x_0 = 0$, $x_1 = 0.25$, $x_2 = 0.5$, $x_3 = 0.75$, $x_4 = 1$; $\xi_1 = 0.1$, $\xi_2 = 0.4$, $\xi_3 = 0.6$, $\xi_4 = 0.9$

6. $f(x) = x^3$, $-1 \le x \le 2$; five subintervals: $x_0 = -1$, $x_1 = -0.4$, $x_2 = 0.2$, $x_3 = 0.8$, $x_4 = 1.4$, $x_5 = 2$; $\xi_1 = -0.5$, $\xi_2 = 0$, $\xi_3 = 0.75$, $\xi_4 = 1$, $\xi_5 = 1.5$

In Exercises 7 through 12, find the exact value of the definite integral by using only the definition of the definite integral; that is, use the method of Example 1 of this section.

7. $\displaystyle\int_0^1 x^2\, dx$

8. $\displaystyle\int_2^7 3x\, dx$

9. $\displaystyle\int_1^2 x^3\, dx$

10. $\displaystyle\int_2^4 x^2\, dx$

11. $\displaystyle\int_1^4 (x^2 + 4x + 5)\, dx$

12. $\displaystyle\int_0^5 (x^3 - 1)\, dx$

In Exercises 13 through 34, evaluate the definite integral by using the fundamental theorem of the calculus.

13. $\displaystyle\int_0^3 (3x^2 - 4x + 1)\, dx$

14. $\displaystyle\int_0^4 (x^3 - x^2 + 1)\, dx$

15. $\displaystyle\int_3^6 (x^2 - 2x)\, dx$

16. $\displaystyle\int_{-1}^3 (3x^2 + 5x - 1)\, dx$

17. $\displaystyle\int_1^2 \frac{x^2 + 1}{x^2}\, dx$

18. $\displaystyle\int_{-3}^5 (y^3 - 4y)\, dy$

19. $\displaystyle\int_0^1 \frac{z}{(z^2 + 1)^3}\, dz$

20. $\displaystyle\int_1^4 \sqrt{x}(2 + x)\, dx$

21. $\displaystyle\int_1^{10} \sqrt{5x - 1}\, dx$

22. $\displaystyle\int_0^{\sqrt{5}} t\sqrt{t^2 + 1}\, dt$

23. $\displaystyle\int_{-2}^0 3w\sqrt{4 - w^2}\, dw$

24. $\displaystyle\int_{-1}^3 \frac{dy}{(y + 2)^3}$

25. $\displaystyle\int_1^2 t^2\sqrt{t^3 + 1}\, dt$

26. $\displaystyle\int_1^3 \frac{x\, dx}{(3x^2 - 1)^3}$

27. $\displaystyle\int_0^1 \frac{(y^2 + 2y)\, dy}{\sqrt[3]{y^3 + 3y^2 + 4}}$

28. $\displaystyle\int_2^4 \frac{w^4 - w}{w^3}\, dw$

29. $\displaystyle\int_0^{15} \frac{w\, dw}{(1 + w)^{3/4}}$

30. $\displaystyle\int_4^5 x^2\sqrt{x - 4}\, dx$

31. $\displaystyle\int_0^3 (x + 2)\sqrt{x + 1}\, dx$

32. $\displaystyle\int_{-2}^1 (x + 1)\sqrt{x + 3}\, dx$

33. $\displaystyle\int_1^{64} \left(\sqrt{t} - \frac{1}{\sqrt{t}} + \sqrt[3]{t}\right) dt$

34. $\displaystyle\int_1^4 \frac{x^5 - x}{3x^3}\, dx$

In Exercises 35 through 40, find the area of the region bounded by the given curve and lines. Draw a figure showing the region and a rectangular element of area. Express the measure of the area as the limit of a Riemann sum, and then with definite integral notation. Evaluate the definite integral by the fundamental theorem of the calculus.

35. $y = 4 - x^2$; x axis

36. $y = x^2 - 2x + 3$; x axis; $x = -2$; $x = 1$

37. $y = 4x - x^2$; x axis; $x = 1$; $x = 3$

38. $y = 6 - x - x^2$; x axis

39. $y = \sqrt{x + 1}$; x axis; y axis; $x = 8$

40. $y = \dfrac{1}{x^2} - x$; x axis; $x = 2$; $x = 3$

7.4 Applications of the definite integral

In Sec. 7.2 we showed how a year's production of a commodity was related to the area of a plane region. This same situation can be formulated by using the concept of the definite integral, as shown in the following illustration.

● ILLUSTRATION 1

If x years have elapsed since an electronic calculator was first introduced, then $f(x)$ units per year should be produced, where

$$f(x) = 60 + 288x^2 \qquad 0 \le x \le 3$$

If N is the number of calculators produced during the second year, then

$$N = \lim_{n \to +\infty} \sum_{i=1}^{n} f(\xi_i)\, \Delta x$$

$$= \int_1^2 f(x)\, dx$$

$$= \int_1^2 (60 + 288x^2)\, dx$$

$$= 60x + 288 \cdot \frac{x^3}{3} \Big]_1^2$$

$$= 60x + 96x^3 \Big]_1^2$$

$$= 120 + 768 - (60 + 96)$$

$$= 732 \qquad\qquad ●$$

We now consider some more applications of the definite integral. The quantity to be found is first expressed as the limit of a Riemann sum, which is a definite integral evaluated by using the fundamental theorem of the calculus.

▌ **EXAMPLE 1** Suppose that during the first 5 years for which a new commodity has been on the market, y units per year are sold when x years have

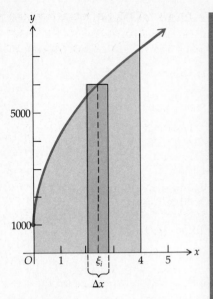

Figure 7.4.1

elapsed since the product was first introduced, and

$$y = 3000\sqrt{x} + 1000 \qquad 0 \le x \le 5 \tag{1}$$

Find the total sales during the first 4 years.

Solution   A sketch of the graph of (1) is shown in Fig. 7.4.1. The number of units sold during the first 4 years is the measure of the area of the shaded region shown in the figure. Let

$$f(x) = 3000\sqrt{x} + 1000$$

The interval $[0, 4]$ is divided into n subintervals of equal length Δx. Let ξ_i be any number in the ith subinterval. If S units are sold during the first 4 years,

$$S = \lim_{n \to +\infty} \sum_{i=1}^{n} f(\xi_i)\, \Delta x$$

$$= \int_0^4 f(x)\, dx$$

$$= \int_0^4 (3000x^{1/2} + 1000)\, dx$$

$$= 3000 \cdot \frac{x^{3/2}}{\frac{3}{2}} + 1000x \Big]_0^4$$

$$= 2000(4)^{3/2} + 1000 \cdot 4$$

$$= 2000(8) + 4000$$

$$= 20,000$$

Therefore 20,000 units are sold during the first 4 years.

EXAMPLE 2   A company manager estimates that the purchase of a particular piece of equipment will result in a savings of operating costs to the company. The rate of operating-cost savings is $f(x)$ dollars per year when the equipment has been in use for x years, and

$$f(x) = 4000x + 1000 \qquad 0 \le x \le 10$$

(a) What is the savings in operating costs for the first 5 years? (b) If the purchase price is $36,000, how many years of use is required for the equipment to pay for itself?

Solution   (a) A sketch of the graph of f on $[0, 5]$ is shown in Fig. 7.4.2. The number of dollars in the savings in operating costs for the first 5 years is the measure of the area of the shaded region in the figure. If S dollars

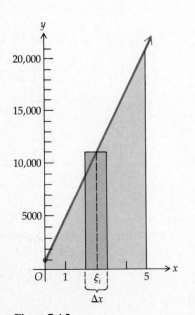

Figure 7.4.2

is this number, then

$$S = \lim_{n \to +\infty} \sum_{i=1}^{n} f(\xi_i) \, \Delta x$$

$$= \int_{0}^{5} f(x) \, dx$$

$$= \int_{0}^{5} (4000x + 1000) \, dx$$

$$= 2000x^2 + 1000x \Big]_{0}^{5}$$

$$= 2000(25) + 1000(5)$$

$$= 55,000$$

Therefore, the savings in operating costs for the first 5 years is $55,000.

(b) Because the purchase price is $36,000, the number of years of use required for the equipment to pay for itself is n, where

$$\int_{0}^{n} f(x) \, dx = 36,000$$

$$\int_{0}^{n} (4000x + 1000) \, dx = 36,000$$

$$2000x^2 + 1000x \Big]_{0}^{n} = 36,000$$

$$2000n^2 + 1000n = 36,000$$

$$2n^2 + n - 36 = 0$$

$$(n - 4)(2n + 9) = 0$$

$$n = 4 \qquad n = -\tfrac{9}{2}$$

Thus 4 years of use are required for the equipment to pay for itself.

The next example shows how the definite integral is used to compute the cost of maintaining an inventory.

EXAMPLE 3 The concession manager of a theatre chain receives a shipment of a certain foodstuff every Monday. Because attendance is low at the beginning of the week and high on the weekend, the demand increases as the week progresses, so that after x days the inventory is y units, where

$$y = 49,000 - 1000x^2 \qquad 0 \le x \le 7 \tag{2}$$

If the daily storage cost is 0.03 cent per unit, find the total cost of maintaining inventory for 7 days.

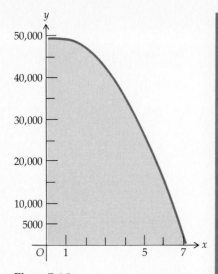

Figure 7.4.3

Solution Figure 7.4.3 shows a sketch of the graph of (2). If A square units is the area of the shaded region in the figure and C cents is the total cost of maintaining inventory for 7 days, then

$$C = 0.03A$$

To compute A, the interval $[0, 7]$ is divided into n subintervals of equal length Δx. If ξ_i is any number in the ith subinterval and $f(x) = 49{,}000 - 1000x^2$,

$$A = \lim_{n \to +\infty} \sum_{i=1}^{n} f(\xi_i) \, \Delta x = \int_0^7 f(x) \, dx$$

$$= \int_0^7 (49{,}000 - 1000x^2) \, dx$$

Therefore

$$C = 0.03 \int_0^7 (49{,}000 - 1000x^2) \, dx$$

$$= 0.03 \left[49{,}000x - 1000 \cdot \frac{x^3}{3} \right]_0^7$$

$$= 0.03 \left[49{,}000 \cdot 7 - 1000 \cdot \frac{7^3}{3} \right]$$

$$= 0.03(7)^3 \left[1000 - \tfrac{1000}{3} \right]$$

$$= 0.01(343)(2000)$$

$$= 6860$$

Thus the total cost of maintaining inventory for 7 days is \$68.60.

Another application of the definite integral is concerned with the **average value** (or **mean value**) of a function on a closed interval. It is a generalization of the arithmetic mean of a finite set of numbers. If $\{f(x_1), f(x_2), \ldots, f(x_n)\}$ is a set of n numbers, then the arithmetic mean is given by

$$\frac{f(x_1) + f(x_2) + \ldots + f(x_n)}{n} = \frac{\displaystyle\sum_{i=1}^{n} f(x_i)}{n}$$

To generalize this definition, consider a division of the closed interval $[a, b]$ into n subintervals of equal length $\Delta x = (b - a)/n$. Let ξ_i be any point in the ith subinterval. Form the sum:

$$\frac{f(\xi_1) + f(\xi_2) + \ldots + f(\xi_n)}{n} = \frac{\displaystyle\sum_{i=1}^{n} f(\xi_i)}{n} \qquad (3)$$

This quotient corresponds to the arithmetic mean of n numbers. Because $\Delta x = (b - a)/n$, then

$$n = \frac{b - a}{\Delta x} \tag{4}$$

Substituting from (4) into (3) we obtain

$$\frac{\displaystyle\sum_{i=1}^{n} f(\xi_i)}{\dfrac{b - a}{\Delta x}} \quad \text{or, equivalently,} \quad \frac{\displaystyle\sum_{i=1}^{n} f(\xi_i)\,\Delta x}{b - a}$$

Taking the limit as $n \to +\infty$ we have, if the limit exists,

$$\lim_{n \to +\infty} \frac{\displaystyle\sum_{i=1}^{n} f(\xi_i)\,\Delta x}{b - a} = \frac{\displaystyle\int_{a}^{b} f(x)\,dx}{b - a}$$

This leads to the following definition.

Definition of the average value of a function	If the function f is integrable on the closed interval $[a, b]$, the **average value** of f on $[a, b]$ is $$\frac{\displaystyle\int_{a}^{b} f(x)\,dx}{b - a}$$

● ILLUSTRATION 2

For the theatre chain of Example 3, the inventory of the particular foodstuff after x days is $f(x)$ units, where

$$f(x) = 49{,}000 - 1000x^2$$

If S units is the average inventory over the 7-day period from Monday to Sunday, then from the definition of average value,

$$S = \frac{\displaystyle\int_{0}^{7} f(x)\,dx}{7 - 0} = \frac{1}{7} \int_{0}^{7} (49{,}000 - 1000x^2)\,dx$$

$$= \frac{1}{7}\left[49{,}000x - 1000 \cdot \frac{x^3}{3} \right]_{0}^{7} = \frac{1}{7}\left[49{,}000 \cdot 7 - 1000 \cdot \frac{7^3}{3} \right]$$

$$= \frac{7^2}{3}(3000 - 1000) = 32{,}667$$

●

EXAMPLE 4 (a) If $f(x) = x^2$, find the average value of f on the interval $[1, 3]$. (b) Find the value of x that gives the average function value. (c) Interpret the result of part (a) geometrically.

Solution (a) If A.V. is the average value of f on $[1, 3]$, we have from the definition

$$\text{A.V.} = \frac{\displaystyle\int_1^3 x^2 \, dx}{3 - 1} = \frac{1}{2} \int_1^3 x^2 \, dx$$

$$= \frac{1}{2}\left[\frac{x^3}{3}\right]_1^3 = \frac{1}{6}(27 - 1)$$

$$= \tfrac{13}{3}$$

(b) Because $f(x) = x^2$, we wish to find the value of x for which

$$x^2 = \tfrac{13}{3}$$

Therefore

$$x = \pm\tfrac{1}{3}\sqrt{39}$$

We reject $-\tfrac{1}{3}\sqrt{39}$ because it is not in the interval $[1, 3]$, and we have

$$x = \tfrac{1}{3}\sqrt{39}$$

(c) In Fig. 7.4.4 there is a sketch of the graph of f on $[1, 3]$ and the line segment from the point $E(\tfrac{1}{3}\sqrt{39}, 0)$ on the x axis to the point $F(\tfrac{1}{3}\sqrt{39}, \tfrac{13}{3})$ on the graph of f. The area of rectangle $AGHB$ having height $\tfrac{13}{3}$ and width 2 is equal to the area of the region $ACDB$. Consequently, the area of the shaded region CGF is equal to the area of the shaded region FDH.

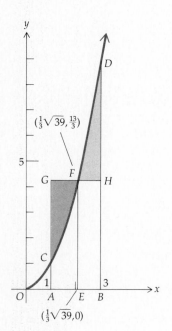

Figure 7.4.4

EXAMPLE 5 The income from sales of a particular commodity x days after its introduction is projected to be $f(x)$ dollars, where

$$f(x) = 24x^2 + 200x$$

Find the average income from sales from the first to the fifth day.

Solution If A dollars is the average income from sales from the first to the fifth day,

$$A = \frac{\displaystyle\int_1^5 (24x^2 + 200x) \, dx}{5 - 1} = \frac{1}{4}\left[24 \cdot \frac{x^3}{3} + 200 \cdot \frac{x^2}{2}\right]_1^5$$

$$= 2x^3 + 25x^2 \bigg]_1^5 = (250 + 625) - (2 + 25)$$

$$= 848$$

Therefore the average income from sales from the first to fifth day is \$848.

An important application of the definite integral involves probability. The probability of a particular event occurring is a number in the closed interval $[0, 1]$. The surer it is that an event will occur, the closer its probability will be to 1.

Suppose that the set of all possible outcomes of a particular situation is the set of all numbers x in some interval I. For instance, x may be the number of hours in the life of a tube for a television set, the number of minutes in the waiting time for a table at a particular restaurant, or the number of inches in a person's height. It is sometimes necessary to determine the probability of x being in some closed subinterval of I. For example, one may wish to find the probability of a television tube lasting between 2000 and 2500 hours, or the probability that a person will have to wait between 20 and 30 minutes for a table at a restaurant, or the probability that someone chosen at random will have a height between 66 and 72 inches. Such problems involve evaluating a definite integral of a function called a **probability density function**.

Probability density functions are obtained from statistical experiments, and the techniques used are beyond the scope of this book. There are two properties that must be satisfied by a probability density function on a closed interval $[a, b]$. They are as follows:

1. $f(x) \geq 0$ for all x in $[a, b]$

2. $$\int_a^b f(x)\, dx = 1$$

● ILLUSTRATION 3

To verify that if $f(x) = \frac{1}{2}x$, f is a probability density function on the closed interval $[0, 2]$, we show that the above two properties hold.

1. $\frac{1}{2}x \geq 0$ for all x in $[0, 2]$

2. $$\int_0^2 \tfrac{1}{2}x\, dx = \frac{x^2}{4} \bigg]_0^2 = 1$$ ●

Following is the definition of the probability that a particular event will occur in an interval.

Definition of $P([c, d])$

> If f is a probability density function on the closed interval $[a, b]$ and $[c, d]$ is a subinterval of $[a, b]$, then the probability that a particular event will occur over the interval $[c, d]$ is denoted by $P([c, d])$, and
>
> $$P([c, d]) = \int_c^d f(x)\, dx$$

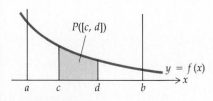

Figure 7.4.5

Figure 7.4.5 shows a sketch of the graph of a probability density function f on $[a, b]$. Because $\int_a^b f(x)\, dx = 1$, the measure of the area of the region, bounded by $y = f(x)$, the x axis, and the lines $x = a$ and $x = b$, is 1. The measure of the area of the shaded region in the figure is $\int_c^d f(x)\, dx$, which is $P([c, d])$.

EXAMPLE 6 At a certain publisher's warehouse, the probability density function for $100x$ percent of the orders being filled per working day is given by

$$f(x) = 12(x^2 - x^3) \qquad 0 \le x \le 1$$

(a) Show that f satisfies the two properties required of a probability density function. (b) Determine the probability that not more than 70 percent of the orders are filled in 1 working day. (c) Find the probability that at least 80 percent of the orders are filled in 1 working day.

Solution   (a) We check the two properties:

1.   If $0 \le x \le 1$, $12x^2 \ge 12x^3$, and so $12(x^2 - x^3) \ge 0$.

2. $\displaystyle\int_0^1 12(x^2 - x^3)\, dx = 12\left[\frac{x^3}{3} - \frac{x^4}{4}\right]_0^1 = 12\left(\frac{1}{3} - \frac{1}{4}\right) = 1$

(b) The probability that not more than 70 percent of the orders are filled in 1 working day is $P([0, 0.70])$, and

$$P([0, 0.70]) = \int_0^{0.70} 12(x^2 - x^3)\, dx$$

$$= 12\left[\frac{x^3}{3} - \frac{x^4}{4}\right]_0^{0.70}$$

$$= 4x^3 - 3x^4\,\Big]_0^{0.70}$$

$$= 4(0.70)^3 - 3(0.70)^4$$

$$= 1.3720 - 0.7203$$

$$= 0.6517$$

(c) The probability that at least 80 percent of the orders are filled in 1 working day is $P([0.80, 1])$, and

$$P([0.80, 1]) = \int_{0.80}^1 12(x^2 - x^3)\, dx$$

$$= 4x^3 - 3x^4\,\Big]_{0.80}^1$$

$$= (4 - 3) - [4(0.80)^3 - 3(0.80)^4]$$

$$= 1 - [2.0480 - 1.2288]$$

$$= 0.1808$$

Exercises 7.4

1. Use the fundamental theorem of the calculus to find the result for part (b) of Exercise 25 in Exercises 7.2.
2. Use the fundamental theorem of the calculus to compute the expected second-month sales for the product of Exercise 26 in Exercises 7.2.
3. Use the fundamental theorem of the calculus to compute the expected first 3 months' sales for the product of Exercise 26 in Exercises 7.2.
4. For the commodity of Example 1, find the total sales for (a) the first year and (b) the fourth year.
5. Suppose that for a 5-minute period a secretary can type at the rate of $f(x)$ words per minute x min from the starting time, where

   $$f(x) = 75 + 10x - 3x^2 \qquad 0 \le x \le 5$$

   How many words does the secretary type during the 5 min?
6. How many words does the secretary in Exercise 6 type during the third minute?
7. The purchase of some new machinery is expected to result in a savings of operating costs so that when the machinery is x years old, the operating cost savings is $f(x)$ dollars per year, where

   $$f(x) = 1000 + 5000x$$

   (a) How much is saved in operating costs during the first 6 years the machinery is in use? (b) If the machinery was purchased at a price of \$67,500, how long will it take for the machinery to pay for itself?
8. It is estimated that x years from now the population of a certain city will be increasing at the rate of $f(x)$ people per year, where

   $$f(x) = 400 + 100x^{3/2} \qquad 0 \le x \le 5$$

   Determine the increase in population for the next 4 years.
9. The book value of a certain piece of equipment is changing at the rate of $f(x)$ dollars per year when the equipment is x years old, and

   $$f(x) = 4000x - 60,000$$

   Determine by how much the equipment depreciates during the third year.
10. If the equipment in Exercise 9 was purchased for \$450,000, how long will it take for it to be fully depreciated?
11. In a certain community, t days since the start of an epidemic, the epidemic is growing at a rate of $f(t)$ people per day, where

    $$f(t) = 2t(50 - 3t)$$

    (a) How many people are infected during the first week of the epidemic? (b) How many new people are infected the eighth day of the epidemic?
12. In a small town a rumor is spreading at the rate of $f(t)$ people per hour t hours since the rumor started, and

    $$f(t) = 40t - 3t^2$$

    (a) How many people hear the rumor during the first 5 hours? (b) How many new people hear the rumor during the sixth hour?
13. The manager of an amusement park expects that during the first month of operation the number of daily admissions will increase so that $f(t)$ admissions per day will occur at t days from the opening where

    $$f(t) = 9800 + 40t$$

    On what day is the 100,000th visitor expected?
14. For the amusement park of Exercise 13, how many admissions are expected (a) the fifth day; (b) during the first 5 days?

15. A merchant receives an annual shipment of 7200 Christmas ornaments on October 23. The sales pattern is essentially the same each year: Over a 60-day period the inventory moves slowly at the beginning, but as Christmas approaches, the demand increases so that x days after October 23 the inventory is y ornaments, where

$$y = 7200 - 2x^2 \qquad 0 \le x \le 60$$

If the daily storage cost is 0.02 cent per ornament, find the total cost of maintaining inventory for 60 days.

16. A candy dealer stocks 9000 Valentine gift boxes at the end of January and plans to sell them over a 15-day period. The sales are low at the beginning of the period but the demand increases as the date gets closer to February 14, so that x days after January 30 the inventory is y boxes, where

$$y = 9000 - 40x^2 \qquad 0 \le x \le 15$$

If the storage cost per box of candy is 0.04 cent per day, find the total cost of maintaining inventory for 15 days.

17. A grocer receives a shipment of 3600 cans of soup that are expected to be sold at a constant rate of 120 cans per day for a 30-day period. (a) If the daily storage cost is 0.01 cent per can, what is the grocer's total cost of maintaining inventory for 30 days? (b) Show that the result of part (a) is the same as the cost of storing 1800 (half of 3600) cans of soup for the entire 30 days.

18. A wholesale distributor has a standing order for 25,000 boxes of detergent that arrive every 20 weeks. These boxes are shipped out by the distributor at a constant rate of 1250 boxes per week. (a) If storage costs are 0.3 cent per box per week, what is the total cost of maintaining inventory for 20 weeks? (b) Show that the result of part (a) is the same as the cost of storing 12,500 (half of 25,000) boxes for the entire 20 weeks.

19. A manufacturing company produces a monthly supply of n items of a certain commodity, and they are stored until they are sold. The items are sold at a constant rate until the inventory becomes zero at the end of the month. If the storage costs are p cents per month per unit, show that the monthly storage costs are $p \cdot \dfrac{n}{2}$, which is the same as the cost of storing $\frac{1}{2}n$ units for the entire month.

In Exercises 20 through 22, find the average value of the function f on the given interval $[a, b]$. In Exercises 20 and 21, find the value of x at which the average value of f occurs, and make a sketch.

20. $f(x) = 9 - x^2$; $[a, b] = [0, 3]$ 21. $f(x) = 8x - x^2$; $[a, b] = [0, 4]$ 22. $f(x) = 3x\sqrt{x^2 - 16}$; $[a, b] = [4, 5]$

23. For the merchant of Exercise 15, determine the average number of ornaments on hand for the 60-day period.

24. For the candy dealer of Exercise 16, determine the average number of Valentine gift boxes on hand for the 15-day period.

25. For the secretary of Exercise 5, find the average typing speed during the first 4 min of the 5-minute period.

26. The demand equation for a certain commodity is $x^2 + 100p^2 = 10,000$, where x units are demanded when p dollars is the price per unit. Find the average total revenue when the number of units demanded takes on all values from 60 to 80.

27. For the x months from June 1 to November 1 the price of a 1-pound can of coffee was $f(x)$ dollars, where

$$f(x) = 3.40 - 0.40x + 0.06x^2 \qquad 0 \le x \le 5$$

Determine the average price for the 3 months from July 1 to October 1.

28. (a) If $f(x) = \frac{1}{18}(2x + 1)$, show that f satisfies the two properties required of a probability density function on the interval $[1, 4]$. (b) Find the probability that an event will occur for x in $[2, 3]$ for the probability density function in part (a).

29. For a particular company the probability density function for $100x$ percent of its weekly orders being filled is given by

$$f(x) = \tfrac{3}{10}(x^2 + 3) \qquad 0 \le x \le 1$$

(a) Show that f satisfies the two properties required of a probability density function. (b) Find the probability that at least 90 percent of the weekly orders will be filled.

30. For a specific part of a video recorder, the probability density function for it lasting $1000x$ hours is given by

$$f(x) = \tfrac{1}{18}(9 - x^2) \qquad 0 \le x \le 3$$

(a) Show that f satisfies the two properties required of a probability density function. (b) Determine the probability that the part will last at least 2000 hours.

31. For the probability density function of Exercise 29, determine the probability that between 70 and 90 percent of the weekly orders will be filled.

32. For the probability density function of Exercise 30, find the probability that the part will last no more than 1000 hours.

7.5 Area of a region in a plane

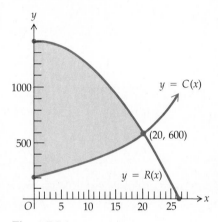

Figure 7.5.1

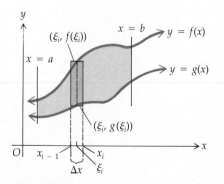

Figure 7.5.2

Suppose that the utilization of a certain piece of machinery yields revenue at the rate of $R(x)$ dollars per month if x months have elapsed since the machinery was new, and

$$R(x) = 1400 - 2x^2$$

Furthermore, assume that the cost of operating and maintaining the equipment is $C(x)$ dollars per month, where

$$C(x) = 200 + x^2$$

Figure 7.5.1 shows sketches of the graphs of R and C.

Observe that when $x = 20$, $R(x) = C(x)$; when $0 \leq x < 20$, $R(x) > C(x)$ and the running of the machinery shows a profit; when $20 < x$, $R(x) < C(x)$, and it is not profitable to use the equipment. The total profit realized from the operation of the equipment for 20 months is related to the area of the region (shaded in Fig. 7.5.1) bounded by the two curves when x is in $[0, 20]$. We proceed now to obtain a method for finding the area of such a region, and we return to this situation in Example 2.

Consider two functions f and g continuous on the closed interval $[a, b]$ and such that $f(x) \geq g(x) \geq 0$ for all x in $[a, b]$. We wish to find the area of the region bounded by the two curves $y = f(x)$ and $y = g(x)$ and the two lines $x = a$ and $x = b$; such a region is shown in Fig. 7.5.2.

Divide the interval $[a, b]$ into n subintervals of equal length Δx. In each subinterval choose a point ξ_i. Consider the rectangle having altitude $[f(\xi_i) - g(\xi_i)]$ units and width Δx units. Such a rectangle is shown in Fig. 7.5.2. There are n such rectangles, one associated with each subinterval. The sum of the measures of the areas of these n rectangles is given by the following Riemann sum:

$$\sum_{i=1}^{n} [f(\xi_i) - g(\xi_i)] \, \Delta x$$

This Riemann sum is an approximation to what we intuitively think of as the number representing the "measure of the area" of the region. The larger the value of n—or, equivalently, the smaller the value of Δx—the better is the approximation. If A square units is the area of the region, we define

$$A = \lim_{n \to +\infty} \sum_{i=1}^{n} [f(\xi_i) - g(\xi_i)] \, \Delta x \tag{1}$$

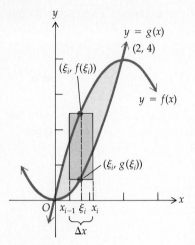

Figure 7.5.3

Because f and g are continuous on $[a, b]$, so also is $(f - g)$; therefore the limit in Eq. (1) exists and is equal to the definite integral

$$\int_a^b [f(x) - g(x)] \, dx$$

EXAMPLE 1   Find the area of the region bounded by the curves $y = x^2$ and $y = -x^2 + 4x$.

Solution   To find the points of intersection of the two curves we solve the equations simultaneously and obtain the points $(0, 0)$ and $(2, 4)$. The region is shown in Fig. 7.5.3.

Let $f(x) = -x^2 + 4x$ and $g(x) = x^2$. Therefore, in the interval $[0, 2]$ the curve $y = f(x)$ is above the curve $y = g(x)$. We draw a vertical rectangular element of area, having altitude $[f(\xi_i) - g(\xi_i)]$ units and width Δx units. The measure of the area of this rectangle then is given by $[f(\xi_i) - g(\xi_i)] \, \Delta x$. The sum of the measures of the areas of n such rectangles is given by the Riemann sum

$$\sum_{i=1}^n [f(\xi_i) - g(\xi_i)] \, \Delta x$$

If A square units is the area of the region, then

$$A = \lim_{n \to +\infty} \sum_{i=1}^n [f(\xi_i) - g(\xi_i)] \, \Delta x$$

and the limit of the Riemann sum is a definite integral. Hence

$$A = \int_0^2 [f(x) - g(x)] \, dx$$

$$= \int_0^2 [(-x^2 + 4x) - x^2] \, dx$$

$$= \int_0^2 (-2x^2 + 4x) \, dx$$

$$= -\tfrac{2}{3}x^3 + 2x^2 \Big]_0^2$$

$$= -\tfrac{16}{3} + 8 - 0$$

$$= \tfrac{8}{3}$$

The area of the region is $\tfrac{8}{3}$ square units.

EXAMPLE 2   The utilization of a certain piece of machinery yields revenue at the rate of $R(x)$ dollars per month if x months have elapsed since the machinery was new, and

$$R(x) = 1400 - 2x^2$$

If the cost of operating and maintaining the equipment is $C(x)$ dollars per month, where

$$C(x) = 200 + x^2$$

find the total profit realized from the operation of the equipment for 20 months.

Solution   As indicated at the beginning of this section, the total profit for 20 months is the number of square units in the area of the shaded region in Fig. 7.5.1. If A square units is this area,

$$A = \lim_{n \to +\infty} \sum_{i=1}^{n} [R(\xi_i) - C(\xi_i)] \, \Delta x = \int_0^{20} [R(x) - C(x)] \, dx$$

$$= \int_0^{20} [(1400 - 2x^2) - (200 + x^2)] \, dx = \int_0^{20} (1200 - 3x^2) \, dx$$

$$= 1200x - x^3 \Big]_0^{20} = 1200(20) - (20)^3 = 24000 - 8000$$

$$= 16{,}000$$

Therefore the total profit is $16,000.

Economic applications of the area of a plane region bounded by two curves are given in Sec. 7.6, where consumers' surplus and producers' surplus are discussed.

In Sec. 7.3 the definition of the area of a plane region bounded above by the curve $y = f(x)$, below by the x axis, and on the sides by the lines $x = a$ and $x = b$ required that $f(x) \geq 0$ on the closed interval $[a, b]$. Then the number of square units in the area is

$$\lim_{n \to +\infty} \sum_{i=1}^{n} f(\xi_i) \, \Delta x$$

Suppose now that $f(x) < 0$ for all x in $[a, b]$. Then each $f(\xi_i)$ is a negative number; so we define the number of square units in the area of the region bounded by $y = f(x)$, the x axis, and the lines $x = a$ and $x = b$ to be

$$\lim_{n \to +\infty} \sum_{i=1}^{n} [-f(\xi_i)] \, \Delta x$$

which equals

$$-\int_a^b f(x) \, dx$$

EXAMPLE 3   Find the area of the region bounded by the curve $y = x^2 - 4x$, the x axis, and the lines $x = 1$ and $x = 3$.

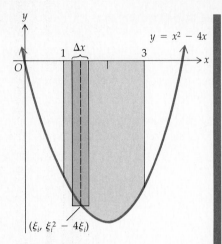

Figure 7.5.4

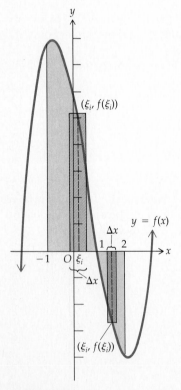

Figure 7.5.5

Solution The region, together with a rectangular element of area, is shown in Fig. 7.5.4.

We take a partition of the interval $[1, 3]$; the width of the ith rectangle is Δx. Because $x^2 - 4x < 0$ on $[1, 3]$, the altitude of the ith rectangle is $-(\xi_i^2 - 4\xi_i) = 4\xi_i - \xi_i^2$. Hence the sum of the measures of the areas of n rectangles is given by

$$\sum_{i=1}^{n} (4\xi_i - \xi_i^2)\, \Delta x$$

The measure of the desired area is given by the limit of this sum as $n \to +\infty$; so if A square units is the area of the region,

$$A = \lim_{n \to +\infty} \sum_{i=1}^{n} (4\xi_i - \xi_i^2)\, \Delta x$$

$$= \int_{1}^{3} (4x - x^2)\, dx$$

$$= 2x^2 - \tfrac{1}{3}x^3 \Big]_{1}^{3}$$

$$= \tfrac{22}{3}$$

Thus the area of the region is $\tfrac{22}{3}$ square units.

EXAMPLE 4 Find the area of the region bounded by the curve $y = x^3 - 2x^2 - 5x + 6$, the x axis, and the lines $x = -1$ and $x = 2$.

Solution The region is shown in Fig. 7.5.5. Let $f(x) = x^3 - 2x^2 - 5x + 6$. Because $f(x) \geq 0$ when x is in the closed interval $[-1, 1]$ and $f(x) \leq 0$ when x is in the closed interval $[1, 2]$, we separate the region into two parts. Let A_1 be the number of square units in the area of the region when x is in $[-1, 1]$, and let A_2 be the number of square units in the area of the region when x is in $[1, 2]$. Then

$$A_1 = \lim_{n \to +\infty} \sum_{i=1}^{n} f(\xi_i)\, \Delta x$$

$$= \int_{-1}^{1} f(x)\, dx$$

$$= \int_{-1}^{1} (x^3 - 2x^2 - 5x + 6)\, dx$$

and

$$A_2 = \lim_{n \to +\infty} \sum_{i=1}^{n} [-f(\xi_i)]\, \Delta x = \int_{1}^{2} -(x^3 - 2x^2 - 5x + 6)\, dx$$

If A square units is the area of the entire region, then

$$A = A_1 + A_2$$

$$= \int_{-1}^{1} (x^3 - 2x^2 - 5x + 6) \, dx - \int_{1}^{2} (x^3 - 2x^2 - 5x + 6) \, dx$$

$$= \left[\tfrac{1}{4}x^4 - \tfrac{2}{3}x^3 - \tfrac{5}{2}x^2 + 6x \right]_{-1}^{1} - \left[\tfrac{1}{4}x^4 - \tfrac{2}{3}x^3 - \tfrac{5}{2}x^2 + 6x \right]_{1}^{2}$$

$$= [(\tfrac{1}{4} - \tfrac{2}{3} - \tfrac{5}{2} + 6) - (\tfrac{1}{4} + \tfrac{2}{3} - \tfrac{5}{2} - 6)]$$
$$- [(4 - \tfrac{16}{3} - 10 + 12) - (\tfrac{1}{4} - \tfrac{2}{3} - \tfrac{5}{2} + 6]$$

$$= \tfrac{32}{3} - (-\tfrac{29}{12})$$

$$= \tfrac{157}{12}$$

The area of the region is therefore $\tfrac{157}{12}$ square units.

● ILLUSTRATION 1

For the function f of Example 4 the definite integral of f from -1 to 1 is

$$\int_{-1}^{1} f(x) \, dx = \int_{-1}^{1} (x^3 - 2x^2 - 5x + 6) \, dx$$

$$= \tfrac{32}{3}$$

This definite integral is the A_1 in Example 4.

The definite integral of f from 1 to 2 is

$$\int_{1}^{2} f(x) \, dx = \int_{1}^{2} (x^3 - 2x^2 - 5x + 6) \, dx$$

$$= -\tfrac{29}{12}$$

This definite integral is the negative of the A_2 in Example 4. This occurs because all of the function values are negative on the interval $[1, 2]$.

Furthermore, the definite integral of f from -1 to 2 is

$$\int_{-1}^{2} f(x) \, dx = \int_{-1}^{2} (x^3 - 2x^2 - 5x + 6) \, dx$$

$$= \tfrac{1}{4}x^4 - \tfrac{2}{3}x^3 - \tfrac{5}{2}x^2 + 6x \Big]_{-1}^{2}$$

$$= (4 - \tfrac{16}{3} - 10 + 12) - (\tfrac{1}{4} + \tfrac{2}{3} - \tfrac{5}{2} - 6)$$
$$= \tfrac{2}{3} - (-\tfrac{91}{12}) = \tfrac{99}{12}$$

Observe that this definite integral is $A_1 + (-A_2)$ because

$$A_1 + (-A_2) = \tfrac{32}{3} + (-\tfrac{29}{12}) = \tfrac{99}{12}$$ ●

The definition, given earlier in this section, that the area of the region bounded above by the curve $y = f(x)$, below by the curve $y = g(x)$, and on the sides by the lines $x = a$ and $x = b$ required $f(x) \geq g(x) \geq 0$. Then the

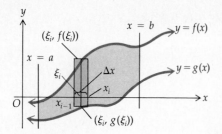

Figure 7.5.6

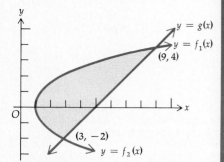

Figure 7.5.7

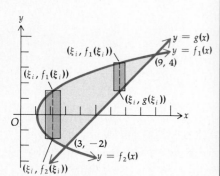

Figure 7.5.8

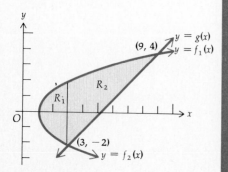

Figure 7.5.9

area is given by

$$\lim_{n \to +\infty} \sum_{i=1}^{n} [f(\xi_i) - g(\xi_i)] \, \Delta x = \int_a^b [f(x) - g(x)] \, dx$$

This definition still applies if $g(x) < 0$ on $[a, b]$ or $f(x)$ and $g(x)$ are both negative on $[a, b]$ as long as $f(x) \geq g(x)$ on $[a, b]$. Figure 7.5.6 shows such a situation.

EXAMPLE 5 Find the area of the region bounded by the parabola $y^2 = 2x - 2$ and the line $y = x - 5$.

Solution The two curves intersect at the points $(3, -2)$ and $(9, 4)$. The region is shown in Fig. 7.5.7.

The equation $y^2 = 2x - 2$ is equivalent to the two equations

$$y = \sqrt{2x - 2} \quad \text{and} \quad y = -\sqrt{2x - 2}$$

with the first equation giving the upper half of the parabola and the second equation giving the bottom half. If we let $f_1(x) = \sqrt{2x - 2}$ and $f_2(x) = -\sqrt{2x - 2}$, the equation of the top half of the parabola is $y = f_1(x)$, and the equation of the bottom half of the parabola is $y = f_2(x)$. If we let $g(x) = x - 5$, the equation of the line is $y = g(x)$.

In Fig. 7.5.8 we see two vertical rectangular elements of area. Each rectangle has the upper base on the curve $y = f_1(x)$. Because the base of the first rectangle is on the curve $y = f_2(x)$, the altitude is $[f_1(\xi_i) - f_2(\xi_i)]$ units. Because the base of the second rectangle is on the curve $y = g(x)$, its altitude is $[f_1(\xi_i) - g(\xi_i)]$ units. If we wish to solve this problem by using vertical rectangular elements of area, we must divide the region into two separate regions, for instance R_1 and R_2, where R_1 is the region bounded by the curves $y = f_1(x)$ and $y = f_2(x)$ and the line $x = 3$, and where R_2 is the region bounded by the curves $y = f_1(x)$ and $y = g(x)$ and the line $x = 3$ (see Fig. 7.5.9).

If A_1 is the number of square units in the area of region R_1, we have

$$A_1 = \lim_{n \to +\infty} \sum_{i=1}^{n} [f_1(\xi_i) - f_2(\xi_i)] \, \Delta x$$

$$= \int_1^3 [f_1(x) - f_2(x)] \, dx$$

$$= \int_1^3 [\sqrt{2x - 2} + \sqrt{2x - 2}] \, dx$$

$$= 2 \int_1^3 \sqrt{2x - 2} \, dx$$

$$= \tfrac{2}{3}(2x - 2)^{3/2} \Big]_1^3$$

$$= \tfrac{16}{3}$$

If A_2 is the number of square units in the area of region R_2, we have

$$A_2 = \lim_{n \to +\infty} \sum_{i=1}^{n} [f_1(\xi_i) - g(\xi_i)]\Delta x$$

$$= \int_3^9 [f_1(x) - g(x)]\, dx$$

$$= \int_3^9 [\sqrt{2x - 2} - (x - 5)]\, dx$$

$$= \left[\tfrac{1}{3}(2x - 2)^{3/2} - \tfrac{1}{2}x^2 + 5x \right]_3^9$$

$$= \left[\tfrac{64}{3} - \tfrac{81}{2} + 45 \right] - \left[\tfrac{8}{3} - \tfrac{9}{2} + 15 \right]$$

$$= \tfrac{38}{3}$$

Hence $A_1 + A_2 = \tfrac{16}{3} + \tfrac{38}{3} = 18$. Therefore the area of the entire region is 18 square units.

EXAMPLE 6 Find the area of the region in Example 5 by taking horizontal rectangular elements of area.

Solution Figure 7.5.10 illustrates the region with a horizontal rectangular element of area.

  If in the equations of the parabola and the line we solve for x,

$$x = \tfrac{1}{2}(y^2 + 2) \quad \text{and} \quad x = y + 5$$

Letting $\phi(y) = \tfrac{1}{2}(y^2 + 2)$ and $\lambda(y) = y + 5$, the equation of the parabola may be written as $x = \phi(y)$ and the equation of the line as $x = \lambda(y)$. Consider the closed interval $[-2, 4]$ on the y axis, and divide the interval into n subintervals each of width Δy. In the ith subinterval $[y_{i-1}, y_i]$, choose a point ξ_i. Then the length of the ith rectangular element is $[\lambda(\xi_i) - \phi(\xi_i)]$ units and the width is Δy units. The measure of the area of the region can be approximated by the Riemann sum

$$\sum_{i=1}^{n} [\lambda(\xi_i) - \phi(\xi_i)]\, \Delta y$$

If A square units is the area of the region, then

$$A = \lim_{n \to +\infty} \sum_{i=1}^{n} [\lambda(\xi_i) - \phi(\xi_i)]\, \Delta y$$

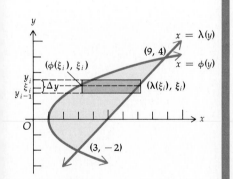

Figure 7.5.10

Because λ and ϕ are continuous on $[-2, 4]$, so too is $(\lambda - \phi)$, and the limit of the Riemann sum is a definite integral:

$$A = \int_{-2}^{4} [\lambda(y) - \phi(y)] \, dy$$

$$= \int_{-2}^{4} [(y + 5) - \tfrac{1}{2}(y^2 + 2)] \, dy$$

$$= \frac{1}{2} \int_{-2}^{4} (-y^2 + 2y + 8) \, dy$$

$$= \tfrac{1}{2} \left[-\tfrac{1}{3}y^3 + y^2 + 8y \right]_{-2}^{4}$$

$$= \tfrac{1}{2}[(-\tfrac{64}{3} + 16 + 32) - (\tfrac{8}{3} + 4 - 16)]$$

$$= 18$$

Comparing the solutions in Examples 5 and 6, we see that in the first case there are two definite integrals to evaluate, whereas in the second case there is only one. In general, if possible, the rectangular elements of area should be constructed so that a single definite integral is obtained.

Exercises 7.5

In Exercises 1 through 20, find the area of the region bounded by the given curves. In each problem do the following: (a) draw a figure showing the region and a rectangular element of area; (b) express the area of the region as the limit of a Riemann sum; (c) find the limit in part (b) by evaluating a definite integral by the fundamental theorem of the calculus.

1. $y = 2x^2$; $y = 2\sqrt{x}$
2. $y = x^3$; $y = \sqrt{x}$
3. $y = x^2 + 1$; $y = x + 1$. Take the elements of area perpendicular to the x axis.
4. $y = 2x^2 + 3$; $y = 2x + 3$. Take the elements of area perpendicular to the x axis.
5. The same region as Exercise 3. Take the elements of area parallel to the x axis.
6. The same region as Exercise 4. Take the elements of area parallel to the x axis.
7. $x^2 = -y$; $y = -4$
8. $y^2 = -x$; $x = -2$; $x = -4$
9. $x^2 + y + 4 = 0$; $y = -8$. Take the elements of area perpendicular to the y axis.
10. The same region as Exercise 9. Take the elements of area parallel to the y axis.
11. $x^3 = 2y^2$; $x = 0$; $y = -2$
12. $y^3 = 4x$; $x = 0$; $y = -2$
13. $y = 2 - x^2$; $y = -x$
14. $y = x^2$; $y = x^4$
15. $y = x^2$; $x^2 = 18 - y$
16. $x = 4 - y^2$; $x = 4 - 4y$
17. $x = y^2 - 2$; $x = 6 - y^2$
18. $x = y^2 - y$; $x = y - y^2$
19. $y = x^3$; $x = y^2$
20. $y^2 = 4x$; $y^2 = 5 - x$

21. Find by integration the area of the triangle having vertices at $(5, 1)$, $(1, 3)$, and $(-1, -2)$.
22. Find by integration the area of the triangle having vertices at $(3, 4)$, $(2, 0)$, and $(0, 1)$.
23. When a certain piece of equipment is x years old, it yields total revenue at the rate of $R(x)$ dollars per year, where

$$R(x) = 15,000 - 80x^2$$

The total cost of operating and maintaining the equipment increases with time so that in x years the total cost is $C(x)$ dollars per year, where

$$C(x) = 3000 + 40x^2$$

(a) Draw a figure showing sketches of the graphs of the functions R and C on the same set of axes. (b) Determine the number of years it is profitable to use the equipment. (c) What is the total profit realized from the operation of the equipment for the number of years determined in part (b)? (d) Show on the figure obtained in part (a) the region whose area is n square units, where n is the number found in part (c).

24. The use of a particular machine x months after its purchase generates total revenue at the rate of $R(x)$ dollars per month, where

$$R(x) = 11,000 - 3x^2$$

The total cost of operating and maintaining the machine increases with time so that in x months the total cost is $C(x)$ dollars per month, where

$$C(x) = 1000 + x^2$$

(a) Draw a figure showing sketches of the graphs of the functions R and C on the same set of axes. (b) How many months is it profitable to operate the machine? (c) Find the total profit realized from the operation of the machine for the number of months determined in part (b). (d) Show on the figure obtained in part (a) the region whose area is n square units, where n is the number found in part (c).

7.6 Consumers' surplus and producers' surplus

Often the price that a consumer pays for a particular commodity is below that which he or she would pay rather than go without it. Let us relate the highest price that a consumer would pay for a commodity rather than go without it to the satisfaction to be obtained from its purchase. Furthermore, we shall relate the price that the consumer actually pays to the satisfaction that is given up in paying that price. The excess of the first satisfaction over the second we call the *surplus satisfaction*. For some commodities the surplus satisfaction is much greater than it is for others. Examples of commodities having a large surplus satisfaction are salt and matches, since for these goods the prices are considerably less than that which most people would pay rather than go without them.

If we wish to give a measure of the combined overall gain of the consumers of a particular commodity, we use the total of the surplus satisfactions afforded by the amounts by which each of the demand prices for that commodity exceeds its selling price. This measure is called the *consumers' surplus*. We now give a discussion leading up to the defintion of consumers' surplus.

In Fig. 7.6.1 the demand curve, having equation $p = f(x)$, is shown. The point A in the figure has coordinates $(\bar{x}, \bar{p})$, where $\bar{p}$ dollars is the market price and $\bar{x}$ is the number of units in the corresponding demand. We are requiring the function f to be continuous on the interval $[0, \bar{x}]$; hence x may assume any value in the interval. Divide the interval into n subintervals, each of length Δx. Thus there are n subintervals of the form $[x_{i-1}, x_i]$, where $i = 1, 2, \ldots, n$. Choose any number ξ_i, with $x_{i-1} \le \xi_i \le x_i$, in each subinterval. When ξ_i units of the commodity are demanded, the price is

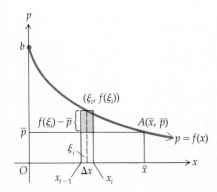

Figure 7.6.1

$f(\xi_i)$ dollars, which is greater than the market price of $\bar{p}$ dollars, and so the surplus satisfaction afforded is $f(\xi_i) - \bar{p}$. For Δx close to zero the surplus satisfaction for each value of x in the interval $[x_{i-1}, x_i]$ is close to $f(\xi_i) - \bar{p}$. The product $[f(\xi_i) - \bar{p}] \Delta x$ is an approximation to the total of the surplus satisfactions afforded when the demand x takes on all values in the interval $[x_{i-1}, x_i]$. The Riemann sum

$$\sum_{i=1}^{n} [f(\xi_i) - \bar{p}] \Delta x$$

is an approximation to what economists intuitively think of as the consumers' surplus. Taking the limit of this Riemann sum as $n \to +\infty$ (the limit exists because of the continuity requirement of the function f on $[0, \bar{x}]$), we have the definite integral

$$\lim_{n \to +\infty} \sum_{i=1}^{n} [f(\xi_i) - \bar{p}] \Delta x = \int_{0}^{\bar{x}} [f(x) - \bar{p}] \, dx$$

The right side of the above may be written as

$$\int_{0}^{\bar{x}} f(x) \, dx - \int_{0}^{\bar{x}} \bar{p} \, dx = \int_{0}^{\bar{x}} f(x) \, dx - \bar{p}\bar{x}$$

which gives the area of the region below the demand curve and above the line $p = \bar{p}$. See Fig. 7.6.2. It is this reasoning that leads to the following definition.

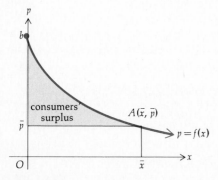

Figure 7.6.2

Definition of consumers' surplus

Suppose that the demand equation, $p = f(x)$ or $x = g(p)$, for a certain commodity is given where p dollars is the price per unit when x units are demanded. Let $\bar{p}$ dollars be the market price and $\bar{x}$ be the number of units in the corresponding market demand. Then the **consumers' surplus** is given by the area of the region below the demand curve and above the line $p = \bar{p}$. Hence, if CS is the number of dollars in the consumers' surplus,

$$CS = \int_{0}^{\bar{x}} f(x) \, dx - \bar{p}\bar{x} \tag{1}$$

or, equivalently,

$$CS = \int_{\bar{p}}^{b} g(p) \, dp \tag{2}$$

where $b = f(0)$.

There are various ways of assigning a market price $\bar{p}$ to a commodity. The market price may be the equilibrium price, or it may be controlled by a monopoly operation, as discussed in Sec. 5.5. It is possible that $\bar{p}$ may be decided on arbitrarily.

EXAMPLE 1   The demand and supply equations for a certain commodity are, respectively,

$$3p + x = 32 \quad \text{and} \quad 3p^2 + p - x = 0$$

where p dollars is the unit price and x units is the quantity. Determine the consumers' surplus if market equilibrium prevails, and draw a sketch showing the region whose area gives the consumers' surplus.

Solution   Solving the demand equation for x and substituting into the supply equation we obtain

$$3p^2 + p - (32 - 3p) = 0$$

$$3p^2 + 4p - 32 = 0$$

$$(3p - 8)(p + 4) = 0$$

from which we get $p = \frac{8}{3}$, and the corresponding value of x is 24. The demand and supply curves intersect at the point of equilibrium $E(24, \frac{8}{3})$. See Figure 7.6.3. If CS is the number of dollars in the consumers' surplus, from (1) we obtain

$$CS = \int_0^{24} \frac{32 - x}{3}\, dx - \tfrac{8}{3} \cdot 24$$

$$= \tfrac{32}{3}x - \tfrac{1}{6}x^2 \Big]_0^{24} - 64$$

$$= 96$$

The consumers' surplus is then $96, and the region whose area gives the consumers' surplus is shaded in Fig. 7.6.3.

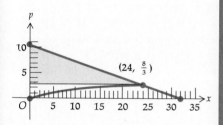

Figure 7.6.3

EXAMPLE 2   Solve Example 1 by using Eq. (2) of the definition of consumers' surplus.

Solution   Solving the demand equation for x we get $x = g(p) = 32 - 3p$. Since $f(x) = \tfrac{1}{3}(32 - x)$, $b = f(0) = \tfrac{32}{3}$. Hence

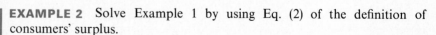

$$CS = \int_{8/3}^{32/3} (32 - 3p)\, dp = 32p - \tfrac{3}{2}p^2 \Big]_{8/3}^{32/3}$$

$$= 32(\tfrac{32}{3}) - \tfrac{3}{2}(\tfrac{32}{3})^2 - 32(\tfrac{8}{3}) + \tfrac{3}{2}(\tfrac{8}{3})^2$$

$$= \tfrac{1024}{3} - \tfrac{512}{3} - \tfrac{256}{3} + \tfrac{32}{3}$$

$$= 96$$

which agrees with the previous answer.

Refer to the supply equation in Example 1, and observe that some producers would be willing to supply the commodity below the equilibrium price of \$2.67 ($\frac{8}{3} = 2.67$). In particular, at a price of \$2, 14 units would be supplied; and at a price of \$1, 4 units would be supplied. Those producers who would be willing to supply the commodity at a price lower than the equilibrium price of \$2.67 will gain by the setting of the price at \$2.67. Suppose that we wish the combined overall gain of the producers of the commodity of Example 1. We would use the total of the amounts by which the market price exceeds each of the supply prices given by the supply equation as x takes on all values from 0 to $\bar{x}$. This measure is called the *producers' surplus*. Before giving the definition of producers' surplus we do as we did with consumers' surplus; we give a preliminary discussion that makes the definition more meaningful.

Figure 7.6.4 shows the supply curve having equation $p = h(x)$ and the point A having coordinates $(\bar{x}, \bar{p})$, where $\bar{x}$ units are supplied when $\bar{p}$ dollars is the market price. The function h is assumed to be continuous on the interval $[0, \bar{x}]$. Divide the interval $[0, \bar{x}]$ into n subintervals of the form $[x_{i-1}, x_i]$, where $i = 1, 2, \ldots, n$, and where each subinterval is of length Δx. In each subinterval choose any number ξ_i, with $x_{i-1} \le \xi_i \le x_i$. For a supply of ξ_i units the supply price is $h(\xi_i)$. The market price, given by $\bar{p}$, is greater than $h(\xi_i)$, and so the gain of the producer at this supply price of $h(\xi_i)$ is $\bar{p} - h(\xi_i)$. For Δx close to zero the producer's gain for each value of x in the interval $[x_{i-1}, x_i]$ is close to $\bar{p} - h(\xi_i)$, and the product $[\bar{p} - h(\xi_i)] \Delta x$ is an approximation to the total gain of the producers when the amount supplied takes on all values in the interval $[x_{i-1}, x_i]$. The Riemann sum

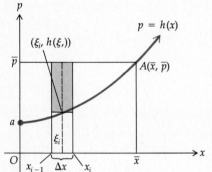

Figure 7.6.4

$$\sum_{i=1}^{n} [\bar{p} - h(\xi_i)] \Delta x$$

is an approximation to what economists intuitively think of as the producers' surplus. Since h is continuous on $[0, \bar{x}]$, the limit of this Riemann sum, as $n \to +\infty$, exists, and we have the definite integral

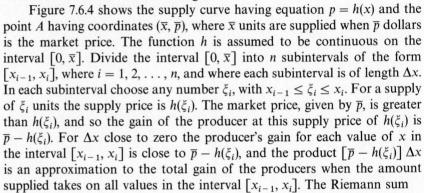

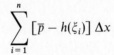

$$\lim_{n \to +\infty} \sum_{i=1}^{n} [\bar{p} - h(\xi_i)] \Delta x = \int_{0}^{\bar{x}} [\bar{p} - h(x)] \, dx$$

$$= \int_{0}^{\bar{x}} \bar{p} \, dx - \int_{0}^{\bar{x}} h(x) \, dx$$

$$= \bar{p}\bar{x} - \int_{0}^{\bar{x}} h(x) \, dx$$

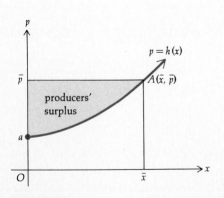

Figure 7.6.5

This definite integral gives the area of the region above the supply curve and below the line $p = \bar{p}$. Refer to Fig. 7.6.5.

Suppose that the supply equation, $p = h(x)$ or $x = \lambda(p)$, for a certain commodity is given where p dollars is the price per unit when x units are supplied. Let $\bar{p}$ dollars be the market price and $\bar{x}$ be the number of units supplied at that price. Then the **producers' surplus** is given by the area of the region above the supply curve and below the line $p = \bar{p}$. Hence, if PS is the number of dollars in the producers' surplus,

$$PS = \bar{p}\bar{x} - \int_0^{\bar{x}} h(x)\, dx \qquad (3)$$

or, equivalently,

$$PS = \int_a^{\bar{p}} \lambda(p)\, dp \qquad (4)$$

where $a = h(0)$.

EXAMPLE 3 For the commodity of Example 1, determine the producers' surplus if market equilibrium prevails. Also draw a sketch showing the region whose area gives the producers' surplus.

Solution Solving the supply equation for p we get

$$p = h(x) = \frac{\sqrt{1 + 12x} - 1}{6}$$

Since the point of equilibrium is $(24, \frac{8}{3})$, $\bar{x} = 24$ and $\bar{p} = \frac{8}{3}$; so if PS is the number of dollars in the producers' surplus, we have from (3)

$$PS = \tfrac{8}{3} \cdot 24 - \int_0^{24} \frac{\sqrt{1 + 12x} - 1}{6}\, dx$$

$$= 64 - \tfrac{1}{6}\left[\tfrac{1}{18}(1 + 12x)^{3/2} - x \right]_0^{24}$$

$$= \tfrac{358}{27} = 13.26$$

Therefore the producers' surplus is $13.26. The region whose area gives the producers' surplus is shaded in Fig. 7.6.6.

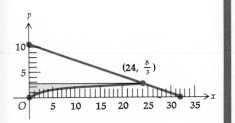

Figure 7.6.6

EXAMPLE 4 For a particular commodity the demand and supply equations are, respectively,

$$100p = 1600 - x^2 \quad \text{and} \quad 400p = x^2 + 2400$$

Determine to the nearest dollar the consumers' surplus and the producers' surplus if market equilibrium prevails. Draw a sketch showing each of the regions whose area determines the consumers' surplus and the producers' surplus.

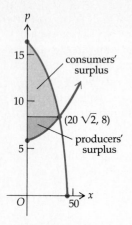

Figure 7.6.7

Solution   The equilibrium price and equilibrium amount are obtained by solving simultaneously the demand and supply equations. We have

$$1600 - x^2 = \tfrac{1}{4}x^2 + 600$$

$$\tfrac{5}{4}x^2 = 1000$$

$$x^2 = 800$$

$$x = 20\sqrt{2}$$

When $x = 20\sqrt{2}$, $p = 8$. Figure 7.6.7 shows the required sketch of the demand curve, the supply curve, and the shaded regions whose areas determine the consumers' surplus and the producers' surplus. If CS is the number of dollars in the consumers' surplus,

$$CS = \int_{8}^{16} 10\sqrt{16 - p}\; dp$$

$$= -\tfrac{20}{3}(16 - p)^{3/2}\Big]_{8}^{16}$$

$$= \tfrac{320}{3}\sqrt{2} = 150.85$$

If PS is the number of dollars in the producers' surplus,

$$PS = \int_{6}^{8} 20\sqrt{p - 6}\; dp$$

$$= \tfrac{40}{3}(p - 6)^{3/2}\Big]_{6}^{8}$$

$$= \tfrac{80}{3}\sqrt{2} = 37.71$$

Hence the consumers' surplus is $150.85 and the producers' surplus is $37.71.

EXAMPLE 5   Under a monopoly in which the producer can control output and hence price, the number of units produced is determined so as to have maximum profit. If the demand equation is

$$10p = (80 - x)^2$$

and the total cost function is given by

$$C(x) = \tfrac{1}{2}x^2 + 100x$$

where $C(x)$ dollars is the total cost of producing x units, determine the consumers' surplus. Draw a sketch showing the region whose area gives the consumers' surplus.

Solution   To find $\bar{p}$ we first find $\bar{x}$ by making use of the fact that the profit will be a maximum when the marginal revenue equals the marginal cost. If R is the total revenue function, we have

$$R(x) = px$$

$$= \frac{x}{10}(80 - x)^2$$

$$= 640x - 16x^2 + \tfrac{1}{10}x^3$$

$$R'(x) = 640 - 32x + \tfrac{3}{10}x^2$$

$$C'(x) = x + 100$$

Setting $R'(x) = C'(x)$ we get

$$\tfrac{3}{10}x^2 - 32x + 640 = x + 100$$

$$3x^2 - 330x + 5400 = 0$$

$$3(x - 20)(x - 90) = 0$$

Because $p = \tfrac{1}{10}(80 - x)^2$, $D_x p = -\tfrac{1}{5}(80 - x)$, and hence $x \leq 80$ for p to be decreasing as x increases. So $\bar{x} = 20$, from which it follows that $\bar{p} = 360$. Figure 7.6.8 shows the demand curve, the line $p = 360$, and the shaded region whose area gives the consumers' surplus. Letting CS be the number of dollars in the consumers' surplus we have

$$CS = \frac{1}{10}\int_0^{20}(80 - x)^2\,dx - 20 \cdot 360$$

$$= -\tfrac{1}{30}(80 - x)^3\Big]_0^{20} - 7200$$

$$= -7200 + \tfrac{51200}{3} - 7200$$

$$= \tfrac{8000}{3}$$

$$= 2666.67$$

The consumers' surplus is therefore $2666.67.

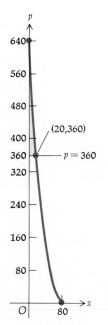

Figure 7.6.8

Exercises 7.6

In Exercises 1 through 14, x units are demanded or supplied when p dollars is the price per unit.

1. The demand equation for a certain commodity is $100p = 2400 - 20x - x^2$, and the market price is $16. Find the consumers' surplus, and draw a sketch showing the region whose area determines the consumers' surplus.

2. Find the consumers' surplus if the demand equation is $10p = 3\sqrt{1600 - 2x}$ and the market demand is 700. Draw a sketch showing the region whose area determines the consumers' surplus.

3. Find the producers' surplus for a commodity whose supply equation is $2x^2 - 300p + 900 = 0$ and whose market price is $9. Draw a sketch showing the region whose area determines the producers' surplus.

4. If the supply equation of a particular article of merchandise is $10p = 2\sqrt{x + 300}$ and the market price is $8, find the producers' surplus. Draw a sketch showing the region whose area determines the producers' surplus.

5. If the demand equation for a particular commodity is $10p = \sqrt{900 - x}$ and the demand is fixed at 500, find the consumers' surplus. Integrate with respect to x.

6. For a particular article of merchandise the demand equation is $x^2 + 100p = 2500$ and the market price is \$16. Find the consumers' surplus by integrating with respect to x.

7. Solve Exercise 5 by integrating with respect to p.

8. Solve Exercise 6 by integrating with respect to p.

9. If the supply equation for a particular commodity is $100p = (x + 20)^2$ and the market price is \$25, find the producers' surplus. Integrate with respect to x.

10. Find the producers' surplus for a commodity whose supply equation is $x^2 - 200p + 900 = 0$ and the market demand is 30. Integrate with respect to x.

11. Solve Exercise 9 by integrating with respect to p.

12. Solve Exercise 10 by integrating with respect to p.

13. The demand and supply equations for a particular commodity are, respectively,

$$x + 25p = 150 \quad \text{and} \quad 200p - x = 300$$

and the market is in equilibrium. Find the consumers' surplus and the producers' surplus without using integration. *Hint:* The required quantities can be determined by areas of right triangles.

14. The demand and supply equations for a particular commodity are, respectively,

$$x = 760 - p^2 \quad \text{and} \quad x = p^2 - 40$$

Find the consumers' surplus and the producers' surplus if the market is in equilibrium.

15. Suppose that the demand equation of Exercise 1 and the supply equation of Exercise 3 apply to the same commodity. Find the consumers' surplus and the producers' surplus if the market is in equilibrium.

16. If the demand equation of Exercise 2 and the supply equation of Exercise 4 apply to the same commodity and the market is in equilibrium, find the consumers' surplus and the producers' surplus.

17. For a certain commodity the demand equation is $30p = 3600 - x^2$, where p dollars is the price per unit when x units are demanded. The total cost function is given by $C(x) = (x + 20)^2$, where $C(x)$ dollars is the total cost of producing x units. Find the consumers' surplus to the nearest dollar if the amount produced, and hence the price, under a monopoly are so determined as to have the maximum total profit.

18. The demand equation for a particular commodity is $10p = 1600 - x^2$, where p dollars is the price per unit when x units are demanded. The total cost function is given by $C(x) = \frac{1}{2}x^2 + 60x$, where $C(x)$ dollars is the total cost of producing x units. Find the consumers' surplus to the nearest dollar if the amount produced, and hence the price, under a monopoly are so determined as to have the maximum total profit.

7.7 Properties of the definite integral (optional)

The fundamental theorem of the calculus was stated as Theorem 7.3.2, and it has been applied to evaluate definite integrals. As stated in Sec. 7.3, a rigorous proof of the fundamental theorem is beyond the scope of this text. However, because of the importance of this theorem, we give a discussion of the proof. We first state some properties of the definite integral that are needed in the development.

In the definition of the definite integral, the closed interval $[a, b]$ is given, and it is assumed that $a < b$. To consider a definite integral of a function f from a to b when $a > b$, or when $a = b$, we have the following definition.

Definition of $\int_a^b f(x)\, dx$

when $a \geq b$

(i) If $a > b$, then if $\int_b^a f(x)\, dx$ exists,

$$\int_a^b f(x)\, dx = -\int_b^a f(x)\, dx \tag{1}$$

(ii) If $f(a)$ exists,

$$\int_a^a f(x)\, dx = 0 \tag{2}$$

● ILLUSTRATION 1

In Example 1 of Sec. 7.3 the definition of the definite integral was used to show that

$$\int_1^3 x^2\, dx = 8\tfrac{2}{3}$$

With this result and Eq. (1) it follows that

$$\int_3^1 x^2\, dx = -8\tfrac{2}{3}$$ ●

● ILLUSTRATION 2

From (2) it follows that

$$\int_1^1 x^2\, dx = 0$$ ●

Another property of the definite integral is given in the following theorem.

Theorem 7.7.1 If f is integrable on a closed interval containing the three numbers a, b, and c, then

$$\int_a^b f(x)\, dx = \int_a^c f(x)\, dx + \int_c^b f(x)\, dx$$

regardless of the order of a, b, and c.

The proof of Theorem 7.7.1 is omitted. However, a geometric interpretation is given in the following illustration for the case $a < c < b$.

● ILLUSTRATION 3

If $f(x) \geq 0$ for all x in $[a, b]$ and $a < c < b$, then Theorem 7.7.1 states that the measure of the area of the region bounded by the curve $y = f(x)$ and the x axis from a to b is equal to the sum of the measures of the areas of the regions from a to c and from c to b. See Fig. 7.7.1.   ●

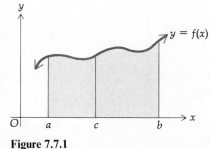

Figure 7.7.1

We need to consider definite integrals having a variable upper limit. Let f be a function continuous on the closed interval $[a, b]$. Then the value of the definite integral depends only on the function f and the numbers a and b, and not on the symbol x, which is used here as the independent variable. From Illustration 1,

$$\int_1^3 x^2 \, dx = 8\tfrac{2}{3}$$

Any symbol instead of x could have been used; for example,

$$\int_1^3 t^2 \, dt = \int_1^3 u^2 \, du = \int_1^3 r^2 \, dr = 8\tfrac{2}{3}$$

If f is continuous on the closed interval $[a, b]$, then by Theorem 7.3.1 and the definition of the definite integral, $\int_a^b f(t) \, dt$ exists. If x is a number in $[a, b]$, then f is continuous on $[a, x]$ because it is continuous on $[a, b]$. Consequently, $\int_a^x f(t) \, dt$ exists and is a unique number whose value depends on x. Therefore $\int_a^x f(t) \, dt$ defines a function F having as its domain all numbers in the closed interval $[a, b]$ and whose function value at any number x in $[a, b]$ is given by

$$F(x) = \int_a^x f(t) \, dt \tag{3}$$

As a notational observation, if the limits of the definite integral are variables, different symbols are used for these limits and for the independent variable in the integrand. Hence in (3), because x is the upper limit, the symbol t is used as the independent variable in the integrand.

If, in (3), $f(t) \geq 0$ for all values of t in $[a, b]$, then the function value $F(x)$ can be interpreted geometrically as the measure of the area of the region bounded by the curve whose equation is $y = f(t)$, the t axis, and the lines $t = a$ and $t = x$. See Fig. 7.7.2. Note that $F(a) = \int_a^a f(t) \, dt$, which by (2) equals 0.

We now wish to show that the function F which is continuous on $[a, b]$ and defined by (3) is differentiable on $[a, b]$; furthermore, we wish to show that if x is any number in $[a, b]$, then

$$F'(x) = f(x) \tag{4}$$

We consider two numbers x_1 and $x_1 + \Delta x$ in $[a, b]$. Then

$$F(x_1) = \int_a^{x_1} f(t) \, dt$$

and

$$F(x_1 + \Delta x) = \int_a^{x_1 + \Delta x} f(t) \, dt$$

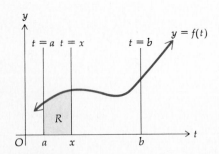

Figure 7.7.2

so that

$$F(x_1 + \Delta x) - F(x_1) = \int_a^{x_1 + \Delta x} f(t)\, dt - \int_a^{x_1} f(t)\, dt \tag{5}$$

By Theorem 7.7.1,

$$\int_a^{x_1} f(t)\, dt + \int_{x_1}^{x_1 + \Delta x} f(t)\, dt = \int_a^{x_1 + \Delta x} f(t)\, dt$$

or, equivalently,

$$\int_a^{x_1 + \Delta x} f(t)\, dt - \int_a^{x_1} f(t)\, dt = \int_{x_1}^{x_1 + \Delta x} f(t)\, dt \tag{6}$$

Substituting from (6) into (5) we get

$$F(x_1 + \Delta x) - F(x_1) = \int_{x_1}^{x_1 + \Delta x} f(t)\, dt$$

or, if we divide by Δx,

$$\frac{F(x_1 + \Delta x) - F(x_1)}{\Delta x} = \frac{\displaystyle\int_{x_1}^{x_1 + \Delta x} f(t)\, dt}{\Delta x}$$

Taking the limit as Δx approaches zero we have

$$\lim_{\Delta x \to 0} \frac{F(x_1 + \Delta x) - F(x_1)}{\Delta x} = \lim_{\Delta x \to 0} \frac{\displaystyle\int_{x_1}^{x_1 + \Delta x} f(t)\, dt}{\Delta x} \tag{7}$$

The left member of (7) is $F'(x_1)$. If we can show that the right member of (7) is $f(x_1)$, we will have shown that (4) holds for any number x in $[a, b]$. A rigorous proof of this fact is too advanced for this text. Instead we give the following informal geometric argument.

We consider $f(t) \geq 0$ and $\Delta x > 0$. Then $\int_{x_1}^{x_1 + \Delta x} f(t)\, dt$ is the measure of the area of the region bounded by the curve $y = f(x)$, the x axis, and the lines $t = x_1$ and $t = x_1 + \Delta x$. See Fig. 7.7.3. Let m units and M units be, respectively, the minimum and maximum values of $f(t)$ on the closed interval $[x_1, x_1 + \Delta x]$. Then

$$m\, \Delta x \leq \int_{x_1}^{x_1 + \Delta x} f(t)\, dt \leq M\, \Delta x$$

or, equivalently,

$$m \leq \frac{\displaystyle\int_{x_1}^{x_1 + \Delta x} f(t)\, dt}{\Delta x} \leq M \tag{8}$$

If Δx approaches zero from the right, then because f is continuous on $[x_1, x_1 + \Delta x]$, both m and M approach $f(x_1)$, and therefore because of

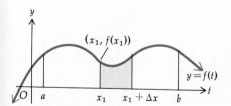

Figure 7.7.3

inequality (8),

$$\lim_{\Delta x \to 0^+} \frac{\displaystyle\int_{x_1}^{x_1+\Delta x} f(t)\, dt}{\Delta x} = f(x_1) \tag{9}$$

A similar informal argument can be given if $\Delta x < 0$. For $\Delta x < 0$, we consider the interval $[x_1 + \Delta x, x_1]$ and let Δx approach zero from the left to obtain

$$\lim_{\Delta x \to 0^-} \frac{\displaystyle\int_{x_1}^{x_1+\Delta x} f(t)\, dt}{\Delta x} = f(x_1) \tag{10}$$

Therefore, from (9) and (10), it follows that the right member of (7) is $f(x_1)$. Thus, from (7) we obtain

$$F'(x_1) = f(x_1)$$

Because x_1 is any number in $[a, b]$, we have (4). We state this result formally as the following theorem.

Theorem 7.7.2   Let the function f be continuous on the closed interval $[a, b]$, and let x be any number in $[a, b]$. If F is the function defined by

$$F(x) = \int_a^x f(t)\, dt$$

then

$$F'(x) = f(x) \tag{11}$$

Theorem 7.7.2 states that the definite integral $\int_a^x f(t)\, dt$, with variable upper limit x, is an antiderivative of f.

EXAMPLE 1   Apply Theorem 7.7.2 to find $D_x \int_1^x \sqrt{9 + t^2}\, dt$.

Solution   With the notation of the statement of Theorem 7.7.2,

$$F(x) = \int_1^x \sqrt{9 + t^2}\, dt \qquad f(t) = \sqrt{9 + t^2}$$

and $a = 1$. Then from the theorem,

$$F'(x) = f(x)$$

Thus

$$D_x \int_1^x \sqrt{9 + t^2}\, dt = \sqrt{9 + x^2}$$

Theorem 7.7.2 is now used to prove the fundamental theorem of the calculus. The theorem states: Let the function f be continuous on the closed interval $[a, b]$, and let g be any antiderivative of f on $[a, b]$. Then

$$\int_a^b f(x)\, dx = g(b) - g(a)$$

Proof If f is continuous at all numbers in $[a, b]$, we know from Theorem 7.7.2 that the definite integral $\int_a^x f(t)\, dt$, with variable upper upper limit x, defines a function F whose derivative on $[a, b]$, is f. Because by hypothesis $g'(x) = f(x)$, it follows from Theorem 6.2.2 that

$$g(x) = \int_a^x f(t)\, dt + k \tag{12}$$

where k is some constant.

Letting $x = b$ and $x = a$, successively, in (12) we get

$$g(b) = \int_a^b f(t)\, dt + k \tag{13}$$

and

$$g(a) = \int_a^a f(t)\, dt + k \tag{14}$$

From (13) and (14),

$$g(b) - g(a) = \int_a^b f(t)\, dt - \int_a^a f(t)\, dt$$

But, by (2), $\int_a^a f(t)\, dt = 0$; so

$$g(b) - g(a) = \int_a^b f(t)\, dt$$

which is what we wished to prove. ∎

We had examples that applied the fundamental theorem of the calculus in Secs. 7.3–7.6. The following two examples utilize the fundamental theorem as well as Theorem 7.7.1.

EXAMPLE 2 Evaluate $\displaystyle\int_{-2}^{2} |x|\, dx$

Solution Because

$$|x| = \begin{cases} x & \text{if } x \geq 0 \\ -x & \text{if } x \leq 0 \end{cases}$$

we first use Theorem 7.7.1. We have

$$\int_{-2}^{2} |x|\, dx = \int_{-2}^{0} |x|\, dx + \int_{0}^{2} |x|\, dx$$

$$= \int_{-2}^{0} (-x)\, dx + \int_{0}^{2} x\, dx$$

$$= \left[-\frac{x^2}{2} \right]_{-2}^{0} + \left[\frac{x^2}{2} \right]_{0}^{2}$$

$$= (-0 + 2) + (2 - 0)$$

$$= 4$$

EXAMPLE 3   Evaluate $\displaystyle\int_{-3}^{4} |x + 2|\, dx$

Solution   If $f(x) = |x + 2|$, instead of finding an antiderivative of f directly, write $f(x)$ as

$$f(x) = \begin{cases} x + 2 & \text{if } x \geq -2 \\ -x - 2 & \text{if } x \leq -2 \end{cases}$$

From Theorem 7.7.1,

$$\int_{-3}^{4} |x + 2|\, dx = \int_{-3}^{-2} |x + 2|\, dx + \int_{-2}^{4} |x + 2|\, dx$$

$$= \int_{-3}^{-2} (-x - 2)\, dx + \int_{-2}^{4} (x + 2)\, dx$$

$$= \left[-\frac{x^2}{2} - 2x \right]_{-3}^{-2} + \left[\frac{x^2}{2} + 2x \right]_{-2}^{4}$$

$$= [(-2 + 4) - (-\tfrac{9}{2} + 6)] + [(8 + 8) - (2 - 4)]$$

$$= 18\tfrac{1}{2}$$

EXAMPLE 4   A particular restaurant that caters to people who work in its vicinity is open from 7 A.M. until 3 P.M. When the restaurant has been open for x hours, the gross receipts for a normal working day are $f(x)$ dollars per hour, where

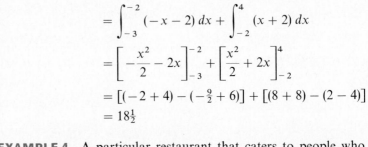

$$f(x) = \begin{cases} 330 - 30(x - 1)^2 & \text{if } 0 \leq x \leq 4 \\ 420 - 90(x - 6)^2 & \text{if } 4 < x \leq 8 \end{cases}$$

(a) Draw a sketch of the graph of f. (b) Interpret the daily gross receipts for a normal working day as the number of square units in the area of a region bounded by the graph in part (a), and find this number.

Solution   (a) A sketch of the graph of f is shown in Fig. 7.7.4. Observe that the peak breakfast period is at 8 A.M. (when $x = 1$) and the peak lunch period is at 1 P.M. (when $x = 6$). Also observe that from 8 A.M. to 11 A.M.

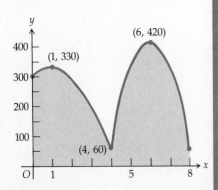

Figure 7.7.4

(from $x = 1$ to $x = 4$) the receipts decrease, and then as lunchtime approaches they increase from 11 A.M. to 1 P.M. (from $x = 4$ to $x = 6$). Then from 1 P.M. to 3 P.M. (from $x = 6$ to $x = 8$) they decrease again. Also notice that f is continuous at $x = 4$ because

$$\lim_{x \to 4^-} f(x) = \lim_{x \to 4^+} f(x) = 60 = f(4)$$

(b) The daily gross receipts for a normal working day is the number of square units in the area of the region shaded in Fig. 7.7.4. If R dollars is this number, then

$$R = \int_0^8 f(x)\, dx = \int_0^4 f(x)\, dx + \int_4^8 f(x)\, dx$$

$$= \int_0^4 [330 - 30(x-1)^2]\, dx + \int_4^8 [420 - 90(x-6)^2]\, dx$$

$$= \left[330x - 10(x-1)^3 \right]_0^4 + \left[420x - 30(x-6)^3 \right]_4^8$$

$$= 1040 + 1200 = 2240$$

Therefore the daily gross receipts are $2240.

Exercises 7.7

In Exercises 1 through 6, use Theorem 7.7.2 to find the indicated derivative.

1. $D_x \int_0^x \sqrt{4 + t^2}\, dt$

2. $D_x \int_2^x \dfrac{dt}{t^4 + 4}$

3. $D_x \int_2^x \dfrac{dt}{\sqrt{t^2 - 1}}$, where $x > 2$

4. $D_x \int_0^x \sqrt{9 - t^2}\, dt$, where $-3 \le x \le 3$

5. $D_x \int_x^3 \dfrac{dt}{3 + t^2}$

6. $D_x \int_x^5 \sqrt{1 + t^2}\, dt$

In Exercises 7 through 14, use Theorem 7.7.1 and the fundamental theorem of the calculus to evaluate the definite integral.

7. $\displaystyle\int_{-1}^2 |x|\, dx$

8. $\displaystyle\int_{-3}^3 |x + 1|\, dx$

9. $\displaystyle\int_{-2}^5 |x - 3|\, dx$

10. $\displaystyle\int_{-1}^4 |3 - x|\, dx$

11. $\displaystyle\int_{-1}^1 \sqrt{|x| - x}\, dx$

12. $\displaystyle\int_{-2}^2 \sqrt{2 + |x|}\, dx$

13. $\displaystyle\int_{-4}^4 f(x)\, dx$ if $f(x) = \begin{cases} x^2 + 1 & \text{if } -4 \le x \le 0 \\ 5 - (x - 2)^2 & \text{if } 0 < x \le 4 \end{cases}$

14. $\displaystyle\int_{-3}^2 f(x)\, dx$ if $f(x) = \begin{cases} 6 - (x + 1)^2 & \text{if } -5 \le x < 0 \\ 2x + 5 & \text{if } 0 < x \le 5 \end{cases}$

In Exercises 15 through 20, find the area of the region bounded by the given curves.

15. $y = |x - 1|$; $y = 0$; $x = -1$; $x = 2$

16. $y = |x + 2|$; $y = 4$

17. $y = 0; x = 4; y = f(x)$ where $f(x) = \begin{cases} x + 1 & \text{if } x \le 0 \\ 9 - 2(x - 2)^2 & \text{if } 0 < x \end{cases}$ **18.** $y = 1; y = f(x)$ where $f(x) = \begin{cases} x^2 & \text{if } x < 0 \\ x^3 & \text{if } 0 \le x \end{cases}$

19. $y = |x|; y = x^2 - 1; x = -1; x = 1$ **20.** $y = |x + 1| + |x|; y = 0; x = -2; x = 3$

21. A certain restaurant is open for lunch and dinner from 11 A.M. until 9 P.M. On a normal Saturday, when the restaurant has been open for x hours, the gross receipts are $f(x)$ dollars per hour, where

$$f(x) = \begin{cases} 400 - 80(x - 2)^2 & \text{if } 0 \le x \le 4 \\ 620 - 60(x - 7)^2 & \text{if } 4 < x \le 10 \end{cases}$$

(a) Draw a sketch of the graph of f. (b) Interpret a normal Saturday's gross receipts as the number of square units in the area of a region bounded by the graph in part (a), and find this number.

Review Exercises for Chapter 7

In Exercises 1 and 2, evaluate the indicated summation by using Theorems 7.1.1 through 7.1.4.

1. $\displaystyle\sum_{i=1}^{20} (i^2 + 2i - 3)$ **2.** $\displaystyle\sum_{i=1}^{30} 2i(i^2 + 3)$

In Exercises 3 through 6, evaluate the limit if it exists.

3. $\displaystyle\lim_{x \to +\infty} \frac{3x - 4}{6x + 5}$ **4.** $\displaystyle\lim_{x \to -\infty} \frac{4x^2 - 5x + 1}{2x^2 + x + 3}$ **5.** $\displaystyle\lim_{t \to +\infty} \frac{t^2 + 1}{t + 1}$ **6.** $\displaystyle\lim_{y \to +\infty} \frac{3y + 1}{y^2 - 2}$

In Exercises 7 and 8, find the exact value of the definite integral by using the definition of the definite integral; do not use the fundamental theorem of the calculus.

7. $\displaystyle\int_0^1 (2x + 5)\,dx$ **8.** $\displaystyle\int_0^2 (3x^2 - 1)\,dx$

In Exercises 9 through 14, evaluate the definite integral by using the fundamental theorem of the calculus.

9. $\displaystyle\int_{-1}^2 (t^3 - 3t)\,dt$ **10.** $\displaystyle\int_1^2 \frac{4\,dx}{3x^3}$ **11.** $\displaystyle\int_1^5 \frac{dx}{\sqrt{3x + 1}}$

12. $\displaystyle\int_{-5}^5 2x\sqrt[3]{x^2 + 2}\,dx$ **13.** $\displaystyle\int_0^2 x^2\sqrt{x^3 + 1}\,dx$ **14.** $\displaystyle\int_{-1}^7 \frac{x^2\,dx}{\sqrt{x + 2}}$

In Exercises 15 through 20, find the area of the region bounded by the given curve and lines. Draw a figure showing the region and a rectangular element of area. Express the measure of the area as the limit of a Riemann sum, and then with definite integral notation. Evaluate the definite integral by the fundamental theorem of the calculus.

15. $y = 9 - x^2$; x axis; y axis; $x = 3$ **16.** $y = \sqrt{x + 1}$; x axis; y axis; $x = 8$

17. $y = 2\sqrt{x - 1}$; x axis; $x = 5$; $x = 17$ **18.** $y = 16 - x^2$; x axis

19. $y = x\sqrt{x + 5}$; x axis; $x = -1$; $x = 4$ **20.** $y = \dfrac{4}{x^2} - x$; x axis; $x = -2$; $x = -1$

21. Find the area of the region bounded by the line $y = x$ and the parabola $y = x^2$.

22. Find the area of the region bounded by the curves $x = y^2$ and $x = y^3$.

23. Find the area of the region bounded by the curve $x^2 + y - 5 = 0$ and the line $y = -4$. Take the elements of area perpendicular to the y axis.

24. Find the area of the region bounded by the curve $y = x^2 - 7x$, the x axis, and the lines $x = 2$ and $x = 4$.

25. Find the area of the region bounded by the loop of the curve $y^2 = x^2(4 - x)$.

26. Find the area of the region bounded by the two parabolas $y^2 = 4x$ and $x^2 = 4y$.

27. If $f(x) = x^2\sqrt{x - 3}$, find the average value of f on $[7, 12]$.

28. (a) Find the average value of the function f defined by $f(x) = 1/x^2$ on the interval $[1, r]$. (b) If A is the average value found in part (a), find $\lim\limits_{r \to +\infty} A$.

29. Suppose that $C(x)$ dollars is the total cost of producing x units of a certain commodity, and $C(x) = \frac{1}{50}x^2 + 500$. Find the average total cost when the number of units produced takes on all values from 0 to 400.

30. The demand equation of a certain commodity is $p^2 + x - 10 = 0$, where $100x$ units of the commodity are demanded when p dollars is the price of 1 unit. Find the average total revenue when the number of units demanded takes on all values from 600 to 800.

31. For the first 5 months that a product was on the market the sales were $f(x)$ units per month x months after the product was first introduced, and

$$f(x) = 50 + 180x^2 \qquad 0 \le x \le 5$$

(a) Interpret the third-month sales as the measure of the area of a plane region, and draw a figure showing the plane region.
(b) Find the third-month sales by computing the area of the plane region in part (a).

32. When some particular machinery is x years old, its book value is changing at the rate of $f(x)$ dollars per year, where

$$f(x) = 3000x - 18,000$$

By how much does the machinery depreciate the first 4 years?

33. For the next 10 years in a particular town it is estimated that x years from now the population will be increasing at the rate of $f(x)$ people per year, where

$$f(x) = 200 + 350x^{4/3} \qquad 0 \le x \le 10$$

What is the expected increase in population for the next 8 years?

34. In the next 5 days how many people are expected to be exposed to a certain virus if it is estimated that t days from now the virus will be spreading at the rate of $f(t)$ people per day, where $f(t) = 9t(8 - t)$?

35. How many people will contact the virus of Exercise 34 during the fifth day?

36. The probability density function for the length of life of a particular electronic component to be $1000x$ hours is given by

$$f(x) = \tfrac{3}{22}(5 - x^2) \qquad 0 \le x \le 2$$

(a) Show that f satisfies the two properties required of a probability density function. (b) What is the probability that the component will last at least 1000 hours?

37. A college bookstore receives a shipment of 600 calculus textbooks on August 10 prior to the beginning of classes on September 9. Over the 30-day period the sales are low at first, and the demand increases as the opening of school approaches so that x days after August 10 the inventory is y books, where

$$y = 600 - \tfrac{2}{3}x^2 \qquad 0 \le x \le 30$$

If the daily storage cost is 0.15 cent per book, find the total cost of maintaining inventory for 30 days.

38. A particular machine yields total revenue at the rate of $R(x)$ dollars per year x years since its purchase, and

$$R(x) = 2500 - 5x^2$$

Furthermore, the total cost of operating and maintaining the machine increases with time so that in x years the total cost is $C(x)$ dollars per year, where

$$C(x) = 880 + 15x^2$$

(a) Draw a figure showing sketches of the graphs of the functions R and C on the same set of axes. (b) How many years is it profitable to operate the machine? (c) What is the total profit realized from the operation of the machine for the number of years determined in part (b)? (d) Show on the figure obtained in part (a) the region whose area is N square units, where N is the number found in part (c).

39. The demand equation for a particular article of merchandise is $100p^2 + x - 2500 = 0$, and the market price is \$3. Find the consumers' surplus to the nearest dollar by integrating (a) with respect to x and (b) with respect to p.

40. If the supply equation for a certain commodity is $64p = (x + 4)^2$ and the market price is \$1, find the producers' surplus to the nearest cent by integrating (a) with respect to x and (b) with respect to p.

41. The demand and supply equations for a particular commodity are, respectively, $2x + p = 12$ and $x^2 - p + 4 = 0$. Find the consumers' surplus and the producers' surplus if the market is in equilibrium.

CHAPTER 8

EXPONENTIAL AND LOGARITHMIC FUNCTIONS

8.1 Types of interest and the number e

To lead up to the definition of e, a number that arises in applications of mathematics to many fields, we obtain its significance to the economist by considering interest on an investment.

If money is loaned at the interest rate of 12 percent per year, then the borrower's debt at the end of a year is $1.12 for each $1 borrowed. In general, if the interest rate is $100i$ percent per year, then for each dollar borrowed the repayment at the end of a year is $(1 + i)$ dollars. If P dollars is borrowed, then the debt is $P(1 + i)$ dollars at the end of a year.

There are different types of interest that we shall consider. One kind is called **simple interest**, which is interest earned only on the original amount that is borrowed. In simple interest no interest is paid on any accrued interest. For example, suppose that interest on $100 is 10 percent simple interest annually. Then the lender would receive $10 at the end of each year. If P dollars is deposited at a simple interest rate of $100i$ percent, the interest at the end of the year is Pi dollars. If no withdrawals are made for n years, the total interest earned is Pni dollars, and if A dollars is the total amount on deposit at the end of n years,

$$A = P + Pni$$

$$A = P(1 + ni) \tag{1}$$

Simple interest is sometimes used for short-term loans of a period of possibly 30, 60, or 90 days. In such cases, in order to simplify calculations, a year is considered as 360 days, and each month is assumed to contain 30 days.

EXAMPLE 1 A loan of $500 is made for a period of 90 days at a simple interest rate of 16 percent annually. Determine the amount to be repaid at the end of 90 days.

Solution We are given $P = 500$, $i = 0.16$, and $n = \frac{90}{360} = \frac{1}{4}$. Hence, if A dollars is the amount to be repaid, we obtain from (1)

$$A = 500[1 + \tfrac{1}{4}(0.16)]$$
$$= 520$$

If during the term of a loan or investment the interest earned each period is added to the principal and then earns interest itself, the interest is called **compound interest**. When the word "interest" is used without an adjective, it is customarily assumed to be compound interest, since this type of interest is the most widely used.

The rate of interest is usually given as an annual rate, but often the interest is computed and then added to the principal more frequently than once a year. If the interest is compounded m times per year, then the annual rate must be divided by m to determine the interest for each period. For example, if $100 is deposited in a savings account that pays 8 percent compounded quarterly, then the number of dollars in the account at the end of the first 3 months' period will be

$$100\left(1 + \frac{0.08}{4}\right) = 100(1.02)$$

The number of dollars in the account at the end of the second 3 months' period will be

$$[100(1.02)](1.02) = 100(1.02)^2$$

At the end of the third 3 months' period the number of dollars in the account will be

$$[100(1.02)^2](1.02) = 100(1.02)^3$$

and so on. At the end of the *n*th 3 months' period the number of dollars in the account will be

$$100(1.02)^n$$

More generally, we have the following theorem.

Theorem 8.1.1 If P dollars is deposited into a savings account that pays an interest rate of $100i$ percent compounded m times per year, and if A_n is the number of dollars in the account at the end of n interest periods, then

$$A_n = P\left(1 + \frac{i}{m}\right)^n \tag{2}$$

The proof of Theorem 8.1.1 involves a process called mathematical induction and is omitted.

If t is the number of years for which P dollars is invested at an interest rate of $100i$ percent compounded m times per year, then the number of interest periods $n = mt$, and letting A dollars be the total amount at t years, we can write formula (2) as

$$A = P\left(1 + \frac{i}{m}\right)^{mt} \tag{3}$$

To facilitate the use of formula (2) or (3), Table 5 in the back of the book gives the value of $(1 + j)^n$, which is the number of dollars in the value of $1 after n periods with an interest rate of $100j$ percent per period.

EXAMPLE 2 Suppose that $400 is deposited into a savings account that pays 8 percent interest per year compounded semiannually. If no withdrawals and no additional deposits are made, find the amount on deposit at the end of 3 years.

Solution The interest is compounded twice a year, and so $m = 2$. Because the time is 3 years, $t = 3$. $P = 400$ and $i = 0.08$, and therefore if A dollars is the amount on deposit at the end of 3 years, we have from formula (3)

$$A = 400\left(1 + \frac{0.08}{2}\right)^6 = 400(1.04)^6$$

From Table 5 in the back of the book, $(1.04)^6 = 1.2653$. Thus

$$A = 400(1.2653) = 506.12$$

The amount on deposit at the end of 3 years is therefore $506.12.

Sometimes a distinction is made between the stated annual rate of interest, called the **nominal rate**, and the **effective rate**, which is the rate that gives the same amount of interest compounded once a year as a nominal rate compounded m times per year. That is, if the effective rate is $100j$ percent and the nominal rate is $100i$ percent, compounded m times per year, then

$$\left(1 + \frac{i}{m}\right)^m = 1 + j \tag{4}$$

● ILLUSTRATION 1

Suppose that $5000 is borrowed at an interest rate of 12 percent compounded monthly, and the loan is to be repaid in one payment at the end of the year. We wish to find the effective rate of interest. Letting j be the effective rate of interest we have from formula (4), with $m = 12$ and $i = 0.12$,

$$1 + j = \left(1 + \frac{0.12}{12}\right)^{12}$$

$$j = (1.01)^{12} - 1$$

Using Table 5 to find $(1.01)^{12} = 1.1268$ we have

$$j = 0.1268 = 12.68 \text{ percent}$$

We use formula (3) to find how much the borrower must repay. If A dollars is the amount to be repaid, we get

$$A = 5000(1.01)^{12}$$
$$= 5000(1.1268)$$
$$= 5634$$

Thus the borrower must repay $5634.   ●

Formula (3) gives the number of dollars in the amount after t years if P dollars is invested at a rate of $100i$ percent compounded m times per year. Let us conceive of a situation where the interest is continuously compounding; that is, consider formula (3) where the number of interest periods per year increases without bound. Then going to the limit in formula (3) we have

$$A = P \lim_{m \to +\infty} \left(1 + \frac{i}{m}\right)^{mt}$$

which may be written as

$$A = P \lim_{m \to +\infty} \left[\left(1 + \frac{i}{m}\right)^{m/i}\right]^{it} \tag{5}$$

Now consider

$$\lim_{m \to +\infty} \left(1 + \frac{i}{m}\right)^{m/i} \tag{6}$$

Let $z = i/m$. Then $m/i = 1/z$, and because "$m \to +\infty$" is equivalent to "$z \to 0^+$", we have

$$\lim_{m \to +\infty} \left(1 + \frac{i}{m}\right)^{m/i} = \lim_{z \to 0^+} (1 + z)^{1/z} \tag{7}$$

Thus the limit in (6) exists if $\lim_{z \to 0^+} (1 + z)^{1/z}$ exists or, even more generally, if the two-sided limit

$$\lim_{z \to 0} (1 + z)^{1/z}$$

exists. You can use a hand calculator to compute values of $(1 + z)^{1/z}$ as z takes on numbers closer and closer to 0. Refer to Table 8.1.1 for values of $(1 + z)^{1/z}$ when z is close to 0 and positive and to Table 8.1.2 for values of $(1 + z)^{1/z}$ when z is close to 0 and negative.

Table 8.1.1

z	0.5	0.1	0.01	0.001	0.0001
$(1 + z)^{1/z}$	2.2500	2.5937	2.7048	2.7169	2.7182

Table 8.1.2

z	-0.5	-0.1	-0.01	-0.001	-0.0001
$(1 + z)^{1/z}$	4.0000	2.8680	2.7320	2.7196	2.7184

These tables lead us to suspect that $\lim_{z \to 0} (1 + z)^{1/z}$ probably exists and is a number that lies between 2.7182 and 2.7184. This is indeed the case, although we will not prove it. The proof of the existence of the limit can be found in more advanced calculus texts. With the assumption that the limit exists, we denote it by the letter e and give a formal definition.

Definition of the number e

$$\lim_{z \to 0} (1 + z)^{1/z} = e \tag{8}$$

The letter e was chosen because of the Swiss mathematician and physicist Leonhard Euler (1707–1783). The number e is an irrational number, and its value can be expressed to any required degree of accuracy. The method for doing this is by using infinite series, which are studied in a more advanced calculus course. The value of e to seven decimal places is 2.7182818. Thus

$$e \approx 2.7182818$$

In the back of the book, Table 4 gives powers of e. Such powers can also be obtained from a hand calculator with an e^x key. The importance of the number e will become apparent as we proceed through this chapter.

Returning now to the discussion of interest compounding continuously, we have, from (5), (7), and (8),

$$A = Pe^{it} \qquad\qquad (9)$$

where A dollars is the amount after t years if P dollars is invested at a rate of $100i$ percent compounded continuously. This value of A is an upper bound for the amount given by (3) when interest is compounded frequently and can be used as an approximation in such a situation. This fact is demonstrated in the following illustration, where we compare the amount at the end of 1 year when interest is compounded continuously with the corresponding amounts obtained when interest is compounded monthly and semimonthly.

● ILLUSTRATION 2

In Illustration 1 we learned that if $5000 is borrowed at an interest rate of 12 percent compounded monthly, and the loan is repaid in one payment at the end of 1 year, then if j is the effective rate of interest,

$$j = 12.68 \text{ percent}$$

and if A dollars is the amount to be repaid,

$$A = 5634$$

Suppose that we have the same problem except that the interest rate of 12 percent is compounded semimonthly instead of monthly. Then we use formula (4) with $m = 24$ and $i = 0.12$, and we have

$$j = \left(1 + \frac{0.12}{24}\right)^{24} - 1$$

$$= (1.005)^{24} - 1$$

Using Table 5 to find $(1.005)^{24} = 1.1272$, we have

$$j = 0.1272$$

$$= 12.72 \text{ percent}$$

Furthermore, from formula (3),

$$A = 5000(1.005)^{24}$$

$$= 5000(1.1272)$$

$$= 5636$$

Now suppose that the interest is compounded continuously at 12 percent. Because $P = 5000$, $i = 0.12$, and $t = 1$, we have from (9)

$$A = 5000e^{0.12}$$

From Table 4, $e^{0.12} = 1.1275$. Thus

$$A = 5000(1.1275)$$
$$= 5637.50$$

Because \$5637.50 is the amount when interest is compounded continuously at 12 percent, it is an upper bound for the amount regardless of how often interest is compounded.

If we let j be the effective rate of interest when interest is compounded continuously at 12 percent, then

$$5000(1 + j) = 5000e^{0.12}$$
$$1 + j = e^{0.12}$$
$$j = 1.1275 - 1$$
$$= 0.1275$$
$$= 12.75 \text{ percent}$$

    •

If in (9), $P = 1$, $i = 1$, and $t = 1$, we get

$$A = e$$

which gives a justification for the economist's interpretation of the number e as the yield on an investment of \$1 for a year at an interest rate of 100 percent compounded continuously.

EXAMPLE 3   A bank advertises that interest on savings accounts is computed at 6 percent per year compounded daily. (a) If \$100 is deposited into a savings account at this bank, find the approximate amount at the end of 1 year by taking the rate as 6 percent compounded continuously. (b) What is the effective rate of interest?

Solution   (a) From (9) with $P = 100$, $i = 0.06$, and $t = 1$, we have

$$A = 100e^{0.06}$$

From Table 4, $e^{0.06} = 1.0618$. Therefore

$$A = 100(1.0618)$$
$$= 106.18$$

Thus the approximate amount on deposit at the end of 1 year is \$106.18.

    (b) If j is the effective rate of interest,

$$100(1 + j) = 106.18$$
$$1 + j = 1.0618$$
$$j = 0.0618$$

Therefore the effective rate of interest is 6.18 percent.

In the above discussions we have been concerned with finding the value in the future of a present sum of money. That is, we determine the amount to which a given principal would grow in a given number of years at various types of interest rates. Often in business one wishes to solve the inverse problem of determining the **present value** (or the value today) of a sum of money that will be available in the future. The present value of A dollars to be received t years hence is the principal to be invested at the present time in order for the investment to amount to A dollars in t years.

Solving (3) for P we obtain

$$P = \frac{A}{\left(1 + \dfrac{i}{m}\right)^{mt}}$$

$$P = A\left(1 + \frac{i}{m}\right)^{-mt} \tag{10}$$

Equation (10) gives the formula for finding the number of dollars in the present value P of A dollars to be received in t years if money is invested at the rate of $100i$ percent per year, compounded m times per year. If interest is compounded only once a year, then $m = 1$, and we have

$$P = A(1 + i)^{-t}$$

EXAMPLE 4   Find the present value of $1000 to be received 3 years from now if money can be invested at an annual rate of 12 percent compounded semiannually.

Solution   Let P be the number of dollars in the present value. From (10) with $A = 1000$, $i = 0.12$, $m = 2$, and $t = 3$,

$$P = 1000\left(1 + \frac{0.12}{2}\right)^{-6}$$

$$= 1000(1.06)^{-6}$$

Table 6 in the back of the book gives the value of $(1 + j)^{-n}$ for various values of j and n. From this table, $(1.06)^{-6} = 0.7050$. Thus

$$P = 1000(0.7050)$$

$$= 705$$

Therefore, receiving $705 now is equivalent to receiving $1000 3 years from now if money is invested at 12 percent compounded semiannually.

Solving (9) for P we obtain

$$P = Ae^{-it} \tag{11}$$

which gives the present value of A dollars payable t years from now if the interest is compounded continuously at the rate $100i$ percent. If $A = 1$ in (11), we have

$$P = e^{-it} \tag{12}$$

which gives the present value of $1 received t years in the future if interest is compounded continuously at the rate of $100i$ percent.

EXAMPLE 5 What is the present value of $1000 payable 3 years from now if money can be invested at an annual rate of 12 percent compounded continuously?

Solution Let P dollars be the present value. From (11) with $A = 1000$, $i = 0.12$ and $t = 3$,

$$P = 1000e^{-0.36}$$

From Table 4, $e^{-0.36} = 0.69768$. Therefore

$$P = 1000(0.69768)$$
$$= 697.68$$

Hence, receiving $697.68 now is equivalent to receiving $1000 3 years from now if money is invested at 12 percent compounded continuously.

Exercises 8.1

1. A loan of $2000 is made at a simple interest rate of 12 percent annually. Determine the amount to be repaid if the period of the loan is (a) 90 days; (b) 6 months; (c) 1 year.
2. Solve Exercise 1 if the loan is for $1500 and the rate is 10 percent annually.
3. Determine the amount at the end of 4 years of an investment of $1000 if the annual interest rate is 8 percent and (a) simple interest is earned; (b) interest is compounded annually; (c) interest is compounded semiannually; (d) interest is compounded quarterly.
4. Find the amount of an investment of $500 at the end of 2 years if the annual interest rate is 6 percent and (a) simple interest is earned; (b) interest is compounded annually; (c) interest is compounded semiannually; (d) interest is compounded monthly.
5. Solve Exercise 4 if the annual interest rate on the $500 investment is 12 percent.
6. A loan of $200 is repaid in one payment at the end of a year. If the interest rate is 12 percent per year, compounded quarterly, determine (a) the total amount repaid and (b) the effective rate of interest.
7. An investment of $5000 earns interest at the rate of 16 percent per year and the interest is paid once at the end of the year. (a) Find the interest earned during the first year if interest is compounded quarterly. (b) What is the effective rate of interest?
8. Work Exercise 6 if the interest rate is 12 percent compounded continuously.
9. Work Exercise 7 if the interest rate is 16 percent compounded continuously.
10. Find the effective rate of interest if interest is at an annual rate of 12 percent compounded (a) semiannually; (b) monthly; (c) continuously.
11. Find the effective rate of interest if interest is at an annual rate of 24 percent compounded (a) semiannually; (b) monthly; (c) continuously.
12. How much should be deposited in a savings account now if it is desired to have $2500 in the account at the end of 5 years if the annual interest rate is 8 percent compounded (a) annually and (b) quarterly?

13. At the end of 4 years a savings account had a balance of $3000. One deposit was made at the beginning of the 4-year period, and no withdrawals were made. How much was the original deposit if the interest rate is 6 percent per year, compounded (a) annually and (b) monthly?

14. A deposit of $1000 is made at a savings bank that advertises that interest on accounts is computed at an annual rate of 7 percent compounded daily. (a) Find the approximate amount at the end of 1 year by taking the rate as 7 percent compounded continuously. (b) What is the effective rate of interest?

15. Solve Exercise 14 if the bank advertises that interest is computed at an annual rate of 9 percent compounded daily, and take the rate as 9 percent compounded continuously.

16. Find the present value of $10,000 payable at the end of 10 years if money is invested at 12 percent per year and interest is compounded (a) annually and (b) continuously.

17. Solve Exercise 16 if the interest rate is 8 percent.

18. Some real estate purchased 8 years ago was sold for $90,000. If the annual interest rate on the investment was determined to be 10 percent compounded annually, what was the purchase price of the real estate to the nearest thousand dollars?

19. A painting was acquired 10 years ago and sold today at auction by the collector for $18,000. If the collector determined that the annual interest rate earned was 24 percent compounded quarterly, what was the purchase price of the painting?

20. A valuable coin was purchased 20 years ago and sold today for $10,000. If it was determined that the annual interest rate was 12 percent compounded continuously, how much, to the nearest hundred dollars, did the seller pay for the coin?

21. A rare book was sold for $2300 by a book dealer who purchased it 4 years ago. If the dealer determined that the annual interest rate was 18 percent compounded continuously, what was the dealer's purchase price to the nearest dollar?

22. In Tables 8.1.1 and 8.1.2 there are values of $(1 + z)^{1/z}$ when $z = 0.001$ and $z = -0.001$. Obtain an approximation of the number e to three decimal places by using these values to find the average value of $(1.001)^{1000}$ and $(0.999)^{-1000}$.

23. Obtain an approximation of the number e to four decimal places by finding the average value of $(1.0001)^{10,000}$ and $(0.9999)^{-10,000}$. Use Tables 8.1.1 and 8.1.2 for the values of $(1 + z)^{1/z}$ when $z = 0.0001$ and $z = -0.0001$.

24. Draw a sketch of the graph of the function defined by

$$f(x) = (1 + x)^{1/x}$$

on the interval $[-0.5, 0.5]$ by assuming that f is continuous on the interval and plotting the points for which x has the values $-0.5, -0.1, -0.01, 0.01, 0.1$, and 0.5. Use Tables 8.1.1 and 8.1.2 for the function values. On the x axis choose the unit of length 0.1, and on the y axis choose the unit of length 1. The y coordinate of the point at which the graph intersects the y axis is the number e.

8.2 Exponential functions

The definition of the power of a positive number when the exponent is a rational number was defined in algebra. In particular, 2^x has been defined for any rational value of x. For instance,

$$2^5 = 2 \cdot 2 \cdot 2 \cdot 2 \cdot 2 \qquad 2^0 = 1 \qquad 2^{-3} = \frac{1}{2^3} \qquad 2^{2/3} = \sqrt[3]{2^2}$$
$$= 32 \qquad\qquad\qquad\qquad = \frac{1}{8} \qquad = \sqrt[3]{4}$$

It is not quite so simple to define 2^x when x is an irrational number. For example, what is meant by $2^{\sqrt{3}}$? The definition of an irrational power of a positive number requires a more advanced approach than we take in this book. However, we can give an intuitive indication that irrational powers of positive numbers can exist by showing how the meaning of $2^{\sqrt{3}}$ can be interpreted. To do this we make use of the following theorem, which is stated without proof.

Theorem 8.2.1 If r and s are rational numbers, then

(i) if $b > 1$: $\qquad r < s$ implies $\quad b^r < b^s$
(ii) if $0 < b < 1$: $\quad r < s$ implies $\quad b^r > b^s$

A decimal approximation for $\sqrt{3}$ can be obtained accurate to any number of decimal places desired. Using four decimal places we have $\sqrt{3} = 1.7321$. Because $1 < 1.7 < 2$, then from Theorem 8.2.1(i) it follows that

$$2^1 < 2^{1.7} < 2^2$$

Because $1.7 < 1.73 < 1.8$, then

$$2^{1.7} < 2^{1.73} < 2^{1.8}$$

Because $1.73 < 1.732 < 1.74$, then

$$2^{1.73} < 2^{1.732} < 2^{1.74}$$

Because $1.732 < 1.7321 < 1.733$, then

$$2^{1.732} < 2^{1.7321} < 2^{1.733}$$

and so on. In each inequality there is a power of 2 for which the exponent is a decimal approximation of the value of $\sqrt{3}$, and in each successive inequality the exponent contains one more decimal place than the exponent in the previous inequality. By following this procedure indefinitely, the difference between the left member of the inequality and the right member of the inequality can be made as small as we please. Hence our intuition leads us to assume that there is a value of $2^{\sqrt{3}}$ that satisfies each successive inequality as the procedure is continued indefinitely. A similar discussion can be given for any irrational power of any positive number. Furthermore, Theorem 8.2.1 is valid if r and s are any real numbers.

We now can define an *exponential function*.

Definition of the exponential function with base b

If $b > 0$ and $b \neq 1$, then the **exponential function with base b** is the function f defined by

$$f(x) = b^x \tag{1}$$

The domain of f is the set of real numbers, and the range of f is the set of positive numbers.

Observe that if $b = 1$, (1) becomes $f(x) = 1^x$. But if x is any real number, then $1^x = 1$, and thus we have a constant function. For this reason we impose the condition that $b \neq 1$ in the above definition.

In the following two illustrations we consider the graphs of the exponential functions with bases 2 and $\frac{1}{2}$, respectively.

● ILLUSTRATION 1

The exponential function with base 2 is the function F such that

$$F(x) = 2^x$$

Table 8.2.1 gives some rational values of x with the corresponding function values.

Table 8.2.1

x	-3	-2	-1	0	1	2	3
2^x	$\frac{1}{8}$	$\frac{1}{4}$	$\frac{1}{2}$	1	2	4	8

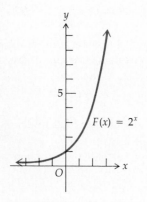

Figure 8.2.1

A sketch of the graph is shown in Fig. 8.2.1; it is drawn by plotting the points whose coordinates are given by Table 8.2.1 and connecting these points with a smooth curve. The function is increasing, which is indicated by the graph and which follows from Theorem 8.2.1(i) with r and s as real numbers and the definition of an increasing function. Observe that

$$\lim_{x \to -\infty} 2^x = 0$$

that is, 2^x approaches zero as x decreases without bound. Furthermore, notice that

$$\lim_{x \to +\infty} 2^x = +\infty$$

that is, 2^x increases without bound as x increases without bound.   ●

● ILLUSTRATION 2

The exponential function with base $\frac{1}{2}$ is the function G such that

$$G(x) = (\tfrac{1}{2})^x$$

In Table 8.2.2 there are some rational values of x with the corresponding function values.

Table 8.2.2

x	-3	-2	-1	0	1	2	3
$(\tfrac{1}{2})^x$	8	4	2	1	$\frac{1}{2}$	$\frac{1}{4}$	$\frac{1}{8}$

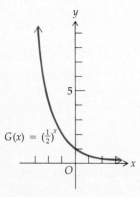

Figure 8.2.2

By plotting the points whose coordinates are given by Table 8.2.2 and connecting these points with a smooth curve, we obtain the sketch of the graph of G shown in Fig. 8.2.2. The function is decreasing, which is indicated

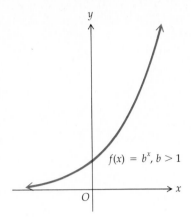

Figure 8.2.3

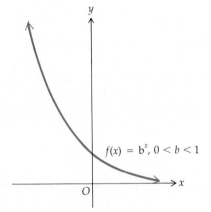

$f(x) = b^x, 0 < b < 1$

Figure 8.2.4

by the graph and which follows from Theorem 8.2.1(ii) with r and s as real numbers and the definition of a decreasing function. Notice that

$$\lim_{x \to +\infty} (\tfrac{1}{2})^x = 0$$

that is, $(\tfrac{1}{2})^x$ approaches zero as x increases without bound. Furthermore,

$$\lim_{x \to -\infty} (\tfrac{1}{2})^x = +\infty$$

that is, $(\tfrac{1}{2})^x$ increases without bound as x decreases without bound. ●

Figure 8.2.3 shows a sketch of the graph of the function f for which $f(x) = b^x$ and $b > 1$. The exponential function with base b, for which $b > 1$, is an increasing function. This fact follows from Theorem 8.2.1(i) with r and s as real numbers and the definition of an increasing function.

In Fig. 8.2.4 there is a sketch of the graph of the exponential function with base b, when $0 < b < 1$. This function is a decreasing function, which follows from Theorem 8.2.1(ii) with r and s as real numbers and the definition of a decreasing function.

The laws of exponents that are valid for rational exponents also hold if the exponents are any real numbers. These laws are summarized in the following theorem.

Theorem 8.2.2 If a and b are any positive numbers, and x and y are any real numbers, then

  (i) $a^x a^y = a^{x+y}$

  (ii) $\dfrac{a^x}{a^y} = a^{x-y}$

  (iii) $(a^x)^y = a^{xy}$

  (iv) $(ab)^x = a^x b^x$

  (v) $\left(\dfrac{a}{b}\right)^x = \dfrac{a^x}{b^x}$

The proofs of properties (i) through (v) of Theorem 8.2.2 for real-number exponents are beyond the scope of this book and are therefore omitted.

EXAMPLE 1 Simplify each of the following by applying laws of exponents:
(a) $2^{\sqrt{3}} \cdot 2^{\sqrt{12}}$; (b) $(7^{\sqrt{5}})^{\sqrt{20}}$

Solution

(a) $2^{\sqrt{3}} \cdot 2^{\sqrt{12}} = 2^{\sqrt{3}} \cdot 2^{2\sqrt{3}}$
$\qquad = 2^{\sqrt{3}+2\sqrt{3}}$
$\qquad = 2^{3\sqrt{3}}$

(b) $(7^{\sqrt{5}})^{\sqrt{20}} = 7^{\sqrt{5} \cdot \sqrt{20}}$
$\qquad = 7^{\sqrt{100}}$
$\qquad = 7^{10}$

From the definition of the exponential function with base b, it follows that the base of an exponential function can be any positive number other than 1. When the base is the number e defined in Sec. 8.1, we have the exponential function with base e, which is often called the *natural exponential function*.

Definition of the natural exponential function

> **The natural exponential function** is the function f defined by
>
> $$f(x) = e^x$$
>
> The domain of the natural exponential function is the set of real numbers, and its range is the set of positive numbers.

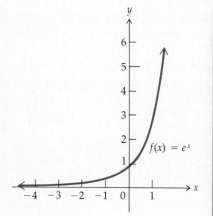

Figure 8.2.5

EXAMPLE 2 Draw a sketch of the graph of the natural exponential function.

Solution Table 8.2.3 gives some values of x and the corresponding function values of the natural exponential function. The approximations of the powers of e are found in Table 4 in the back of the book.

Table 8.2.3

x	0	0.5	1	1.5	2	2.5	-0.5	-1	-2
e^x	1	1.6	2.7	4.5	7.4	12.2	0.6	0.4	0.1

The points whose coordinates are given by Table 8.2.3 are plotted, and these points are connected with a smooth curve to yield the sketch of the graph of the natural exponential function shown in Fig. 8.2.5.

Mathematical models involving powers of e occur in many fields. Some models involve what are called *exponential growth* or *exponential decay*. A function defined by an equation of the form

$$f(t) = Be^{kt} \qquad t \geq 0 \tag{2}$$

where B and k are positive constants, is said to have **exponential growth**. In Sec. 8.6 you will learn that when the rate of growth of a quantity is proportional to its size, the quantity grows exponentially. To draw a sketch of the graph of (2) we note that $f(0) = B$ and $f(t)$ is always positive. Furthermore,

$$\lim_{t \to +\infty} f(t) = \lim_{t \to +\infty} Be^{kt} = +\infty$$

Thus $f(t)$ increases without bound as t increases without bound. See Fig. 8.2.6.

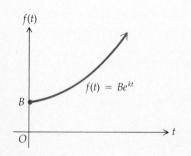

Figure 8.2.6

A particular example of exponential growth occurs if $1000 is invested at 12 percent compounded continuously. If $f(t)$ dollars is the amount of such an investment after t years, we have from formula (9) of Sec. 8.1

$$f(t) = 1000e^{0.12t}$$

Thus the amount of the investment grows exponentially. The following example involves exponential growth in biology.

EXAMPLE 3 In a particular bacterial culture, if $f(t)$ bacteria are present at t min, then

$$f(t) = Be^{0.04t} \tag{3}$$

where B is a constant. If 1500 bacteria are present initially, how many bacteria will be present after 1 hour?

Solution There are 1500 bacteria present initially, and so $f(0) = 1500$. Therefore, from (3),

$$f(0) = Be^{0.04(0)}$$
$$1500 = Be^{0}$$
$$1500 = B$$

From (3), with $B = 1500$, we have

$$f(t) = 1500e^{0.04t} \tag{4}$$

The number of bacteria present after 1 hour is $f(60)$. From (4),

$$f(60) = 1500e^{0.04(60)}$$
$$= 1500e^{2.4}$$

From Table 4, $e^{2.4} = 11.023$. Therefore

$$f(60) = 1500(11.023)$$
$$= 16,535$$

Hence 16,535 bacteria are present in the culture after 1 hour.

A function defined by an equation of the form

$$f(t) = Be^{-kt} \qquad t \geq 0 \tag{5}$$

where B and k are positive constants, is said to have **exponential decay**. Exponential decay occurs when the rate of decrease of a quantity is proportional to its size, as will be shown in Sec. 8.6. For instance, it is known from experiments that the rate of decay of radium is proportional to the amount of radium present at a given instant. A sketch of the graph of (5) appears in Fig. 8.2.7. Observe that

$$\lim_{t \to +\infty} f(t) = \lim_{t \to +\infty} Be^{-kt} = 0$$

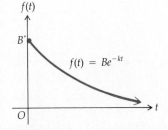

Figure 8.2.7

The next example involves exponential decay where the value of some equipment is decreasing exponentially.

EXAMPLE 4   If $V(t)$ dollars is the value of a certain piece of equipment t years after its purchase, then

$$V(t) = Be^{-0.20t} \tag{6}$$

where B is a constant. If the equipment was purchased for \$8000, what will be its value in 2 years?

Solution   Because the equipment was purchased for \$8000, $V(0) = 8000$. Therefore, from (6),

$$V(0) = Be^{-0.20(0)}$$

$$8000 = Be^0$$

$$8000 = B$$

Replacing B by 8000 in (6) we have

$$V(t) = 8000e^{-0.20t} \tag{7}$$

The number of dollars in the value of the equipment after 2 years is $V(2)$. From (7),

$$V(2) = 8000e^{-0.20(2)}$$

$$= 8000e^{-0.40}$$

From Table 4, $e^{-0.40} = 0.670320$. Therefore

$$V(2) = 8000(0.670320)$$

$$= 5362.56$$

Thus the value of the equipment in 2 years will be \$5362.56.

Another mathematical model involving powers of e is given by the function defined by

$$f(t) = A(1 - e^{-kt}) \qquad t \geq 0 \tag{8}$$

where A and k are positive constants. This function describes **bounded growth**. Because

$$\lim_{t \to +\infty} f(t) = \lim_{t \to +\infty} \left(A - \frac{A}{e^{kt}} \right) = A - 0 = A$$

then as t increases without bound, $f(t)$ approaches A. Also note that

$$f(0) = A(1 - e^0) = 0$$

From this information we obtain the sketch of the graph of f shown in Fig. 8.2.8. This graph is sometimes called a **learning curve**. The appropriateness of the name is apparent when $f(t)$ represents the competence at which

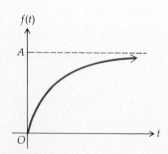

$f(t)$

Figure 8.2.8

a person performs a job. As a person's experience increases, the competence increases rapidly at first and then slows down as additional experience has little effect on the skill at which the task is performed.

EXAMPLE 5 A typical worker at a certain factory can produce $f(t)$ units per day after starting the job, where

$$f(t) = 50(1 - e^{-kt}) \tag{9}$$

If a worker can produce 37 units per day after 4 days, how many units per day can the worker produce after 7 days?

Solution We are given that $f(4) = 37$. Thus

$$37 = 50(1 - e^{-4k})$$

$$\tfrac{37}{50} = 1 - e^{-4k}$$

$$e^{-4k} = 1 - 0.74$$

$$e^{-4k} = 0.26 \tag{10}$$

In Table 4 we wish to find a value of x for which $e^{-x} = 0.26$. In the table we observe that $e^{-1.35} = 0.259240$. Thus

$$e^{-1.35} = 0.26 \tag{11}$$

Therefore, from (10) and (11), we have

$$e^{-4k} = e^{-1.35}$$

$$-4k = -1.35$$

$$k = 0.34$$

Taking $k = 0.34$ in (9) we have

$$f(t) = 50(1 - e^{-0.34t}) \tag{12}$$

We wish to find $f(7)$. From (12),

$$f(7) = 50(1 - e^{-0.34(7)})$$

$$= 50(1 - e^{-2.38})$$

From Table 4, $e^{-2.38} = 0.093$. Therefore

$$f(7) = 50(1 - 0.093)$$

$$= 45$$

Hence the worker can produce 45 units per day after 7 days.

Bounded growth is also described by a function defined by

$$f(t) = A - Be^{-kt} \qquad t \geq 0 \tag{13}$$

where A, B, and k are positive constants. For this function $f(0) = A - B$. A sketch of the graph appears in Fig. 8.2.9. In Sec. 8.7 we show that bounded

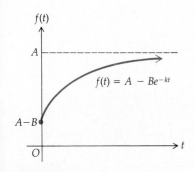

Figure 8.2.9

growth occurs when a quantity grows at a rate proportional to the difference between a fixed number and its size, where the fixed number serves as an upper bound.

Consider now the growth of a population that is affected by the environment imposing an upper bound on its size. For instance, space or reproduction may be factors that are limited by the environment. In such cases a mathematical model of the form (2) does not apply because the population does not increase beyond a certain point. A model that takes into account environmental factors is given by the function defined by

$$f(t) = \frac{A}{1 + Be^{-Akt}} \qquad t \geq 0 \tag{14}$$

where A, B, and k are positive constants. In Fig. 8.2.10 there is a sketch of the graph of this function. It is called a curve of **logistic growth**. Observe that when t is small, the graph is similar to the one for exponential growth in Fig. 8.2.6, and as t increases, the curve is analogous to that shown in Fig. 8.2.9 for bounded growth.

An application of logistic growth in economics is the distribution of information about a particular product. Logistic growth is used by biologists to describe the spread of a disease and by sociologists to describe the spread of a rumor or a joke.

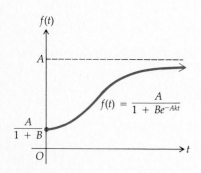

Figure 8.2.10

EXAMPLE 6 In a certain community the spread of a particular flu germ was such that t weeks after its outbreak $f(t)$ persons had contracted the flu, where

$$f(t) = \frac{45,000}{1 + 224e^{-0.9t}} \tag{15}$$

How many people had the flu (a) at the outbreak; (b) after 3 weeks; (c) after 10 weeks? (d) If the epidemic continues indefinitely, how many people will contract the flu?

Solution (a) The number of people who had the flu at the outbreak is $f(0)$, and

$$f(0) = \frac{45,000}{1 + 224e^{-0.9(0)}} = \frac{45,000}{1 + 224} = 200$$

(b) After 3 weeks the number of people who had the flu is $f(3)$, and

$$f(3) = \frac{45,000}{1 + 224e^{-0.9(3)}} = \frac{45,000}{1 + 224e^{-2.7}}$$

$$= \frac{45,000}{1 + 224(0.067206)} = \frac{45,000}{16.054}$$

$$= 2803$$

(c) After 10 weeks the number of people who had the flu is $f(10)$, and

$$f(10) = \frac{45,000}{1 + 224e^{-9}} = \frac{45,000}{1 + 224(0.0001234)}$$

$$= \frac{45,000}{1.02764} = 43,790$$

(d) $\displaystyle\lim_{t \to +\infty} f(t) = \lim_{t \to +\infty} \frac{45,000}{1 + 224e^{-0.9t}} = \frac{45,000}{1 + 0} = 45,000$

Therefore approximately 45,000 people will contract the flu if the epidemic continues indefinitely.

Exercises 8.2

In Exercises 1 through 14, draw a sketch of the graph of the given exponential function. In Exercises 9 through 14, use Table 4 in the back of the book or a hand calculator for powers of e.

1. $f(x) = 3^x$
2. $g(x) = 4^x$
3. $F(x) = 3^{-x}$
4. $G(x) = 4^{-x}$
5. $g(x) = (\frac{1}{3})^x$
6. $f(x) = 10^x$
7. $f(x) = 2^{x+1}$
8. $g(x) = 3^{x-1}$
9. $G(x) = e^{2x}$
10. $F(x) = e^{-x}$
11. $f(x) = 10e^{0.2x}$
12. $g(x) = 100e^{0.1x}$
13. $F(x) = 100e^{-0.1x}$
14. $G(x) = 10e^{-0.2x}$

15. Draw sketches of the graphs of $y = 3^x$ and $x = 3^y$ on the same coordinate axes.
16. Draw sketches of the graphs of $y = e^x$ and $x = e^y$ on the same coordinate axes.
17. The function f defined by $f(x) = e^{-x^2}$ is important in statistics. Draw a sketch of the graph of f by assuming that f is continuous and plotting the points for which x has the values -2, $-\frac{3}{2}$, -1, $-\frac{1}{2}$, 0, $\frac{1}{2}$, 1, $\frac{3}{2}$, and 2. Use Table 4 or a hand calculator for powers of e.

In Exercises 18 through 21, simplify the given expression by applying laws of exponents. When variables appear, the base is a positive number.

18. (a) $x^2 \cdot x^5$; (b) $(x^2)^5$; (c) $x^5 \div x^2$; (d) $x^2 \div x^5$
19. (a) $y^6 \cdot y^{-2}$; (b) $(y^6)^{-2}$; (c) $y^6 \div y^{-2}$; (d) $y^{-2} \div y^{-6}$
20. (a) $3^{\sqrt{2}} \cdot 3^{\sqrt{50}}$; (b) $(e^{\sqrt{2}})^{\sqrt{50}}$
21. (a) $2^{\sqrt{12}} \cdot 2^{\sqrt{27}}$; (b) $(e^{\sqrt{12}})^{\sqrt{27}}$

22. The value of a particular machine t years after its purchase is $V(t)$ dollars, where

$$V(t) = ke^{-0.30t}$$

and k is a constant. If the machine was purchased 8 years ago for $10,000, what is its value now?

23. An historically important abstract painting was purchased in 1922 for $200, and its value has doubled every 10 years since its purchase. (a) If $f(t)$ dollars is the value t years after its purchase, define $f(t)$. (b) What was the value of the painting in 1982?

24. If $P(h)$ pounds per square foot is the atmospheric pressure at altitude h ft above sea level, then

$$P(h) = ke^{-0.00003h}$$

where k is a constant. If the atmospheric pressure at sea level is 2116 pounds per square foot, find the atmospheric pressure outside of an airplane that is at altitude 10,000 ft.

25. If $f(t)$ grams of a radioactive substance are present after t seconds, then

$$f(t) = ke^{-0.3t}$$

where k is a constant. If 100 g of the substance are present initially, how much is present after 5 sec?

26. Suppose that $f(t)$ is the number of bacteria present in a certain culture at t min, and

$$f(t) = ke^{0.035t}$$

where k is a constant. If 5000 bacteria are present after 10 min have elapsed, how many bacteria were present initially?

27. In 1975 it was estimated that for the succeeding 20 years the population of a particular city is expected to be $f(t)$ people t years from 1975, where

$$f(t) = C \cdot 10^{kt}$$

where C and k are constants. If the population in 1975 was 1000 and in 1980 it was 4000, what is the expected population in 1990?

28. After t hours of practice typing it was determined that a certain person could type $f(t)$ words per minute, where

$$f(t) = 90(1 - e^{-0.03t})$$

(a) Draw a sketch of the graph of f, and observe the behavior of f as t increases without bound. (b) How many words per minute can the person type after 30 hours of practice? (c) How many words per minute can the person eventually be expected to type?

29. The efficiency of a typical worker at a certain factory is given by the function defined by

$$f(t) = 100 - 60e^{-0.2t}$$

where the worker can complete $f(t)$ units of work per day after being on the job for t months. (a) Draw a sketch of the graph of f, and observe the behavior of f as t increases without bound. (b) How many units per day can be completed by a beginning worker? (c) How many units per day can be completed by a worker having 6 months' experience? (d) How many units per day can the typical worker eventually be expected to complete?

30. The resale value of a certain piece of equipment is $f(t)$ dollars t years after its purchase, where

$$f(t) = 1200 + 8000e^{-0.25t}$$

(a) Draw a sketch of the graph of f, and observe the behavior of f as t increases without bound. (b) What is the value of the equipment when it is purchased? (c) What is the value of the equipment 10 years after its purchase? (d) What is the anticipated scrap value of the equipment after a long period of time?

31. One day on a college campus when there were 5000 people in attendance, a particular student heard that a certain controversial speaker was going to make an unscheduled appearance. This information was told to friends who in turn related it to others. After t min elapsed, $f(t)$ people had heard the rumor, where

$$f(t) = \frac{5000}{1 + 4999e^{-0.5t}}$$

How many people had heard the rumor (a) after 10 min and (b) after 20 min? (c) Find $\lim_{t \to +\infty} f(t)$.

32. In a particular town of population A, 20 percent of the residents heard a radio announcement about a local political scandal. After t hours, $f(t)$ people had heard about the scandal, where

$$f(t) = \frac{A}{1 + Be^{-Akt}}$$

If 50 percent of the population heard about the scandal after 1 hour, how long was it until 80 percent of the population heard about it?

33. In a community in which A people were susceptible to a particular virus, this virus spread in such a way that t weeks after its appearance, $f(t)$ persons had caught the virus, where

$$f(t) = \frac{A}{1 + Be^{-Akt}}$$

If 10 percent of those susceptible had the virus initially, and 25 percent had been infected after 3 weeks, what percent of those susceptible had been infected after 6 weeks?

8.3 Logarithmic functions

Suppose we wish to find the number of years that it will take for $900 to accumulate to $1500 if money is invested at 10 percent compounded continuously. If T is the number of years to be determined, then from formula (9) in Sec. 8.1 with $P = 900$, $A = 1500$, $i = 0.10$, and $t = T$ we have

$$1500 = 900e^{0.10T}$$

In this equation the unknown T appears in an exponent. At the present we cannot solve such an equation by using functions so far considered. However, the concept of a logarithm will give us the means to do so. We will develop this concept and then return to the problem in Example 4.

When $b > 1$, the exponential function with base b is an increasing function, and when $0 < b < 1$, it is a decreasing function. Furthermore, this function is continuous on its domain, which is the set of real numbers. Because the range of the exponential function is the set of positive numbers, it follows from a theorem, called the intermediate-value theorem, that for any positive number y there is a real number x such that

$$y = b^x$$

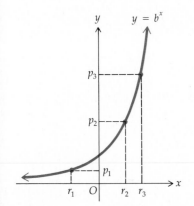

Figure 8.3.1

See Fig. 8.3.1, which shows a sketch of the graph of $y = b^x$ when $b > 1$; in this figure a positive number p_i ($i = 1, 2, 3$) is selected on the y axis, and the unique real number r_i is the corresponding point on the x axis such that

$$p_i = b^{r_i}$$

Thus we are able to make the following definition.

Definition of the logarithmic function to the base b

Let b be any positive number except 1. If x is any positive number, there is exactly one real number y such that $x = b^y$. This unique number y is the value of the **logarithmic function to the base b** at x, and is denoted by $\log_b x$. Therefore

$$y = \log_b x \quad \text{if and only if} \quad x = b^y \tag{1}$$

$\log_b x$ is read as "the logarithm of x to the base b."

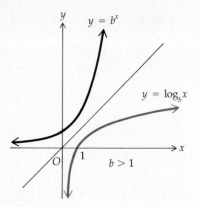

Figure 8.3.2

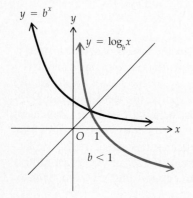

Figure 8.3.3

Because the range of the exponential function with base b is the set of positive numbers, the domain of the logarithmic function to the base b is the set of positive numbers. The range of the logarithmic function to the base b is the set of real numbers, because this is the domain of the exponential function with base b. Figure 8.3.2 shows in color a sketch of the graph of the logarithmic function to the base b, where $b > 1$. It is the reflection across the line $y = x$ of the graph of the exponential function with base $b(b > 1)$; that is, if the paper were folded along the line $y = x$, one graph would lie on the other graph. A sketch of the graph of the logarithmic function to the base b, where $0 < b < 1$, is shown in color in Fig. 8.3.3. Observe that it is the reflection across the line $y = x$ of the graph of the exponential function with base $b(0 < b < 1)$.

The sketches of the graphs in Figs. 8.3.2 and 8.3.3 indicate certain properties of the logarithmic function to the base b. We list these properties, and they are valid, but you should be aware that we have not given formal proofs of them.

1. The function is continuous on its domain of positive numbers.
2. If $b > 1$, the function is increasing; if $0 < b < 1$, the function is decreasing.
3. If $b > 1$, $\log_b x$ is positive if $x > 1$, and $\log_b x$ is negative if $0 < x < 1$. If $0 < b < 1$, $\log_b x$ is negative if $x > 1$, and $\log_b x$ is positive if $0 < x < 1$. Furthermore, $\log_b x$ is not defined if x is negative.
4. $\log_b 1 = 0$
5. If $b > 1$, $\lim\limits_{x \to 0^+} \log_b x = -\infty$; and if $0 < b < 1$, $\lim\limits_{x \to 0^+} \log_b x = +\infty$.

The two equations appearing in statement (1) are equivalent.

● ILLUSTRATION 1

$$4^2 = 16 \quad \text{is equivalent to} \quad \log_4 16 = 2$$
$$2^3 = 8 \quad \text{is equivalent to} \quad \log_2 8 = 3$$
$$(\tfrac{1}{9})^{1/2} = \tfrac{1}{3} \quad \text{is equivalent to} \quad \log_{1/9} \tfrac{1}{3} = \tfrac{1}{2}$$
$$5^{-2} = \tfrac{1}{25} \quad \text{is equivalent to} \quad \log_5 \tfrac{1}{25} = -2$$

● ILLUSTRATION 2

$$\log_{10} 10{,}000 = 4 \quad \text{is equivalent to} \quad 10^4 = 10{,}000$$
$$\log_8 2 = \tfrac{1}{3} \quad \text{is equivalent to} \quad 8^{1/3} = 2$$
$$\log_7 1 = 0 \quad \text{is equivalent to} \quad 7^0 = 1$$
$$\log_9 \tfrac{1}{3} = -\tfrac{1}{2} \quad \text{is equivalent to} \quad 9^{-1/2} = \tfrac{1}{3}$$
$$\log_6 6 = 1 \quad \text{is equivalent to} \quad 6^1 = 6$$

EXAMPLE 1 Find the value of each of the following logarithms: (a) $\log_7 49$; (b) $\log_5 \sqrt{5}$; (c) $\log_6 \tfrac{1}{6}$; (d) $\log_3 81$; (e) $\log_{10} 0.001$.

Solution In each part we let y represent the given logarithm and obtain an equivalent equation in exponential form. We then solve for y by making use of the fact that if $b > 0$ and $b \neq 1$, $b^y = b^n$ implies $y = n$.

(a) Let $\log_7 49 = y$. This equation is equivalent to $7^y = 49$. Because $49 = 7^2$, we have

$$7^y = 7^2$$

Therefore $y = 2$; that is, $\log_7 49 = 2$.

(b) Let $\log_5 \sqrt{5} = y$. Therefore $5^y = \sqrt{5}$ or, equivalently,

$$5^y = 5^{1/2}$$

Hence $y = \frac{1}{2}$; that is, $\log_5 \sqrt{5} = \frac{1}{2}$.

(c) Let $\log_6 \frac{1}{6} = y$. Thus $6^y = \frac{1}{6}$ or, equivalently,

$$6^y = 6^{-1}$$

Therefore $y = -1$; that is, $\log_6 \frac{1}{6} = -1$.

(d) Let $\log_3 81 = y$. Thus $3^y = 81$ or, equivalently,

$$3^y = 3^4$$

Hence $y = 4$; that is, $\log_3 81 = 4$.

(e) Let $\log_{10} 0.001 = y$. Then $10^y = 0.001$. Because $10^{-3} = 0.001$, we have

$$10^y = 10^{-3}$$

Therefore $y = -3$; that is, $\log_{10} 0.001 = -3$.

If we eliminate y between the equations in (1), we obtain

$$b^{\log_b x} = x \tag{2}$$

Eliminating x between the same pair of equations we get

$$\log_b b^y = y$$

If in this equation we replace the variable y by x, we obtain the equivalent equation

$$\log_b b^x = x \tag{3}$$

Equations (2) and (3) are identities, and because of these identities, the exponential function with base b and the logarithmic function to the base b are *inverse functions*. In general, two functions f and g for which

$$f(g(x)) = x \quad \text{and} \quad g(f(x)) = x \tag{4}$$

are said to be **inverse functions**. If in Eqs. (4) we let f be the exponential function with base b and g be the logarithmic function to the base b, we obtain Eqs. (2) and (3), respectively.

The following theorem gives three important properties of logarithms. The proofs are based on corresponding properties of exponents.

Theorem 8.3.1   If $b > 0, b \neq 1, M > 0, N > 0$, and r is any real number, then

$$\text{(i)} \quad \log_b MN = \log_b M + \log_b N \tag{5}$$

$$\text{(ii)} \quad \log_b \frac{M}{N} = \log_b M - \log_b N \tag{6}$$

$$\text{(iii)} \quad \log_b M^r = r \log_b M \tag{7}$$

Proof of Part (i)

Let

$$u = \log_b M \quad \text{and} \quad v = \log_b N \tag{8}$$

The exponential forms of Eqs. (8) are, respectively,

$$M = b^u \quad \text{and} \quad N = b^v$$

Therefore

$$MN = b^u \cdot b^v$$
$$= b^{u+v}$$

The logarithmic form of this equation is

$$\log_b MN = u + v \tag{9}$$

Substituting the values of u and v from Eqs. (8) into (9) we obtain

$$\log_b MN = \log_b M + \log_b N \qquad\qquad \blacksquare$$

The proofs of parts (ii) and (iii) are similar and are omitted.

EXAMPLE 2   Express each of the following in terms of logarithms of x, y, and z, where the variables represent positive numbers: (a) $\log_b x^2 y^3 z^4$; (b) $\log_b \dfrac{x}{yz^2}$; (c) $\log_b \sqrt[5]{\dfrac{xy^2}{z^3}}$.

Solution   (a) Using (5) we have

$$\log_b x^2 y^3 z^4 = \log_b x^2 + \log_b y^3 + \log_b z^4$$

Applying (7) to each of the logarithms in the right member we obtain

$$\log_b x^2 y^3 z^4 = 2 \log_b x + 3 \log_b y + 4 \log_b z$$

(b) From (6) it follows that

$$\log_b \frac{x}{yz^2} = \log_b x - \log_b yz^2$$

Applying (5) to the second logarithm in the right member we have

$$\log_b \frac{x}{yz^2} = \log_b x - (\log_b y + \log_b z^2)$$

$$= \log_b x - \log_b y - 2 \log_b z$$

(c) From (7) it follows that

$$\log_b \sqrt[5]{\frac{xy^2}{z^3}} = \frac{1}{5} \log_b \frac{xy^2}{z^3}$$

Applying (6) to the right member we obtain

$$\log_b \sqrt[5]{\frac{xy^2}{z^3}} = \frac{1}{5}(\log_b xy^2 - \log_b z^3)$$

$$= \frac{1}{5}(\log_b x + \log_b y^2 - \log_b z^3)$$

$$= \frac{1}{5}(\log_b x + 2 \log_b y - 3 \log_b z)$$

$$= \frac{1}{5}\log_b x + \frac{2}{5}\log_b y - \frac{3}{5}\log_b z$$

EXAMPLE 3 Write each of the following expressions as a single logarithm with a coefficient of 1: (a) $\log_b x + 2 \log_b y - 3 \log_b z$; (b) $\frac{1}{3}(\log_b 4 - \log_b 3 + 2 \log_b x - \log_b y)$.

Solution

(a) $\log_b x + 2 \log_b y - 3 \log_b z = (\log_b x + \log_b y^2) - \log_b z^3$

$$= \log_b xy^2 - \log_b z^3$$

$$= \log_b \frac{xy^2}{z^3}$$

(b) $\frac{1}{3}(\log_b 4 - \log_b 3 + 2 \log_b x - \log_b y)$

$$= \frac{1}{3}[(\log_b 4 + \log_b x^2) - (\log_b 3 + \log_b y)]$$

$$= \frac{1}{3}[\log_b 4x^2 - \log_b 3y]$$

$$= \frac{1}{3}\log_b \frac{4x^2}{3y}$$

$$= \log_b \sqrt[3]{\frac{4x^2}{3y}}$$

When logarithms are used mainly as an aid to computation, the base is taken as 10. Such logarithms are called **common logarithms**. Table 2 in the back of the book gives values of common logarithms to four places. A hand calculator with a log key can also be used for these values.

Suppose that we have a logarithmic function whose base is the number e, defined in Sec. 8.1. This function is called the *natural logarithmic function* and is given a formal definition.

Definition of the natural logarithmic function

> If x is any positive number, there is exactly one real number y such that $x = e^y$. This unique number y is the value of the **natural logarithmic function** at x and is denoted by $\ln x$. Therefore
>
> $$y = \ln x \quad \text{if and only if} \quad x = e^y \tag{10}$$

We read $\ln x$ as "the natural logarithm of x." You will become aware of the importance of the natural logarithmic function in calculus when studying Sec. 8.4.

A sketch of the graph of the natural logarithmic function appears in Fig. 8.3.4. It has, of course, the appearance of the graph of the logarithmic function to the base b when $b > 1$, as shown in Fig. 8.3.2. In Fig. 8.3.4 observe that e is the number whose natural logarithm is 1; that is

$$\ln e = 1 \tag{11}$$

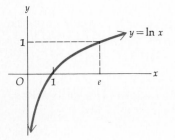

Figure 8.3.4

This follows from statement (10) because

$$\ln e = 1 \quad \text{is equivalent to} \quad e = e^1$$

If in (2) $b = e$, we have

$$e^{\ln x} = x \tag{12}$$

and if $b = e$ in (3), we have

$$\ln e^x = x \tag{13}$$

Because of (12) and (13) the natural exponential function and the natural logarithmic function are inverses.

If in Theorem 8.3.1 we replace b by e, we have the following three properties, where $M > 0$, $N > 0$, and r is any real number:

$$\ln MN = \ln M + \ln N \tag{14}$$

$$\ln \frac{M}{N} = \ln M - \ln N \tag{15}$$

$$\ln M^r = r \ln M \tag{16}$$

In the following example we use natural logarithms to solve the problem presented at the beginning of this section. In the solution we use Table 3 in the back of the book, where values of natural logarithms to four places are given. A hand calculator with an ln key could also be used.

EXAMPLE 4 Determine the number of years that it will take for $900 to accumulate to $1500 if money is invested at 10 percent compounded continuously.

Solution If T is the number of years to be determined, then from formula (9) in Sec. 8.1 with $P = 900$, $A = 1500$, $i = 0.10$, and $t = T$ we have

$$900e^{0.10T} = 1500$$

$$e^{0.10T} = \frac{1500}{900}$$

$$e^{0.10T} = \frac{5}{3}$$

Taking the natural logarithm on both sides of the equation we have

$$\ln e^{0.10T} = \ln \frac{5}{3}$$

On the left side of the above equation we use (16) and on the right side we use (15), and we have

$$0.10T(\ln e) = \ln 5 - \ln 3$$

From Table 3, $\ln 5 = 1.6094$ and $\ln 3 = 1.0986$; and because $\ln e = 1$ we have

$$0.10T(1) = 1.6094 - 1.0986$$

$$0.10T = 0.5108$$

$$T = 5.108$$

Because 5.108 years is 5 years, 1 month, and 9 days, it will take that long for $900 to accumulate to $1500 if money is invested at 10 percent compounded continuously.

EXAMPLE 5 Solve Example 4 if money is invested at 10 percent compounded semiannually.

Solution If T is the number of years to be determined, then from formula (3) in Sec. 8.1 with $P = 900$, $A = 1500$, $i = 0.10$, $m = 2$, and $t = T$ we have

$$1500 = 900\left(1 + \frac{0.10}{2}\right)^{2T}$$

$$(1.05)^{2T} = \frac{5}{3}$$

Taking the natural logarithm on both sides of the equation we obtain

$$\ln(1.05)^{2T} = \ln \frac{5}{3}$$

$$2T \ln(1.05) = \ln 5 - \ln 3$$

$$T = \frac{\ln 5 - \ln 3}{2 \ln(1.05)}$$

From Table 3, $\ln 5 = 1.6094$, $\ln 3 = 1.0986$, and $\ln 1.05 = 0.0488$. Therefore

$$T = \frac{1.6094 - 1.0986}{2(0.0488)}$$

$$= \frac{0.5108}{0.0976}$$

$$= 5.234$$

Because 5.234 years is 5 years, 2 months, and 24 days, it will take that long for $900 to accumulate to $1500 if money is invested at 10 percent compounded semiannually.

A relationship between logarithms to the base b and natural logarithms can be obtained by starting with the equation

$$y = \log_b x \tag{17}$$

An equivalent equation in exponential form is

$$b^y = x$$

Take the natural logarithm on both sides and we have

$$\ln b^y = \ln x$$

$$y \ln b = \ln x$$

$$y = \frac{\ln x}{\ln b}$$

From (17), $y = \log_b x$. Therefore

$$\log_b x = \frac{\ln x}{\ln b} \tag{18}$$

Equation (18) can be used to compute the logarithm of a number to any base if one can determine the required natural logarithms.

If in (18) we take $x = e$, we have

$$\log_b e = \frac{\ln e}{\ln b}$$

or, because $\ln e = 1$,

$$\log_b e = \frac{1}{\ln b} \tag{19}$$

EXAMPLE 6 Compute the value of the given logarithm: (a) $\log_2 e$; (b) $\log_5 10$.

Solution (a) From (19),

$$\log_2 e = \frac{1}{\ln 2} = \frac{1}{0.6931} = 1.443$$

(b) From (18),

$$\log_5 10 = \frac{\ln 10}{\ln 5} = \frac{2.3026}{1.6094} = 1.431$$

Exercises 8.3

In Exercises 1 through 6, draw a sketch of the graph of the given function.

1. $f(x) = \log_{10} x$
2. $g(x) = \log_2 x$
3. $g(x) = \ln(x + 1)$
4. $f(x) = \ln(x - 1)$
5. $F(x) = \ln(-x)$
6. $G(x) = -\ln x$

In Exercises 7 through 12, find the value of the given logarithm.

7. (a) $\log_8 64$; (b) $\log_4 64$; (c) $\log_{64} 8$; (d) $\log_2 \frac{1}{64}$
8. (a) $\log_9 81$; (b) $\log_3 81$; (c) $\log_{27} 81$; (d) $\log_{81} 3$
9. (a) $\log_2 1$; (b) $\log_2 2$; (c) $\log_2 4$; (d) $\log_2 \frac{1}{2}$
10. (a) $\log_2 \frac{1}{8}$; (b) $\log_4 \frac{1}{8}$; (c) $\log_{1/2} 8$; (d) $\log_{1/4} 8$
11. (a) $\log_3 \frac{1}{81}$; (b) $\log_{27} \frac{1}{81}$; (c) $\log_{1/3} 81$; (d) $\log_{81} \frac{1}{27}$
12. (a) $\log_{10} 1$; (b) $\log_{10} 10$; (c) $\log_{100} 10$; (d) $\log_{10} 0.001$

In Exercises 13 through 16, express each of the logarithms in terms of logarithms of x, y, and z, where the variables represent positive numbers.

13. (a) $\log_b xyz$; (b) $\ln \dfrac{x^4 y}{z^2}$
14. (a) $\log_b x^2 y^3 z^4$; (b) $\ln \dfrac{x^3 z}{y^4}$

15. (a) $\log_b \sqrt[3]{yz^2}$; (b) $\ln \sqrt[5]{\dfrac{x^3 y^4}{z^2}}$
16. (a) $\log_b \sqrt[4]{xy^3}\, z$; (b) $\ln \sqrt[3]{\dfrac{x^2}{yz^2}}$

In Exercises 17 through 20, write the given expression as a single logarithm with a coefficient of 1.

17. (a) $4 \log_{10} x + \frac{1}{2} \log_{10} y$; (b) $5 \ln x + \frac{1}{2} \ln y - \frac{1}{3} \ln z$
18. (a) $3 \log_2 x - \frac{1}{3} \log_2 y + 4 \log_2 y + 1$; (b) $\frac{3}{4} \ln x - 6 \ln y - \frac{4}{5} \ln z$
19. (a) $\frac{2}{3} \log_b x - 4 \log_b y + \log_b z - 1$; (b) $\ln \pi + \ln h + 2 \ln r - \ln 3$
20. (a) $\log_{10} 2 + \log_{10} \pi + \frac{1}{2} \log_{10} t - \frac{1}{2} \log_{10} g$; (b) $\frac{1}{4} \ln x^3 + \frac{1}{4} \ln y - \frac{1}{2} \ln z$

21. The supply equation for a particular commodity is $p = 20 + 10 \ln(1 + x)$, where x units are supplied when p dollars is the price per unit. (a) Draw a sketch of the supply curve. (b) Find the price at which 10 units would be supplied.
22. The demand equation of a certain commodity is $x = \ln 10 - \ln p$, where x units are demanded when p dollars is the price per unit. (a) Draw a sketch of the demand curve. (b) Find the highest price anyone would pay for the commodity.
23. How long will it take for $500 to accumulate to $900 if money is invested at 9 percent compounded continuously?
24. Solve Exercise 23 if money is invested at 12 percent compounded continuously.
25. Solve Exercise 23 if money is invested at 9 percent compounded annually.
26. Solve Exercise 23 if money is invested at 12 percent compounded monthly.
27. How long will it take for an investment to double itself if interest is paid at the rate of 8 percent compounded continuously?
28. How long will it take for an investment to triple itself if interest is paid at the rate of 12 percent compounded continuously?
29. Solve Exercise 27 if interest is paid at the rate of 8 percent compounded (a) annually and (b) quarterly.

30. Solve Exercise 28 if interest is paid at the rate of 12 percent compounded (a) annually and (b) semiannually.

31. In the bacterial culture of Example 3 in Sec. 8.2, after how many minutes will there be 3000 bacteria present?

32. After how many years since its purchase was the value of the machine in Exercise 22 of Exercises 8.2 $4000?

33. When is the value of the abstract painting in Exercise 23 of Exercises 8.2 expected to be $18,000?

34. Refer to Exercise 24 in Exercises 8.2. At what altitude is the atmospheric pressure 500 pounds per square foot?

35. After how many seconds will there be present only 1 g of the radioactive substance of Exercise 25 in Exercises 8.2?

36. After how many days on the job will the worker of Example 5 in Sec. 8.2 produce 25 units per day?

37. After how many months on the job will a typical worker at the factory in Exercise 29 of Exercises 8.2 complete 90 units per day?

38. How long after its purchase will the resale value of the equipment in Exercise 30 of Exercises 8.2 be $5500?

39. For the flu epidemic of Example 6 in Sec. 8.2, after how many weeks had 22,500 people, one-half the population of the community, contracted the flu?

40. On the college campus of Exercise 31 in Exercises 8.2, after how many minutes had one-half of those in attendance heard the rumor?

In Exercises 41 through 44, compute the value of the given logarithm.

41. $\log_{10} e$ **42.** $\log_5 e$ **43.** $\log_2 10$ **44.** $\log_4 100$

8.4 Derivatives of logarithmic functions and integrals yielding the natural logarithmic function

To obtain the derivative of the logarithmic function to the base b, we apply the definition of a derivative given by (2) in Sec. 2.5.

If $f(x) = \log_b x$, then

$$f'(x) = \lim_{\Delta x \to 0} \frac{f(x + \Delta x) - f(x)}{\Delta x}$$

$$= \lim_{\Delta x \to 0} \frac{\log_b(x + \Delta x) - \log_b x}{\Delta x}$$

$$= \lim_{\Delta x \to 0} \frac{1}{\Delta x} \log_b \frac{x + \Delta x}{x}$$

$$= \lim_{\Delta x \to 0} \frac{1}{\Delta x} \log_b \left(1 + \frac{\Delta x}{x}\right) \tag{1}$$

In (1) let $z = \dfrac{\Delta x}{x}$. Then $\dfrac{1}{\Delta x} = \dfrac{1}{xz}$, and "$\Delta x \to 0$" is equivalent to "$z \to 0$." Then (1) can be written in the form

$$f'(x) = \lim_{z \to 0} \frac{1}{xz} \log_b(1 + z)$$

$$= \lim_{z \to 0} \frac{1}{x} \left[\frac{1}{z} \log_b(1 + z)\right]$$

$$= \lim_{z \to 0} \frac{1}{x} \log_b(1 + z)^{1/z}$$

$$= \frac{1}{x} \lim_{z \to 0} \log_b(1 + z)^{1/z} \tag{2}$$

Because the logarithmic function is continuous, (2) can be written as

$$f'(x) = \frac{1}{x} \log_b \lim_{z \to 0} (1 + z)^{1/z}$$

From the definition of the number e given by (8) in Sec. 8.1 we have

$$f'(x) = \frac{1}{x} \log_b e \tag{3}$$

The following theorem follows from (3) and the chain rule.

> **Theorem 8.4.1**   If u is a differentiable function of x,
>
> $$D_x(\log_b u) = \frac{\log_b e}{u} \cdot D_x u \tag{4}$$

If the base b is e, then we have the natural logarithmic function, and because

$$\log_e e = \ln e = 1$$

we have the following theorem.

> **Theorem 8.4.2**   If u is a differentiable function of x,
>
> $$D_x(\ln u) = \frac{1}{u} \cdot D_x u \tag{5}$$

By comparing formulas (4) and (5) you can see that when the base is the number e, we obtain the simplest formula for the derivative of a logarithmic function.

From Theorem 8.4.2, with $u = x$, we have

$$D_x(\ln x) = \frac{1}{x}$$

● ILLUSTRATION 1

We compute $D_x(\ln x^2)$ by two methods.
   (a) From Theorem 8.4.2, with $u = x^2$,

$$D_x(\ln x^2) = \frac{1}{x^2}(2x) = \frac{2}{x}$$

(b) If Theorem 8.3.1(iii) is applied before finding the derivative, we have

$$\ln x^2 = 2 \ln x$$

Thus

$$D_x(\ln x^2) = 2 \cdot \frac{1}{x} = \frac{2}{x} \qquad \bullet$$

EXAMPLE 1   Find the derivative of the given function. (a) $f(x) = \log_{10}(3x^2 - 7)$; (b) $g(x) = \ln \dfrac{x}{x^2 + 1}$

Solution   (a) From Theorem 8.4.1 with $b = 10$ and $u = 3x^2 - 7$,

$$f'(x) = \frac{\log_{10} e}{3x^2 - 7} \cdot D_x(3x^2 - 7)$$

$$= \frac{6x \log_{10} e}{3x^2 - 7}$$

(b) From Theorem 8.4.2 with $u = \dfrac{x}{x^2 + 1}$

$$g'(x) = \frac{1}{\dfrac{x}{x^2 + 1}} \cdot D_x\!\left(\frac{x}{x^2 + 1}\right)$$

$$= \frac{x^2 + 1}{x} \cdot \frac{x^2 + 1 - 2x(x)}{(x^2 + 1)^2}$$

$$= \frac{1 - x^2}{x(x^2 + 1)}$$

EXAMPLE 2   Find the derivative: (a) $y = \ln(2x - 1)^3$; (b) $z = \left[\ln(2x - 1)\right]^3$.

Solution

(a) $D_x y = \dfrac{1}{(2x - 1)^3} \cdot D_x\!\left[(2x - 1)^3\right]$

$$= \frac{1}{(2x - 1)^3}\left[3(2x - 1)^2(2)\right]$$

$$= \frac{6}{2x - 1}$$

(b) $D_x z = 3\left[\ln(2x - 1)\right]^2 \cdot D_x\!\left[\ln(2x - 1)\right]$

$$= 3\left[\ln(2x - 1)\right]^2 \cdot \frac{1}{2x - 1} \cdot 2$$

$$= \frac{6\left[\ln(2x - 1)\right]^2}{2x - 1}$$

The following illustration shows the computation of the derivatives in Examples 1(b) and 2(a) if the properties of logarithms given in Theorem 8.3.1 are applied.

● ILLUSTRATION 2

(a) If Theorem 8.3.1(ii) is applied before finding the derivative in Example 1(b), we have

$$g(x) = \ln x - \ln(x^2 + 1)$$

$$g'(x) = \frac{1}{x} - \frac{1}{x^2 + 1} \cdot 2x$$

$$= \frac{x^2 + 1 - 2x^2}{x(x^2 + 1)}$$

$$= \frac{1 - x^2}{x(x^2 + 1)}$$

(b) If Theorem 8.3.1(iii) is applied before finding the derivative in Example 2(a), we have

$$y = 3 \ln(2x - 1)$$

$$D_x y = 3 \cdot \frac{1}{2x - 1} \cdot 2$$

$$= \frac{6}{2x - 1}$$ ●

EXAMPLE 3 The demand equation for a particular commodity is

$$x = 5000 - 1000 \ln(p + 40)$$

where x units are demanded when p dollars is the price per unit. (a) Find the price elasticity of demand when the price is $60 per unit. (b) Use the result of part (a) to find the approximate change in the demand when the price of $60 is increased by 5 percent.

Solution   (a) If η is the price elasticity of demand, then from the definition (Eq. (3) in Sec. 5.4),

$$\eta = \frac{p}{x} \cdot \frac{dx}{dp} \tag{6}$$

Because

$$x = 5000 - 1000 \ln(p + 40) \tag{7}$$

$$\frac{dx}{dp} = -\frac{1000}{p + 40} \tag{8}$$

When $p = 60$, from (7),

$$x = 5000 - 1000 \ln 100$$
$$= 5000 - 1000(4.6052)$$
$$= 5000 - 4605.2$$
$$= 395$$

and from (8)

$$\frac{dx}{dp} = -\frac{1000}{100}$$
$$= -10$$

Thus, when $p = 60$, from (6) we have

$$\eta = \tfrac{60}{395}(-10)$$
$$= -1.52$$

The result of $\eta = -1.52$ means that when the unit price is \$60, an increase of 1 percent in the unit price will cause an approximate decrease of 1.52 percent in the demand, or a decrease of 1 percent in the unit price will cause an approximate increase of 1.52 percent in the demand.

(b) If the price of \$60 is increased by 5 percent, there is an approximate decrease in the demand of $5(1.52)$ percent $= 7.6$ percent.

In the discussion that follows we need to make use of $D_x(\ln |x|)$. To find this by using Theorem 8.4.2, $\sqrt{x^2}$ is substituted for $|x|$, and so

$$D_x(\ln |x|) = D_x(\ln \sqrt{x^2})$$
$$= \frac{1}{\sqrt{x^2}} \cdot D_x(\sqrt{x^2})$$
$$= \frac{1}{\sqrt{x^2}} \cdot \frac{x}{\sqrt{x^2}}$$
$$= \frac{x}{x^2}$$
$$= \frac{1}{x}$$

From this result and the chain rule, the next theorem follows.

Theorem 8.4.3   If u is a differentiable function of x,

$$D_x(\ln |u|) = \frac{1}{u} \cdot D_x u$$

The following example illustrates how the properties of the natural logarithmic function, given in Theorem 8.3.1, can simplify the work involved when differentiating complicated expressions involving products, quotients, and powers.

EXAMPLE 4 Given

$$y = \frac{\sqrt[3]{x+1}}{(x+2)\sqrt{x+3}}$$

find $D_x y$.

Solution From the given equation,

$$|y| = \left| \frac{\sqrt[3]{x+1}}{(x+2)\sqrt{x+3}} \right| = \frac{|\sqrt[3]{x+1}|}{|x+2||\sqrt{x+3}|}$$

Taking the natural logarithm and applying the properties of logarithms we obtain

$$\ln|y| = \tfrac{1}{3}\ln|x+1| - \ln|x+2| - \tfrac{1}{2}\ln|x+3|$$

Differentiating on both sides implicitly with respect to x and applying Theorem 8.4.3 we get

$$\frac{1}{y}D_x y = \frac{1}{3(x+1)} - \frac{1}{x+2} - \frac{1}{2(x+3)}$$

Multiplying on both sides by y we obtain

$$D_x y = y \cdot \frac{2(x+2)(x+3) - 6(x+1)(x+3) - 3(x+1)(x+2)}{6(x+1)(x+2)(x+3)}$$

Replacing y by its given value we obtain $D_x y$ equal to

$$\frac{(x+1)^{1/3}}{(x+2)(x+3)^{1/2}} \cdot \frac{2x^2 + 10x + 12 - 6x^2 - 24x - 18 - 3x^2 - 9x - 6}{6(x+1)(x+2)(x+3)}$$

and so

$$D_x y = \frac{-7x^2 - 23x - 12}{6(x+1)^{2/3}(x+2)^2(x+3)^{3/2}}$$

The process illustrated in Example 4 is called **logarithmic differentiation**, which was developed in 1697 by Johann Bernoulli (1667–1748).

From Theorem 8.4.3 we obtain the following theorem for indefinite integration.

Theorem 8.4.4

$$\int \frac{1}{u}\,du = \ln|u| + C$$

From Theorems 8.4.4 and 6.2.8 for n any real number,

$$\int u^n \, du = \begin{cases} \dfrac{u^{n+1}}{n+1} + C & \text{if } n \neq -1 \\ \ln |u| + C & \text{if } n = -1 \end{cases} \tag{9}$$

EXAMPLE 5   Evaluate

$$\int \frac{x^2 \, dx}{x^3 + 1}$$

Solution

$$\int \frac{x^2 \, dx}{x^3 + 1} = \frac{1}{3} \int \frac{3x^2 \, dx}{x^3 + 1} = \frac{1}{3} \ln |x^3 + 1| + C$$

EXAMPLE 6   Evaluate

$$\int_0^2 \frac{x^2 + 2}{x + 1} \, dx$$

Solution   Because $(x^2 + 2)/(x + 1)$ is an improper fraction, we divide the numerator by the denominator and obtain

$$\frac{x^2 + 2}{x + 1} = x - 1 + \frac{3}{x + 1}$$

Therefore

$$\int_0^2 \frac{x^2 + 2}{x + 1} \, dx = \int_0^2 \left(x - 1 + \frac{3}{x + 1} \right) dx$$

$$= \tfrac{1}{2}x^2 - x + 3 \ln |x + 1| \Big]_0^2$$

$$= 2 - 2 + 3 \ln 3 - 3 \ln 1$$

$$= 3 \ln 3 - 3 \cdot 0$$

$$= 3 \ln 3$$

The answer in Example 6 also can be written as $\ln 27$ because, by Theorem 8.3.1(iii), $3 \ln 3 = \ln 3^3$.

EXAMPLE 7   Evaluate

$$\int \frac{\ln x}{x} \, dx$$

Solution   Let $u = \ln x$; then $du = dx/x$; therefore

$$\int \frac{\ln x}{x} \, dx = \int u \, du = \frac{1}{2} u^2 + C = \frac{1}{2} (\ln x)^2 + C$$

Exercises 8.4

In Exercises 1 through 16, differentiate the given function and simplify the result.

1. (a) $f(x) = \ln(4 + 5x)$; (b) $g(x) = \ln\sqrt{4 + 5x}$
2. (a) $f(x) = \ln(8 - 2x)$; (b) $g(x) = \ln(8 - 2x)^5$
3. (a) $f(t) = \ln(3t + 1)^2$; (b) $g(t) = [\ln(3t + 1)]^2$
4. (a) $F(x) = \ln(1 + 4x^2)$; (b) $G(x) = \ln\sqrt{1 + 4x^2}$
5. (a) $f(x) = \ln(3x^2 - 2x + 1)^2$; (b) $g(x) = \log_2(3x^2 - 2x + 1)$
6. (a) $f(x) = \ln(x^2 + 2)^3$; (b) $g(x) = \log_{10}(x^2 + 2)$

7. $f(t) = \log_{10}\dfrac{t}{t + 1}$ 8. $h(x) = \dfrac{\log_{10} x}{x}$

9. $f(x) = \sqrt{\log_b x}$ 10. $g(t) = \log_{10}[\log_{10}(t + 1)]$

11. $h(w) = \ln\sqrt[3]{\dfrac{w + 1}{w^2 + 1}}$ 12. $f(r) = r \ln r$

13. $g(y) = \ln(\ln y)$ 14. $F(x) = \sqrt{x + 1} - \ln(1 + \sqrt{x + 1})$

15. $g(x) = \ln|x^3 + 1|$ 16. $f(t) = \ln\left|\dfrac{3t}{t^2 + 4}\right|$

In Exercises 17 through 20, find $D_x y$ by logarithmic differentiation.

17. $y = x^2(x^2 - 1)^3(x + 1)^4$ 18. $y = \dfrac{x^5(x + 2)}{x - 3}$ 19. $y = \dfrac{x^3 + 2x}{\sqrt[5]{x^7 + 1}}$ 20. $y = \dfrac{x\sqrt{x + 1}}{\sqrt[3]{x - 1}}$

In Exercises 21 and 22, find $D_x y$ by implicit differentiation.

21. $\ln xy + x + y = 2$ 22. $\ln\dfrac{y}{x} + xy = 1$

In Exercises 23 through 30, evaluate the indefinite integral.

23. $\displaystyle\int \dfrac{dx}{3 - 2x}$ 24. $\displaystyle\int \dfrac{dx}{7x + 10}$ 25. $\displaystyle\int \dfrac{3x}{x^2 + 4}\,dx$ 26. $\displaystyle\int \dfrac{x}{2 - x^2}\,dx$

27. $\displaystyle\int \dfrac{3x^2}{5x^3 - 1}\,dx$ 28. $\displaystyle\int \dfrac{2x - 1}{x(x - 1)}\,dx$ 29. $\displaystyle\int \dfrac{2x^3}{x^2 - 4}\,dx$ 30. $\displaystyle\int \dfrac{5 - 4y^2}{3 - 2y}\,dy$

In Exercises 31 through 34, evaluate the definite integral.

31. $\displaystyle\int_3^5 \dfrac{2x}{x^2 - 5}\,dx$ 32. $\displaystyle\int_4^5 \dfrac{x}{4 - x^2}\,dx$ 33. $\displaystyle\int_2^4 \dfrac{dx}{x \ln^2 x}$ 34. $\displaystyle\int_2^4 \dfrac{\ln x}{x}\,dx$

35. The marginal revenue function for a commodity is given by $R'(x) = 12/(x + 2)$, where $R(x)$ dollars is the total revenue when x units are sold. Find the demand equation.

36. A particular company has determined that when its weekly advertising expense is x dollars, then if S dollars is its total weekly income from sales, $S = 4000 \ln x$. (a) Determine the rate of change of sales income with respect to advertising expense when $800 is the weekly advertising budget. (b) If the weekly advertising budget is increased to $950, what is the approximate increase in the total weekly income from sales?

37. The demand equation for a certain salad dressing is $x = 20 - 10 \ln p$, where x bottles are demanded when p dollars is the price per bottle. (a) Find the price elasticity of demand when the price is $1 per bottle. (b) Use the result of part (a) to determine the approximate change in the demand when the price of $1 is increased by 10 percent.

38. A manufacturer of electric generators began operations on Jan. 1, 1973. During the first year there were no sales because the company concentrated on product development and research. After the first year the sales have increased steadily according to the equation $y = x \ln x$, where x is the number of years during which the company has been operating and y is the number of millions of dollars in the sales volume. (a) Draw a sketch of the graph of the equation. Determine the rate at which the sales were increasing (b) on Jan. 1, 1977 and (c) on Jan. 1, 1983.

39. The demand equation for a particular article of merchandise is $p(x + 2) = 40 - x$, where x units are demanded when p dollars is the price per unit. Find the consumers' surplus if the market demand is 10.

40. The demand equation for a certain commodity is $p = 10/(x + 1)$, where x units are demanded when p dollars is the price per unit. Find the consumers' surplus if the market price is $4.

41. The demand and supply equations for a particular commodity are, respectively,

$$p = \frac{600 - 2x}{x + 100} \quad \text{and} \quad 200p = 300 + x$$

where x units are demanded or supplied when p dollars is the price per unit. If the market is in equilibrium, find the consumers' surplus to the nearest dollar.

42. In a telegraph cable, the measure of the speed of the signal is directly proportional to $x^2 \ln(1/x)$, where x is the ratio of the measure of the radius of the core of the cable to the measure of the thickness of the cable's winding. Find the value of x for which the speed of the signal is greatest.

43. In biology an equation sometimes used to describe the restricted growth of a population is the Gompertz growth equation

$$\frac{dy}{dt} = ky \ln \frac{a}{y}$$

where a and k are positive constants. Find the general solution of this differential equation.

In Exercises 44 and 45, draw a sketch of the curve having the given equation.

44. $y = x \ln x$ **45.** $y = x - \ln x$

8.5 Differentiation and integration of exponential functions

To obtain the derivative of the exponential function with base b, we start with the equation

$$y = b^x \tag{1}$$

Then from (1) in Sec. 8.3,

$$x = \log_b y$$

On both sides of this equation we differentiate implicitly with respect to x and get

$$1 = \frac{\log_b e}{y} \cdot D_x y$$

Thus

$$D_x y = y \cdot \frac{1}{\log_b e} \tag{2}$$

From (19) in Sec. 8.3,

$$\frac{1}{\log_b e} = \ln b \tag{3}$$

Substituting from (1) and (3) into (2) we get

$$D_x(b^x) = b^x \ln b \tag{4}$$

From (4) and the chain rule we have the following theorem.

Theorem 8.5.1 If u is a differentiable function of x,

$$D_x(b^u) = b^u \ln b \, D_x u \tag{5}$$

● ILLUSTRATION 1

If $y = 10^{3x}$, then from Theorem 8.5.1,

$$D_x y = 10^{3x} \ln 10(3) = 3(\ln 10)10^{3x} \qquad ●$$

If in (4) the base b is e, we have the natural exponential function, and from (4) with $b = e$ we obtain

$$D_x(e^x) = e^x \ln e$$

$$D_x(e^x) = e^x \tag{6}$$

From (6) and the chain rule we have the following theorem.

Theorem 8.5.2 If u is a differentiable function of x,

$$D_x(e^u) = e^u \, D_x u \tag{7}$$

● ILLUSTRATION 2

If $y = e^{3x}$, then from Theorem 8.5.2,

$$D_x y = e^{3x}(3) = 3e^{3x} \qquad ●$$

By comparing formulas (5) and (7) you see again that when the base is the number e we obtain the simplest formula for the derivative of an exponential function.

EXAMPLE 1 Find the derivative of the given function: (a) $f(x) = 3^{x^2}$; (b) $g(x) = e^{1/x^2}$.

Solution (a) From Theorem 8.5.1 with $b = 3$ and $u = x^2$,

$$f'(x) = 3^{x^2} \ln 3 \cdot D_x(x^2)$$
$$= 3^{x^2} \ln 3(2x)$$
$$= 2(\ln 3)x3^{x^2}$$

(b) From Theorem 8.5.2 with $u = 1/x^2$,

$$g'(x) = e^{1/x^2} \cdot D_x\left(\frac{1}{x^2}\right)$$

$$= e^{1/x^2}\left(-\frac{2}{x^3}\right)$$

$$= -\frac{2e^{1/x^2}}{x^3}$$

From Theorem 8.5.2, if $f(x) = ke^x$, where k is a constant, then $f'(x) = ke^x$. Thus the derivative of this function is itself. The only other function we have previously encountered that has this property is the constant function $f(x) = 0$; actually, this is the special case of $f(x) = ke^x$ when $k = 0$. It can be proved that the most general function which is its own derivative is given by $f(x) = ke^x$ (see Exercise 42).

From Theorems 8.5.1 and 8.5.2 we obtain the following two theorems for indefinite integration.

Theorem 8.5.3   If b is any positive number other than 1,

$$\int b^u \, du = \frac{b^u}{\ln b} + C$$

● ILLUSTRATION 3

From Theorem 8.5.3 with $b = 2$ and $u = x$,

$$\int 2^x \, dx = \frac{2^x}{\ln 2} + C$$ ●

Theorem 8.5.4

$$\int e^u \, du = e^u + C$$

● ILLUSTRATION 4

From Theorem 8.5.4 with $u = 2x$ and $du = 2dx$ we have

$$\int e^{2x} \, dx = \frac{1}{2}\int e^{2x}(2 \, dx)$$

$$= \frac{1}{2}e^{2x} + C$$ ●

EXAMPLE 2 Find $\int \sqrt{10^{3x}} \, dx$

Solution $\int \sqrt{10^{3x}} \, dx = \int 10^{3x/2} \, dx$. Let $u = \frac{3}{2}x$; then $du = \frac{3}{2} \, dx$; thus $\frac{2}{3} \, du = dx$. Therefore

$$\int 10^{3x/2} \, dx = \int 10^u \left(\tfrac{2}{3} \, du\right)$$

$$= \frac{2}{3} \cdot \frac{10^u}{\ln 10} + C$$

$$= \frac{2 \cdot 10^{3x/2}}{3 \ln 10} + C$$

EXAMPLE 3 Find $\int \dfrac{e^{\sqrt{x}}}{\sqrt{x}} \, dx$

Solution Let $u = \sqrt{x}$; then $du = \frac{1}{2}x^{-1/2} \, dx$; so

$$\int \frac{e^{\sqrt{x}}}{\sqrt{x}} \, dx = 2 \int e^u \, du$$

$$= 2e^u + C$$

$$= 2e^{\sqrt{x}} + C$$

From Eq. (12) of Sec. 8.3, if $x > 0$,

$$e^{\ln x} = x$$

Thus if $x > 0$ and n is any real number,

$$x^n = (e^{\ln x})^n$$

$$x^n = e^{n \ln x} \tag{8}$$

In a more advanced approach to a discussion of exponential functions (8) is given as the definition of a real number exponent. Equation (8) and a table of powers of e (or a hand calculator) can be used to compute a^n if n is any irrational number. The following example shows the procedure.

EXAMPLE 4 Compute the value of $2^{\sqrt{3}}$ to two decimal places.

Solution From (8),

$$2^{\sqrt{3}} = e^{\sqrt{3} \ln 2}$$

$$= e^{1.732(0.6931)}$$

$$= e^{1.200}$$

$$= 3.32$$

Because x^n has been defined for any real number n, we now prove Theorem 2.6.2 (the derivative of a power function) when the exponent is any real number. From (8), if $x > 0$,

$$x^n = e^{n \ln x}$$

Therefore

$$D_x(x^n) = e^{n \ln x} D_x(n \ln x)$$

$$= e^{n \ln x} \left(\frac{n}{x} \right)$$

$$= x^n \cdot \frac{n}{x}$$

$$\boxed{D_x(x^n) = nx^{n-1}} \tag{9}$$

Equation (9) enables us to find the derivative of a variable to a constant power. Previously in this section we learned how to differentiate a constant to a variable power. We now consider the derivative of a function whose function value is a variable to a variable power.

EXAMPLE 5   If $y = x^x$, where $x > 0$, find dy/dx.

Solution   From (8), if $x > 0$, $x^x = e^{x \ln x}$. Therefore

$$y = e^{x \ln x}$$

$$\frac{dy}{dx} = e^{x \ln x} D_x(x \ln x)$$

$$= e^{x \ln x} \left(x \cdot \frac{1}{x} + \ln x \right)$$

$$= x^x (1 + \ln x)$$

The method of logarithmic differentiation can also be used to find the derivative of a function whose function value is a variable to a variable power, as shown in the following example.

EXAMPLE 6   Find dy/dx in Example 5 by using logarithmic differentiation.

Solution   We are given $y = x^x$ with $x > 0$. We take the natural logarithm on both sides of the equation and obtain

$$\ln y = \ln x^x$$

$$\ln y = x \ln x$$

Differentiating on both sides of the above equation with respect to x gives

$$\frac{1}{y} \cdot \frac{dy}{dx} = x \cdot \frac{1}{x} + \ln x$$

$$\frac{dy}{dx} = y(1 + \ln x)$$

$$\frac{dy}{dx} = x^x (1 + \ln x)$$

Exercises 8.5

In Exercises 1 through 12, find the derivative of the given function.

1. (a) $f(x) = e^{5x}$; (b) $g(x) = 2^{5x}$
2. (a) $f(x) = e^{-7x}$; (b) $g(x) = 10^{-7x}$
3. (a) $f(x) = e^{-3x^2}$; (b) $g(x) = b^{-3x^2}, b > 0$
4. (a) $f(x) = e^{x^2-3}$; (b) $g(x) = b^{x^2-3}, b > 0$

5. $f(t) = \dfrac{e^t}{t}$
6. $g(x) = e^{e^x}$
7. $h(x) = \dfrac{e^x - e^{-x}}{e^x + e^{-x}}$
8. $f(w) = \dfrac{e^{2w}}{w^2}$

9. $g(x) = 10^{x^2-2x}$
10. $f(x) = (x^3 + 3)2^{-7x}$
11. $f(x) = \ln(e^x + e^{-x})$
12. $g(x) = \ln\dfrac{e^{4x} - 1}{e^{4x} + 1}$

In Exercises 13 and 14, find $D_x y$ by implicit differentiation.

13. $e^x + e^y = e^{x+y}$
14. $ye^{2x} + xe^{2y} = 1$

In Exercises 15 through 28, evaluate the indefinite integral.

15. (a) $\displaystyle\int e^{2x}\,dx$; (b) $\displaystyle\int 3^{2x}\,dx$
16. (a) $\displaystyle\int e^{-4x}\,dx$; (b) $\displaystyle\int 10^{-4x}\,dx$

17. (a) $\displaystyle\int e^{2-5x}\,dx$; (b) $\displaystyle\int 10^{2-5x}\,dx$
18. (a) $\displaystyle\int e^{2x+1}\,dx$; (b) $\displaystyle\int 5^{2x+1}\,dx$

19. $\displaystyle\int x^2 e^{2x^3}\,dx$
20. $\displaystyle\int 3xe^{4x^2}\,dx$
21. $\displaystyle\int \dfrac{1 + e^{2x}}{e^x}\,dx$
22. $\displaystyle\int \dfrac{e^{2x}}{e^x + 3}\,dx$

23. $\displaystyle\int \dfrac{e^{3x}}{(1 - 2e^{3x})^2}\,dx$
24. $\displaystyle\int e^{3x}e^{2x}\,dx$
25. $\displaystyle\int 3^t e^t\,dt$
26. $\displaystyle\int x^2 10^{x^3}\,dx$

27. $\displaystyle\int 5^{x^4+2x}(2x^3 + 1)\,dx$
28. $\displaystyle\int 2^{z \ln z}(\ln z + 1)\,dz$

In Exercises 29 through 31, evaluate the definite integral.

29. $\displaystyle\int_0^3 \dfrac{e^x + e^{-x}}{2}\,dx$
30. $\displaystyle\int_1^2 \dfrac{e^x}{e^x + e}\,dx$
31. $\displaystyle\int_0^2 xe^{4-x^2}\,dx$

In Exercises 32 through 37, find dy/dx.

32. $y = x^{x^2}; x > 0$
33. $y = x^{\sqrt{x}}; x > 0$
34. $y = (\ln x)^{\ln x}; x > 1$
35. $y = x^{\ln x}; x > 0$
36. $y = x^{e^x}; x > 0$
37. $y = (x)^{x^x}; x > 0$

In Exercises 38 through 41, compute the value to two decimal places by first expressing it as a power of e.

38. $2^{\sqrt{2}}$
39. $(\sqrt{2})^{\sqrt{2}}$
40. $(\sqrt{2})^e$
41. 3^π

42. Prove that the most general function that is equal to its derivative is given by $f(x) = ke^x$. (*Hint:* Let $y = f(x)$ and solve the differential equation $dy/dx = y$.)

43. Find the rate at which the value of the abstract painting in Exercise 23 of Exercises 8.2 was increasing in 1982.

44. A company estimates that in t years the number of its employees will be $N(t)$, where $N(t) = 1000(0.8)^{t/2}$. (a) How many employees does the company expect to have in 4 years? (b) At what rate is the number of employees expected to be changing in 4 years?

45. A company has learned that when it initiates a new sales campaign, the number of sales per day increases. However, the number of extra daily sales decreases as the impact of the campaign wears off. For a specific campaign the company has determined that if $S(t)$ is the number of extra daily sales as a result of the campaign and t is the number of days that have elapsed since the campaign ended, then $S(t) = 1000(3^{-t/2})$. Find the rate at which the extra daily sales is decreasing when (a) $t = 4$ and (b) $t = 10$.

46. The distributor of a particular article of merchandise has determined that the number of units sold depends on the advertising budget. Let $S(x)$ be the number of units sold when the advertising budget is x dollars. It is estimated that $S(x)$ is increasing at the rate of $10e^{-0.02x}$ units per dollar increase in the advertising budget; that is, $S'(x) = 10e^{-0.02x}$. If 250 units are sold without any advertising, find an equation defining $S(x)$.

47. The demand equation for a certain commodity is $p = 10e^{-x}$, where x units are demanded when p dollars is the price per unit. Find the consumers' surplus when the market price is \$1.

48. The supply equation for a particular commodity is $p = 20e^{x/3}$, where x units are supplied when p dollars is the price per unit. If the market price is \$6, find the producers' surplus.

8.6 Laws of growth and decay

Mathematical models involving exponential growth and exponential decay were discussed in Sec. 8.2. In this section we show that these models arise when the rate of change of the amount of a quantity with respect to time is proportional to the amount of the quantity present at a given instant. In such cases if the time is represented by t units, and y units is the amount of the quantity present at any time, then

$$\frac{dy}{dt} = ky$$

where k is a constant. If y increases as t increases, then $k > 0$, and we have the **law of natural growth**; the solution of the differential equation gives a model of exponential growth. If y decreases as t increases, then $k < 0$, and we have the **law of natural decay**; the solution of the differential equation gives a model of exponential decay.

In the following Example 1 the rate of increase of the population of a community is proportional to the actual population at any given instant, and the solution involves exponential growth. In Example 2 the daily sales of a commodity decrease at a rate proportional to the daily sales, and the solution involves exponential decay. In Sec. 8.1 we discussed an investment for which interest is compounded continuously, which is an example of exponential growth; you will see that this situation is the case where the amount of an investment is increasing at a rate proportional to its size. Exponential growth occurs in biology under certain circumstances when the rate of growth of a culture of bacteria is proportional to the number of bacteria present at any specific time. In a chemical reaction we have exponential decay when the rate of decay of a substance is proportional to the quantity of the substance present.

EXAMPLE 1 The rate of increase of the population of a certain city is proportional to the population. If the population in 1930 was 50,000 and in 1960 it was 75,000, what is the expected population in 1990?

Solution Let t be the number of years in the time since 1930. Let A be the population in t years. We have the initial conditions given in Table 8.6.1.

Table 8.6.1

t	0	30	60
A	50,000	75,000	A_{60}

The differential equation is

$$\frac{dA}{dt} = kA$$

Separating the variables we obtain

$$\frac{dA}{A} = k \, dt$$

Integrating, we have

$$\int \frac{dA}{A} = k \int dt$$

$$\ln |A| = kt + \overline{c}$$

$$|A| = e^{kt + \overline{c}} = e^{\overline{c}} \cdot e^{kt}$$

Letting $e^{\overline{c}} = C$ we have $|A| = Ce^{kt}$, and because A is nonnegative, we can omit the absolute-value bars, thereby giving

$$A = Ce^{kt}$$

Because $A = 50,000$ when $t = 0$, we obtain $C = 50,000$. Thus

$$A = 50,000e^{kt} \tag{1}$$

Because $A = 75,000$ when $t = 30$, we get

$$75,000 = 50,000e^{30k}$$

$$e^{30k} = \tfrac{3}{2} \tag{2}$$

When $t = 60$, $A = A_{60}$. Therefore

$$A_{60} = 50,000e^{60k}$$

$$A_{60} = 50,000(e^{30k})^2 \tag{3}$$

Substituting from (2) into (3) we obtain

$$A_{60} = 50,000(\tfrac{3}{2})^2$$
$$= 112,500$$

Therefore the expected population in 1990 is 112,500.

In the above example, because the population is increasing with time, we have a case of the law of growth. If a population decreases with time, which could occur if the death rate is greater than the birth rate, then we have a case of the law of decay (see Exercise 3). In the next example there is another situation involving the law of decay.

EXAMPLE 2 A particular commodity was promoted by a substantial advertising campaign, and just before the promotion ceased, the number of sales per day was 10,000 units. Immediately following the stoppage of the advertising, the daily sales decreased at a rate proportional to the daily sales. If 10 days after the advertising ceased the number of sales per day was 8000 units, find the daily sales 20 days after the advertising stopped.

Solution If S units is the daily sales t days after the advertising ceased, then

$$\frac{dS}{dt} = kS \tag{4}$$

When $t = 0$, $S = 10,000$, and when $t = 10$, $S = 8000$. We wish to find S when $t = 20$. Let this value of S be S_{20}. These conditions are stated in Table 8.6.2.

As in Example 1, the general solution of differential equation (4) is

$$S = Ce^{kt}$$

Because $S = 10,000$ when $t = 0$, then $C = 10,000$. Therefore

$$S = 10,000e^{kt} \tag{5}$$

Because $S = 8000$ when $t = 10$, we have

$$8000 = 10,000e^{10k}$$

$$e^{10k} = 0.8 \tag{6}$$

When $t = 20$, $S = S_{20}$, and so

$$S_{20} = 10,000e^{20k}$$

$$S_{20} = 10,000(e^{10k})^2 \tag{7}$$

Substituting from (6) into (7) we obtain

$$S_{20} = 10,000(0.8)^2$$

$$= 6400$$

Thus the daily sales 20 days after the advertising stopped is 6400 units.

Table 8.6.2

t	0	10	20
S	10,000	8000	S_{20}

EXAMPLE 3 Suppose that the GNP (gross national product) of a certain country has a rate of increase that is proportional to the GNP. If the GNP on January 1, 1978, was \$80 billion and on January 1, 1982, it was \$96 billion, when is it expected to be \$128 billion?

Solution Let t be the number of years that have elapsed since January 1, 1978. Let x be the number of billions of dollars in the GNP at t years. Table 8.6.3 gives the initial conditions. The differential equation is

$$\frac{dx}{dt} = kx$$

As in Example 1 the general solution is

$$x = Ce^{kt}$$

Table 8.6.3

t	0	4	T
x	80	96	128

Because $x = 80$ when $t = 0$, it follows that $C = 80$. Therefore

$$x = 80e^{kt}$$

Because $x = 96$ when $t = 4$, we have

$$96 = 80e^{4k}$$

$$1.2 = e^{4k}$$

$$\ln 1.2 = 4k$$

$$k = \tfrac{1}{4} \ln 1.2$$

$$k = \tfrac{1}{4}(0.1823)$$

$$k = 0.0456$$

Thus

$$x = 80e^{0.0456t}$$

Therefore

$$128 = 80e^{0.0456T}$$

$$e^{0.0456T} = 1.6$$

$$0.0456T = \ln 1.6$$

$$T = \frac{0.4700}{0.0456} = 10.3$$

Because 10.3 years is approximately 10 years and 4 months, it follows that the GNP is expected to be \$128 billion on May 1, 1988.

Now consider an investment of P dollars that increases at a rate proportional to its size. This is the law of natural growth. Then if A dollars is the amount at t years,

$$\frac{dA}{dt} = kA$$

$$\int \frac{dA}{A} = k \int dt$$

$$\ln |A| = kt + \overline{c}$$

$$A = Ce^{kt}$$

When $t = 0$, $A = P$; so $C = P$. Therefore we have

$$A = Pe^{kt} \tag{8}$$

Comparing (8) with Eq. (9) in Sec. 8.1 we see that they are the same if $k = i$. So if the amount of an investment increases at a rate proportional to its size, then the interest is compounded continuously, and the annual interest rate is the constant of proportionality.

● ILLUSTRATION 1

If P dollars is invested at a rate of 14 percent per year compounded continuously and A dollars is the amount of the investment at t years, then

$$\frac{dA}{dt} = 0.14A$$

and $A = P$ when $t = 0$. Therefore

$$A = Pe^{0.14t}$$ ●

EXAMPLE 4   If an amount of money doubles itself in 6 years at interest compounded continuously, what is the annual rate of interest?

Solution   From (8) with $A = 2P$ and $t = 6$,

$$2P = Pe^{6k}$$

$$e^{6k} = 2$$

$$\ln e^{6k} = \ln 2$$

$$6k = \ln 2$$

$$k = \tfrac{1}{6}(0.6931)$$

$$k = 0.115$$

Therefore the annual rate of interest is 11.5 percent.

EXAMPLE 5   In a certain bacterial culture the rate of growth of bacteria is proportional to the number present. If there are 1000 bacteria present initially and the number doubles in 20 min, how long will it take before there will be 1,000,000 bacteria present?

Table 8.6.4

t	0	20	T
A	1000	2000	1,000,000

Solution   Let t be the number of minutes in the time from now, and let A be the number of bacteria present at t min. Table 8.6.4 gives the initial conditions. The differential equation is

$$\frac{dA}{dt} = kA$$

As in Example 1, the general solution is

$$A = Ce^{kt}$$

When $t = 0$, $A = 1000$; hence $C = 1000$, which gives

$$A = 1000e^{kt}$$

From the condition that $A = 2000$ when $t = 20$ we obtain

$$e^{20k} = 2$$

$$20k = \ln 2$$

$$k = \tfrac{1}{20} \ln 2$$

$$k = 0.03466$$

Thus we have

$$A = 1000e^{0.03466t}$$

Replacing t by T and A by 1,000,000 we get

$$1,000,000 = 1000e^{0.03466T}$$

$$e^{0.03466T} = 1000$$

$$0.03466T = \ln 1000$$

$$T = \frac{6.9078}{0.03466}$$

$$T = 199.30$$

Therefore there will be 1,000,000 bacteria present in 3 hours, 19 min, 18 sec.

In problems involving the law of natural decay, the **half-life** of a substance is the time required for half of it to decay.

EXAMPLE 6 The rate of decay of radium is proportional to the amount present at any time. If 60 mg of radium are present now and its half-life is 1690 years, how much radium will be present 100 years from now?

Solution Let t be the number of years in the time from now. Let A be the number of milligrams of radium present at t years. We have the initial conditions given in Table 8.6.5. The differential equation is

$$\frac{dA}{dt} = kA$$

As in Example 1 the general solution is

$$A = Ce^{kt}$$

Because $A = 60$ when $t = 0$, we obtain $60 = C$. Therefore

$$A = 60e^{kt} \qquad (9)$$

Because $A = 30$ when $t = 1690$, we get $30 = 60e^{1690k}$, or

$$0.5 = e^{1690k}$$

So

$$\ln 0.5 = 1690k$$

and

$$k = \frac{\ln 0.5}{1690} = \frac{-0.6931}{1690} = -0.000410$$

Substituting this value of k into (9) we obtain

$$A = 60e^{-0.000410t}$$

Table 8.6.5

t	0	1690	100
A	60	30	A_{100}

When $t = 100$, $A = A_{100}$, and we have

$$A_{100} = 60e^{-0.0410} = 57.6$$

Therefore there will be 57.6 mg of radium present 100 years from now.

EXAMPLE 7 There are 100 million liters of fluoridated water in the reservoir containing a city's water supply, and the water contains 700 kg of fluoride. To decrease the fluoride content, fresh water runs into the reservoir at the rate of 3 million liters per day, and the mixture of water and fluoride, kept uniform, runs out of the reservoir at the same rate. How many kilograms of fluoride are in the reservoir 60 days after the pure water started to flow into the reservoir?

Solution Let t be the number of days that have elapsed since the pure water started to flow into the reservoir. Let x be the number of kilograms of fluoride in the reservoir at t days.

Because 100 million liters of fluoridated water are in the tank at all times, at t days the number of kilograms of fluoride per million liters is $x/100$. Three million liters of the fluoridated water run out of the reservoir each day; so the reservoir loses $3(x/100)$ kg of fluoride per day. Because $D_t x$ is the rate of change of x with respect to t, and x is decreasing as t increases, we have the differential equation

$$\frac{dx}{dt} = -\frac{3x}{100}$$

Table 8.6.6

t	0	60
x	700	x_{60}

We also have the initial conditions given in Table 8.6.6. Separating the variables and integrating we have

$$\int \frac{dx}{x} = -0.03 \int dt$$

$$\ln |x| = -0.03t + \bar{c}$$

$$x = Ce^{-0.03t}$$

When $t = 0$, $x = 700$; so $C = 700$. Letting $t = 60$ and $x = x_{60}$ we have

$$x_{60} = 700e^{-1.8}$$
$$= 700(0.1653)$$
$$= 115.71$$

Hence there are 115.71 kg of fluoride in the reservoir 60 days after the pure water started to flow into the reservoir.

Exercises 8.6

1. The rate of natural increase of the population of a certain city is proportional to the population. If the population increases from 40,000 to 60,000 in 40 years, when will the population be 80,000?

2. The population of a particular city doubled in the 60 years from 1890 to 1950. If the rate of natural increase of the population at any time is proportional to the population at the time, and the population in 1950 was 60,000, estimate the population in the year 2000.

3. The population of a town is decreasing at a rate proportional to its size. In 1970 the population was 50,000 and in 1980 it was 44,000. What is the expected population in 1990?

4. For the commodity of Example 2, how many days after the advertising ceased is the number of daily sales expected to be 6000 units?

5. After the pre-opening and opening-day publicity of a certain exploitation film stopped, the attendance decreased at a rate proportional to its size. If the opening day's attendance at a specific theatre was 5000 and the attendance on the third day was 2000, what is the expected attendance on the sixth day?

6. For the country of Example 3, what is the expected GNP on January 1, 1991?

7. The GNP of a particular country has a rate of increase that is proportional to the GNP. If the GNP on January 1, 1971, was 60 billion dollars and twice that amount on January 1, 1981, when is it expected to be three times that amount?

8. If an amount of money invested doubles itself in 10 years at interest compounded continuously, how long will it take for the original amount to triple itself?

9. If an amount of money triples itself in 9 years at interest compounded continuously, what is the annual rate of interest?

10. If the purchasing power of a dollar is decreasing at the rate of 10 percent annually compounded continuously, how long will it take for the purchasing power to be 50 cents?

11. After an automobile is 1 year old, its rate of depreciation at any time is proportional to its value at that time. If an automobile was purchased on June 1, 1981, and its values on June 1, 1982, and June 1, 1983, were, respectively, $7000 and $5800, what is its expected value on June 1, 1987?

12. Suppose that the value of a certain antique collection increases with age and its rate of appreciation at any time is proportional to its value at that time. If the value of the collection was $25,000 10 years ago and its present value is $35,000, in how many years is its value expected to be $50,000?

13. Bacteria grown in a certain culture increase at a rate proportional to the number present. If 1000 bacteria are present initially and the number doubles in 30 min, how many bacteria will there be in 2 hours?

14. In a certain bacterial culture where the rate of growth of bacteria is proportional to the number present, the number triples in 1 hour. If at the end of 4 hours there were 10 million bacteria, how many bacteria were present initially?

15. The winter mortality rate of a certain species of wildlife in a particular geographical region is proportional to the number of the species present at any time. There were 2400 of the species present in the region on December 21, the first day of winter; and 30 days later there were 2000. How many of the species were expected to survive the winter? That is, how many will be living 90 days after December 21?

16. When a simple electric circuit, containing no capacitors but having inductance and resistance, has the electromotive force removed, the rate of decrease of the current is proportional to the current. The current is i amperes t sec after the cutoff, and $i = 40$ when $t = 0$. If the current dies down to 15 amperes in 0.01 sec, find i in terms of t.

17. In a certain chemical reaction the rate of conversion of a substance is proportional to the amount of the substance still unreacted at that time. After 10 min one-third of the original amount of the substance has been reacted and 20 g has been reacted after 15 min. What was the original amount of the substance?

18. Sugar decomposes in water at a rate proportional to the amount still unchanged. If there were 50 kg of sugar present initially and at the end of 5 hours this is reduced to 20 kg, how long will it take until 90 percent of the sugar is decomposed?

19. If the half-life of radium is 1690 years, what percent of the amount present now will be remaining after (a) 100 years and (b) 1000 years?

20. Thirty percent of a radioactive substance disappears in 15 years. Find the half-life of the substance.

21. There are 100 liters of brine in a tank, and the brine contains 70 kg of dissolved salt. Fresh water runs into the tank at the rate of 3 liters/min, and the mixture, kept uniform by stirring, runs out at the same rate. How many kilograms of salt are there in the tank at the end of 1 hour?

22. A tank contains 200 liters of brine in which there are 3 kg of salt per liter. It is desired to dilute this solution by adding brine containing 1 kg of salt per liter, which flows into the tank at the rate of 4 liters/min and runs out at the same rate. When will the tank contain $1\frac{1}{2}$ kg of salt per liter?

23. Professor Willard Libby of the University of California at Los Angeles was awarded the Nobel prize in chemistry for discovering a method of determining the date of death of a once-living object. Professor Libby made use of the fact that the tissue of a

living organism is composed of two kinds of carbons, a radioactive carbon-14 (commonly written ^{14}C) and a stable carbon-12 (^{12}C), in which the ratio of the amount of ^{14}C to the amount of ^{12}C is approximately constant. When the organism dies, the law of natural decay applies to ^{14}C. If it is determined that the amount of ^{14}C in a piece of charcoal is only 15 percent of its original amount and the half-life of ^{14}C is 5600 years, when did the tree from which the charcoal came die?

24. Refer to Exercise 23. Suppose that after finding a fossil an archaeologist determines that the amount of ^{14}C present in the fossil is 25 percent of its original amount. Using the fact that the half-life of ^{14}C is 5600 years, what is the age of the fossil?

8.7 Additional applications of exponential functions

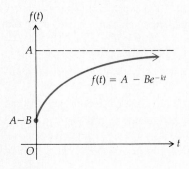

Figure 8.7.1

In Sec. 8.2 we discussed the mathematical model describing bounded growth and given by a function defined by

$$f(t) = A - Be^{-kt} \qquad t \geq 0 \tag{1}$$

where A, B, and k are positive constants. The graph of such a function is called a learning curve and is shown in Fig. 8.7.1. Such a model arises when a quantity grows at a rate proportional to the difference between a fixed number and its size, where the fixed number serves as an upper bound. This fact is shown in the following illustration.

● ILLUSTRATION 1

Suppose that a quantity increases at a rate proportional to the difference between a fixed number A and its size. Then if time is represented by t units and y units is the amount of the quantity present at any time,

$$\frac{dy}{dt} = k(A - y) \tag{2}$$

Separating the variables we obtain

$$\frac{dy}{A - y} = k \, dt$$

Integrating, we have

$$\int \frac{dy}{A - y} = k \int dt$$

$$-\ln |A - y| = kt + C$$

$$\ln |A - y| = -kt - C$$

$$|A - y| = e^{-C}e^{-kt}$$

Let $e^{-C} = B$. Because $A - y$ is nonnegative, we can omit the absolute-value bars. Thus we have

$$A - y = Be^{-kt}$$

$$y = A - Be^{-kt}$$

If y is replaced by $f(t)$, we get

$$f(t) = A - Be^{-kt}$$

which is the equation in (1).   ●

EXAMPLE 1 A new employee is performing her job more efficiently each day in such a way that if y is the number of units produced per day after t days on the job,

$$\frac{dy}{dt} = k(80 - y)$$

The employee produces 20 units the day she begins work and 50 units per day after being on the job 10 days. (a) How many units per day does she produce after being on the job 30 days? (b) Show that after being on the job 60 days she is producing just 1 unit less than her full potential.

Solution The given differential equation is similar to (2) in Illustration 1 with $A = 80$. We proceed as we did there by separating the variables and integrating. The solution takes the following form.

$$\frac{dy}{dt} = k(80 - y)$$

$$\int \frac{dy}{80 - y} = k \int dt$$

$$-\ln|80 - y| = kt + C$$

$$\ln|80 - y| = -kt - C$$

$$80 - y = e^{-C}e^{-kt}$$

$$y = 80 - Be^{-kt} \tag{3}$$

Because she produces 20 units the day she begins work, $y = 20$ when $t = 0$. Substituting these values in (3) we have

$$20 = 80 - Be^0$$

$$20 = 80 - B$$

$$B = 60$$

Setting $B = 60$ in (3) we obtain

$$y = 80 - 60e^{-kt} \tag{4}$$

After being on the job 10 days she produces 50 units; therefore $y = 50$ when $t = 10$. Thus from (4),

$$50 = 80 - 60e^{-k(10)}$$

$$e^{-10k} = 0.5$$

$$-10k = \ln 0.5$$

$$-10k = -0.6931$$

$$k = 0.069$$

Substituting this value of k into (4) we have

$$y = 80 - 60e^{-0.069t} \tag{5}$$

(a) Let $y = y_{30}$ when $t = 30$. Hence from (5),

$$
\begin{aligned}
y_{30} &= 80 - 60e^{-0.069(30)} \\
&= 80 - 60e^{-2.07} \\
&= 80 - 60(0.126) \\
&= 80 - 7.56 \\
&= 72.44
\end{aligned}
$$

Therefore she is producing 72 units per day after being on the job 30 days.

(b) Let $y = y_{60}$ when $t = 60$. Thus from (5),

$$
\begin{aligned}
y_{60} &= 80 - 60e^{-0.069(60)} \\
&= 80 - 60e^{-4.14} \\
&= 80 - 60(0.016) \\
&= 80 - 0.96 \\
&= 79.04
\end{aligned}
$$

After being on the job 60 days she is producing 79 units per day. Because

$$\lim_{t \to +\infty} (80 - 60e^{-0.069t}) = 80$$

her full potential is 80 units per day. Thus after 60 days she is producing just 1 unit less than her full potential.

EXAMPLE 2   Newton's law of cooling states that the rate at which a body changes temperature is proportional to the difference between its temperature and that of the surrounding medium. If a body is in air of temperature $35°$ and the body cools from $120°$ to $60°$ in 40 min, find the temperature of the body after 100 min.

Solution   Let t be the number of minutes in the time that has elapsed since the body started to cool. Let x be the number of degrees in the temperature of the body at t minutes.

Table 8.7.1 gives the initial conditions. From Newton's law of cooling we have

$$\frac{dx}{dt} = k(x - 35)$$

Separating the variables we obtain

$$\frac{dx}{x - 35} = k \, dt$$

Table 8.7.1

t	0	40	100
x	120	60	x_{100}

Thus

$$\int \frac{dx}{x - 35} = k \int dt$$

$$\ln |x - 35| = kt + \bar{c}$$

$$x - 35 = e^{kt + \bar{c}}$$

$$x = e^{\bar{c}} e^{kt} + 35$$

$$x = C e^{kt} + 35$$

When $t = 0$, $x = 120$; so $C = 85$. Therefore

$$x = 85 e^{kt} + 35$$

When $t = 40$, $x = 60$, and we obtain

$$60 = 85 e^{40k} + 35$$

$$40k = \ln \tfrac{5}{17}$$

$$k = \tfrac{1}{40}(\ln 5 - \ln 17)$$

$$k = \tfrac{1}{40}(1.6094 - 2.8332)$$

$$k = -0.0306$$

So

$$x = 85 e^{-0.0306t} + 35$$

Then

$$x_{100} = 85 e^{-3.06} + 35 = 39$$

Therefore the temperature of the body is $39°$ after 100 min.

In the next example there is a probability density function of the form

$$f(x) = k e^{-kx} \qquad x \geq 0$$

where $k > 0$. In this case the variable is said to be exponentially distributed on the interval $[0, +\infty)$. Recall that there are two conditions that must be satisfied by the probability density function on $[0, +\infty)$. The first condition holds because $f(x) \geq 0$ for all x in $[0, +\infty)$. To verify the second condition that the definite integral of f on $[0, +\infty)$ is 1 involves an improper integral. Such integrals are treated in Sec. 9.5, and in that section we refer to this probability density function and show that the integral does have a value of 1.

EXAMPLE 3 For a particular kind of battery the probability density function for x hours to be the life of a battery selected at random is given by

$$f(x) = \tfrac{1}{60} e^{-x/60} \qquad x \geq 0$$

Find the probability that the life of a battery selected at random will be (a) between 50 and 60 hours and (b) between 15 and 25 hours.

Solution (a) The probability that the life of a battery selected at random will be between 50 and 60 hours is $P([50, 60])$ and

$$P([50, 60]) = \int_{50}^{60} \frac{1}{60} e^{-x/60} \, dx$$

$$= -\int_{50}^{60} e^{-x/60} \left(-\frac{dx}{60} \right)$$

$$= -e^{-x/60} \Big]_{50}^{60}$$

$$= -e^{-60/60} + e^{-50/60}$$

$$= -e^{-1} + e^{-0.83}$$

$$= -0.37 + 0.44$$

$$= 0.07$$

(b) The probability that the life of a battery selected at random will be between 15 and 25 hours is $P([15, 25])$, and

$$P([15, 25]) = \int_{15}^{25} \frac{1}{60} e^{-x/60} \, dx$$

$$= -e^{-x/60} \Big]_{15}^{25}$$

$$= -e^{-25/60} + e^{-15/60}$$

$$= -e^{-0.42} + e^{-0.25}$$

$$= -0.66 + 0.78$$

$$= 0.12$$

EXAMPLE 4 The demand equation for a new kind of appliance is given by

$$x = 5000e^{-0.04p}$$

where x units are demanded when p dollars is the price per unit. Find the value of p for which the total sales revenue will be the greatest. Also find the greatest total sales revenue.

Solution Let R dollars be the total sales revenue. Because $R = px$,

$$R = 5000pe^{-0.04p} \qquad p \in [0, +\infty)$$

$$D_p R = 5000e^{-0.04p} + 5000p(-0.04)e^{-0.04p}$$

$$= 5000e^{-0.04p}(1 - 0.04p)$$

Setting $D_p R = 0$ we get

$$1 - 0.04p = 0$$

$$p = 25$$

Table 8.7.2

	R	$D_p R$	Conclusion
$0 < p < 25$		$+$	R is increasing
$p = 25$	45,985	0	R has a relative maximum value
$25 < p$		$-$	R is decreasing

From Table 8.7.2 we can conclude that on the interval $[0, +\infty)$ R has a relative maximum value at 25. Because R is continuous on $[0, +\infty)$ and R has only one relative extremum on the interval, it follows that R has an absolute maximum value at 25. Therefore the greatest total sales revenue is $45,985, which occurs when the price is $25 per unit.

The next example involves some real estate whose value increases over a period of time. The real estate will be worth more the longer it is kept. Eventually, however, there may be a time when money invested at the current interest rate will have a more rapid growth than the value of the real estate. If it is planned to sell the real estate at the time when its present value is greatest, then the most profit will be realized.

EXAMPLE 5 An investment company estimated that a certain piece of real estate will have a value of $(100{,}000 + 20{,}000n)$ dollars in n years. If the company can receive an interest rate of 12 percent compounded quarterly, when should the investment company plan to sell the real estate so that its present value will be greatest?

Solution Let P dollars be the present value of the real estate. Using Eq. (10) in Sec. 8.1 with $A = 100{,}000 + 20{,}000n$, $i = 0.12$, $m = 4$, and $t = n$, we have

$$P = (100{,}000 + 20{,}000n)(1 + 0.03)^{-4n}$$

The present value will be greatest for the value of n that makes it an absolute maximum. Differentiating, we get

$$
\begin{aligned}
D_n P &= 20{,}000(1.03)^{-4n} + (100{,}000 + 20{,}000n)(1.03)^{-4n}(\ln 1.03)(-4) \\
&= (1.03)^{-4n}[20{,}000 + (-400{,}000 - 80{,}000n)(0.0296)] \\
&= (1.03)^{-4n}[20{,}000 - 11{,}800 - 2370n] \\
&= (1.03)^{-4n}(8200 - 2370n)
\end{aligned}
$$

Setting $D_n P = 0$ we have

$$8200 - 2370n = 0$$

$$n = \frac{8200}{2370}$$

$$n = 3.46$$

Because $n \geq 0$, and since $D_n P > 0$ when $0 < n < 3.46$ and $D_n P < 0$ when $3.46 < n$, it follows from the first-derivative test that P has a relative maximum value when $n = 3.46$. Because P is continuous on $[0, +\infty)$ and there is only one relative extremum on the interval, P has an absolute maximum value when $n = 3.46$. Therefore the company should plan to sell the real estate in 3.46 years, that is, in 3 years and 6 months.

EXAMPLE 6   Solve Example 5 if the company can receive an interest rate of 12 percent compounded continuously.

Solution   We apply Eq. (11) in Sec. 8.1 with $A = 100,000 + 20,000n$, $i = 0.12$, and $t = n$. Then if P dollars is the present value of the real estate,

$$P = (100,000 + 20,000n)e^{-0.12n}$$

Therefore

$$D_n P = 20,000e^{-0.12n} + (100,000 + 20,000n)e^{-0.12n}(-0.12)$$
$$= e^{-0.12n}[20,000 + (-12,000 - 2400n)]$$
$$= e^{-0.12n}(8000 - 2400n)$$

Setting $D_n P = 0$ we obtain

$$8000 - 2400n = 0$$

$$n = \frac{8000}{2400}$$

$$n = 3.33$$

Because $D_n P > 0$ when $0 < n < 3.33$ and $D_n P < 0$ when $3.33 < n$, then P has a relative maximum value when $n = 3.33$. This value of n gives P an absolute maximum value, because P is continuous for $n \geq 0$ and there is only one relative extremum. Thus the company should plan to sell the real estate in 3.33 years, that is, in 3 years and 4 months.

Exercises 8.7

1. Suppose that a student has 3 hours to cram for an examination and during this time wishes to memorize a set of 60 facts. According to psychologists, the rate at which a person can memorize a set of facts is proportional to the number of facts remaining to be memorized. Thus if the student memorizes y facts in t min,

$$\frac{dy}{dt} = k(60 - y)$$

It is assumed that initially zero facts are memorized. If the student memorizes 15 facts in the first 20 min, how many facts will the student memorize in (a) 1 hour and (b) 3 hours?

2. A new worker on an assembly line can do a particular task in such a way that if y units are completed per day after t days on the assembly line, then

$$\frac{dy}{dt} = k(90 - y)$$

On the day the worker starts, 60 units are completed, and after being on the job 5 days, the worker does 75 units per day. (a) How many units per day are completed after the worker is on the job 9 days? (b) Show that the worker is producing at almost full potential after 30 days.

3. Under the conditions of Example 2, after how many minutes will the temperature of the body be $45°$?

4. If a body in air at a temperature of $0°$ cools from $200°$ to $100°$ in 40 min, how many more minutes will it take for the body to cool to $50°$? Use Newton's law of cooling given in Example 2.

5. If a thermometer is taken from a room in which the temperature is $75°$ into the open, where the temperature is $35°$, and the reading of the thermometer is $65°$ after 30 sec, (a) how long after the removal will the reading be $50°$? (b) What is the thermometer reading 3 min after the removal? Use Newton's law of cooling given in Example 2.

6. For a certain light bulb, the probability density function that x hours will be the life of a bulb selected at random is given by

$$f(x) = \tfrac{1}{40}e^{-x/40} \qquad x \geq 0$$

Find the probability that the life of a bulb selected at random will be (a) between 40 and 60 hours and (b) between 10 and 30 hours.

7. For a particular appliance, the probability density function that it will need servicing x months after it is purchased is given by

$$f(x) = 0.02e^{-0.02x} \qquad x \geq 0$$

If the appliance is guaranteed for a year, what is the probability that a customer selected at random will need to have servicing during the 1-year warranty period?

8. In a certain city, the probability density function for x min to be the length of a telephone call selected at random is given by

$$f(x) = \tfrac{1}{3}e^{-x/3} \qquad x \geq 0$$

Find the probability that a telephone call selected at random will last (a) between 3 and 4 min and (b) between 1 and 2 min.

9. The total cost of producing x units of a commodity is $C(x)$ dollars and $C(x) = 40e^{x/4}$. Find the function giving (a) the average cost and (b) the marginal cost. (c) Find the absolute minimum average unit cost. (d) Verify that the average cost and marginal cost are equal when the average cost has its least value.

10. A certain article of merchandise has the demand equation $p = 10e^{-x/400}$, where x units are demanded when p dollars is the price per unit. Find (a) the total revenue function and (b) the marginal revenue function. (c) Find the absolute maximum total revenue.

11. An advertising agency determined statistically that if a breakfast food manufacturer increases its budget for television commercials by x thousand dollars, there will be an increase in the total profit of $25x^2e^{-0.2x}$ hundred dollars. (a) What should be the advertising budget increase for the manufacturer to have the greatest profit? (b) What will be the corresponding increase in the company's profit?

12. A manufacturer has determined that if $100x$ units of a particular commodity are produced each week, the marginal cost is given by $2^{x/2}$ and the marginal revenue is given by $8(2^{-x/2})$, where both the production cost and revenue are in thousands of dollars. If the weekly fixed costs amount to $2000, find to the nearest dollar the maximum weekly profit that can be obtained.

13. A dealer of fine arts has the current value of a painting appraised at $55,000 and feels that the painting will increase in value by $10,000 each year. If the dealer can invest money at an interest rate of 12 percent per year compounded quarterly, how long should the dealer plan to keep the painting so that its present value will be greatest?

14. A woman owns two apartment houses. The larger apartment house can be sold for $250,000, and it is increasing in value at the rate of $25,000 per year. The smaller apartment house can be sold for $200,000, and it is increasing in value at the rate of $20,000 per year. The woman wishes to sell one apartment house and keep the other. If she invests her money at an interest rate of 8 percent compounded annually, which apartment house should she sell now, and how long should she plan to keep the other one so that her present value will be greatest?

15. Solve Exercise 13 if the dealer can receive an interest rate of 12 percent compounded continuously.

16. Solve Exercise 14 if the woman invests her money at an interest rate of 8 percent compounded continuously.

8.8 Annuities

In Sec. 8.1 we considered problems involving the determination of the value at the end of t years of an investment of P dollars at a given rate of interest. In these problems we assume that no additional deposits are made other than the interest. Frequently there are business situations in which additional principal deposits are made. When equal payments are made at equal periods of time, we have what is called an **annuity**. If the equal payments are made at the ends of the payment periods, the annuity is called an **ordinary annuity**. If each of the equal payments is made at the beginning of the payment period, the annuity is called an **annuity due**.

Let us consider an ordinary annuity for which equal deposits of P dollars are made at the end of each year for a term of t years at an interest rate of $100i$ percent compounded annually. At the end of the first year the number of dollars in the amount of the annuity will be P. At the end of the second year the first deposit of P dollars has earned interest for 1 year, and hence its value will be $P(1 + i)$ dollars. Also, at the end of the second year an additional deposit of P dollars is made. Therefore the number of dollars in the amount of the annuity at the end of 2 years will be

$$P(1 + i) + P$$

Similarly, the number of dollars in the amount at the end of 3 years will be

$$[P(1 + i) + P](1 + i) + P = P(1 + i)^2 + P(1 + i) + P$$

Continuing on, the number of dollars in the amount at the end of 4 years will be

$$[P(1 + i)^2 + P(1 + i) + P](1 + i) + P$$
$$= P(1 + i)^3 + P(1 + i)^2 + P(1 + i) + P$$

Assuming that the number of dollars in the amount of the annuity at the end of k years will be

$$P(1 + i)^{k-1} + P(1 + i)^{k-2} + \ldots + P(1 + i) + P$$

then the number of dollars in the amount at the end of $(k + 1)$ years will be

$$[P(1 + i)^{k-1} + P(1 + i)^{k-2} + \ldots + P(1 + i) + P](1 + i) + P$$
$$= P(1 + i)^k + P(1 + i)^{k-1} + \ldots + P(1 + i)^2 + P(1 + i) + P$$

We have therefore proved by mathematical induction that if A dollars is the amount of the annuity at the end of t years, then

$$A = P[(1 + i)^{t-1} + (1 + i)^{t-2} + \ldots + (1 + i)^2 + (1 + i) + 1] \tag{1}$$

The expression on the right side of (1) is called a *geometric progression*. To find a formula for computing the sum we multiply each term on both sides of Eq. (1) by $(1 + i)$ and obtain

$$(1 + i)A = P[(1 + i)^t + (1 + i)^{t-1} + \ldots + (1 + i)^2 + (1 + i)] \tag{2}$$

Subtracting the terms in Eq. (1) from the corresponding terms in Eq. (2) we get

$$(1 + i)A - A = P[(1 + i)^t - 1]$$

and solving for A we obtain

$$A = P\left[\frac{(1 + i)^t - 1}{i}\right] \tag{3}$$

If for an ordinary annuity the interest is compounded m times per year and the payments of P dollars are made m times per year, then the number of interest periods in t years will be mt, which we shall denote by n, and the interest rate per period will be denoted by $100j$ percent, where $j = i/m$. In this case the amount of the annuity after t years (or, equivalently, n periods) is given by

$$A = P\left[\frac{(1 + j)^n - 1}{j}\right] \tag{4}$$

The computation involved in using formulas (3) and (4) is simplified by using Table 7 in the back of the book, which gives the value of $[(1 + j)^n - 1]/j$, which is the number of dollars in the amount of an ordinary annuity of equal payments of $1 after n periods with an interest rate of $100j$ percent per period.

Examples in determining the size of the equal payments in an ordinary annuity arise from the consideration of *sinking funds*. A **sinking fund** is a fund created to pay for an obligation that will arise at a future date. The following example illustrates such a situation.

EXAMPLE 1 A loan of $500 with interest at an annual rate of 12 percent compounded semiannually must be paid off in one payment 2 years hence. To provide for this, quarterly payments are to be placed into a sinking fund that pays 8 percent compounded quarterly. How much is the quarterly payment?

Solution Let A dollars be the amount to be paid in 2 years. Then

$$A = 500(1.06)^4 = 500(1.2625)$$
$$= 631.25$$

Let P dollars be the quarterly payment into the sinking fund. Then using formula (4), where $j = 0.08/4 = 0.02$ and $n = 4 \cdot 2 = 8$, we have

$$631.25 = P\left[\frac{(1.02)^8 - 1}{0.02}\right]$$

From Table 7 we find $[(1.02)^8 - 1]/0.02 = 8.5830$, and so

$$631.25 = P(8.5830)$$

$$P = 73.55$$

Therefore the quarterly payments into the sinking fund should be \$73.55.

We now wish to obtain a formula for finding the present value of an ordinary annuity of equal payments of P dollars for n periods with an interest rate of $100j$ percent per period. Let V_1 dollars be the present value of the first payment of P dollars to be made at the end of the first period. We have, then,

$$P = V_1(1 + j)$$

Hence

$$V_1 = \frac{P}{1 + j}$$

If V_2 dollars is the present value of the second payment of P dollars to be made at the end of the second period, we have

$$P = V_2(1 + j)^2$$

from which we obtain

$$V_2 = \frac{P}{(1 + j)^2}$$

Continuing on, if V_k dollars is the present value of the kth payment of P dollars to be made at the end of the kth period, we have

$$V_k = \frac{P}{(1 + j)^k}$$

Letting V dollars be the present value of the annuity we have $V = V_1 + V_2 + \ldots + V_n$, and so

$$V = \frac{P}{1 + j} + \frac{P}{(1 + j)^2} + \frac{P}{(1 + j)^3} + \ldots + \frac{P}{(1 + j)^n} \tag{5}$$

Multiplying each term in Eq. (5) by $1/(1 + j)$ we get

$$\frac{V}{1 + j} = \frac{P}{(1 + j)^2} + \frac{P}{(1 + j)^3} + \ldots + \frac{P}{(1 + j)^{n+1}} \tag{6}$$

Subtracting the terms of Eq. (5) from the corresponding terms of Eq. (6) we obtain

$$\frac{V}{1 + j} - V = \frac{P}{(1 + j)^{n+1}} - \frac{P}{1 + j}$$

Multiplying each term of the above equation by $(1 + j)$ gives us

$$V - (1 + j)V = P[(1 + j)^{-n} - 1]$$

and solving for V we get

$$V = P\left[\frac{1 - (1 + j)^{-n}}{j}\right] \tag{7}$$

Table 8 in the back of the book gives values of $[1 - (1 + j)^{-n}]/j$, which is the number of dollars in the present value of an ordinary annuity of equal payments of $1 for n periods with an interest rate of $100j$ percent per period.

If the interest rate is $100i$ percent per year and the equal payments are made in 1-year periods, then in formula (7) $j = i$ and $n = t$, and we have

$$V = P\left[\frac{1 - (1 + i)^{-t}}{i}\right] \tag{8}$$

EXAMPLE 2 The rent for an apartment is $450 per month, payable 1 month in advance. If the tenant wishes to pay a year's rent in advance at an annual rate of 12 percent compounded monthly, what amount should be paid?

Solution This problem involves equal payments of $450 for 12 periods. Because the payments are made at the beginning of the payment period, this is an annuity due. We can work the problem as one involving an ordinary annuity by considering the first payment separately and then the remaining 11 payments as constituting an ordinary annuity (payments at the end of each of the payment periods) for 11 periods. Letting V dollars be the present value of the ordinary annuity and S dollars be the amount that should be paid now, we have

$$S = 450 + V$$

Using formula (7) with $P = 450$, $j = 0.12/12 = 0.01$, and $n = 11$ we have

$$S = 450 + 450\left[\frac{1 - (1.01)^{-11}}{0.01}\right]$$

From Table 8 we find $[1 - (1.01)^{-11}]/0.01 = 10.3676$, and substituting this into the above gives

$$S = 450 + 450(10.3676)$$
$$= 5115.42$$

Thus the tenant should pay $5115.42 now to cover the year's rent.

EXAMPLE 3 A company is considering whether to bid on a contract that will produce a net return of $3000 semiannually for 5 years. However, if the contract is obtained, the company must invest $24,000 in special equipment, which will have a scrap value of $5000 at the end of 5 years. If money is worth 10 percent per year compounded semiannually, should the company bid on the contract?

Solution   We first find the present value of $5000, which is the scrap value of the equipment at the end of 5 years. If P dollars is this present value, then

$$P = 5000(1.05)^{-10}$$
$$= 5000(0.6139)$$
$$= 3070$$

The present value of the cost of the equipment is then $24,000 − $3,070 = $20,930. Letting V dollars be the present value of the returns, we use formula (7), where $P = 3000$, $j = 0.10/2 = 0.05$, and $n = 5 \cdot 2 = 10$, which gives

$$V = 3000 \left[\frac{1 - (1.05)^{-10}}{0.05} \right]$$

$$= 3000(7.7217)$$
$$= 23,165$$

The present value of the returns is then $23,165, and this is greater than the present value of the cost of the equipment. Therefore the company should bid on the contract.

Let us now extend the concept of equal payments at equal periods of time to that of a continuous investment for which interest is compounded continuously. Suppose that the number of dollars per year in the investment at any time t years from the present is $f(t)$, where f is a continuous function. We wish to find a suitable formula for determining A, the number of dollars in the amount T years from now at an interest rate of $100i$ percent compounded continuously.

Because the investment at t years is $f(t)$ dollars and the rate is $100i$ percent compounded continuously for a period of $(T - t)$ years, from Eq. (9) in Sec. 8.1 we see that it will amount to $f(t)e^{i(T-t)}$ dollars at T years. Let the function g be defined by $g(t) = f(t)e^{i(T-t)}$. Divide the closed interval $[0, T]$ into n subintervals each of width Δt. Let ξ_i be any number in the ith subinterval $[t_{i-1}, t_i]$. We define A to be

$$\lim_{n \to +\infty} \sum_{i=1}^{n} g(\xi_i) \, \Delta t$$

This is the limit of a Riemann sum, and because g is continuous on $[0, T]$, the limit exists and is the definite integral of g from 0 to T. Recalling that $g(t) = f(t)e^{i(T-t)}$, we have the formula

$$A = \int_0^T f(t)e^{i(T-t)} \, dt \tag{9}$$

Observe that A is a function of T. Often it is necessary to use integration by parts to apply formula (9). This topic is discussed in Sec. 9.1. However, if $f(t)$ is constant or a power of e, we can apply integration techniques we already know.

Suppose that $f(t)$ is a constant P; then from formula (9),

$$A = \int_0^T Pe^{i(T-t)} \, dt$$

$$= P\left[\frac{-e^{i(T-t)}}{i}\right]_0^T$$

$$= P\left(\frac{-1}{i} + \frac{e^{iT}}{i}\right)$$

$$A = P\left(\frac{e^{iT} - 1}{i}\right) \tag{10}$$

Compare formula (10) with formula (3), and you will notice that they are the same except that the $(1 + i)$ in (3) is replaced by e^i in (10).

EXAMPLE 4   Find the amount in 10 years of a continuous investment of $500 per year at an annual rate of 11 percent compounded continuously.

Solution   If A dollars is the amount, then from (9),

$$A = \int_0^{10} 500e^{0.11(10-t)} \, dt$$

$$= 500\left[\frac{-e^{0.11(10-t)}}{0.11}\right]_0^{10}$$

$$= 500\left[\frac{e^{1.1} - 1}{0.11}\right]$$

$$= \frac{50,000}{11}(3.0042 - 1)$$

$$= 9110$$

Therefore the amount in 10 years will be $9110.

Observe in Example 4 that instead of using the integral we could have obtained the third line of the solution by substituting directly into (10).

We now obtain a formula for finding the present value of the amount A dollars given in (10). From Eq. (12) in Sec. 8.1 we know that the present value of 1 dollar t years from now is e^{-it}, and therefore the present value of $f(t)$ dollars received t years in the future is $f(t)e^{-it}$. By an argument similar to the one preceding (9) we conclude that if V dollars is the present value of the amount earned in T years from a continuous flow of income at t years of $f(t)$ dollars per year invested at a rate of $100i$ percent compounded continuously, then

$$V = \int_0^T f(t)e^{-it} \, dt \tag{11}$$

If in formula (11) $f(t)$ is a constant P, then

$$V = \int_0^T Pe^{-it}\, dt$$

$$= P\left[\frac{-e^{-it}}{i}\right]_0^T$$

$$= P\left(\frac{-e^{-iT}}{i} + \frac{1}{i}\right)$$

$$V = P\left(\frac{1 - e^{-iT}}{i}\right) \qquad\qquad (12)$$

By comparing formula (12) with formula (8), notice that they have the same relationship that formula (10) has with formula (3).

EXAMPLE 5 In Example 3, suppose that instead of the contract paying $3000 semiannually for 5 years, it offers a continuous flow of income of $6000 per year. Furthermore, suppose that money is worth 10 percent per year compounded continuously. Find the present value of the cost of the special equipment and the present value of the returns under these conditions.

Solution The scrap value of the equipment at the end of 5 years is $5000. The present value of this $5000 is obtained from Eq. (11) of Sec. 8.1 with $A = 5000$, $i = 0.1$, and $t = 5$. Thus if P dollars is the present value,

$$P = 5000e^{-0.5}$$

$$= 5000(0.6065)$$

$$= 3033$$

Therefore the present value of the cost of the special equipment is $24,000 − $3033 = $20,967.

If V dollars is the present value of the returns, then from (11) with $f(t) = 6000$, $i = 0.1$, and $T = 5$,

$$V = \int_0^5 6000e^{-0.1t}\, dt$$

$$= 6000(-10)e^{-0.1t}\Big]_0^5$$

$$= -60,000(e^{-0.5} - 1)$$

$$= -60,000(0.6065 - 1)$$

$$= -60,000(-0.3935)$$

$$= 23,610$$

Hence the present value of the returns is $23,610.

Exercises 8.8

1. At the end of every 6 months, $200 is deposited in a savings account. If the annual interest rate is 8 percent compounded semi-annually, what is the total amount in the account at the end of 4 years?

2. Equal deposits are made into a savings account at the end of each year for 10 years so that there will be $40,000 in the account at the end of the tenth year. What should be the amount of each deposit if interest is compounded annually at the rate of 10 percent?

3. It is desired to have $20,000 in a savings account at the end of 10 years. If equal deposits are made at the end of every 3 months, what should be the amount of each deposit if interest is compounded quarterly at the annual rate of 8 percent?

4. If $500 is deposited in a savings account at the end of every 3 months, and the interest rate is 8 percent per year compounded quarterly, what is the total amount in the account at the end of 2 years?

5. A loan of $2000 with interest at 16 percent per year compounded quarterly must be paid off in one payment 4 years from now. To provide for this, quarterly payments are to be placed into a sinking fund that pays an annual rate of 12 percent compounded quarterly. How much is the quarterly payment?

6. Solve Exercise 5 if the loan is for $5000 and the sinking fund earns interest at only 8 percent compounded quarterly.

7. A man makes a downpayment of $20,000 cash on a house and agrees to pay $4000 at the end of each 6-month period for the next 15 years. If money is worth 12 percent per year compounded semiannually, what is the present value of the purchase price of the house?

8. A deposit of $400 is made every 6 months into a fund that pays interest at an annual rate of 12 percent compounded semiannually. How much is in the fund just after the twentieth deposit?

9. A life insurance policy has a present cash value of $12,000. However, the policy provides for a cash payment of $2000 at the end of 10 years, and the remainder is to be paid in 10 equal consecutive annual payments starting 1 year later. If money is worth 10 percent per year compounded annually, what is the annual payment?

10. What is the present value of a contract that pays $200 at the end of each quarter for 4 years and an additional $1000 at the end of the last quarter if money is worth 16 percent per year compounded quarterly?

11. A company wishes to make semiannual payments into a sinking fund to provide for the replacement in 5 years of a piece of equipment that will cost $10,000. If the present piece of equipment has a scrap value of $500 at the end of 5 years, what equal deposits should be made, starting in 6 months, if the fund earns an annual rate of 8 percent compounded semiannually?

12. A business may be purchased for either $100,000 cash or a $60,000 downpayment with six equal annual payments of $10,000. What is the approximate yearly rate of interest, compounded annually, if the second alternative is chosen?

13. A special piece of equipment will produce earnings of $50,000 at the end of each year for 5 years. The purchase price of the machine is $200,000, and in order to buy it a company must obtain a loan to be repaid in five equal annual payments at a yearly rate of 10 percent compounded annually. Should the company buy the equipment if at the end of 5 years there is no scrap value?

14. In Exercise 13, suppose that the company pays cash for the equipment and that at the end of 5 years it has a scrap value of $30,000. Furthermore, suppose that the company could invest its money at a yearly rate of 10 percent compounded annually. In this case should the company buy the equipment?

15. In Exercise 13, suppose that instead of equal annual earnings of $50,000 per year the earnings are $40,000 at the end of the first year, $40,000 at the end of the second year, $50,000 at the end of the third year, $60,000 at the end of the fourth year, and $60,000 at the end of the fifth year. Furthermore, suppose that the company pays cash for the equipment and that at the end of 5 years it has a scrap value of $30,000. If money can be invested at a rate of 10 percent per year compounded annually, should the company buy the equipment?

16. A company is considering whether to bid on a 5-year contract that will produce an estimated net return of $1000 at the end of the first year, $5000 at the end of the second year, $10,000 at the end of the third year, $10,000 at the end of the fourth year, and $5000 at the end of the fifth year. To fulfill the contract the company must invest $5000 at the beginning of each of the 5 years. If the company can invest its money at an interest rate of 10 percent per year compounded annually, should it bid on the contract?

17. Work Exercise 1 if instead of depositing $200 at the end of each 6 months the $400 per year is a continuous investment and the interest rate is 8 percent per year compounded continuously.

18. Work Exercise 4 if instead of depositing $500 at the end of every 3 months the $2000 per year is a continuous investment and the interest rate is 8 percent per year compounded continuously.

19. In Exercise 5, suppose that the interest on the loan is 16 percent per year compounded continuously. Furthermore, suppose that the payments are to be a continuous flow of x dollars per year for 4 years into a sinking fund that pays 12 percent per year compounded continuously. Find x.

20. In Exercise 6, suppose that the loan of $5000 has interest at 16 percent per year compounded continuously. Furthermore, suppose that the payments are to be a continuous flow of x dollars per year for 4 years into a sinking fund that pays an annual interest rate of 8 percent compounded continuously. Find x.

21. In Exercise 9, if the remainder is to be paid at a continuous flow of income of x dollars per year for 10 years and the annual interest rate is 10 percent compounded continuously, find x.

22. Work Exercise 10 if instead of the contract paying $200 at the end of each quarter for 4 years it offers a continuous flow of income of $800 per year for 4 years and money is worth 16 percent per year compounded continuously.

23. Suppose that a continuous flow of income decreases for a period of T years and at t years the number of dollars in the income per year is ae^{-bt}, where a and b are constants. (a) Find the present value of this income if the annual interest rate is $100i$ percent compounded continuously. (b) Show that the answer in part (a) is the same as the present value of a continuous flow of income of a dollars per year if the annual rate of interest is $100(i + b)$ percent compounded continuously.

Review Exercises for Chapter 8

In Exercises 1 through 14, differentiate the given function.

1. $f(x) = \ln(5x + 3)$

2. $f(x) = \ln(x^2 - 2x)$

3. $g(x) = \ln\sqrt{x^2 + 1}$

4. $f(t) = \ln(3t - 1)^2$

5. $f(x) = \log_{10} x^2$

6. $g(x) = \log_2 \dfrac{x + 1}{x - 1}$

7. $f(r) = e^{r^2}$

8. $g(y) = 10^{-3y}$

9. $g(t) = t^2 2^t$

10. $F(x) = x^2 e^{2x}$

11. $f(x) = \dfrac{e^x}{e^x + e^{-x}}$

12. $f(x) = \ln\sqrt{\dfrac{2x + 1}{x - 3}}$

13. $G(x) = x^{2x}; x > 0$

14. $h(t) = t^{3/\ln t}$

In Exercises 15 through 20, evaluate the indefinite integral.

15. $\displaystyle\int \dfrac{3e^{2x}\, dx}{1 + e^{2x}}$

16. $\displaystyle\int e^{2x^2 - 4x}(x - 1)\, dx$

17. $\displaystyle\int (e^{3t} + 2^{3t})\, dt$

18. $\displaystyle\int \dfrac{10^{\ln x^2}\, dx}{x}$

19. $\displaystyle\int \dfrac{xe^{6x^2}\, dx}{1 + e^{6x^2}}$

20. $\displaystyle\int (w + 1)e^w 7^{we^w}\, dw$

In Exercises 21 through 26, evaluate the definite integral.

21. $\displaystyle\int_0^2 x^2 e^{x^3}\, dx$

22. $\displaystyle\int_0^1 (e^{2x} + 1)^2\, dx$

23. $\displaystyle\int_1^8 \dfrac{t^{1/3}\, dt}{t^{4/3} + 4}$

24. $\displaystyle\int_e^{e^2} \dfrac{dy}{y(\ln y)}$

25. Find $D_x y$ if $ye^x + xe^y + x + y = 0$.

26. A loan of $1000 is repaid in one payment at the end of a year. If the interest rate is 12 percent per year compounded monthly, find (a) the total amount repaid and (b) the effective rate of interest.

27. Find the effective rate of interest if interest is computed at an annual rate of 16 percent compounded (a) semiannually; (b) quarterly; (c) continuously.

28. Work Exercise 26 if the interest rate is 12 percent compounded continuously.

29. A house purchased 10 years ago was sold for $100,000. If the annual interest rate was determined to be 20 percent compounded quarterly, what was the purchase price of the house to the nearest thousand dollars?

30. Interest on a savings account is computed at 8 percent per year compounded continuously. If one wishes to have $1000 in the account at the end of a year by making a single deposit now, what should be the amount of the deposit?

31. How long will it take for an investment to double itself if interest is earned at the rate of 12 percent per year compounded (a) quarterly and (b) continuously?

32. How long will it take for a deposit of $500 into a savings account to accumulate to $600 if interest is computed at 8 percent per year compounded (a) quarterly and (b) continuously?

33. If A mg of radium are present after t years, then $A = ke^{-0.0004t}$, where k is a constant. Furthermore, 60 mg of radium are present now. (a) How much radium will be present 100 years from now? (b) How long will it take until there are only 50 mg of radium present?

34. In t min there will be $f(t)$ bacteria present in a certain culture, where $f(t) = ke^{0.03t}$ and k is a constant. If 60,000 bacteria are present initially, (a) how many will be present in 15 min, and (b) after how many minutes will there be 200,000 present?

35. In a small town an epidemic spread in such a way that $f(t)$ persons had contracted the disease t weeks after its outbreak, where

$$f(t) = \frac{10,000}{1 + 599e^{-0.8t}}$$

How many people had the disease (a) initially, (b) after 6 weeks, and (c) after 12 weeks? (d) If the epidemic continues indefinitely, how many people will contract the disease?

36. In the town of Exercise 35, after how many weeks will 5000 people, one-half of the town's population, contract the disease?

37. A particular article of merchandise has the demand equation $pe^{x/200} = 20$, where x units are demanded when p dollars is the price per unit. Find (a) the total revenue function and (b) the marginal revenue function. (c) Find the absolute maximum total revenue.

38. The demand equation of a certain commodity is $xe^p = 200$, where x units are demanded when p dollars is the price per unit. (a) Find the price elasticity of demand when $p = 10$. (b) From the result of part (a) find an approximate change in the demand if the price of $10 is decreased by 2 percent.

39. On a certain college campus the probability density function for the length of a telephone call to be t min is given by

$$f(t) = 0.4e^{-0.4t} \qquad t \geq 0$$

What is the probability that a telephone call selected at random will last (a) between 2 and 3 min, (b) 2 min or less, and (c) at most 3 min?

40. The rate of natural increase of the population of a certain city is proportional to the population. If the population doubles in 60 years, and if the population in 1950 was 60,000, estimate the population in the year 2000.

41. The rate of decay of a radioactive substance is proportional to the amount present. If half of a given deposit of the substance disappears in 1900 years, how long will it take for 95 percent of the deposit to disappear?

42. The charge of electricity on a spherical surface leaks off at a rate proportional to the charge. Initially the charge of electricity was 8 coulombs, and one-fourth leaks off in 15 min. When will there be only 2 coulombs remaining?

43. The rate of bacterial growth in a certain culture is proportional to the number present, and the number doubles in 20 min. If at the end of 1 hr there were 1,500,000 bacteria, how many bacteria were present initially?

44. A tank contains 100 liters of fresh water, and brine containing 2 kg of salt per liter flows into the tank at the rate of 3 liters/min. If the mixture, kept uniform by stirring, flows out at the same rate, how many kilograms of salt are there in the tank at the end of 30 min?

45. A tank contains 60 gal of salt water with 120 lb of dissolved salt. Salt water with 3 lb of salt per gallon flows into the tank at the rate of 2 gal/min, and the mixture, kept uniform by stirring, flows out at the same rate. How long will it be before there are 100 lb of salt in the tank?

46. Refer to Exercise 23 in Exercises 8.6. A paleontologist discovered an insect preserved inside a transparent amber, which is hardened tree pitch, and the amount of ^{14}C present in the insect was determined to be 2 percent of its original amount. Use the fact that the half life of ^{14}C is 5600 years to determine the age of the insect at the time of discovery.

47. A student studying a foreign language has 50 verbs to memorize. The rate at which the student can memorize these verbs is proportional to the number of verbs remaining to be memorized; that is, if the student memorizes y verbs in t min,

$$\frac{dy}{dt} = k(50 - y)$$

Assume that initially no verbs are memorized, and suppose that 20 verbs are memorized in the first 30 minutes. How many verbs will the student memorize in (a) 1 hour and (b) 2 hours? (c) After how many hours will the student have only one verb left to memorize?

48. Find the demand equation for a commodity if the marginal revenue function is given by $R'(x) = 10/(x + 1)$, where $R(x)$ dollars is the total revenue when x units are sold.

49. At the end of each year equal deposits are placed into a savings account yielding an interest rate of 10 percent per year compounded annually. (a) If the amount of each deposit is $500, how much is in the account at the end of 8 years? (b) If it is desired to have $6000 in the account at the end of the eighth year, what should be the amount of each deposit?

50. Determine the present value of a contract that pays $1000 at the end of each quarter for 3 years and an additional $5000 at the end of the last quarter if money is worth 12 percent per year compounded quarterly.

51. Work part (a) of Exercise 49 if instead of depositing $500 at the end of each year the $500 per year is a continuous investment and the interest rate is 10 percent per year compounded continuously.

52. The demand equation for a particular commodity is $4^{x/2}p = 10$, where x units are demanded when the price per unit is p dollars. If the market price is $5, find the consumers' surplus. Draw a sketch showing the demand curve and the region whose area gives the consumers' surplus.

53. Use Newton's law of cooling, given in Example 2 of Sec. 8.7, to determine the current temperature of a body in air of temperature $40°$ if 30 min ago the body's temperature was $150°$ and 10 min ago it was $90°$.

CHAPTER 9

TOPICS ON INTEGRATION

9.1 Integration by parts

The standard integration formulas that you learned in previous chapters and that are used frequently are listed below.

$$\int du = u + C$$

$$\int a\,du = au + C \qquad \text{where } a \text{ is any constant}$$

$$\int [f(u) + g(u)]\,du = \int f(u)\,du + \int g(u)\,du$$

$$\int u^n\,du = \frac{u^{n+1}}{n+1} + C \qquad n \neq -1$$

$$\int \frac{du}{u} = \ln |u| + C$$

$$\int a^u\,du = \frac{a^u}{\ln a} + C$$

$$\int e^u\,du = e^u + C$$

A method of integration that is quite useful is **integration by parts**. It depends on the formula for the differential of a product. If u and v are functions of a variable x, then

$$d(uv) = u\,dv + v\,du$$

or, equivalently,

$$u\,dv = d(uv) - v\,du \tag{1}$$

Integrating on both sides of (1) we have

$$\int u\,dv = uv - \int v\,du \tag{2}$$

Formula (2) is called the **formula for integration by parts**. This formula expresses the integral $\int u\,dv$ in terms of another integral, $\int v\,du$. By a suitable choice of u and dv it may be easier to evaluate the second integral than the first.

● ILLUSTRATION 1

We wish to evaluate

$$\int xe^x\,dx$$

If we let $u = x$ and $dv = e^x \, dx$, then we find du and v as follows:

$$u = x \qquad dv = e^x \, dx$$

$$du = dx \qquad v = \int e^x \, dx$$

$$= e^x + C_1$$

From formula (2),

$$\int xe^x \, dx = x(e^x + C_1) - \int (e^x + C_1) \, dx$$

$$= xe^x + C_1 x - \int e^x \, dx - \int C_1 \, dx$$

$$= xe^x + C_1 x - e^x - C_1 x + C_2$$

$$= xe^x - e^x + C_2 \qquad\qquad \bullet$$

In Illustration 1 observe that the first constant of integration C_1 does not appear in the final result. This is true in general, and we prove it as follows: By writing $v + C_1$ in formula (2) we have

$$\int u \, dv = u(v + C_1) - \int (v + C_1) \, du$$

$$= uv + C_1 u - \int v \, du - C_1 \int du$$

$$= uv + C_1 u - \int v \, du - C_1 u$$

$$= uv - \int v \, du$$

Therefore it is not necessary to write C_1 when finding v from dv.

● ILLUSTRATION 2

If you have the integral

$$\int x^3 e^{x^2} \, dx$$

then to determine the substitutions for u and dv, bear in mind that to find v you must be able to integrate dv. This suggests letting $dv = xe^{x^2} \, dx$, and then $u = x^2$. We integrate to find

$$v = \int xe^{x^2}\, dx$$

$$= \frac{1}{2}\int e^{x^2}(2x\, dx)$$

$$= \tfrac{1}{2}\, e^{x^2}$$

Remember it is not necessary to write the arbitrary constant of integration when finding v from dv. We find the differential of u:

$$u = x^2$$

$$du = 2x\, dx$$

From formula (2),

$$\int x^3 e^{x^2}\, dx = x^2\left(\frac{1}{2}\, e^{x^2}\right) - \int \frac{1}{2}\, e^{x^2}(2x\, dx)$$

$$= \tfrac{1}{2}x^2 e^{x^2} - \tfrac{1}{2}e^{x^2} + C \qquad \bullet$$

In applying integration by parts to a specific integral, one pair of choices for u and dv may work while another pair may not. This situation occurs in Illustration 3.

● ILLUSTRATION 3

In Illustration 1, if instead of the choices of u and dv as shown, suppose that we let

$$u = e^x \qquad \text{and} \quad dv = x\, dx$$

Then

$$du = e^x\, dx \quad \text{and} \quad v = \tfrac{1}{2}x^2$$

Thus

$$\int xe^x\, dx = \frac{1}{2}\, x^2 e^x - \frac{1}{2}\int x^2 e^x\, dx$$

The integral on the right is more complicated than the integral with which we started, thereby indicating that these are not desirable choices for u and dv. $\qquad \bullet$

EXAMPLE 1   Find $\displaystyle\int x \ln x\, dx$.

Solution   Let $u = \ln x$ and $dv = x\, dx$. Then

$$du = \frac{dx}{x} \qquad v = \frac{x^2}{2}$$

So

$$\int x \ln x \, dx = \frac{x^2}{2} \ln x - \int \frac{x^2}{2} \cdot \frac{dx}{x}$$

$$= \frac{x^2}{2} \ln x - \frac{1}{2} \int x \, dx$$

$$= \tfrac{1}{2}x^2 \ln x - \tfrac{1}{4}x^2 + C$$

It may happen that a particular integral requires repeated applications of integration by parts. This is illustrated in the following example.

EXAMPLE 2 Find $\int x^2 e^x \, dx$

Solution Let $u = x^2$ and $dv = e^x \, dx$. Then

$$du = 2x \, dx \quad \text{and} \quad v = e^x$$

We have, then,

$$\int x^2 e^x \, dx = x^2 e^x - 2 \int x e^x \, dx$$

We now apply integration by parts to the integral on the right. Let

$$\bar{u} = x \quad \text{and} \quad d\bar{v} = e^x \, dx$$

Then

$$d\bar{u} = dx \quad \text{and} \quad \bar{v} = e^x$$

So we obtain

$$\int x e^x \, dx = x e^x - \int e^x \, dx$$

$$= x e^x - e^x + \bar{C}$$

Therefore

$$\int x^2 e^x \, dx = x^2 e^x - 2(x e^x - e^x + \bar{C})$$

$$= x^2 e^x - 2x e^x + 2e^x + C$$

Integration by parts is often used when the integrand contains logarithms or products of two types of functions, such as in Example 2, which involves the product of a polynomial function and an exponential function. In the next example the integrand is the product of a polynomial function and a logarithmic function.

EXAMPLE 3 Find a formula for $\int x^r \ln x \, dx$ if r is any real number.

Solution We distinguish two cases: $r \neq -1$ and $r = -1$.

Case 1: $r \neq -1$. Let $u = \ln x$ and $dv = x^r \, dx$. Then

$$du = \frac{1}{x} \, dx \quad \text{and} \quad v = \frac{x^{r+1}}{r+1}$$

Therefore

$$\int x^r \ln x \, dx = \frac{x^{r+1}}{r+1} \ln x - \frac{1}{r+1} \int x^r \, dx$$

$$= \frac{x^{r+1}}{r+1} \ln x - \frac{x^{r+1}}{(r+1)^2} + C$$

Case 2: $r = -1$. The integral becomes

$$\int \frac{\ln x}{x} \, dx = \int w \, dw \qquad \text{where } w = \ln x$$

Hence we obtain

$$\int \frac{\ln x}{x} \, dx = \tfrac{1}{2} w^2 + C$$

$$= \tfrac{1}{2} (\ln x)^2 + C$$

Combining Cases 1 and 2 we get the formula

$$\int x^r \ln x \, dx = \begin{cases} \dfrac{x^{r+1}}{r+1} \ln x - \dfrac{x^{r+1}}{(r+1)^2} + C & \text{if } r \neq -1 \\[2ex] \tfrac{1}{2} (\ln x)^2 + C & \text{if } r = -1 \end{cases}$$

EXAMPLE 4 An investment made now will produce income continuously of $200t$ dollars per year, where t is the number of years from the present. If the income is deposited in a bank account with interest at 8 percent compounded continuously, find to the nearest dollar the amount in the account at the end of 6 years.

Solution We apply formula (9) of Sec. 8.8 with $f(t) = 200t$, $i = 0.08$, and $T = 6$. Therefore, if A dollars is the amount in the account at the end of 6 years,

$$A = 200 \int_0^6 t e^{0.08(6-t)} \, dt$$

To evaluate the integral we use integration by parts, with

$$u = t \qquad dv = e^{0.08(6-t)}$$

$$du = dt \qquad v = -\frac{e^{0.08(6-t)}}{0.08}$$

Therefore

$$A = 200\left[-\frac{e^{0.08(6-t)}}{0.08} t \right]_0^6 + \frac{1}{0.08} \int_0^6 e^{0.08(6-t)} \, dt$$

$$= 200\left[-\frac{e^{0.08(6-t)}}{0.08} t - \frac{e^{0.08(6-t)}}{0.0064} \right]_0^6$$

$$= 200\left(-\frac{6}{0.08} - \frac{1}{0.0064} + \frac{e^{0.48}}{0.0064} \right)$$

$$= \frac{200}{0.0064} (-0.48 - 1 + 1.6161)$$

$$= 31,250(0.1361)$$

$$= 4253$$

Thus the amount in the account at the end of 6 years is $4253.

Exercises 9.1

In Exercises 1 through 24, evaluate the indefinite integral.

1. $\displaystyle\int xe^{3x} \, dx$

2. $\displaystyle\int xe^{-2x} \, dx$

3. $\displaystyle\int x3^x \, dx$

4. $\displaystyle\int x10^{x/2} \, dx$

5. $\displaystyle\int \ln x \, dx$

6. $\displaystyle\int \log_{10} x \, dx$

7. $\displaystyle\int x \log_{10} x \, dx$

8. $\displaystyle\int \ln 2x^2 \, dx$

9. $\displaystyle\int x^2 \ln x \, dx$

10. $\displaystyle\int x^2 \log_2 x \, dx$

11. $\displaystyle\int (\ln x)^2 \, dx$

12. $\displaystyle\int (\ln x)^3 \, dx$

13. $\displaystyle\int xa^x \, dx$

14. $\displaystyle\int x \log_a x \, dx$

15. $\displaystyle\int x^2 e^{-2x} \, dx$

16. $\displaystyle\int x^3 e^x \, dx$

17. $\displaystyle\int \frac{xe^x}{(x+1)^2} \, dx$

18. $\displaystyle\int \frac{\ln(x+1)}{\sqrt{x+1}} \, dx$

19. $\displaystyle\int \frac{\ln x}{x^2} \, dx$

20. $\displaystyle\int \frac{\ln x}{x^3} \, dx$

21. $\displaystyle\int \frac{x^3 \, dx}{\sqrt{1-x^2}}$

22. $\displaystyle\int x^3 \sqrt{1-x^2} \, dx$

23. $\displaystyle\int e^{3\sqrt{x}} \, dx$

24. $\displaystyle\int (2^x + x)^2 \, dx$

In Exercises 25 through 28, evaluate the definite integral.

25. $\displaystyle\int_0^2 x^2 3^x \, dx$

26. $\displaystyle\int_0^1 x^2 e^{-2x} \, dx$

27. $\displaystyle\int_{-1}^2 \ln(x+2) \, dx$

28. $\displaystyle\int_1^3 x^2 (\ln x)^2 \, dx$

29. Find the area of the region bounded by the curve $y = \ln x$, the x axis, and the line $x = e^2$.

30. Find the area of the region bounded by the curve $y = 2xe^{-x/2}$, the x axis, and the line $x = 4$.

31. During the morning after arriving at work at 8 A.M., an employee can complete $f(x)$ units per hour after being on the job for x hours, where

$$f(x) = 50xe^{-0.4x}$$

How many units does the worker complete by 11 A.M.?

32. The marginal cost function is C', and $C'(x) = \ln x$, where $x \geq 1$. Find the total cost function if $C(x)$ dollars is the total cost of producing x units and $C(1) = 5$.

33. For a certain commodity the demand equation is $p = \ln(10 - x)$, and the supply equation is $p = \ln(2x + 1)$. The market price is the equilibrium price at which x units are demanded and supplied when p dollars is the price per unit. Determine (a) the consumers' surplus and (b) the producers' surplus. Draw a sketch showing each of the regions whose areas determine the consumers' surplus and the producers' surplus.

34. The supply equation for a particular article of merchandise is $p = 2\ln(x + 2)$, where x units are supplied when p dollars is the price per unit. If the market price is \$4, determine the producers' surplus.

35. Find to the nearest dollar the amount in 5 years of a continuous investment of $100t$ dollars per year, where t is the number of years from the present, if the interest rate is 12 percent compounded continuously. Use formula (9) of Sec. 8.8.

36. A continuous flow of income of $500t$ dollars per year is produced from an investment made now, where t is the number of years from the present. If the income for 5 years is to be deposited into a bank account with interest at 10 percent compounded continuously, find to the nearest dollar the present value of this income. Use formula (11) of Sec. 8.8.

37. The income per year at t years of a continuous flow of income is $1000t(2^{-t})$ dollars. Find to the nearest dollar the present value of this income over a period of 5 years if the interest rate is 8 percent compounded continuously. Use formula (11) of Sec. 8.8.

38. (a) Derive the following formula where r is any real number:

$$\int x^r e^x \, dx = x^r e^x - r \int x^{r-1} e^x \, dx$$

(b) Use the formula derived in (a) to find $\int x^4 e^x \, dx$.

39. For a certain kind of bathing suit the probability density function that a particular store will sell $1000x$ bathing suits during the month of July is given by

$$f(x) = \tfrac{1}{9} x e^{-x/3} \qquad x \geq 0$$

(a) Find the probability that the store will sell no more than 5000 bathing suits in July. (b) What is the probability that the store will sell between 9000 and 12,000 bathing suits in July?

40. (a) Derive the following formula, where r and q are any real numbers:

$$\int x^r (\ln x)^q \, dx = \begin{cases} \dfrac{x^{r+1}(\ln x)^q}{r+1} - \dfrac{q}{r+1} \displaystyle\int x^r (\ln x)^{q-1} \, dx & \text{if } r \neq -1 \\[3ex] \dfrac{(\ln x)^{q+1}}{q+1} + C & \text{if } r = -1 \end{cases}$$

(b) Use the formula derived in (a) to find $\int x^4 (\ln x)^2 \, dx$.

9.2 Integration of rational functions by partial fractions

In Sec. 8.2 we had the mathematical model given by the function defined by

$$f(t) = \frac{A}{1 + Be^{-Akt}} \qquad t \geq 0 \tag{1}$$

where A, B, and k are positive constants. The graph of this function is a curve of logistic growth and is shown in Fig. 8.2.10 and repeated in this section in Fig. 9.2.1.

The function in (1) is a model of the growth of a population when the environment imposes an upper bound on its size. It is also a model describing the spread of a disease or a rumor as well as the distribution of information. The model arises when a quantity grows at a rate that is jointly proportional to the amount present and the difference between a fixed number A and the amount present. Thus if time is represented by t units, and if y units is the amount of the quantity present at any time,

$$\frac{dy}{dt} = ky(A - y) \tag{2}$$

If the variables are separated in this differential equation, we obtain

$$\frac{dy}{y(A - y)} = k \, dt \tag{3}$$

To integrate the expression on the left-hand side of (3) requires the use of partial fractions. We shall now develop the process of integration by partial fractions and then in Illustration 2 show that differential equation (2) has the function in (1) as its general solution.

In Sec. 1.4 a rational function was defined as one that can be expressed as the quotient of two polynomial functions. That is, the function H is a rational function if $H(x) = P(x)/Q(x)$, where $P(x)$ and $Q(x)$ are polynomials. We saw previously that if the degree of the numerator is not less than the degree of the denominator, we have an improper fraction, and in that case we divide the numerator by the denominator until we obtain a proper fraction, one in which the degree of the numerator is less than the degree of the denominator. For example,

$$\frac{x^4 - 10x^2 + 3x + 1}{x^2 - 4} = x^2 - 6 + \frac{3x - 23}{x^2 - 4}$$

So if we wish to integrate

$$\int \frac{x^4 - 10x^2 + 3x + 1}{x^2 - 4} \, dx$$

the problem is reduced to integrating

$$\int (x^2 - 6) \, dx + \int \frac{3x - 23}{x^2 - 4} \, dx$$

In general, then, we are concerned with the integration of expressions of the form

$$\int \frac{P(x)}{Q(x)} \, dx$$

where the degree of $P(x)$ is less than the degree of $Q(x)$.

To do this it is often necessary to write $P(x)/Q(x)$ as the sum of **partial fractions**. The denominators of the partial fractions are obtained by factoring $Q(x)$ into a product of linear and quadratic factors. Sometimes it may be difficult to find these factors of $Q(x)$; however, a theorem from advanced algebra states that theoretically this can always be done.

After $Q(x)$ has been factored into a product of linear and quadratic factors, the method of determining the partial fractions depends on the nature of these factors. There are various cases, but we shall consider here only the two in which the factors of $Q(x)$ are linear.

Case 1: The factors of $Q(x)$ are all linear, and none is repeated. That is,

$$Q(x) = (a_1x + b_1)(a_2x + b_2) \ldots (a_nx + b_n)$$

where no two of the factors are identical. In this case we write

$$\frac{P(x)}{Q(x)} \equiv \frac{A_1}{a_1x + b_1} + \frac{A_2}{a_2x + b_2} + \ldots + \frac{A_n}{a_nx + b_n} \tag{4}$$

where $A_1, A_2, \ldots, A_n$ are constants to be determined.

Note that we used $\equiv$ (read as "identically equal") instead of $=$ in (4). This is because (1) is an identity in x.

The following illustration shows how the values of A_i are found.

● ILLUSTRATION 1

To evaluate

$$\int \frac{(x-1)\,dx}{x^3 - x^2 - 2x}$$

we factor the denominator and have

$$\frac{x-1}{x^3 - x^2 - 2x} \equiv \frac{x-1}{x(x-2)(x+1)}$$

So

$$\frac{x-1}{x(x-2)(x+1)} \equiv \frac{A}{x} + \frac{B}{x-2} + \frac{C}{x+1} \tag{5}$$

Equation (5) is an identity for all x (except $x = 0, 2, -1$). From (5),

$$x - 1 \equiv A(x-2)(x+1) + Bx(x+1) + Cx(x-2) \tag{6}$$

Equation (6) is an identity that is true for all values of x, including 0, 2, and -1. We wish to find the constants A, B, and C. Substituting 0 for x in (6) we obtain

$$-1 = -2A \quad \text{or} \quad A = \tfrac{1}{2}$$

Substituting 2 for x in (6) we get

$$1 = 6B \quad \text{or} \quad B = \tfrac{1}{6}$$

Substituting -1 for x in (6) we obtain

$$-2 = 3C \quad \text{or} \quad C = -\tfrac{2}{3}$$

There is another method for finding the values of A, B, and C. If on the right side of (6) we combine terms,

$$x - 1 \equiv (A + B + C)x^2 + (-A + B - 2C)x - 2A \tag{7}$$

For (7) to be an identity, the coefficients on the left must equal the corresponding coefficients on the right. Hence

$$A + B + C = 0$$
$$-A + B - 2C = 1$$
$$-2A = -1$$

Solving these equations simultaneously we get $A = \frac{1}{2}$, $B = \frac{1}{6}$, and $C = -\frac{2}{3}$. Substituting these values in (5) we get

$$\frac{x-1}{x(x-2)(x+1)} \equiv \frac{\frac{1}{2}}{x} + \frac{\frac{1}{6}}{x-2} + \frac{-\frac{2}{3}}{x+1}$$

So the given integral can be expressed as follows:

$$\int \frac{x-1}{x^3 - x^2 - 2x}\, dx = \frac{1}{2}\int \frac{dx}{x} + \frac{1}{6}\int \frac{dx}{x-2} - \frac{2}{3}\int \frac{dx}{x+1}$$
$$= \frac{1}{2}\ln|x| + \frac{1}{6}\ln|x-2| - \frac{2}{3}\ln|x+1| + \frac{1}{6}\ln C$$
$$= \frac{1}{6}(3\ln|x| + \ln|x-2| - 4\ln|x+1| + \ln C)$$
$$= \frac{1}{6}\ln\left|\frac{Cx^3(x-2)}{(x+1)^4}\right|$$

Case 2: The factors of $Q(x)$ are all linear, and some are repeated.

Suppose that $(a_i x + b_i)$ is a p-fold factor. Then, corresponding to this factor there will be the sum of p partial fractions

$$\frac{A_1}{(a_i x + b_i)^p} + \frac{A_2}{(a_i x + b_i)^{p-1}} + \cdots + \frac{A_{p-1}}{(a_i x + b_i)^2} + \frac{A_p}{a_i x + b_i}$$

where $A_1, A_2, \ldots, A_p$ are constants to be determined.

Example 1 following illustrates this case and the method of determining each A_i.

EXAMPLE 1 Find

$$\int \frac{(x^3 - 1)\, dx}{x^2(x-2)^3}$$

Solution The fraction in the integrand is written as a sum of partial fractions as follows:

$$\frac{x^3 - 1}{x^2(x-2)^3} \equiv \frac{A}{x^2} + \frac{B}{x} + \frac{C}{(x-2)^3} + \frac{D}{(x-2)^2} + \frac{E}{x-2} \tag{8}$$

The above is an identity for all x except $x = 0, 2$. Multiplying on both sides of (8) by the lowest common denominator we get

$$x^3 - 1 \equiv A(x-2)^3 + Bx(x-2)^3 + Cx^2 + Dx^2(x-2) + Ex^2(x-2)^2 \quad (9)$$

We substitute 2 for x in (9) and obtain

$$7 = 4C \quad \text{or} \quad C = \tfrac{7}{4}$$

Substituting 0 for x in (9) we get

$$-1 = -8A \quad \text{or} \quad A = \tfrac{1}{8}$$

We substitute these values for A and C in (9) and expand the powers of the binomials, and we have

$$x^3 - 1 \equiv \tfrac{1}{8}(x^3 - 6x^2 + 12x - 8) + Bx(x^3 - 6x^2 + 12x - 8) + \tfrac{7}{4}x^2$$
$$+ Dx^3 - 2Dx^2 + Ex^2(x^2 - 4x + 4)$$

$$x^3 - 1 \equiv (B + E)x^4 + (\tfrac{1}{8} - 6B + D - 4E)x^3$$
$$+ (-\tfrac{3}{4} + 12B + \tfrac{7}{4} - 2D + 4E)x^2 + (\tfrac{3}{2} - 8B)x - 1$$

Equating the coefficients of like powers of x we obtain

$$B + E = 0$$
$$\tfrac{1}{8} - 6B + D - 4E = 1$$
$$-\tfrac{3}{4} + 12B + \tfrac{7}{4} - 2D + 4E = 0$$
$$\tfrac{3}{2} - 8B = 0$$

Solving, we get

$$B = \tfrac{3}{16} \qquad D = \tfrac{5}{4} \qquad E = -\tfrac{3}{16}$$

Therefore, from (8),

$$\frac{x^3 - 1}{x^2(x - 2)^3} \equiv \frac{\tfrac{1}{8}}{x^2} + \frac{\tfrac{3}{16}}{x} + \frac{\tfrac{7}{4}}{(x - 2)^3} + \frac{\tfrac{5}{4}}{(x - 2)^2} + \frac{-\tfrac{3}{16}}{x - 2}$$

Thus

$$\int \frac{x^3 - 1}{x^2(x - 2)^3}\, dx$$

$$= \frac{1}{8} \int \frac{dx}{x^2} + \frac{3}{16} \int \frac{dx}{x} + \frac{7}{4} \int \frac{dx}{(x - 2)^3} + \frac{5}{4} \int \frac{dx}{(x - 2)^2} - \frac{3}{16} \int \frac{dx}{x - 2}$$

$$= -\frac{1}{8x} + \frac{3}{16} \ln |x| - \frac{7}{8(x - 2)^2} - \frac{5}{4(x - 2)} - \frac{3}{16} \ln |x - 2| + C$$

$$= \frac{-11x^2 + 17x - 4}{8x(x - 2)^2} + \frac{3}{16} \ln \left| \frac{x}{x - 2} \right| + C$$

EXAMPLE 2 Find

$$\int \frac{du}{u^2 - a^2}$$

Solution

$$\frac{1}{u^2 - a^2} \equiv \frac{A}{u - a} + \frac{B}{u + a}$$

Multiplying by $(u - a)(u + a)$ we get

$$1 \equiv A(u + a) + B(u - a)$$

$$1 \equiv (A + B)u + Aa - Ba$$

Equating coefficients we have

$$A + B = 0$$

$$Aa - Ba = 1$$

Solving simultaneously we get

$$A = \frac{1}{2a} \quad \text{and} \quad B = -\frac{1}{2a}$$

Therefore

$$\int \frac{du}{u^2 - a^2} = \frac{1}{2a} \int \frac{du}{u - a} - \frac{1}{2a} \int \frac{du}{u + a}$$

$$= \frac{1}{2a} \ln |u - a| - \frac{1}{2a} \ln |u + a| + C$$

or, equivalently,

$$\int \frac{du}{u^2 - a^2} = \frac{1}{2a} \ln \left| \frac{u - a}{u + a} \right| + C$$

The type of integral of the above example occurs frequently enough for it to be listed as a formula.

$$\int \frac{du}{u^2 - a^2} = \frac{1}{2a} \ln \left| \frac{u - a}{u + a} \right| + C \tag{10}$$

Furthermore

$$\int \frac{du}{a^2 - u^2} = -\int \frac{du}{u^2 - a^2}$$

$$= -\frac{1}{2a} \ln \left| \frac{u - a}{u + a} \right| + C$$

$$= \frac{1}{2a} \ln \left| \frac{u + a}{u - a} \right| + C$$

This is also listed as a formula.

$$\int \frac{du}{a^2 - u^2} = \frac{1}{2a} \ln \left| \frac{u + a}{u - a} \right| + C \tag{11}$$

We now return to the discussion started at the beginning of this section and show that differential equation (2) describes logistic growth.

● ILLUSTRATION 2

We have the differential equation

$$\frac{dy}{dt} = ky(A - y)$$

Separating the variables and integrating gives

$$\int \frac{dy}{y(A - y)} = k \int dt \tag{12}$$

To evaluate the integral on the left we use partial fractions.

$$\frac{1}{y(A - y)} \equiv \frac{D}{y} + \frac{E}{A - y}$$

$$1 \equiv D(A - y) + Ey \tag{13}$$

Substituting 0 for y in (13) we get

$$1 = DA \quad \text{or} \quad D = \frac{1}{A}$$

We substitute A for y in (13) and obtain

$$1 = EA \quad \text{or} \quad E = \frac{1}{A}$$

Therefore

$$\frac{1}{y(A - y)} = \frac{\frac{1}{A}}{y} + \frac{\frac{1}{A}}{A - y}$$

Thus

$$\int \frac{dy}{y(A - y)} = \frac{1}{A} \int \frac{dy}{y} + \frac{1}{A} \int \frac{dy}{A - y}$$

$$= \frac{1}{A} \ln |y| - \frac{1}{A} \ln |A - y| + C$$

$$= \frac{1}{A} \ln \left| \frac{y}{A - y} \right| + C \tag{14}$$

Substituting from (14) in (12) we have

$$\frac{1}{A} \ln \left| \frac{y}{A - y} \right| + C = kt$$

$$\ln \left| \frac{y}{A - y} \right| = Akt - AC$$

$$\frac{y}{A - y} = e^{-AC} e^{Akt}$$

$$y = e^{-AC} e^{Akt} (A - y)$$

$$y = Ae^{-AC} e^{Akt} - ye^{-AC} e^{Akt}$$

$$y + ye^{-AC} e^{Akt} = Ae^{-AC} e^{Akt}$$

$$y(1 + e^{-AC} e^{Akt}) = Ae^{-AC} e^{Akt}$$

$$y = \frac{Ae^{-AC} e^{Akt}}{1 + e^{-AC} e^{Akt}}$$

We divide numerator and denominator by $e^{-AC} e^{Akt}$ to obtain

$$y = \frac{A}{e^{AC} e^{-Akt} + 1}$$

We let $e^{AC} = B$, and because $e^{AC} > 0$, then B is positive. Thus if $y = f(t)$, we have

$$f(t) = \frac{A}{1 + Be^{-Akt}}$$

which is the function in (1). ●

In Fig. 9.2.1 we have a curve of logistic growth, which is the graph of (1). The point of inflection of this graph occurs where $f(t) = \frac{1}{2}A$. You are asked to show this in Exercise 24. Thus if (1) is a model for the spread of a rumor, where A is the total number of people who will eventually hear the rumor, then the rumor is spreading most rapidly when $\frac{1}{2}A$ people have heard it. Similarly, if (1) is a model for an epidemic, then the disease is spreading fastest when half of those people susceptible to the disease have contracted it.

EXAMPLE 3   A particular pond can support up to 1200 fish so that the quantity of fish in the pond grows at a rate that is jointly proportional to the number present and the difference between 1200 and the number present. Ten weeks ago the pond contained 60 fish and it now contains 400 fish. When the pond contains 600 fish, the growth rate is greatest. When will this situation occur?

Solution   Let y be the number of fish in the pond at t weeks, where time is measured from 10 weeks ago. Thus when $t = 0$, $y = 60$, and when $t = 10$, $y = 400$; let $t = \bar{t}$ when $y = 600$. These sets of values appear in Table 9.2.1. The differential equation is

$$\frac{dy}{dt} = ky(1200 - y)$$

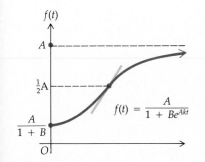

Figure 9.2.1

Table 9.2.1

t	0	10	$\bar{t}$
y	60	400	600

As in Illustration 2, the general solution to this differential equation is

$$y = \frac{1200}{1 + Be^{-1200kt}} \tag{15}$$

Because $y = 60$ when $t = 0$, we have from (15)

$$60 = \frac{1200}{1 + B}$$

$$1 + B = 20$$

$$B = 19$$

Setting $B = 19$ in (15) we obtain

$$y = \frac{1200}{1 + 19e^{-1200kt}} \tag{16}$$

In (16), substitute 10 for t and 400 for y to get

$$400 = \frac{1200}{1 + 19e^{-12,000k}}$$

$$1 + 19e^{-12,000k} = 3$$

$$e^{-12,000k} = \tfrac{2}{19}$$

$$-12,000k = \ln 2 - \ln 19$$

$$-12,000k = 0.6931 - 2.9445$$

$$-12,000k = -2.2514$$

$$1200k = 0.22514$$

Substituting 0.22514 for $1200k$ in (16) we have

$$y = \frac{1200}{1 + 19e^{-0.22514t}}$$

In this equation let $y = 600$ and $t = \bar{t}$, and we obtain

$$600 = \frac{1200}{1 + 19e^{-0.22514\bar{t}}}$$

$$1 + 19e^{-0.22514\bar{t}} = 2$$

$$e^{-0.22514\bar{t}} = \frac{1}{19}$$

$$-0.22514\bar{t} = \ln 1 - \ln 19$$

$$-0.22514\bar{t} = -2.9445$$

$$\bar{t} = \frac{2.9445}{0.22514}$$

$$\bar{t} = 13.08$$

Because time is measured from 10 weeks ago, we conclude that the pond will contain 600 fish 3 weeks from now.

EXAMPLE 4 A company that began transacting business on April 1, 1983, has estimated that during its first 6 years of operation the total income from sales will increase at the rate of

$$\frac{t^3 + t^2 + 3t + 1}{t^2 + t}$$

millions of dollars, where t is the number of years that the company has been doing business. If the total income from sales for the year ending March 31, 1984, was \$8 million, what is the total income from sales expected for the year ending March 31, 1988?

Table 9.2.2

t	1	5
S	8	S_5

Solution Let S millions of dollars be the total income from sales for the year ending t years from April 1, 1983. We have the initial conditions given in Table 9.2.2. We also have the differential equation

$$\frac{dS}{dt} = \frac{t^3 + t^2 + 3t + 1}{t^2 + t}$$

from which we obtain

$$S = \int \frac{t^3 + t^2 + 3t + 1}{t^2 + t}\, dt \qquad (17)$$

Observe that the integrand is an improper fraction. We divide the numerator by the denominator.

$$\frac{t^3 + t^2 + 3t + 1}{t^2 + t} \equiv t + \frac{3t + 1}{t^2 + t} \qquad (18)$$

Substituting from (18) into (17) we have

$$S = \int t\, dt + \int \frac{3t + 1}{t^2 + t}\, dt \qquad (19)$$

To evaluate the second integral on the right-hand side of (19), factor the denominator and then write the fraction as a sum of partial fractions.

$$\frac{3t + 1}{t(t + 1)} \equiv \frac{A}{t} + \frac{B}{t + 1}$$

Therefore

$$3t + 1 \equiv A(t + 1) + Bt$$

$$3t + 1 \equiv (A + B)t + A$$

Equating coefficients we have

$$A + B = 3 \quad \text{and} \quad A = 1$$

Therefore

$$A = 1 \quad \text{and} \quad B = 2$$

Hence

$$\frac{3t + 1}{t(t + 1)} \equiv \frac{1}{t} + \frac{2}{t + 1} \tag{20}$$

We substitute from (20) into (19) and obtain

$$S = \int t \, dt + \int \frac{dt}{t} + 2 \int \frac{dt}{t + 1}$$
$$S = \tfrac{1}{2}t^2 + \ln |t| + 2 \ln |t + 1| + C$$
$$S = \tfrac{1}{2}t^2 + \ln |t(t + 1)^2| + C \tag{21}$$

Because $S = 8$ when $t = 1$, we substitute these values in (21), and we have

$$8 = \tfrac{1}{2} + \ln 4 + C$$
$$C = 7.5 - \ln 4$$

In (21) we replace C by its value and get

$$S = \tfrac{1}{2}t^2 + \ln |t(t + 1)^2| + 7.5 - \ln 4$$

Because $S = S_5$ when $t = 5$,

$$S_5 = \tfrac{1}{2}(25) + \ln(5 \cdot 36) + 7.5 - \ln 4$$
$$= 20 + \ln \frac{5 \cdot 36}{4}$$
$$= 20 + \ln 45$$
$$= 20 + 3.8067$$
$$= 23.8067$$

Therefore the total income from sales expected for the year ending March 31, 1988, is $23,806,700.

In chemistry, the **law of mass action** affords an application of integration that leads to the use of partial fractions. Under certain conditions it is found that a substance A reacts with a substance B to form a third substance C in such a way that the rate of change of the amount of C is jointly proportional to the amounts of A and B remaining at any given time.

Suppose that initially there are α grams of A and β grams of B and that r grams of A combine with s grams of B to form $(r + s)$ grams of C. If x is the number of grams of substance C present at t units of time, then C contains $rx/(r + s)$ grams of A and $sx/(r + s)$ grams of B. The number of grams of substance A remaining is then $\alpha - rx/(r + s)$, and the number of grams of substance B remaining is $\beta - sx/(r + s)$. Therefore the law of mass action gives

$$\frac{dx}{dt} = K\left(\alpha - \frac{rx}{r + s}\right)\left(\beta - \frac{sx}{r + s}\right)$$

where K is the constant of proportionality. This equation can be written as

$$\frac{dx}{dt} = \frac{Krs}{(r + s)^2}\left(\frac{r + s}{r}\alpha - x\right)\left(\frac{r + s}{s}\beta - x\right) \tag{22}$$

Letting

$$k = \frac{Krs}{(r + s)^2} \qquad a = \frac{r + s}{r}\alpha \qquad b = \frac{r + s}{s}\beta$$

Eq. (22) becomes

$$\frac{dx}{dt} = k(a - x)(b - x) \tag{23}$$

We can separate the variables in (23) and get

$$\frac{dx}{(a - x)(b - x)} = k\,dt$$

If $a = b$, then the left side of the above equation can be integrated by the power formula. If $a \neq b$, partial fractions can be used for the integration.

EXAMPLE 5   A chemical reaction causes a substance A to combine with a substance B to form a substance C so that the law of mass action is obeyed. If in Eq. (23) $a = 8$ and $b = 6$, and 2 g of substance C is formed in 10 min, how many grams of C are formed in 15 min?

Table 9.2.3

t	0	10	15
x	0	2	x_{15}

Solution   Letting x grams be the amount of substance C present at t minutes, we have the initial conditions shown in Table 9.2.3. Equation (23) becomes

$$\frac{dx}{dt} = k(8 - x)(6 - x)$$

Separating the variables we have

$$\int \frac{dx}{(8 - x)(6 - x)} = k\int dt \tag{24}$$

Writing the integrand as the sum of partial fractions gives

$$\frac{1}{(8 - x)(6 - x)} \equiv \frac{A}{8 - x} + \frac{B}{6 - x}$$

$$1 \equiv A(6 - x) + B(8 - x)$$

Substituting 6 for x gives $B = \frac{1}{2}$, and substituting 8 for x gives $A = -\frac{1}{2}$. Hence (24) is written as

$$-\frac{1}{2} \int \frac{dx}{8-x} + \frac{1}{2} \int \frac{dx}{6-x} = k \int dt$$

Integrating, we have

$$\tfrac{1}{2} \ln |8-x| - \tfrac{1}{2} \ln |6-x| + \tfrac{1}{2} \ln |C| = kt$$

$$\ln \left| \frac{6-x}{C(8-x)} \right| = -2kt$$

$$\frac{6-x}{8-x} = Ce^{-2kt} \tag{25}$$

Substituting $x = 0$, $t = 0$ in (25) gives $C = \frac{3}{4}$. Hence

$$\frac{6-x}{8-x} = \frac{3}{4} e^{-2kt} \tag{26}$$

Substituting $x = 2$, $t = 10$ in (26) we have

$$\tfrac{4}{6} = \tfrac{3}{4} e^{-20k}$$

$$e^{-20k} = \tfrac{8}{9}$$

Substituting $x = x_{15}$, $t = 15$ into (26) we get

$$\frac{6 - x_{15}}{8 - x_{15}} = \frac{3}{4} e^{-30k}$$

$$4(6 - x_{15}) = 3(e^{-20k})^{3/2}(8 - x_{15})$$

$$24 - 4x_{15} = 3(\tfrac{8}{9})^{3/2}(8 - x_{15})$$

$$24 - 4x_{15} = \frac{16\sqrt{2}}{9}(8 - x_{15})$$

$$x_{15} = \frac{54 - 32\sqrt{2}}{9 - 4\sqrt{2}}$$

$$x_{15} = 2.6$$

Therefore 2.6 g of substance C will be formed in 15 min.

Exercises 9.2

In Exercises 1 through 16, evaluate the indefinite integral.

1. $\displaystyle\int \frac{dx}{x^2 - 4}$

2. $\displaystyle\int \frac{5x - 2}{x^2 - 4}\, dx$

3. $\displaystyle\int \frac{x^2\, dx}{x^2 + x - 6}$

4. $\displaystyle\int \frac{(4x - 2)\, dx}{x^3 - x^2 - 2x}$

5. $\displaystyle\int \frac{4w - 11}{2w^2 + 7w - 4}\, dw$

6. $\displaystyle\int \frac{9t^2 - 26t - 5}{3t^2 - 5t - 2}\, dt$

7. $\displaystyle\int \frac{6x^2 - 2x - 1}{4x^3 - x}\, dx$

8. $\displaystyle\int \frac{x^2 + x + 2}{x^2 - 1}\, dx$

9. $\displaystyle\int \frac{dx}{x^3 + 3x^2}$

10. $\displaystyle\int \frac{x^2 + 4x - 1}{x^3 - x} \, dx$

11. $\displaystyle\int \frac{dx}{x^2(x + 1)^2}$

12. $\displaystyle\int \frac{3x^2 - x + 1}{x^3 - x^2} \, dx$

13. $\displaystyle\int \frac{x^2 - 3x - 7}{(2x + 3)(x + 1)^2} \, dx$

14. $\displaystyle\int \frac{dt}{(t + 2)^2(t + 1)}$

15. $\displaystyle\int \frac{3z + 1}{(z^2 - 4)^2} \, dz$

16. $\displaystyle\int \frac{2x^4 - 2x + 1}{2x^5 - x^4} \, dx$

In Exercises 17 through 22, evaluate the definite integral.

17. $\displaystyle\int_1^2 \frac{x - 3}{x^3 + x^2} \, dx$

18. $\displaystyle\int_0^4 \frac{(x - 2) \, dx}{2x^2 + 7x + 3}$

19. $\displaystyle\int_1^3 \frac{x^2 - 4x + 3}{x(x + 1)^2} \, dx$

20. $\displaystyle\int_1^4 \frac{(2x^2 + 13x + 18) \, dx}{x^3 + 6x^2 + 9x}$

21. $\displaystyle\int_1^2 \frac{5x^2 - 3x + 18}{9x - x^3} \, dx$

22. $\displaystyle\int_0^1 \frac{(3x^2 + 7x) \, dx}{(x + 1)(x^2 + 5x + 6)}$

23. Find the area of the region in the first quadrant bounded by the curve $(x + 2)^2 y = 4 - x$.

24. The general solution of the differential equation $\dfrac{dy}{dt} = ky(A - y)$ is

$$y = \frac{A}{1 + Be^{-Akt}} \tag{27}$$

where A, B, and k are positive constants. Show that $\dfrac{dy}{dt}$ is a maximum when $y = \frac{1}{2}A$, and therefore the graph of (27) has a point of inflection where $y = \frac{1}{2}A$.

25. In a certain town there are 4800 people susceptible to a particular disease. The rate of growth of an epidemic of this disease is jointly proportional to the number of people infected and the number of susceptible people who are not infected. Initially 300 people are infected and 1200 are infected after 10 days. (a) How many people are infected after 20 days? (b) When will the disease be spreading fastest? That is, when will 2400 people be infected?

26. A certain environment can support up to 2,000,000 bacteria so that the rate of bacterial growth is jointly proportional to the number present and the difference between 2,000,000 and the number present. Twenty minutes ago there were 1000 bacteria present in the environment and now there are 2000 bacteria. (a) How many bacteria will be present 1 hour from now? (b) When will the bacterial growth be greatest? That is, when will there be 1,000,000 bacteria present?

27. In a community of 7500 people, the rate at which a rumor spreads is jointly proportional to the number of people who have heard the rumor and the number of people who have not heard it. Initially 300 people heard the rumor and 1500 had heard it after 2 hours. (a) How many people heard the rumor after 3 hours? (b) When will the rumor be spreading most rapidly? That is, when will 3750 people have heard it?

28. A manufacturer who began operations 4 years ago has determined that income from sales has increased steadily at the rate of

$$\frac{t^3 + 3t^2 + 6t + 7}{t^2 + 3t + 2}$$

millions of dollars per year, where t is the number of years that the company has been operating. It is estimated that the total income from sales will increase at the same rate for the next 2 years. If the total income from sales for the year just ended was $6 million, what is the total income from sales expected for the period ending 1 year from now? Give the answer to the nearest $100.

29. Suppose in Example 5 that $a = 5$ and $b = 4$ and 1 g of substance C is formed in 5 min. How many grams of C are formed in 10 min?

30. Suppose in Example 5 that $a = 6$ and $b = 3$ and 1 g of substance C is formed in 4 min. How long will it take for 2 g of substance C to be formed?

31. At any instant the rate at which a substance dissolves is jointly proportional to the amount of the substance present at that instant and the difference between the concentration of the substance in solution at that instant and the concentration of the substance in a saturated solution. A quantity of insoluble material is mixed with 10 lb of salt initially, and the salt is dissolving in a tank containing 20 gal of water. If 5 lb of salt dissolves in 10 min and the concentration of salt in a saturated solution is 3 lb/gal, how much salt will dissolve in 20 min?

9.3 Approximate integration

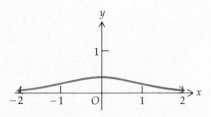

Figure 9.3.1

There is an important function in statistics called the **standardized normal probability density function** that is defined by

$$N(x) = \frac{1}{\sqrt{2\pi}} e^{-x^2/2}$$

A sketch of the graph of N is shown in Fig. 9.3.1. It is the bell-shaped curve well-known to statisticians. Observe that the curve is symmetric with respect to the y axis; that is, the portion of the curve for $x \leq 0$ is the mirror image with respect to the y axis of the portion of the curve for $x \geq 0$. Table 9.3.1 gives values of $N(x)$ for some values of x.

Table 9.3.1

x	0	± 0.5	± 1.0	± 1.5	± 2.0	± 2.5	± 3.0	± 3.5	± 4.0
$N(x)$	0.40	0.35	0.24	0.13	0.05	0.02	0.004	0.001	0.0001

Observe that as either x increases without bound or decreases without bound, $N(x)$ rapidly approaches 0. For instance, $N(3) = 0.004$ and $N(4) = 0.0001$. It can be proved that the area bounded by the curve $y = N(x)$ and the x axis on the interval $(-\infty, +\infty)$ is 1, although it is difficult to do so. To compute the probability $P([a, b])$ from this probability density function, it is necessary to evaluate the definite integral

$$\frac{1}{\sqrt{2\pi}} \int_a^b e^{-x^2/2} \, dx \tag{1}$$

which is the measure of the area of the region bounded above by the curve $y = N(x)$, below by the x axis, and on the sides by the lines $x = a$ and $x = b$. However, the definite integral in (1) cannot be evaluated by the fundamental theorem of the calculus because we cannot find an antiderivative of the integrand by using elementary functions. There are other definite integrals that cannot be evaluated by the fundamental theorem for the same reason. Examples of such integrals are $\int_0^1 \sqrt{1 + x^3} \, dx$ and the ones in Exercises 27 through 30. We can compute an approximate value of such a definite integral by one of two methods discussed in this section. These devices often give fairly good accuracy. We proceed now to develop these processes and in Example 5 at the end of this section we return to the standardized normal probability density function and compute an approximate value of a probability by each of the two methods.

The first process is called the **trapezoidal rule**. Let f be a function continuous on a closed interval $[a, b]$. The definite integral of f from a to b is the limit of a Riemann sum; that is,

$$\int_a^b f(x)\,dx = \lim_{n \to +\infty} \sum_{i=1}^{n} f(\xi_i)\,\Delta x$$

The geometric intepretation of the Riemann sum

$$\sum_{i=1}^{n} f(\xi_i)\,\Delta x$$

is that it is equal to the sum of the measures of the areas of the rectangles lying above the x axis plus the negatives of the measures of the areas of the rectangles lying below the x axis (see Fig. 7.3.2). To approximate the measure of the area of a region we shall use trapezoids instead of rectangles.

For the definite integral $\int_a^b f(x)\,dx$ we divide the interval $[a,b]$ into n subintervals, each of length $\Delta x = (b - a)/n$. This gives the following $(n + 1)$ points: $x_0 = a$, $x_1 = a + \Delta x$, $x_2 = a + 2\,\Delta x$, ..., $x_i = a + i\,\Delta x$, ..., $x_{n-1} = a + (n - 1)\,\Delta x$, $x_n = b$. Then the definite integral may be expressed as the sum of n definite integrals as follows:

$$\int_a^b f(x)\,dx = \int_a^{x_1} f(x)\,dx + \int_{x_1}^{x_2} f(x)\,dx + \ldots$$

$$+ \int_{x_{i-1}}^{x_i} f(x)\,dx + \ldots + \int_{x_{n-1}}^{b} f(x)\,dx \qquad (2)$$

To interpret (2) geometrically, refer to Fig. 9.3.2, in which $f(x) \geq 0$ for all x in $[a, b]$; however, (2) holds for any function that is continuous on $[a, b]$.

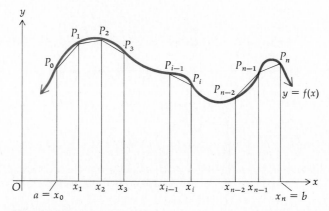

Figure 9.3.2

Then the integral $\int_a^{x_1} f(x)\, dx$ is the measure of the area of the region bounded by the x axis, the lines $x = a$ and $x = x_1$, and the portion of the curve from P_0 to P_1. This integral may be approximated by the measure of the area of the trapezoid formed by the lines $x = a$, $x = x_1$, $P_0 P_1$, and the x axis. By a formula from geometry the measure of the area of this trapezoid is

$$\tfrac{1}{2}[f(x_0) + f(x_1)]\, \Delta x$$

Similarly, the other integrals on the right side of (2) may be approximated by the measure of the area of a trapezoid. For the ith integral,

$$\int_{x_{i-1}}^{x_i} f(x)\, dx \approx \tfrac{1}{2}[f(x_{i-1}) + f(x_i)]\, \Delta x \tag{3}$$

So, using (3) for each of the integrals on the right side of (1) we have

$$\int_a^b f(x)\, dx \approx \tfrac{1}{2}[f(x_0) + f(x_1)]\, \Delta x + \tfrac{1}{2}[f(x_1) + f(x_2)]\, \Delta x + \dots$$

$$+ \tfrac{1}{2}[f(x_{n-2}) + f(x_{n-1})]\, \Delta x + \tfrac{1}{2}[f(x_{n-1}) + f(x_n)]\, \Delta x$$

Thus

$$\int_a^b f(x)\, dx \approx \tfrac{1}{2}\, \Delta x[f(x_0) + 2f(x_1) + 2f(x_2) + \dots + 2f(x_{n-1}) + f(x_n)] \tag{4}$$

Formula (4) is known as the *trapezoidal rule* and is stated formally in the following theorem.

The trapezoidal rule

> **Theorem 9.3.1**   If the function f is continuous on the closed interval $[a, b]$ and the numbers $a = x_0, x_1, x_2, \dots, x_n = b$ divide the interval $[a, b]$ into n subintervals of equal length, then
>
> $$\int_a^b f(x)\, dx \approx \frac{b - a}{2n}[f(x_0) + 2f(x_1) + 2f(x_2) + \dots + 2f(x_{n-1}) + f(x_n)]$$

EXAMPLE 1   Compute

$$\int_0^3 \frac{dx}{16 + x^2}$$

by using the trapezoidal rule with $n = 6$. Express the result to three decimal places.

Table 9.3.2

i	x_i	$f(x_i)$	k_i	$k_i \cdot f(x_i)$
0	0	0.0625	1	0.0625
1	0.5	0.0615	2	0.1230
2	1	0.0588	2	0.1176
3	1.5	0.0548	2	0.1096
4	2	0.0500	2	0.1000
5	2.5	0.0450	2	0.0900
6	3	0.0400	1	0.0400

$$\sum_{i=0}^{6} k_i f(x_i) = 0.6427$$

Solution   Because $[a, b] = [0, 3]$ and $n = 6$,

$$\Delta x = \frac{b-a}{n} = \frac{3}{6} = 0.5 \quad \text{and} \quad \frac{b-a}{2n} = 0.25$$

Therefore

$$\int_0^3 \frac{dx}{16 + x^2}$$

$$\approx 0.25 \left[f(x_0) + 2f(x_1) + 2f(x_2) + 2f(x_3) + 2f(x_4) + 2f(x_5) + f(x_6) \right]$$

where $f(x) = 1/(16 + x^2)$. The computation of the sum in brackets in the above is shown in Table 9.3.2. So

$$\int_0^3 \frac{dx}{16 + x^2} \approx 0.25(0.6427) = 0.161$$

To find the exact value of this definite integral requires the use of a function from trigonometry (the inverse tangent function). The exact value to four decimal places is 0.1609.

By increasing the value of n when using the trapezoidal rule, the better is the approximation of the definite integral. For large n it is desirable to use a computer to do the calculations. The following formula is equivalent to the one in Theorem 9.3.1 and is easier to program for a computer.

$$\int_a^b f(x)\,dx \approx \left(\frac{1}{2}[f(a) - f(b)] + \sum_{i=1}^{n} f(x_i) \right) \Delta x$$

To consider the accuracy of the approximation of a definite integral by the trapezoidal rule, we prove first that as n increases without bound, the limit of the approximation by the trapezoidal rule is the exact value of the definite integral. Let

$$T = \tfrac{1}{2} \Delta x [f(x_0) + 2f(x_1) + \cdots + 2f(x_{n-1}) + f(x_n)]$$

Then

$$T = [f(x_1) + f(x_2) + \cdots + f(x_n)]\,\Delta x + \tfrac{1}{2}[f(x_0) - f(x_n)]\,\Delta x$$

or, equivalently,

$$T = \sum_{i=1}^{n} f(x_i)\,\Delta x + \tfrac{1}{2}[f(a) - f(b)]\,\Delta x$$

Therefore, if $n \to +\infty$,

$$\lim_{n \to +\infty} T = \lim_{n \to +\infty} \sum_{i=1}^{n} f(x_i)\, \Delta x + \lim_{n \to +\infty} \tfrac{1}{2}[f(a) - f(b)]\left(\frac{b-a}{n}\right)$$

$$= \int_a^b f(x)\, dx + 0$$

Thus we can make the difference between T and the value of the definite integral as small as we please by taking n sufficiently large.

The following theorem, which is proved in numerical analysis, gives a method for estimating the error obtained when using the trapezoidal rule. The error is denoted by ε_T.

Theorem 9.3.2 Let the function f be continuous on the closed interval $[a, b]$, and f' and f'' both exist on $[a, b]$. If

$$\varepsilon_T = \int_a^b f(x)\, dx - T$$

where T is the approximate value of $\int_a^b f(x)\, dx$ found by the trapezoidal rule, then there is some number η in $[a, b]$ such that

$$\varepsilon_T = -\tfrac{1}{12}(b-a)f''(\eta)(\Delta x)^2 \qquad\qquad (5)$$

EXAMPLE 2 Find the bounds for the error in the result of Example 1.

Solution We first find the absolute minimum and absolute maximum values of $f''(x)$ on $[0, 3]$.

$$f(x) = (16 + x^2)^{-1}$$

$$f'(x) = -2x(16 + x^2)^{-2}$$

$$f''(x) = 8x^2(16 + x^2)^{-3} - 2(16 + x^2)^{-2} = (6x^2 - 32)(16 + x^2)^{-3}$$

$$f'''(x) = -6x(6x^2 - 32)(16 + x^2)^{-4} + 12x(16 + x^2)^{-3}$$

$$= 24x(16 - x^2)(16 + x^2)^{-4}$$

Because $f'''(x) > 0$ for all x in the open interval $(0, 3)$, then f'' is increasing on the open interval $(0, 3)$. Therefore the absolute minimum value of f'' on $[0, 3]$ is $f''(0)$, and the absolute maximum value of f'' on $[0, 3]$ is $f''(3)$.

$$f''(0) = -\frac{1}{128} \quad \text{and} \quad f''(3) = \frac{22}{15{,}625}$$

Taking $\eta = 0$ on the right side of (5) we get

$$-\frac{3}{12}\left(-\frac{1}{128}\right)\frac{1}{4} = \frac{1}{2048}$$

Taking $\eta = 3$ on the right side of (5) we have

$$-\frac{3}{12}\left(\frac{22}{15,625}\right)\frac{1}{4} = -\frac{11}{125,000}$$

Therefore if ε_T is the error in the result of Example 1.

$$-\frac{11}{125,000} \le \varepsilon_T \le \frac{1}{2048}$$

$$-0.0001 \le \varepsilon_T \le 0.0005$$

If in Theorem 9.3.2 $f(x) = mx + b$, then $f''(x) = 0$ for all x. Therefore $\varepsilon_T = 0$; so the trapezoidal rule gives the exact value of the definite integral of a linear function.

Another method for approximating the value of a definite integral is provided by **Simpson's rule**, sometimes referred to as the **parabolic rule**. For a given subdivision of the closed interval $[a, b]$, Simpson's rule usually gives a better approximation than the trapezoidal rule. In the trapezoidal rule, successive points on the graph of $y = f(x)$ are connected by segments of straight lines, whereas in Simpson's rule the points are connected by segments of parabolas. Before Simpson's rule is developed, we state a theorem that will be needed.

Theorem 9.3.3 If $P_0(x_0, y_0)$, $P_1(x_1, y_1)$, and $P_2(x_2, y_2)$ are three non-collinear points on the parabola having the equation $y = Ax^2 + Bx + C$, where $y_0 \ge 0$, $y_1 \ge 0$, $y_2 \ge 0$, $x_1 = x_0 + h$, and $x_2 = x_0 + 2h$, then the measure of the area of the region bounded by the parabola, the x axis, and the lines $x = x_0$ and $x = x_2$ is given by

$$\tfrac{1}{3}h(y_0 + 4y_1 + y_2) \tag{6}$$

The proof of Theorem 9.3.3 consists of showing that the area of the shaded region in Fig. 9.3.3 is given by (6). To do this it is necessary to find an equation of the parabola through the points P_0, P_1, and P_2. If this equation is $y = f(x)$, then the area is given by $\int_{x_0}^{x_0 + 2h} f(x)\,dx$. The details of the proof are omitted.

We now proceed to obtain Simpson's rule. Let the function f be continuous on the closed interval $[a, b]$. Consider a subdivision of the interval $[a, b]$ into $2n$ subintervals ($2n$ is used instead of n because we want an even number of subintervals). The length of each subinterval is given by $\Delta x = (b - a)/2n$. Let the points on the curve $y = f(x)$ having these partitioning

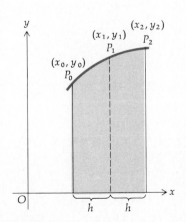

Figure 9.3.3

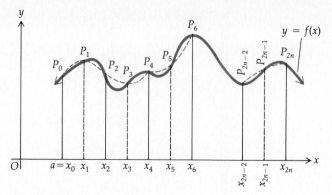

Figure 9.3.4

points as abscissas be denoted by $P_0(x_0, y_0)$, $P_1(x_1, y_1), \ldots, P_{2n}(x_{2n}, y_{2n})$; see Fig. 9.3.4, where $f(x) \geq 0$ for all x in $[a, b]$.

We approximate the segment of the curve $y = f(x)$ from P_0 to P_2 by the segment of the parabola with a vertical axis and through P_0, P_1, and P_2. Then, by Theorem 9.3.3, the measure of the area of the region bounded by this parabola, the x axis, and the lines $x = x_0$ and $x = x_2$, with $h = \Delta x$, is given by

$$\tfrac{1}{3} \Delta x (y_0 + 4y_1 + y_2) \quad \text{or} \quad \tfrac{1}{3} \Delta x [f(x_0) + 4f(x_1) + f(x_2)]$$

In a similar manner we approximate the segment of the curve $y = f(x)$ from P_2 to P_4 by the segment of the parabola with a vertical axis and through P_2, P_3, and P_4. The measure of the area of the region bounded by this parabola, the x axis, and the lines $x = x_2$ and $x = x_4$ is given by

$$\tfrac{1}{3} \Delta x (y_2 + 4y_3 + y_4) \quad \text{or} \quad \tfrac{1}{3} \Delta x [f(x_2) + 4f(x_3) + f(x_4)]$$

This process is continued until there are n such regions, and the measure of the area of the last region is given by

$$\tfrac{1}{3} \Delta x (y_{2n-2} + 4y_{2n-1} + y_{2n}) \quad \text{or} \quad \tfrac{1}{3} \Delta x [f(x_{2n-2}) + 4f(x_{2n-1}) + f(x_{2n})]$$

The sum of the measures of the areas of these regions approximates the measure of the area of the region bounded by the curve whose equation is $y = f(x)$, the x axis, and the lines $x = a$ and $x = b$. The measure of the area of this region is given by the definite integral $\int_a^b f(x)\, dx$. So we have as an approximation to the definite integral

$$\tfrac{1}{3} \Delta x [f(x_0) + 4f(x_1) + f(x_2)] + \tfrac{1}{3} \Delta x [f(x_2) + 4f(x_3) + f(x_4)] + \ldots$$
$$+ \tfrac{1}{3} \Delta x [f(x_{2n-4}) + 4f(x_{2n-3}) + f(x_{2n-2})] + \tfrac{1}{3} \Delta x [f(x_{2n-2}) + 4f(x_{2n-1}) + f(x_{2n})]$$

Thus

$$\int_a^b f(x)\, dx \approx \tfrac{1}{3} \Delta x [f(x_0) + 4f(x_1) + 2f(x_2) + 4f(x_3) + 2f(x_4) + \ldots$$
$$+ 2f(x_{2n-2}) + 4f(x_{2n-1}) + f(x_{2n})] \qquad (7)$$

where $\Delta x = (b - a)/2n$.

Formula (7) is known as *Simpson's rule*. It is stated formally in the next theorem.

Simpson's rule

Theorem 9.3.4 If the function f is continuous on the closed interval $[a, b]$, $2n$ is an even integer, and the numbers $a = x_0, x_1, x_2, \ldots, x_{2n-1}$, $x_{2n} = b$ divide the interval $[a, b]$ into n subintervals of equal length, then

$$\int_a^b f(x)\, dx \approx \frac{b-a}{6n}\left[f(x_0) + 4f(x_1) + 2f(x_2) + 4f(x_3) + 2f(x_4) + \ldots\right.$$

$$\left. + 2f(x_{2n-2}) + 4f(x_{2n-1}) + f(x_{2n})\right]$$

EXAMPLE 3 Use Simpson's rule to approximate the value of

$$\int_0^1 \frac{dx}{x+1}$$

with $2n = 4$. Give the result to four decimal places.

Solution Applying Simpson's rule with $2n = 4$ we have

$$\Delta x = \frac{b-a}{2n} = \frac{1}{4} \quad \text{and} \quad \frac{b-a}{6n} = \frac{1}{12}$$

Therefore, if $f(x) = 1/(x+1)$,

$$\int_0^1 \frac{dx}{x+1} \approx \tfrac{1}{12}\left[f(x_0) + 4f(x_1) + 2f(x_2) + 4f(x_3) + f(x_4)\right] \tag{8}$$

The computation of the expression in brackets on the right side of (8) is shown in Table 9.3.3.

Table 9.3.3

i	x_i	$f(x_i)$	k_i	$k_i \cdot f(x_i)$
0	0	1.00000	1	1.00000
1	0.25	0.80000	4	3.20000
2	0.5	0.66667	2	1.33334
3	0.75	0.57143	4	2.28572
4	1	0.50000	1	0.50000

$$\sum_{i=0}^{4} k_i f(x_i) = 8.31906$$

Substituting the sum from Table 9.3.3 in (8) we get

$$\int_0^1 \frac{dx}{x+1} \approx \frac{1}{12}(8.31906) = 0.69325^+$$

Rounding off the result to four decimal places gives

$$\int_0^1 \frac{dx}{x+1} \approx 0.6933$$

The exact value of this definite integral is found as follows:

$$\int_0^1 \frac{dx}{x+1} = \ln|x+1|\Big]_0^1 = \ln 2 - \ln 1 = \ln 2$$

From a table of natural logarithms the value of ln 2 to four decimal places is 0.6931, which agrees with our approximation in the first three places. And the error in our approximation is -0.0002.

In applying Simpson's rule, the larger we take the value of $2n$, the smaller will be the value of Δx; so geometrically it seems evident that the greater will be the accuracy of the approximation, because a parabola passing through three points of a curve that are close to each other will be close to the curve throughout the subinterval of width $2\,\Delta x$.

When n is large, you will want a computer to do the calculations. The following formula, which is equivalent to the one in Theorem 9.3.4, is easier to program for a computer.

$$\int_a^b f(x)\,dx \approx \left(\left[f(a) - f(b)\right] + \sum_{i=1}^{n}\left[4f(x_{2i-1}) + 2f(x_{2i})\right]\right)\frac{1}{3}\,\Delta x$$

The following theorem, which is proved in numerical analysis, gives a method for determining the error in applying Simpson's rule. The error is denoted by ε_S.

Theorem 9.3.5   Let the function f be continuous on the closed interval $[a, b]$, and f', f'', f''', and $f^{(iv)}$ all exist on $[a, b]$. If

$$\varepsilon_S = \int_a^b f(x)\,dx - S$$

where S is the approximate value of $\int_a^b f(x)\,dx$ found by Simpson's rule, then there is some number η in $[a, b]$ such that

$$\varepsilon_S = -\tfrac{1}{180}(b-a)f^{(iv)}(\eta)(\Delta x)^4 \tag{9}$$

EXAMPLE 4 Find the bounds for the error in Example 3.

Solution

$$f(x) = (x + 1)^{-1}$$
$$f'(x) = -1(x + 1)^{-2}$$
$$f''(x) = 2(x + 1)^{-3}$$
$$f'''(x) = -6(x + 1)^{-4}$$
$$f^{(iv)}(x) = 24(x + 1)^{-5}$$
$$f^{(v)}(x) = -120(x + 1)^{-6}$$

Because $f^{(v)}(x) < 0$ for all x in $[0, 1]$, $f^{(iv)}$ is decreasing on $[0, 1]$. Thus the absolute minimum value of $f^{(iv)}$ is at the right endpoint 1, and the absolute maximum value of $f^{(iv)}$ on $[0, 1]$ is at the left endpoint 0.

$$f^{(iv)}(0) = 24 \quad \text{and} \quad f^{(iv)}(1) = \tfrac{3}{4}$$

Substituting 0 for η in the right side of (9) we get

$$-\tfrac{1}{180}(b - a)f^{(iv)}(0)(\Delta x)^4 = -\tfrac{1}{180}(24)(\tfrac{1}{4})^4 = -\tfrac{1}{1920} = -0.00052$$

Substituting 1 for η in the right side of (9) we have

$$-\frac{1}{180}(b - a)f^{(iv)}(1)(\Delta x)^4 = -\frac{1}{180} \cdot \frac{3}{4}\left(\frac{1}{4}\right)^4 = -\frac{1}{61,440} = -0.00002$$

So

$$-0.00052 \le \varepsilon_S \le -0.00002 \tag{10}$$

The inequality (10) agrees with the discussion in Example 3 regarding the error in the approximation of the definite integral by Simpson's rule because $-0.00052 < -0.0002 < -0.00002$.

If $f(x)$ is a polynomial of degree three or less, then $f^{(iv)}(x) \equiv 0$ and therefore $\varepsilon_S = 0$. In other words, Simpson's rule gives an exact result for a polynomial of the third degree or lower. This statement is geometrically obvious if $f(x)$ is of the second or first degree because in the first case the graph of $y = f(x)$ is a parabola, and in the second case the graph is a straight line.

EXAMPLE 5 For the standardized normal probability density function, determine the probability that a random choice of x will be in the interval $[0, 2]$. Approximate the value of the definite integral by (a) the trapezoidal rule with $n = 4$ and (b) Simpson's rule with $2n = 4$.

Solution The standardized normal probability density function is given by

$$N(x) = \frac{1}{\sqrt{2\pi}}e^{-x^2/2}$$

The probability that a random choice of x will be in the interval $[0, 2]$ is $P([0, 2])$, and

$$P([0, 2]) = \frac{1}{\sqrt{2\pi}} \int_0^2 e^{-x^2/2} \, dx \tag{11}$$

(a) We approximate the integral in (11) by the trapezoidal rule with $n = 4$. Because $[a, b] = [0, 2]$, $\Delta x = (b - a)/n = (2 - 0)/4 = \frac{1}{2}$. Therefore, with $f(x) = e^{-x^2/2}$,

$$\int_0^2 e^{-x^2/2} \, dx \approx \tfrac{1}{4}[f(0) + 2f(\tfrac{1}{2}) + 2f(1) + 2f(\tfrac{3}{2}) + f(2)]$$

$$= \tfrac{1}{4}[e^0 + 2e^{-1/8} + 2e^{-1/2} + 2e^{-9/8} + e^{-2}]$$
$$= \tfrac{1}{4}[1 + 2(0.8825) + 2(0.6065) + 2(0.3246) + 0.1353]$$
$$= \tfrac{1}{4}(4.7625)$$
$$= 1.191$$

Thus

$$P([0, 2]) \approx \frac{1}{\sqrt{2\pi}}(1.191) = \frac{1.191}{2.507} = 0.475$$

(b) If Simpson's rule with $2n = 4$ is used to approximate the integral in (11), we have

$$\int_0^2 e^{-x^2/2} \, dx \approx \tfrac{1}{6}[f(0) + 4f(\tfrac{1}{2}) + 2f(1) + 4f(\tfrac{3}{2}) + f(2)]$$

$$= \tfrac{1}{6}[e^0 + 4e^{-1/8} + 2e^{-1/2} + 4e^{-9/8} + e^{-2}]$$
$$= \tfrac{1}{6}[1 + 4(0.8825) + 2(0.6065) + 4(0.3246) + 0.1353]$$
$$= \tfrac{1}{6}(7.1767)$$
$$= 1.196$$

Therefore

$$P([0, 2]) \approx \frac{1}{\sqrt{2\pi}}(1.196) = \frac{1.196}{2.507} = 0.477$$

The exact value of $P([0, 2])$ in Example 5 is the measure of the area of the shaded region in Fig. 9.3.5. This value is less than 0.5, because 0.5 is one-half of 1, which is the measure of the area of the region bounded by the curve and the x axis on the interval $(-\infty, +\infty)$.

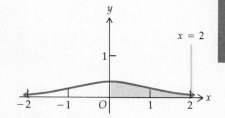

Figure 9.3.5

Exercises 9.3

In Exercises 1 through 10, compute the approximate value of the given definite integral by the trapezoidal rule for the indicated value of n. Express the result to three decimal places. In Exercises 1 through 4, find the exact value of the definite integral, and compare the result with the approximation.

1. $\displaystyle\int_1^2 \frac{dx}{x}$; $n = 5$

2. $\displaystyle\int_2^{10} \frac{dx}{1 + x}$; $n = 8$

3. $\displaystyle\int_0^2 x^3 \, dx$; $n = 4$

4. $\int_0^2 x\sqrt{4 - x^2}\, dx$; $n = 8$ **5.** $\int_0^2 \dfrac{dx}{1 + x^2}$; $n = 8$ **6.** $\int_0^1 \dfrac{dx}{\sqrt{1 + x^2}}$; $n = 5$

7. $\int_0^2 e^{-x^2}\, dx$; $n = 4$ **8.** $\int_2^3 \ln(1 + x^2)\, dx$; $n = 4$ **9.** $\int_0^2 \sqrt{1 + x^4}\, dx$; $n = 6$

10. $\int_0^1 \sqrt{1 + x^3}\, dx$; $n = 4$

In Exercises 11 through 16, find bounds for the error in the approximation of the indicated exercise.

11. Exercise 1   **12.** Exercise 2   **13.** Exercise 3
14. Exercise 4   **15.** Exercise 7   **16.** Exercise 8

In Exercises 17 through 22, approximate the definite integral by Simpson's rule, using the indicated value of $2n$. Express the answer to three decimal places. In Exercises 17 through 20, find the exact value of the definite integral, and compare the result with the approximation.

17. $\int_0^2 x^3\, dx$; $2n = 4$ **18.** $\int_0^2 x^2\, dx$; $2n = 4$ **19.** $\int_{-1}^0 \dfrac{dx}{1 - x}$; $2n = 4$

20. $\int_1^2 \dfrac{dx}{x + 1}$; $2n = 8$ **21.** $\int_0^2 \dfrac{dx}{1 + x^2}$; $2n = 8$ **22.** $\int_0^1 \dfrac{dx}{x^2 + x + 1}$; $2n = 4$

In Exercises 23 through 26, find bounds for the error in the approximation of the indicated exercise.

23. Exercise 17   **24.** Exercise 18   **25.** Exercise 19   **26.** Exercise 20

Each of the definite integrals in Exercises 27 through 30 cannot be evaluated exactly in terms of elementary functions. Use Simpson's rule, with the indicated value of $2n$, to find an approximate value of the definite integral. Express the result to three decimal places.

27. $\int_1^{1.8} \sqrt{1 + x^3}\, dx$; $2n = 4$ **28.** $\int_0^1 \sqrt[3]{1 - x^2}\, dx$; $2n = 4$

29. $\int_0^2 \dfrac{dx}{\sqrt{1 + x^3}}$; $2n = 8$ **30.** $\int_0^2 \sqrt{1 + x^4}\, dx$; $2n = 6$

31. Show that the exact value of the integral $\int_0^2 \sqrt{4 - x^2}\, dx$ is π by interpreting it as the measure of one-fourth the area of a circular region. Approximate the definite integral by the trapezoidal rule with $n = 8$. Give the result to three decimal places, and compare the value so obtained with the exact value.

32. Show that the exact value of $4 \int_0^1 \sqrt{1 - x^2}\, dx$ is π by interpreting it as the measure of the area of a circular region. Use Simpson's rule with $2n = 6$ to get an approximate value of the definite integral to three decimal places. Compare the results.

33. For the standardized normal probability density function determine the probability that a random choice of x will be in the interval $[0, 1]$. Approximate the value of the definite integral by (a) the trapezoidal rule with $n = 4$ and (b) Simpson's rule with $2n = 4$.

34. For the standardized normal probability density function, determine the probability that a random choice of x will be in the interval $[-3, 3]$. Approximate the value of the definite integral by (a) the trapezoidal rule with $n = 6$ and (b) Simpson's rule with $2n = 6$.

35. The demand equation for a particular commodity is $p = 2\sqrt{100 - x^3}$, and the market price is \$12. Find the consumers' surplus, and draw a sketch showing the region whose area determines the consumers' surplus. Evaluate the definite integral by Simpson's rule, with $2n = 4$. Express the result to two decimal places.

36. Find the area of the region enclosed by the loop of the curve whose equation is $y^2 = 8x^2 - x^5$. Evaluate the definite integral by Simpson's rule, with $2n = 8$. Express the result to three decimal places.

37. Find the producers' surplus for a commodity whose supply equation is $p = \sqrt{x^4 + 68}$ and whose market price is $18. Draw a sketch showing the region whose area determines the producers' surplus. Evaluate the definite integral by Simpson's rule, with $2n = 4$, and express the result to two decimal places.

38. The error function, denoted by erf, is defined by

$$\operatorname{erf}(x) = \frac{2}{\sqrt{\pi}} \int_0^x e^{-t^2} \, dt$$

Find an approximate value of erf(1) to four decimal places by using Simpson's rule with $2n = 10$.

39. For the error function defined in Exercise 38, find an approximate value of erf(10) to four decimal places by using Simpson's rule, with $2n = 10$. *Note:* $\lim\limits_{x \to +\infty} \operatorname{erf}(x) = 1$, and your result is close to 1.

9.4 Use of a table of integrals

We have presented various techniques of integration, and you have seen how they are useful for evaluating many integrals. However, there may be occasions when these procedures are either not sufficient or else lead to a complicated integration. In such cases you may wish to use a **table of integrals**. Fairly complete tables of integrals appear in mathematics handbooks, and shorter tables are found in most calculus textbooks. You should be cautioned not to rely too heavily on tables when evaluating integrals. A mastery of integration skills is essential, because it may be necessary to employ some of the techniques to express the integrand in a form that is found in a table. Therefore you should acquire proficiency in recognizing which technique to apply to a given integral. Furthermore, development of computational skills is important in all branches of mathematics, and the exercises involving integration provide a good training ground. For these reasons you are advised to use a table of integrals only after you have mastered integration.

A short table of integrals appears on the endpapers of this book. The formulas used in the examples and exercises of this section appear in this table. Observe that in the table there are various headings indicating the form of the integrand. The first heading is *Some Elementary Forms*, and the five formulas listed here are included in those given at the beginning of Sec. 9.1. The second heading is *Rational Forms Containing a + bu*. The first example utilizes one of these formulas.

EXAMPLE 1 Find

$$\int \frac{x \, dx}{(4 - x)^3}$$

Solution Formula 10 in the table of integrals is

$$\int \frac{u \, du}{(a + bu)^3} = \frac{1}{b^2} \left[\frac{a}{2(a + bu)^2} - \frac{1}{a + bu} \right] + C$$

Using this formula with $u = x$, $a = 4$, and $b = -1$ we have

$$\int \frac{x\,dx}{(4-x)^3} = \frac{1}{(-1)^2}\left[\frac{4}{2(4-x)^2} - \frac{1}{4-x}\right] + C$$

$$= \frac{2}{(4-x)^2} - \frac{1}{4-x} + C$$

EXAMPLE 2 Find

$$\int \frac{dx}{6 - 2x^2}$$

Solution   Formula 25 in the table is

$$\int \frac{du}{a^2 - u^2} = \frac{1}{2a}\ln\left|\frac{u+a}{u-a}\right| + C \tag{1}$$

Observe that this formula can be used if the coefficient of x^2 in the given integral is 1 instead of 2. Thus we write

$$\int \frac{dx}{6 - 2x^2} = \frac{1}{2}\int \frac{dx}{3 - x^2} \tag{2}$$

On the right side of (2) we apply (1) with $u = x$ and $a = \sqrt{3}$, and we have

$$\int \frac{dx}{6 - 2x^2} = \frac{1}{2}\cdot\frac{1}{2\sqrt{3}}\ln\left|\frac{x+\sqrt{3}}{x-\sqrt{3}}\right| + C$$

$$= \frac{\sqrt{3}}{12}\ln\left|\frac{x+\sqrt{3}}{x-\sqrt{3}}\right| + C$$

EXAMPLE 3 Find

$$\int \frac{dx}{8x^2 + 4x}$$

Solution

$$\int \frac{dx}{8x^2 + 4x} = \frac{1}{4}\int \frac{dx}{x(2x+1)} \tag{3}$$

The integral on the right side of (3) is of the form

$$\int \frac{du}{u(a+bu)}$$

where $u = x$, $a = 1$, and $b = 2$. Formula (11) in the table is

$$\int \frac{du}{u(a+bu)} = \frac{1}{a}\ln\left|\frac{u}{a+bu}\right| + C$$

Using this formula we have

$$\frac{1}{4} \int \frac{dx}{x(2x + 1)} = \frac{1}{4} \cdot \frac{1}{1} \ln \left| \frac{x}{1 + 2x} \right| + C$$

$$= \frac{1}{4} \ln \left| \frac{x}{2x + 1} \right| + C$$

EXAMPLE 4 Find

$$\int \frac{dx}{\sqrt{x^2 + 2x - 3}}$$

Solution　Formula 27 in the table is

$$\int \frac{du}{\sqrt{u^2 \pm a^2}} = \ln \left| u + \sqrt{u^2 \pm a^2} \right| + C \tag{4}$$

We may be able to apply this formula to the given integral if by completing the square under the radical sign we obtain an expression of the form $u^2 \pm a^2$. To complete the square of $x^2 + 2x$ we add 1, and so we also subtract 1. Thus we write

$$x^2 + 2x - 3 = (x^2 + 2x + 1) - 1 - 3 = (x + 1)^2 - 4$$

Therefore

$$\int \frac{dx}{\sqrt{x^2 + 2x - 3}} = \int \frac{dx}{\sqrt{(x + 1)^2 - 4}}$$

The integral is of the form

$$\int \frac{du}{\sqrt{u^2 - a^2}}$$

where $u = x + 1$ and $a = 2$. Hence, from (4),

$$\int \frac{dx}{\sqrt{(x + 1)^2 - 4}} = \ln \left| (x + 1) + \sqrt{(x + 1)^2 - 4} \right| + C$$

$$= \ln \left| x + 1 + \sqrt{x^2 + 2x - 3} \right| + C$$

EXAMPLE 5 Find

$$\int x^2 \sqrt{4x^2 + 1}\, dx$$

Solution　Formula 29 in the table, with the $+$ sign, is

$$\int u^2 \sqrt{u^2 + a^2}\, du = \frac{u}{8} (2u^2 + a^2)\sqrt{u^2 + a^2} - \frac{a^4}{8} \ln \left| u + \sqrt{u^2 + a^2} \right| + C \tag{5}$$

We can apply this formula by writing the given integral as follows:

$$\int x^2 \sqrt{4x^2 + 1} \, dx = \int x^2 \sqrt{4\left(x^2 + \frac{1}{4}\right)} \, dx$$

$$= 2 \int x^2 \sqrt{x^2 + \frac{1}{4}} \, dx$$

From (5), with $u = x$ and $a = \frac{1}{2}$, we have

$$2 \int x^2 \sqrt{x^2 + \frac{1}{4}} \, dx$$

$$= 2\left[\frac{x}{8}\left(2x^2 + \frac{1}{4}\right)\sqrt{x^2 + \frac{1}{4}} - \frac{\frac{1}{16}}{8} \ln \left|x + \sqrt{x^2 + \frac{1}{4}}\right|\right] + C$$

$$= \frac{x}{16}(8x^2 + 1)\sqrt{x^2 + \frac{1}{4}} - \frac{1}{64} \ln \left|x + \sqrt{x^2 + \frac{1}{4}}\right| + C$$

Some of the formulas in the table express one integral in terms of a simpler integral of the same form. These are formulas 16, 19, 21, 22, 23, 52, 53, 54, and 55. Such formulas are called **reduction formulas**. The next example shows how they are applied.

EXAMPLE 6   Find $\int x^3 e^x \, dx$

Solution   Formula 52 is

$$\int u^n e^u \, du = u^n e^u - n \int u^{n-1} e^u \, du \tag{6}$$

Using this formula with $u = x$ and $n = 3$ we have

$$\int x^3 e^x \, dx = x^3 e^x - 3 \int x^2 e^x \, dx \tag{7}$$

We apply (6) to the integral on the right side of (7) with $n = 2$, and we obtain

$$\int x^3 e^x \, dx = x^3 e^x - 3\left[x^2 e^x - 2 \int x e^x \, dx\right] \tag{8}$$

For the integral on the right side of (8) we can continue to use (6) with $n = 1$, or we can apply formula 51 in the table. We have in either case

$$\int x^3 e^x \, dx = x^3 e^x - 3x^2 e^x - 6e^x(x - 1) + C$$

$$= x^3 e^x - 3x^2 e^x - 6xe^x - 6e^x + C$$

It is not necessary to use a table of integrals to find the integral in Example 6. It can be evaluated by using integration by parts. As a matter of fact, this integral appears as Exercise 16 in Exercises 9.1, the section devoted to integration by parts.

Exercises 9.4

In Exercises 1 through 22, use the table of integrals on the endpapers to evaluate the integral.

In Exercises 1 through 4, the integrand is a rational form containing $a + bu$. Use one of the formulas 6 through 13.

1. $\int \dfrac{x\,dx}{2 + 3x}$
2. $\int \dfrac{x\,dx}{(5 - 2x)^3}$
3. $\int \dfrac{x^2\,dx}{(6 - x)^2}$
4. $\int \dfrac{dx}{x(7 + 3x)}$

In Exercises 5 through 8, the integrand is a form containing $\sqrt{a + bu}$. Use one of the formulas 14 through 23.

5. $\int x\sqrt{1 + 2x}\,dx$
6. $\int x^2\sqrt{1 + 2x}\,dx$
7. $\int \dfrac{\sqrt{1 + 2x}}{x}\,dx$
8. $\int \dfrac{dx}{x^2\sqrt{1 + 2x}}$

In Exercises 9 and 10, the integrand is a form containing $a^2 \pm u^2$. Use one of the formulas 24 through 26.

9. $\int \dfrac{dx}{4 - x^2}$
10. $\int \dfrac{dx}{x^2 - 25}$

In Exercises 11 through 14, the integrand is a form containing $\sqrt{u^2 \pm a^2}$. Use one of the formulas 27 through 38.

11. $\int \dfrac{dx}{\sqrt{x^2 + 6x}}$
12. $\int \sqrt{4x^2 + 1}\,dx$
13. $\int \dfrac{\sqrt{9x^2 + 4}}{x}\,dx$
14. $\int \dfrac{dx}{(x - 1)^2\sqrt{x^2 - 2x - 3}}$

In Exercises 15 and 16, the integrand is a form containing $\sqrt{a^2 - u^2}$. Use one of the formulas 39 through 48.

15. $\int \dfrac{\sqrt{9 - 4x^2}}{x}\,dx$
16. $\int \dfrac{dx}{x^2\sqrt{25 - 9x^2}}$

In Exercises 17 through 22, the integrand is a form containing an exponential or logarithmic function. Use one of the formulas 49 through 58.

17. $\int x^4 e^x\,dx$
18. $\int x^3 2^x\,dx$
19. $\int x^2 e^{4x}\,dx$

20. $\int x^2 \ln x\,dx$
21. $\int x^3 \ln(3x)\,dx$
22. $\int 5x^2 e^{-2x}\,dx$

In Exercises 23 through 30, use the table of integrals on the endpapers to evaluate the definite integral.

23. $\int_1^2 \dfrac{dx}{x(5 - x^2)}$
24. $\int_0^3 \dfrac{x\,dx}{(1 + x)^2}$
25. $\int_0^3 \dfrac{x^2\,dx}{\sqrt{x^2 + 16}}$
26. $\int_0^2 \dfrac{dx}{(9 + 4x^2)^{3/2}}$

27. $\int_1^2 x^4 \ln x\,dx$
28. $\int_0^1 x^2 e^{-x}\,dx$
29. $\int_3^4 \sqrt{x^2 + 2x - 15}\,dx$
30. $\int_3^5 x^2\sqrt{x^2 - 9}\,dx$

9.5 Improper integrals

When defining the definite integral $\int_a^b f(x)\,dx$, we assumed the function f to exist on the closed interval $[a, b]$. We now extend the definition of the definite integral to consider an infinite interval of integration, and we also discuss a definite integral in which the integrand has an infinite discontinuity on a finite closed interval. Either integral is called an **improper integral**.

In Sec. 8.7 we referred to a probability density function of the form

$$f(x) = ke^{-kx} \qquad x \geq 0$$

where k is a constant. We indicated there that in this section we would verify that the definite integral of f on $[0, +\infty)$ is 1.

We first evaluate the definite integral of f on an interval $[0, b]$. This definite integral is the measure of the area of the shaded region shown in Fig. 9.5.1.

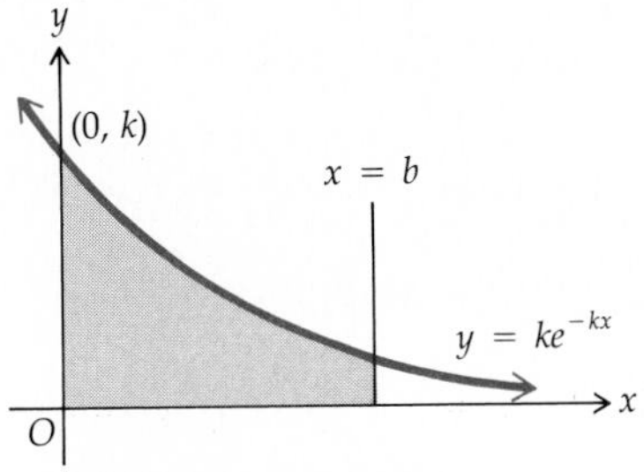

$$\int_0^b ke^{-kx}\, dx = -\int_0^b e^{-kx}(-k\, dx)$$

$$= -e^{-kx}\Big]_0^b$$

$$= 1 - e^{-kb}$$

If we let b increase without bound, then

$$\lim_{b \to +\infty} \int_0^b ke^{-kx}\, dx = \lim_{b \to +\infty} (1 - e^{-kb})$$

$$\lim_{b \to +\infty} \int_0^b ke^{-kx}\, dx = 1 \qquad\qquad (1)$$

We conclude from (1) that no matter how large a value we take for b, the area of the region shown in Fig. 9.5.1 will always be less than 1 square unit.

Equation (1) states that

$$\left| \int_0^b ke^{-kx}\, dx - 1 \right|$$

can be made as small as we please by taking b sufficiently large. In place of (1) we write

$$\int_0^{+\infty} ke^{-kx}\, dx = 1$$

In general we have the following definition.

Figure 9.5.1

Definition of $\displaystyle\int_a^{+\infty} f(x)\, dx$

If f is continuous for all $x \geq a$, then

$$\int_a^{+\infty} f(x)\, dx = \lim_{b \to +\infty} \int_a^b f(x)\, dx$$

if this limit exists.

If the lower limit of integration is infinite, we have the following definition.

Definition of $\displaystyle\int_{-\infty}^{b} f(x)\, dx$

> If f is continuous for all $x \leq b$, then
>
> $$\int_{-\infty}^{b} f(x)\, dx = \lim_{a \to -\infty} \int_{a}^{b} f(x)\, dx$$
>
> if this limit exists.

Finally, we have the case when both limits of integration are infinite.

Definition of $\displaystyle\int_{-\infty}^{+\infty} f(x)\, dx$

> If f is continuous for all values of x, and c is any real number, then
>
> $$\int_{-\infty}^{+\infty} f(x)\, dx = \lim_{a \to -\infty} \int_{a}^{c} f(x)\, dx + \lim_{b \to +\infty} \int_{c}^{b} f(x)\, dx$$
>
> if these limits exist.

When the above definition is applied, c is usually taken as 0. In the three definitions, if the limits exist, we say that the improper integral is **convergent**. If the limits do not exist, we say that the improper integral is **divergent**.

EXAMPLE 1 Evaluate

$$\int_{-\infty}^{2} \frac{dx}{(4 - x)^2}$$

if it converges.

Solution

$$\int_{-\infty}^{2} \frac{dx}{(4 - x)^2} = \lim_{a \to -\infty} \int_{a}^{2} \frac{dx}{(4 - x)^2}$$

$$= \lim_{a \to -\infty} \left[\frac{1}{4 - x} \right]_{a}^{2}$$

$$= \lim_{a \to -\infty} \left(\frac{1}{2} - \frac{1}{4 - a} \right)$$

$$= \tfrac{1}{2} - 0$$

$$= \tfrac{1}{2}$$

EXAMPLE 2 Evaluate if they exist:

(a) $\displaystyle\int_{-\infty}^{+\infty} x\,dx$; (b) $\displaystyle\lim_{r \to +\infty} \int_{-r}^{r} x\,dx$.

Solution (a) From the defintion of $\displaystyle\int_{-\infty}^{+\infty} f(x)\,dx$ with $c = 0$ we have

$$\int_{-\infty}^{+\infty} x\,dx = \lim_{a \to -\infty} \int_{a}^{0} x\,dx + \lim_{b \to +\infty} \int_{0}^{b} x\,dx$$

$$= \lim_{a \to -\infty} \left[\tfrac{1}{2}x^2\right]_{a}^{0} + \lim_{b \to +\infty} \left[\tfrac{1}{2}x^2\right]_{0}^{b}$$

$$= \lim_{a \to -\infty} \left(-\tfrac{1}{2}a^2\right) + \lim_{b \to +\infty} \tfrac{1}{2}b^2$$

Because neither of these two limits exists, the improper integral diverges.

$$\text{(b)} \quad \lim_{r \to +\infty} \int_{-r}^{r} x\,dx = \lim_{r \to +\infty} \left[\tfrac{1}{2}x^2\right]_{-r}^{r}$$

$$= \lim_{r \to +\infty} \left(\tfrac{1}{2}r^2 - \tfrac{1}{2}r^2\right)$$

$$= \lim_{r \to +\infty} 0$$

$$= 0$$

Example 2 illustrates why we do not use the limit in (b) to determine the convergence of an improper integral where both limits of integration are infinite. That is, the improper integral in (a) is divergent, but the limit in (b) exists and is zero.

An interesting application of improper integrals arises in the field of economics. In Sec. 8.8 we discussed a continuous flow of income from which interest is compounded continuously, and formula (11) of that section is

$$V = \int_{0}^{T} f(t)e^{-it}\,dt$$

where $f(t)$ dollars per year is the income at any time t years, $100i$ percent is the annual rate of interest compounded continuously, and V dollars is the present value of the amount earned in T years. If the income continues indefinitely, then if V dollars is the present value of all future income,

$$V = \int_{0}^{+\infty} f(t)e^{-it}\,dt \tag{2}$$

EXAMPLE 3 A contract produces a continuous flow of income that is decreasing with time, and at t years the number of dollars per year in the income is $8000 \cdot 3^{-t}$. Find the present value of this income if it continues indefinitely with an interest rate of 12 percent compounded continuously.

Solution If V dollars is the present value, then from (2), with $f(t) = 8000 \cdot 3^{-t}$ and $i = 0.12$, we have

$$V = \int_0^{+\infty} (8000 \cdot 3^{-t}) e^{-0.12t} \, dt$$

$$= 8000 \lim_{b \to +\infty} \int_0^b (3e^{0.12})^{-t} \, dt$$

$$= 8000 \lim_{b \to +\infty} \left[\frac{-(3e^{0.12})^{-t}}{\ln(3e^{0.12})} \right]_0^b$$

$$= 8000 \lim_{b \to +\infty} \left[\frac{-(3e^{0.12})^{-b} + 1}{\ln 3 + 0.12} \right]$$

$$= 8000 \left(\frac{0 + 1}{\ln 3 + 0.12} \right)$$

$$= \frac{8000}{1.0986 + 0.12}$$

$$= 6565$$

Therefore the present value of the income is \$6565.

EXAMPLE 4 Is it possible to assign a finite number to represent the measure of the area of the region bounded by the graphs of the equations $y = 1/x$, $y = 0$, and $x = 1$?

Solution The region is shown in Fig. 9.5.2. Let L be the number we wish to assign to the measure of the area, if possible. Let A be the measure of the area of the region bounded by the graphs of the equations $y = 1/x$, $y = 0$, $x = 1$, and $x = b$, where $b > 1$. Then

$$A = \lim_{n \to +\infty} \sum_{i=1}^{n} \frac{1}{\xi_i} \Delta x = \int_1^b \frac{1}{x} \, dx$$

So we shall let $L = \lim_{b \to +\infty} A$ if this limit exists. But

$$\lim_{b \to +\infty} A = \lim_{b \to +\infty} \int_1^b \frac{1}{x} \, dx$$

$$= \lim_{b \to +\infty} \left[\ln b - \ln 1 \right]$$

$$= +\infty$$

Therefore it is not possible to assign a finite number to represent the measure of the area of the region.

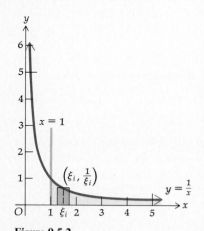

Figure 9.5.2

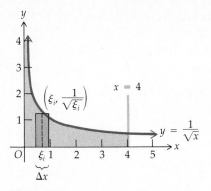

Figure 9.5.3

Figure 9.5.3 shows the region bounded by the curve whose equation is $y = 1/\sqrt{x}$, the x axis, the y axis, and the line $x = 4$. If it is possible to assign a finite number to represent the measure of the area of this region, it would be given by

$$\lim_{n \to +\infty} \sum_{i=1}^{n} \frac{1}{\sqrt{\xi_i}} \Delta x$$

If this limit exists, it is the definite integral denoted by

$$\int_0^4 \frac{dx}{\sqrt{x}} \tag{3}$$

However, the integrand is discontinuous at the lower limit, zero. Furthermore, $\lim_{x \to 0^+} 1/\sqrt{x} = +\infty$, and so we state that the integrand has an infinite discontinuity at the lower limit. Such an integral is improper, and its existence can be determined from the following definition.

Definition of an improper integral for which the integrand has an infinite discontinuity at the lower limit of integration

> If f is continuous at all x in the interval half-open on the left $(a, b]$, and if $\lim_{x \to a^+} f(x) = \pm\infty$, then
>
> $$\int_a^b f(x)\, dx = \lim_{\epsilon \to 0^+} \int_{a+\epsilon}^b f(x)\, dx$$
>
> if this limit exists.

● ILLUSTRATION 1

We determine if it is possible to assign a finite number to represent the measure of the area of the region shown in Fig. 9.5.3. From the discussion preceding the above definition, the measure of the area of the given region will be the improper integral (3) if it exists. By the definition,

$$\int_0^4 \frac{dx}{\sqrt{x}} = \lim_{\epsilon \to 0^+} \int_\epsilon^4 \frac{dx}{\sqrt{x}}$$

$$= \lim_{\epsilon \to 0^+} 2x^{1/2} \Big]_\epsilon^4$$

$$= \lim_{\epsilon \to 0^+} [4 - 2\sqrt{\epsilon}]$$

$$= 4 - 0$$

$$= 4$$

Therefore we assign 4 to the measure of the area of the given region.   ●

If the integrand has an infinite discontinuity at the upper limit of integration, we use the following definition to determine the existence of the improper integral.

Definition of an improper integral for which the integrand has an infinite discontinuity at the upper limit of integration

If f is continuous at all x in the interval half-open on the right $[a, b)$, and if $\lim\limits_{x \to b^-} f(x) = \pm\infty$, then

$$\int_a^b f(x)\, dx = \lim_{\epsilon \to 0^+} \int_a^{b-\epsilon} f(x)\, dx$$

if this limit exists.

If there is an infinite discontinuity at an interior point of the interval of integration, the existence of the improper integral is determined from the following definition.

Definition of an improper integral for which the integrand has an infinite discontinuity at an interior point of the interval of integration

If f is continuous at all x in the interval $[a, b]$ except c, where $a < c < b$, and if $\lim\limits_{x \to c} |f(x)| = +\infty$, then

$$\int_a^b f(x)\, dx = \lim_{\epsilon \to 0^+} \int_a^{c-\epsilon} f(x)\, dx + \lim_{\delta \to 0^+} \int_{c+\delta}^b f(x)\, dx$$

if these limits exist.

If $\int_a^b f(x)\, dx$ is an improper integral, it is convergent if the corresponding limit exists; otherwise it is divergent.

EXAMPLE 5 Evaluate

$$\int_0^2 \frac{dx}{(x-1)^2}$$

if it is convergent.

Solution The integrand has an infinite discontinuity at 1. Applying the definition we have

$$\int_0^2 \frac{dx}{(x-1)^2} = \lim_{\epsilon \to 0^+} \int_0^{1-\epsilon} \frac{dx}{(x-1)^2} + \lim_{\delta \to 0^+} \int_{1+\delta}^2 \frac{dx}{(x-1)^2}$$

$$= \lim_{\epsilon \to 0^+} \left[-\frac{1}{x-1} \right]_0^{1-\epsilon} + \lim_{\delta \to 0^+} \left[-\frac{1}{x-1} \right]_{1+\delta}^2$$

$$= \lim_{\epsilon \to 0^+} \left[\frac{1}{\epsilon} - 1 \right] + \lim_{\delta \to 0^+} \left[-1 + \frac{1}{\delta} \right]$$

Because neither of these limits exist, the improper integral is divergent.

● ILLUSTRATION 2

Suppose that in evaluating the integral in Example 5 we had failed to note the infinite discontinuity of the integrand at 1. We would have obtained

$$-\frac{1}{x-1}\Bigg]_0^2 = -\frac{1}{1} + \frac{1}{-1} = -2$$

This is obviously an incorrect result. Because the integrand $1/(x-1)^2$ is never negative, the integral from 0 to 2 could not possibly be a negative number. ●

EXAMPLE 6 Evaluate

$$\int_0^{+\infty} \frac{e^{-\sqrt{x}}}{\sqrt{x}}\, dx$$

if it is convergent.

Solution For this integral there is an infinite upper limit and an infinite discontinuity at the lower limit. We proceed as follows.

$$\int_0^{+\infty} \frac{e^{-\sqrt{x}}}{\sqrt{x}}\, dx = \lim_{\epsilon \to 0^+} \int_\epsilon^1 \frac{e^{-\sqrt{x}}}{\sqrt{x}}\, dx + \lim_{b \to +\infty} \int_1^b \frac{e^{-\sqrt{x}}}{\sqrt{x}}\, dx \qquad (4)$$

To evaluate the indefinite integral we let $u = -\sqrt{x}$; then $du = -dx/2\sqrt{x}$. We then have

$$\int \frac{e^{-\sqrt{x}}}{\sqrt{x}}\, dx = -2 \int (e^{-\sqrt{x}})\left(\frac{-dx}{2\sqrt{x}}\right) = -2 \int e^u\, du$$

$$= -2e^u + C = -2e^{-\sqrt{x}} + C$$

So in (4) we have

$$\int_0^{+\infty} \frac{e^{-\sqrt{x}}}{\sqrt{x}}\, dx = \lim_{\epsilon \to 0^+}\left[-2e^{-\sqrt{x}}\right]_\epsilon^1 + \lim_{b \to +\infty}\left[-2e^{-\sqrt{x}}\right]_1^b$$

$$= \lim_{\epsilon \to 0^+}\left(-2e^{-1} + 2e^{-\sqrt{\epsilon}}\right) + \lim_{b \to +\infty}\left(-2e^{-\sqrt{b}} + 2e^{-1}\right)$$

$$= -2e^{-1} + 2 - 0 + 2e^{-1}$$

$$= 2$$

Exercises 9.5

In Exercises 1 through 22, determine whether the improper integral is convergent or divergent. If it is convergent, evaluate it.

1. $\displaystyle\int_{-\infty}^1 e^x\, dx$

2. $\displaystyle\int_5^{+\infty} \frac{dx}{\sqrt{x-1}}$

3. $\displaystyle\int_0^{+\infty} \frac{dx}{\sqrt{e^x}}$

4. $\displaystyle\int_1^{+\infty} 2^{-x}\, dx$

5. $\displaystyle\int_1^{+\infty} \ln x\, dx$

6. $\displaystyle\int_0^1 \ln x\, dx$

7. $\displaystyle\int_0^1 \frac{dx}{\sqrt{1-x}}$

8. $\displaystyle\int_{-2}^0 \frac{dx}{2x+3}$

9. $\displaystyle\int_0^2 \frac{dx}{(x-1)^{2/3}}$ **10.** $\displaystyle\int_{-\infty}^0 x5^{-x^2}\,dx$ **11.** $\displaystyle\int_e^{+\infty} \frac{dx}{x(\ln x)^2}$ **12.** $\displaystyle\int_1^e \frac{dx}{x\ln x}$

13. $\displaystyle\int_{-\infty}^{+\infty} xe^{-x^2}\,dx$ **14.** $\displaystyle\int_{-\infty}^{+\infty} e^{-|x|}\,dx$ **15.** $\displaystyle\int_{-1}^1 \frac{dx}{x^2}$ **16.** $\displaystyle\int_{-2}^2 \frac{dx}{x^3}$

17. $\displaystyle\int_{-\infty}^0 \frac{dx}{x+3}$ **18.** $\displaystyle\int_0^{+\infty} \frac{dx}{x-5}$ **19.** $\displaystyle\int_0^{+\infty} \frac{dx}{\sqrt{x}e^{\sqrt{x}}}$ **20.** $\displaystyle\int_{-\infty}^0 x^2 e^x\,dx$

21. $\displaystyle\int_2^3 \frac{dy}{\sqrt[3]{y-2}}$ **22.** $\displaystyle\int_0^2 \frac{x\,dx}{1-x}$

23. For the light bulb of Exercise 6 in Exercises 8.7, find the probability that the life of a bulb selected at random will be 60 or more hours.

24. In the city of Exercise 8 in Exercises 8.7, what is the probability that the length of a telephone call selected at random will last 5 or more minutes?

25. Determine whether it is possible to assign a finite number to represent the measure of the area of the region bounded by the curve whose equation is $y = 1/(e^x + e^{-x})$ and the x axis. If a finite number can be assigned, find it.

26. Determine whether it is possible to assign a finite number to represent the measure of the area of the region bounded by the x axis, the line $x = 2$, and the curve whose equation is $y = 1/(x^2 - 1)$. If a finite number can be assigned, find it.

27. Determine whether it is possible to assign a finite number to represent the measure of the area of the region bounded by the curve whose equation is $y = 1/\sqrt{x}$, the line $x = 1$, and the x and y axes. If a finite number can be assigned, find it.

28. Suppose that the owner of a piece of business property holds a permanent lease on the property so that the rent is paid perpetually. If the annual rent is $12,000 and money is worth 10 percent compounded continuously, find the present value of all future rent payments.

29. A continuous flow of income is decreasing with time, and at t years the number of dollars in the annual income is $1000 \cdot 2^{-t}$. Find the present value of this income if it continues indefinitely with an interest rate of 8 percent compounded continuously.

30. The British Consol is a bond with no maturity (that is, it never comes due), and it affords the holder an annual lump-sum payment. By finding the present value of a flow of payments of R dollars annually, and using the current interest rate $100i$ percent compounded continuously, show that the fair selling price of a British Consol is R/i dollars.

31. The continuous flow of profit for a company is increasing with time, and at t years the number of dollars in the profit per year is directly proportional to t. Show that the present value of the company is inversely proportional to i^2, where $100i$ percent is the interest rate compounded continuously. $\left(\textit{Hint: } \text{If } c \text{ is a positive constant, } \displaystyle\lim_{t \to +\infty} \frac{t}{e^{ct}} = 0.\right)$

32. Show that the improper integral $\displaystyle\int_{-\infty}^{+\infty} \frac{x\,dx}{(1+x^2)^2}$ is convergent and that the improper integral $\displaystyle\int_{-\infty}^{+\infty} \frac{x\,dx}{1+x^2}$ is divergent.

33. Evaluate, if they exist:

     (a) $\displaystyle\int_{-1}^1 \frac{dx}{x}$ (b) $\displaystyle\lim_{r \to 0^+} \left[\int_{-1}^{-r} \frac{dx}{x} + \int_r^1 \frac{dx}{x}\right]$

34. Show that the improper integral $\displaystyle\int_1^{+\infty} \frac{dx}{x^n}$ is convergent if $n > 1$ and divergent if $n \leq 1$.

Review Exercises for Chapter 9

In Exercises 1 through 10, evaluate the indefinite integral.

1. $\displaystyle\int xe^{-4x}\,dx$ **2.** $\displaystyle\int x4^x\,dx$ **3.** $\displaystyle\int \frac{e^x\,dx}{\sqrt{4-e^x}}$ **4.** $\displaystyle\int x^3 e^{3x}\,dx$

5. $\displaystyle\int \frac{5x^2 - 3}{x^3 - x}\, dx$ **6.** $\displaystyle\int \frac{x^2 + 1}{(x - 1)^3}\, dx$ **7.** $\displaystyle\int \frac{2t^3 + 11t + 8}{t^3 + 4t^2 + 4t}\, dt$ **8.** $\displaystyle\int \frac{dw}{w \ln w(\ln w - 1)}$

9. $\displaystyle\int \ln(x + 1)\, dx$ **10.** $\displaystyle\int x^2 \ln x\, dx$

In Exercises 11 through 18, evaluate the definite integral.

11. $\displaystyle\int_1^2 (\ln x)^2\, dx$ **12.** $\displaystyle\int_0^2 \frac{(1 - x)\, dx}{x^2 + 3x + 2}$ **13.** $\displaystyle\int_1^2 \frac{t + 2}{(t + 1)^2}\, dt$ **14.** $\displaystyle\int_0^1 \frac{dx}{e^x - e^{-x}}$

15. $\displaystyle\int_0^{16} \sqrt{4 - \sqrt{x}}\, dx$ **16.** $\displaystyle\int_1^{10} \log_{10} \sqrt{ex}\, dx$ **17.** $\displaystyle\int_1^2 \frac{2x^2 + x + 4}{x^3 + 4x^2}\, dx$ **18.** $\displaystyle\int_0^1 w^3 \sqrt{1 + w^2}\, dw$

In Exercises 19 through 26, determine whether the improper integral is convergent or divergent. If it is convergent, evaluate it.

19. $\displaystyle\int_{-2}^0 \frac{dx}{2x + 1}$ **20.** $\displaystyle\int_0^{+\infty} \frac{dx}{\sqrt{e^x}}$ **21.** $\displaystyle\int_{-\infty}^0 \frac{dx}{(x - 2)^2}$ **22.** $\displaystyle\int_2^4 \frac{x\, dx}{\sqrt{x - 2}}$

23. $\displaystyle\int_0^1 \frac{(\ln x)^2}{x}\, dx$ **24.** $\displaystyle\int_{-\infty}^3 3^t\, dt$ **25.** $\displaystyle\int_0^{+\infty} \frac{3^{-\sqrt{r}}}{\sqrt{r}}\, dr$ **26.** $\displaystyle\int_{-\infty}^{+\infty} \frac{dx}{4x^2 + 4x + 1}$

In Exercises 27 and 28, find an approximate value for the given integral by using the trapezoidal rule with $n = 4$. Express the result to three decimal places.

27. $\displaystyle\int_0^2 \sqrt{1 + x^2}\, dx$ **28.** $\displaystyle\int_1^{9/5} \sqrt{1 + x^3}\, dx$

29. Find an approximate value for the integral of Exercise 27 by using Simpson's rule with $2n = 4$. Express the result to three decimal places.
30. Find an approximate value for the integral of Exercise 28 by using Simpson's rule with $2n = 4$. Express the result to three decimal places.

In Exercises 31 through 34, use the table of integrals in the endpapers to evaluate the indefinite or definite integral.

31. $\displaystyle\int \sqrt{9x^2 + 1}\, dx$ **32.** $\displaystyle\int \frac{dt}{\sqrt{4t^2 - 9}}$ **33.** $\displaystyle\int_4^5 w^2 \sqrt{w^2 - 16}\, dw$ **34.** $\displaystyle\int_0^1 x^3 e^{2x}\, dx$

35. Find to the nearest dollar the amount in 4 years of a continuous investment of $300t$ dollars per year, where t is the number of years from the present, if the interest rate is 10 percent per year compounded continuously. Use formula (9) of Sec. 8.8.
36. The revenue produced by a particular piece of equipment for 10 years is estimated to be a continuous flow of income of $4000 - 50t$ dollars per year, where t is the number of years from the present. If the income is to be deposited into a bank account with interest at 12 percent per year compounded continuously, find to the nearest dollar the present value of this income. Use formula (11) of Sec. 8.8.
37. A pond can support up to a maximum of 10,000 fish so that the rate of growth of the fish population is jointly proportional to the number of fish present and the difference between 10,000 and the number present. The pond initially contained 400 fish, and 6 weeks later there were 3000 fish. (a) How many fish did the pond contain after 8 weeks? (b) When was the growth rate greatest? That is, after how many weeks did the pond contain 5000 fish?
38. In a town of population 12,000 the growth rate of a flu epidemic is jointly proportional to the number of people who have the flu and the number of people who do not have it. Five days ago 400 people in the town had the flu, and today 1000 people have it. How many people are expected to have the flu tomorrow? (b) In how many days from now will the epidemic be spreading the fastest? That is, when will half the population have the flu?

39. Two chemicals A and B react to form a chemical C, and the rate of change of the amount of C is jointly proportional to the amounts of A and B remaining at any given time. Initially there are 60 kg of chemical A and 60 kg of chemical B, and to form 5 kg of C, 3 kg of A and 2 kg of B are required. After 1 hour, 15 kg of C are formed. (a) If x kg of C are formed in t hours, find an expression for x in terms of t. (b) Find the amount of C after 3 hours.

40. The demand equation for a particular commodity is $p = \ln(20 - x)$, where x units are demanded when the price per unit is p dollars. If the market price is \$5, find the consumers' surplus. Draw a sketch showing the demand curve and the region whose area gives the consumers' surplus.

41. On the college campus of Exercise 39 in Review Exercises for Chapter 8, what is the probability that the length of a telephone call selected at random will last 4 or more minutes?

42. Determine whether it is possible to assign a finite number to represent the measure of the area of the region in the first quadrant below the curve having the equation $y = e^{-x}$. If a finite number can be assigned, find it.

43. Assuming a continuous flow of income for a particular business, and assuming that at t years from now the number of dollars in the income per year is expected to be $1000t - 300$, what is the present value of all expected future income if the interest rate is 8 percent per year compounded continuously? (See hint for Exercise 31 in Exercises 9.5.)

CHAPTER 10

DIFFERENTIAL CALCULUS OF FUNCTIONS OF SEVERAL VARIABLES

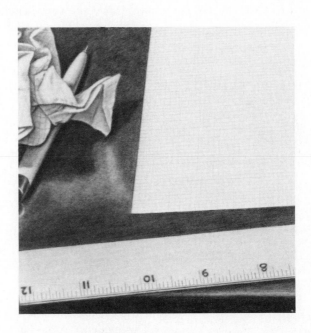

10.1 R^3, the three-dimensional number space

In this chapter we are concerned with functions of more than one variable. In our treatment of functions of two variables we wish to deal with their graphs, and to do this we need a geometric three-dimensional space. This section is devoted to a discussion of such a space.

In Chapter 1 we introduced the number line R^1, the one-dimensional number space, and the number plane R^2, the two-dimensional number space. We identified the real numbers in R^1 with points on a horizontal axis and the real number pairs in R^2 with points in a geometric plane. In an analogous fashion we now introduce the set of all ordered triples of real numbers.

Definition of R^3, the three-dimensional number space

> The set of all ordered triples of real numbers is called the **three-dimensional number space** and is denoted by R^3. Each ordered triple (x, y, z) is called a *point* in the three-dimensional number space.

To represent R^3 in a geometric three-dimensional space we consider the distances of a point from three mutually perpendicular planes. The planes are formed by first considering three mutually perpendicular lines that intersect at a point called the origin, denoted by the letter O. These lines, called the coordinate axes, are designated as the x axis, the y axis, and the z axis. Usually the x axis and the y axis are taken in a horizontal plane, and the z axis is vertical. A positive direction is selected on each axis. If the positive directions are chosen as in Fig. 10.1.1, the coordinate system is called a **right-handed system**. This terminology follows from the fact that if the right hand is placed so the thumb is pointed in the positive direction of the x axis and the index finger is pointed in the positive direction of the y axis, then the middle finger is pointed in the positive direction of the z axis. If the middle finger is pointed in the negative direction of the z axis, then the coordinate system is called **left handed**. A left-handed system is shown in Fig. 10.1.2. In general we use a right-handed system. The three axes determine three coordinate planes: the xy plane containing the x and y axes, the xz plane containing the x and z axes, and the yz plane containing the y and z axes.

An ordered triple of real numbers (x, y, z) is associated with each point P in a geometric three-dimensional space. The directed distance of P from the yz plane is called the x *coordinate*, the directed distance of P from the xz plane is called the y *coordinate*, and the z *coordinate* is the directed distance of P from the xy plane. These three coordinates are called **the rectangular cartesian coordinates** of the point, and there is a one-to-one correspondence (called a *rectangular cartesian coordinate system*) between all such ordered triples of real numbers and the points in a geometric three-dimensional space. Hence we identify R^3 with the geometric three-dimensional space, and we call an ordered triple (x, y, z) a point. The point $(3, 2, 4)$ is shown in Fig. 10.1.3, and the point $(4, -2, -5)$ is shown in Fig.

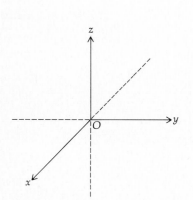

Figure 10.1.1

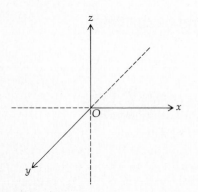

Figure 10.1.2

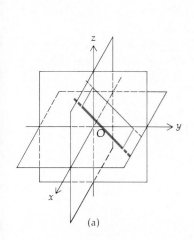

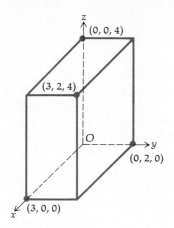

Figure 10.1.3

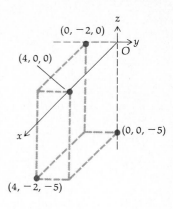

Figure 10.1.4

10.1.4. The three coordinate planes divide the space into eight parts, called **octants**. The first octant is the one in which all three coordinates are positive.

A line is parallel to a plane if and only if the distance from any point on the line to the plane is the same.

● ILLUSTRATION 1

A line parallel to the yz plane, a line parallel to the xz plane, and a line parallel to the xy plane appear in Fig. 10.1.5(a), (b), and (c), respectively. ●

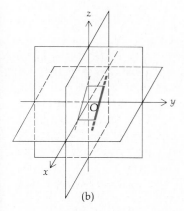

We consider all lines lying in a given plane as being parallel to the plane, in which case the distance from any point on the line to the plane is zero. The next theorem follows immediately.

Theorem 10.1.1

  (i) A line is parallel to the yz plane if and only if all points on the line have equal x coordinates.
  (ii) A line is parallel to the xz plane if and only if all points on the line have equal y coordinates.
  (iii) A line is parallel to the xy plane if and only if all points on the line have equal z coordinates.

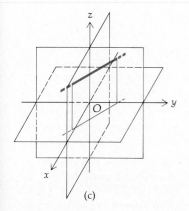

Figure 10.1.5

In three-dimensional space, if a line is parallel to each of two intersecting planes, it is parallel to the line of intersection of the two planes. Also, if a given line is parallel to a second line, then the given line is parallel to any plane containing the second line. Theorem 10.1.2 follows from these two facts from solid geometry and from Theorem 10.1.1.

> **Theorem 10.1.2**
>
> (i) A line is parallel to the x axis if and only if all points on the line have equal y coordinates and equal z coordinates.
> (ii) A line is parallel to the y axis if and only if all points on the line have equal x coordinates and equal z coordinates.
> (iii) A line is parallel to the z axis if and only if all points on the line have equal x coordinates and equal y coordinates.

● ILLUSTRATION 2

A line parallel to the x axis, a line parallel to the y axis, and a line parallel to the z axis are shown in Fig. 10.1.6(a) (b), and (c), respectively.   ●

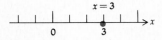

Figure 10.1.7

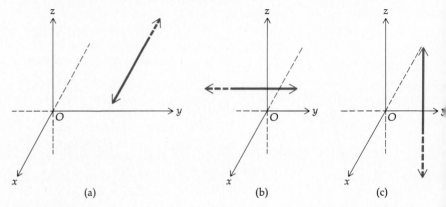

(a)   (b)   (c)

Figure 10.1.6

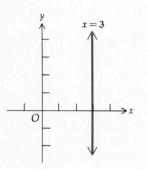

Figure 10.1.8

Definition of the graph of an equation in R^3

> **The graph of an equation in R^3** is the set of all points (x, y, z) whose coordinates are numbers satisfying the equation.

The graph of an equation in R^3 is called a **surface**. A plane is a particular surface. A sphere, discussed later, is another particular surface.

Consider the equation $x = 3$. In R^1 this is an equation of a point that is 3 units to the right of the origin; in R^2 the equation $x = 3$ is an equation of a line that is 3 units to the right of the y axis; and in R^3 this equation is that of a plane that is 3 units in front of the yz plane. See Figs. 10.1.7, 10.1.8, and 10.1.9.

A plane parallel to the yz plane has an equation of the form $x = k$, where k is a constant. Figure 10.1.9 shows a sketch of the plane having the equation $x = 3$. A plane parallel to the xz plane has an equation of the form $y = k$, and a plane parallel to the xy plane has an equation of the form $z = k$. Figures 10.1.10 and 10.1.11 show sketches of the planes having the equations $y = -5$ and $z = 6$, respectively.

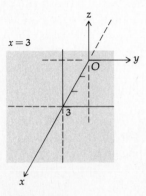

Figure 10.1.9

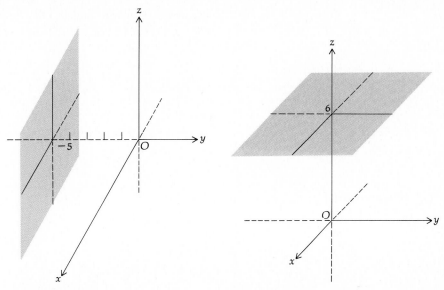

Figure 10.1.10   **Figure 10.1.11**

Theorem 1.2.1 states that in R^2 the graph of the general equation of the first degree in x and y is a straight line. Although we will not prove it here, in R^3 the graph of the general first-degree equation in x, y, and z, $Ax + By + Cz + D = 0$, is a plane.

A plane is determined by three noncollinear points, by a line and a point not on the line, by two intersecting lines, or by two parallel lines. To draw a sketch of a plane from its equation it is convenient to find the points at which the plane intersects each of the coordinate axes. The x coordinate of the point at which the plane intersects the x axis is called the **x intercept** of the plane, the y coordinate of the point at which the plane intersects the y axis is called the **y intercept** of the plane, and the **z intercept** of the plane is the z coordinate of the point at which the plane intersects the z axis. In Examples 1 and 2 that follow, we want a sketch of the portion of a plane in the first octant. To do this we first draw the lines of intersection of the given plane with the three coordinate planes. These lines are called the **traces** of the given plane in the coordinate planes. To obtain the traces we can first find the intercepts and then in the coordinate planes draw lines through the points corresponding to these intercepts. An equation of the trace in the yz plane can be found by setting $x = 0$ in the equation of the given plane. Similarly, an equation of the trace in the xz plane can be found by setting $y = 0$, and an equation of the trace in the xy plane is obtained by setting $z = 0$. Observe that $x = 0$, $y = 0$, and $z = 0$ are equations of the yz plane, the xz plane, and the xy plane, respectively.

EXAMPLE 1 Draw a sketch of the portion of the plane

$$2x + 4y + 3z = 8$$

in the first octant.

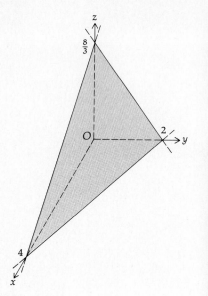

Figure 10.1.12

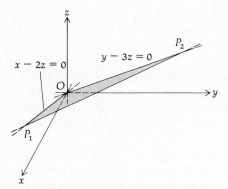

Figure 10.1.13

Solution   By substituting zero for y and z in the given equation we obtain $x = 4$; so the x intercept of the plane is 4. In a similar manner we obtain the y intercept and the z intercept, which are 2 and $\frac{8}{3}$, respectively. By plotting the points corresponding to these intercepts and connecting them with lines we have the traces of the given plane in the coordinate planes. Thus we obtain the required sketch shown in Fig. 10.1.12.

EXAMPLE 2   Draw a sketch of the portion of the plane

$$3x + 2y - 6z = 0$$

in the first octant.

Solution   Because the given equation is satisfied when x, y, and z are all zero, the plane intersects each of the coordinate axes at the origin. If we set $x = 0$ in the given equation, we obtain $y - 3z = 0$, which is the trace of the given plane in the yz plane. The trace of the given plane in the xz plane is obtained by setting $y = 0$, and we get $x - 2z = 0$. Drawing a sketch of each of these two traces and drawing a line segment from a point on one of the traces to a point on the other, we obtain Fig. 10.1.13.

To represent a line in R^3 we consider simultaneously the equations of two planes containing the line. This is analogous to representing a point in R^2 by considering simultaneously the equations of two lines containing the point. The two simultaneous equations $x = 5$ and $y = -2$ in R^2 represent the point $(5, -2)$, which is the intersection of the two lines, as shown in Fig. 10.1.14. However, in R^3 the same two equations represent the line of intersection of the two planes, one of which is 5 units in front of the yz plane while the other is 2 units to the left of the xz plane; see Fig. 10.1.15. In R^2 an equation of the form

$$x^2 + y^2 = r^2$$

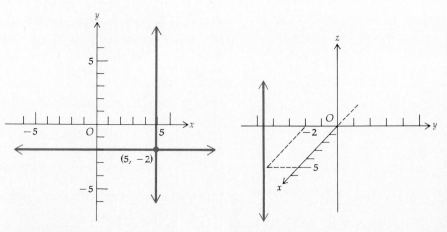

Figure 10.1.14   **Figure 10.1.15**

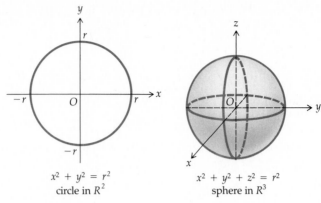

$x^2 + y^2 = r^2$
circle in R^2

$x^2 + y^2 + z^2 = r^2$
sphere in R^3

Figure 10.1.16   **Figure 10.1.17**

is an equation of the circle having its center at the origin and radius r. See Fig. 10.1.16. In R^3 an equation of the form

$$x^2 + y^2 + z^2 = r^2$$

is an equation of the sphere having its center at the origin and radius r. See Fig. 10.1.17. We shall use this surface to illustrate certain results in subsequent sections.

Another surface we shall refer to later is the one defined by the equation

$$z = x^2 + y^2$$

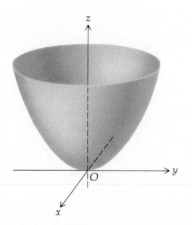

Figure 10.1.18

To obtain a graph of this equation we find the traces of the surface in the coordinate planes. The trace in the xy plane is found by using the equation $z = 0$ simultaneously with the equation of the surface. We obtain $x^2 + y^2 = 0$, which is the origin. The traces in the xz and yz planes are found by using the equations $y = 0$ and $x = 0$, respectively, with the equation $z = x^2 + y^2$. We obtain the parabolas $z = x^2$ and $z = y^2$. The cross section of the surface in a plane $z = k$, parallel to the xy plane, is a circle with its center on the z axis and radius $\sqrt{k}$. With this information we have the sketch of the surface shown in Fig. 10.1.18. The surface is called a *paraboloid*.

Exercises 10.1

In Exercises 1 through 5, the given points A and B are opposite vertices of a rectangular parallelepiped having its faces parallel to the coordinate planes. In each exercise, (a) draw a sketch of the figure; (b) find the coordinates of the other six vertices.

1. $A(0, 0, 0)$; $B(7, 2, 3)$ 2. $A(1, 1, 1)$; $B(3, 4, 2)$ 3. $A(-1, 1, 2)$; $B(2, 3, 5)$
4. $A(2, -1, -3)$; $B(4, 0, -1)$ 5. $A(1, -1, 0)$; $B(3, 3, 5)$

6. The vertex opposite one corner of a room is 18 ft east, 15 ft south, and 12 ft up from the first corner. (a) Draw a sketch of the figure; (b) find the coordinates of all eight vertices of the room.
7. Follow the instructions of Exercise 6 if the vertex opposite one corner of the room is 14 ft west, 16 ft north, and 10 ft up from the first corner.

8. Follow the instructions of Exercise 6 if the vertex opposite one corner of a room is 12 ft west, 20 ft south, and 11 ft down from the first corner.

In Exercises 9 through 12, draw a sketch of the graph of the given equation in (a) R^1, (b) R^2, (c) R^3.

9. $x = 3$ **10.** $y = 5$ **11.** $y = -4$ **12.** $x = -2$

In Exercises 13 and 14, draw a sketch of the graph of the two simultaneous equations in (a) R^2, (b) R^3.

13. $x = 6, y = 3$ **14.** $x = -3, y = 4$

In Exercises 15 and 16, draw a sketch of the graph of the two simultaneous equations in R^3.

15. $x = 4, z = 6$ **16.** $y = 5, z = 7$

In Exercises 17 and 18, describe in words the graph of the simultaneous equations in R^3.

17. $x = 0, y = 0$ **18.** $y = 0, z = 0$

In Exercises 19 through 30, draw a sketch of the given plane.

19. $4x + 6y + 3z = 12$ **20.** $5x + 3y + 7z = 15$ **21.** $3x - 3y - z + 6 = 0$
22. $2x - y + 2z - 6 = 0$ **23.** $y + 2z - 4 = 0$ **24.** $3x + 2z - 6 = 0$
25. $5x + 2y - 10 = 0$ **26.** $4x - 4y - 2z = 9$ **27.** $4x + 3y - 12z = 0$
28. $z = 5$ **29.** $x - y = 0$ **30.** $2y - 3z = 0$

10.2 Functions of more than one variable

In the preceding chapters we have been concerned with the calculus of functions of one variable. We now generalize the notion of a function to functions of more than one independent variable. Such functions often occur in practical situations. For example, the cost of a particular product may be dependent upon the cost of labor, the price of materials, and overhead expenses. The number of items of a commodity sold may depend on the selling price of the item and the amount of advertising. The demand for a product is dependent upon its price and could also be influenced by the price of competing products as well as the income of potential consumers. A person's approximate body surface area depends on the person's weight and height. According to the ideal gas law, the volume occupied by a confined gas is directly proportional to its temperature and inversely proportional to its pressure.

● ILLUSTRATION 1

A clothing store sells two kinds of sweaters that are similar but are made by different manufacturers. The cost to the store of the first kind is $40 and the cost of the second kind is $50. It has been determined by experience that if the selling price of the first kind is x dollars and the selling price of the second kind is y dollars, then the number sold weekly of the first kind is $3200 - 50x + 25y$, and the number sold weekly of the second kind is $400 - 25y + 25x$. The gross profit on the sale of a sweater of the first kind is $(x - 40)$ dollars, and on the sale of a sweater of the second kind it is $(y - 50)$ dollars. If each of these amounts is multiplied by the number of sweaters sold, the sum of the products gives the gross profit. Thus, if P dollars is the weekly gross profit,

$$P = (x - 40)(3200 - 50x + 25y) + (y - 50)(400 - 25y + 25x)$$

$$P = 50xy - 50x^2 - 25y^2 + 3950x + 650y - 148{,}000 \qquad (1)$$

Equation (1) expresses P as a function of the two independent variables x and y. The notation used for functions of several variables is similar to that for functions of one variable. For instance, (1) can be written as

$$P(x, y) = 50xy - 50x^2 - 25y^2 + 3950x + 650y - 148{,}000 \qquad (2) \quad \bullet$$

A thorough study of the calculus of functions of several variables belongs to a more advanced text than this one. Hence our discussion will be more informal than our treatment of functions of one variable.

An equation of the form

$$z = f(x, y) \qquad (3)$$

defines a function of two independent variables if for each ordered pair (x, y) of real numbers in the domain of f there is one and only one real number z in the range of f that satisfies Eq. (3). In a similar manner, an equation of the form

$$w = f(x, y, z)$$

defines a function of three independent variables, and an equation of the form

$$u = f(x, y, z, w)$$

defines a function of four independent variables. In this chapter we will be concerned mainly with functions of two independent variables.

● ILLUSTRATION 2

The function f of two variables x and y is defined by

$$f(x, y) = \sqrt{25 - x^2 - y^2}$$

The domain of f is the set of all ordered pairs (x, y) for which $25 - x^2 - y^2 \geq 0$. This is the set of all points in the xy plane on the circle $x^2 + y^2 = 25$ and in the interior region bounded by the circle. In Fig. 10.2.1 there is a sketch showing as a shaded region the set of points in the domain of f.

If $z = f(x, y)$, then we can write

$$z = \sqrt{25 - (x^2 + y^2)}$$

From this equation we see that $0 \leq z \leq 5$; therefore the range of f is the set of all real numbers in the closed interval $[0, 5]$. ●

● ILLUSTRATION 3

The function g of two variables x and y is defined by

$$g(x, y) = \ln(x^2 - y)$$

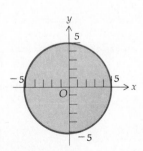

Figure 10.2.1

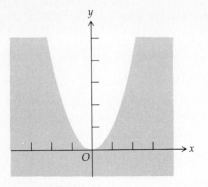

Figure 10.2.2

The domain of g is the set of all points in R^2 for which $x^2 - y > 0$, that is, $y < x^2$. This is the set of all points in the xy plane below the parabola $y = x^2$. Figure 10.2.2 is a sketch showing as a shaded region the set of points in the domain of g. ●

● ILLUSTRATION 4

If the store of Illustration 1 sells sweaters of the first kind for $90 and sweaters of the second kind for $100, then the number of dollars in the weekly gross profit is $P(90, 100)$, and from (2),

$$P(90, 100) = 50(90)(100) - 50(90)^2 - 25(100)^2 + 3950(90) + 650(100) - 148{,}000$$
$$= 450{,}000 - 405{,}000 - 250{,}000 + 355{,}500 + 65{,}000 - 148{,}000$$
$$= 67{,}500 \qquad ●$$

● ILLUSTRATION 5

Let f be the function of Illustration 2; that is,

$$f(x, y) = \sqrt{25 - x^2 - y^2}$$

Then

$$f(3, -4) = \sqrt{25 - (3)^2 - (-4)^2} = \sqrt{25 - 9 - 16} = 0$$
$$f(-2, 1) = \sqrt{25 - (-2)^2 - (1)^2} = \sqrt{25 - 4 - 1} = 2\sqrt{5}$$
$$f(u, 3v) = \sqrt{25 - u^2 - (3v)^2} = \sqrt{25 - u^2 - 9v^2} \qquad ●$$

EXAMPLE 1   The function g is defined by $g(x, y, z) = x^2 - 5xz + yz^2$. Find (a) $g(1, 4, -2)$; (b) $g(2a, -b, 3c)$; (c) $g(x^2, y^2, z^2)$; (d) $g(y, z, -x)$.

Solution

(a) $g(1, 4, -2)\ \ = 1^2 - 5(1)(-2) + 4(-2)^2 = 1 + 10 + 16 = 27$

(b) $g(2a, -b, 3c) = (2a)^2 - 5(2a)(3c) + (-b)(3c)^2$
$\qquad\qquad\qquad = 4a^2 - 30ac - 9bc^2$

(c) $g(x^2, y^2, z^2)\ \ = (x^2)^2 - 5(x^2)(z^2) + (y^2)(z^2)^2 = x^4 - 5x^2z^2 + y^2z^4$

(d) $g(y, z, -x)\ \ = y^2 - 5y(-x) + z(-x)^2 = y^2 + 5xy + x^2z$

A **polynomial function** of two variables x and y is a function f such that $f(x, y)$ is the sum of terms of the form cx^ny^m, where c is a real number and n and m are nonnegative integers. The **degree** of the polynomial function is determined by the largest sum of the exponents of x and y appearing in any one term. Hence the function f defined by

$$f(x, y) = 6x^3y^2 - 5xy^3 + 7x^2y - 2x^2 + y$$

is a polynomial function of degree 5.

A **rational function** of two variables is a function h such that $h(x, y) = f(x, y)/g(x, y)$, where f and g are two polynomial functions. For example, the function f defined by

$$f(x, y) = \frac{x^2 y^2}{x^2 + y^2}$$

is a rational function.

The graph of a function f of a single variable consists of the set of points (x, y) in R^2 for which $y = f(x)$. Similarly, the *graph of a function of two variables* is a set of points in R^3.

Definition of the graph of a function of two variables

> If f is a function of two variables, then the **graph** of f is the set of all points (x, y, z) in R^3 for which (x, y) is a point in the domain of f and $z = f(x, y)$.

The graph of a function f of two variables is a surface that is a set of points in three-dimensional space whose cartesian coordinates are given by ordered triples of real numbers (x, y, z). Because the domain of f is a set of points in the xy plane, and because for each ordered pair (x, y) in the domain of f there corresponds a unique value of z, no line perpendicular to the xy plane can intersect the graph of f in more than one point.

● ILLUSTRATION 6

The function of Illustration 2 is the function f that is defined by

$$f(x, y) = \sqrt{25 - x^2 - y^2}$$

Then the graph of f is the set of all points (x, y, z) in R^3 for which

$$z = \sqrt{25 - x^2 - y^2}$$

So the graph of f is the hemisphere on and above the xy plane having its center at the origin and a radius of 5. A sketch of this hemisphere is shown in Fig. 10.2.3. ●

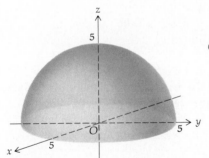

Figure 10.2.3

● ILLUSTRATION 7

If g is the function defined by

$$g(x, y) = x^2 + y^2$$

the graph of g is the set of all points (x, y, z) in R^3 for which

$$z = x^2 + y^2$$

This equation was discussed at the end of Sec. 10.1, and its graph is the paraboloid shown in Fig. 10.1.18, repeated here as Fig. 10.2.4. ●

Another useful method of representing a function of two variables geometrically is similar to that of representing a three-dimensional landscape by a two-dimensional topographical map. Suppose that the surface

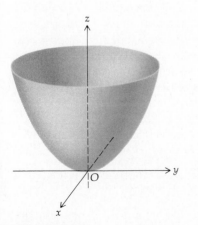

Figure 10.2.4

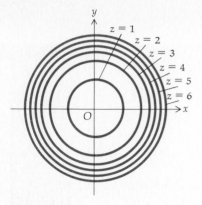

Figure 10.2.5

$z = f(x, y)$ is intersected by the plane $z = k$, and the curve of intersection is projected onto the xy plane. This projected curve has $f(x, y) = k$ as an equation, and the curve is called the **level curve** (or **contour curve**) of the function f at k. Each point on the level curve corresponds to the unique point on the surface that is k units above it if k is positive, or k units below it if k is negative. By considering different values for the constant k we obtain a set of level curves called a **contour map**. The set of all possible values of k is the range of the function f, and each level curve, $f(x, y) = k$, in the contour map consists of the points (x, y) in the domain of f having equal function values of k. For example, for the function g of Illustration 7 the level curves are circles with the center at the origin. The particular level curves for $z = 1, 2, 3, 4, 5,$ and 6 are shown in Fig. 10.2.5.

A contour map shows the variation of z with x and y. The level curves are usually shown for values of z at constant intervals, and the values of z are changing more rapidly when the level curves are close together than when they are far apart; that is, when the level curves are close together, the surface is steep, and when the level curves are far apart, the elevation of the surface is changing slowly. On a two-dimensional topographical map of a landscape, a general notion of its steepness is obtained by considering the spacing of its level curves. Also on a topographical map, if the path of a level curve is followed, the elevation remains constant.

To illustrate a use of level curves, suppose that the temperature at any point of a flat metal plate is given by the function f; that is, if t degrees is the temperature, then at the point (x, y), $t = f(x, y)$. Then the curves having equations of the form $f(x, y) = k$, where k is a constant, are curves on which the temperature is constant. These are the level curves of f and are called **isothermals**. Furthermore, if V volts gives the amount of electric potential at any point (x, y) of the xy plane, and $V = f(x, y)$, then the level curves of f are called **equipotential curves** because the electric potential at each point of such a curve is the same.

For an application of level curves in economics, consider the productivity (or output) of an industry that is dependent upon several inputs. Among the inputs may be the number of machines used in production, the number of person-hours available, the amount of working capital to be had, the quantity of material used, and the amount of land available. Suppose that the amounts of the inputs are given by x and y, the amount of the output is given by z, and $z = f(x, y)$. Such a function is called a **production function**, and the level curves of f, having equations of the form $f(x, y) = k$, where k is a constant, are called **constant product curves**.

EXAMPLE 2   Let f be the production function for which $f(x, y) = 2x^{1/2}y^{1/2}$. Draw a contour map of f showing the constant product curves at 8, 6, 4, and 2.

Solution   The contour map consists of the curves that are the intersection of the surface

$$z = 2x^{1/2}y^{1/2} \tag{4}$$

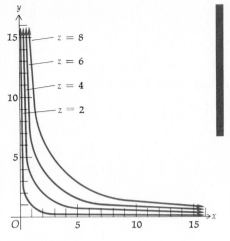

Figure 10.2.6

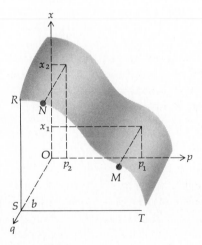

Figure 10.2.7

with the planes $z = k$, where $k = 8, 6, 4, 2$, and 1. Substituting $z = 8$ in (4) we obtain $4 = x^{1/2}y^{1/2}$ or, equivalently,

$$xy = 16 \qquad x > 0 \quad \text{and} \quad y > 0 \tag{5}$$

The curve in the xy plane represented by (5) is a branch of a hyperbola lying in the first quadrant. With each of the numbers 6, 4, and 2 we also obtain a branch of a hyperbola in the first quadrant. These are the constant product curves, and they are shown in Fig. 10.2.6.

In Sec. 1.6 we discussed a demand equation giving the relationship between x and p, where x units are demanded when p dollars is the price of one unit of a commodity. In addition to the price of the given commodity, the demand often will depend on the prices of other commodities related to the given one. In particular, let us consider two related commodities for which p dollars is the price per unit of x units of the first commodity and q dollars is the price per unit of y units of the second commodity. Then the demand equations for these two commodities can be written, respectively, as

$$\alpha(x, p, q) = 0 \quad \text{and} \quad \beta(y, p, q) = 0$$

or, solving the first equation for x and the second equation for y, as

$$x = f(p, q) \tag{6}$$

and

$$y = g(p, q) \tag{7}$$

The functions f and g in Eqs. (6) and (7) are the demand functions, and the graphs of these functions are surfaces. Under normal circumstances x, y, p, and q are nonnegative, and so the surfaces are restricted to the first octant. These surfaces are called **demand surfaces**. Recalling that p dollars is the price of one unit of x units of the first commodity, we note that if the variable q is held constant, then x decreases as p increases and x increases as p decreases. This is illustrated in Fig. 10.2.7, which is a sketch of the demand surface for an equation of type (6) under normal circumstances. The plane $q = b$ intersects the surface in section RST. For any point on the curve RT, q is the constant b. Referring to the points $M(p_1, b, x_1)$ and $N(p_2, b, x_2)$, we see that $x_2 > x_1$ if and only if $p_2 < p_1$; that is, x decreases as p increases and x increases as p decreases.

When q is constant, therefore, as p increases, x decreases; but y may either increase or decrease. If y increases, then a decrease in the demand for one commodity, caused by an increase in its price, results in an increase in the demand for the other, and the two commodities are said to be **substitutes** (for example, butter and margarine). Now if, when q is constant, y decreases as p increases, then a decrease in the demand for one commodity, caused by an increase in its price, results in a decrease in the demand for the other, and the two commodities are said to be **complementary** (for example, tires and gasoline).

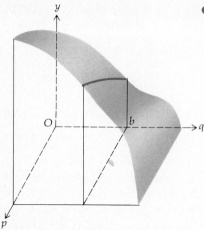

Figure 10.2.8

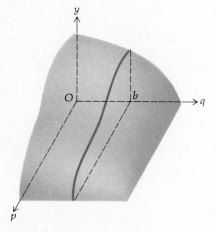

Figure 10.2.9

● ILLUSTRATION 8

Figures 10.2.8 and 10.2.9 each show a sketch of a demand surface for an equation of type (7). In Fig. 10.2.8 we see that when q is constant, y increases as p increases, and so the two commodities are substitutes. In Fig. 10.2.9 the two commodities are complementary because when q is constant, y decreases as p increases.   ●

Observe that in Figs. 10.2.7 and 10.2.8, which show the demand surfaces for equations of types (6) and (7), respectively, the p and q axes are interchanged and the vertical axis in Fig. 10.2.7 is labeled x and in Fig. 10.2.8 it is labeled y.

It is possible that for a fixed value of q, y may increase for some values of p and decrease for other values of p. For example, if the demand surface of Eq. (7) is that shown in Fig. 10.2.10, then for $q = b$, when $p = a_1$, y is increasing, and when $p = a_2$, y is decreasing. This of course means that if the price of the second commodity is held constant, then for some prices of the first commodity the two commodities will be substitutes, and for other prices of the first commodity the two commodities will be complementary. These relationships between the two commodities that are determined by the demand surface having equation $y = g(p, q)$ will correspond to similar relationships determined by the demand surface having equation $x = f(p, q)$ for the same fixed values of p and q. An economic example might be one in which investors allocate funds between the stock market and real estate. As stock prices climb, they invest in real estate. Yet once stock prices seem to reach a crash level, investors begin to decrease the amount of real estate purchases with any increase in stock prices in anticipation of a collapse that will affect the real estate market and its values.

If the demand equations for the two commodities are given by equations (6) and (7), then under normal circumstances f and g are restricted so that it is possible to solve the two equations for p and q in terms of x and y, giving

$$p = F(x, y) \quad \text{and} \quad q = G(x, y) \tag{8}$$

Furthermore, if the demand equations are given by Eqs. (8), it must be possible to find the functions f and g so that they may be written in the form of Eqs. (6) and (7).

Suppose the functions f and g defined by Eqs. (6) and (7) are linear so that those equations are of the form

$$x = mp + cq + x_0 \quad \text{and} \quad y = nq + dp + y_0$$

where the constants m and n are negative (or else zero in a trivial case) and x_0 and y_0 are positive or zero. If the constants c and d are both positive, the commodities are substitutes, since if p is held constant, then an increase in q causes an increase in x but a decrease in y, or if q is a fixed constant, an increase in p causes an increase in y but a decrease in x. By a similar discussion, if the constants c and d are both negative, we see that the commodities are complementary.

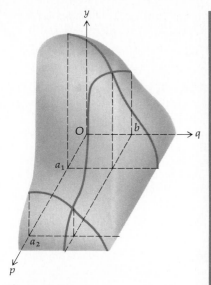

Figure 10.2.10

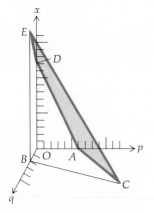

Figure 10.2.11

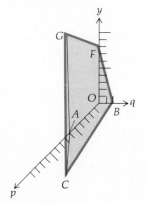

Figure 10.2.12

EXAMPLE 3 Suppose that x units of one commodity and y units of a second commodity are demanded when the prices per unit are p dollars and q dollars, respectively, and the demand equations are

$$x = -2p + 3q + 12 \tag{9}$$

and

$$y = -4q + p + 8 \tag{10}$$

Determine whether the commodities are substitutes or complementary, and draw sketches of the two demand surfaces.

Solution Because the coefficient of q in Eq. (9) is positive and the coefficient of p in Eq. (10) is positive, the two commodities are substitutes. A sketch of the demand surface of (9) is shown in Fig. 10.2.11. To draw this sketch we first determine from both equations the permissible values of p and q. Because x and y must be positive or zero, p and q must satisfy the inequalities

$$-2p + 3q + 12 \geq 0 \quad \text{and} \quad -4q + p + 8 \geq 0$$

Also, p and q are each nonnegative. Hence the values of p and q are restricted to the quadrilateral $AOBC$. The required demand surface, then, is the portion in the first octant of the plane defined by (9), which is above $AOBC$. This is the shaded quadrilateral $ADEC$ in the figure. In Fig. 10.2.12 there is a sketch of the demand surface defined by (10). This demand surface is the shaded quadrilateral $BFGC$, which is the portion in the first octant of the plane defined by (10) that is above the quadrilateral $AOBC$.

For nonlinear demand functions of two variables it is often more convenient to represent them geometrically by means of contour maps than by surfaces. The following example gives such a case.

EXAMPLE 4 Suppose that x units of one commodity and y units of a second commodity are demanded when the prices per unit are p dollars and q dollars, respectively, and the demand equations are

$$x = \frac{8}{pq} \quad \text{and} \quad y = \frac{12}{pq}$$

draw sketches of the contour maps of the two demand functions showing the level curves of each function at 6, 4, 2, 1, and $\frac{1}{2}$. Are the commodities substitutes or complementary?

Solution Let the two demand functions be f and g, so that

$$x = f(p, q) = \frac{8}{pq}$$

and

$$y = g(p, q) = \frac{12}{pq}$$

Sketches of the contour maps of f and g, showing the level curves of these functions at the required numbers, are shown in Figs. 10.2.13 and 10.2.14, respectively. If q is held constant, we see that as p increases both x and y decrease, and so the commodities are complementary.

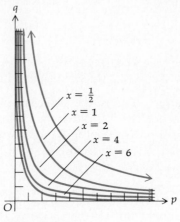

Figure 10.2.13   **Figure 10.2.14**

Exercises 10.2

1. Given: $f(x, y) = 3x + 2y - 5$. Find: (a) $f(3, -1)$; (b) $f(-4, 2)$; (c) $f(a + 1, b - 2)$; (d) $f(x + 1, y - 2)$; (e) $f(2x, 3y)$; (f) $f(x + h, y) - f(x, y)$; (g) $f(x, y + h) - f(x, y)$.

2. Given: $g(x, y) = x^2 - 3y + 5$. Find: (a) $g(-2, 3)$; (b) $g(4, 7)$; (c) $g(4r, 2s)$; (d) $g(4x, 2y)$; (e) $g(x - 2, 2 - y)$; (f) $g(x + h, y) - g(x, y)$; (g) $g(x, y + h) - g(x, y)$.

3. Given:
$$f(x, y) = \frac{x + y}{x - y}$$
Find: (a) $f(-3, 4)$; (b) $f(x^2, y^2)$; (c) $[f(x, y)]^2$; (d) $f(-x, y) - f(x, -y)$; (e) the domain of f.

4. Given: $g(x, y) = \sqrt{x^2 + y^2 - 25}$. Find: (a) $g(-4, 3)$; (b) $g(10, -5)$; (c) $g(x^2, y^2)$; (d) $[g(x, y)]^2$; (e) $g(-x, y) - g(x, -y)$; (f) the domain of g; (g) the range of g.

5. Given: $g(x, y, z) = \sqrt{4 - x^2 - y^2 - z^2}$. Find: (a) $g(1, -1, -1)$; (b) $g(-a, 2b, \frac{1}{2}c)$; (c) $g(y, -x, -y)$; (d) $[g(x, y, z)]^2 - [g(x + 2, y + 2, z)]^2$; (e) the domain of g; (f) the range of g.

6. Given:
$$f(x, y, z) = \frac{x + y + z}{x - y + z}$$
Find: (a) $f(1, -1, 1)$; (b) $f(-3, 2, -5)$; (c) $f(-y, z, -x)$; (d) $f(r^2, 2rs, s^2)$; (e) the domain of f.

In Exercises 7 through 14, draw a sketch showing as a shaded region in R^2 the set of points in the domain of f.

7. $f(x, y) = \sqrt{16 - x^2 - y^2}$ 8. $f(x, y) = \dfrac{\sqrt{16 - x^2 - y^2}}{x}$ 9. $f(x, y) = \sqrt{x^2 + y^2 - 16}$ 10. $f(x, y) = \dfrac{x^2 - y^2}{x - y}$

11. $f(x, y) = \ln(x^2 - 4y)$ 12. $f(x, y) = \ln(4y^2 - x)$ 13. $f(x, y) = \ln(xy - 1)$ 14. $f(x, y) = \sqrt{x + y}$

In Exercises 15 and 16, determine the domain and range of the given function.

15. $g(x, y, z) = |x|e^{y/z}$ 16. $h(x, y, z) = \ln(x^2 + y^2 + z^2 - 1)$

In Exercises 17 through 20, draw a sketch of a contour map of the function f showing the level curves of f at the given numbers.

17. $f(x, y) = 4x^2 + 4y^2$ at 16, 12, 8, 4, 1, 0

18. $f(x, y) = \frac{1}{4}(x^2 + y^2)$ at 5, 4, 3, 2, 1, 0

19. $f(x, y) = 16 - x^2 - y^2$ at 8, 4, 0, -4, -8

20. $f(x, y) = \sqrt{100 - x^2 - y^2}$ at 8, 6, 4, 2, 0

21. Suppose that f is a production function for which $f(x, y) = 6xy$ and $f(x, y)$ units are produced when x machines and y person-hours are used in production. Draw a contour map of f showing the constant product curves at 30, 24, 18, 12, and 6.

22. The production function f for a certain commodity has function values $f(x, y) = 4x^{1/3}y^{2/3}$, where x and y give the amounts of two inputs. Draw a contour map of f showing the constant product curves at 16, 12, 8, 4, and 2.

23. The temperature at a point (x, y) of a flat metal plate is t degrees, and $t = 2x^2 + 2y^2$. Draw the isothermals for $t = 12, 8, 4, 2,$ and 0.

24. The electric potential at a point (x, y) of the xy plane is V volts, and $V = 4/\sqrt{9 - x^2 - y^2}$. Draw the equipotential curves for $V = 16, 12, 8, 4,$ and 2.

In Exercises 25 through 30, x units of one commodity and y units of a second commodity are demanded when the prices per unit are p dollars and q dollars, respectively. Determine from the given demand equations if the commodities are substitutes, complementary, or neither, and draw sketches of the two demand surfaces.

25. $x = -p - 3q + 6$ and $y = -2q - p + 8$

26. $x = -4p + 2q + 6$ and $y = 5p - q + 10$

27. $x = 6 - 3p - 2q$ and $y = 4 + 2p - q$

28. $x = -7q - p + 7$ and $y = 18 - 3q - 9p$

29. $x = -3p + 5q + 15$ and $y = 2p - 4q + 10$

30. $x = 9 - 3p + q$ and $y = 10 - 2p - 5q$

In Exercises 31 and 32, x units of one commodity and y units of a second commodity are demanded when the prices per unit are p dollars and q dollars, respectively. From the given demand equations, find the two demand functions, and draw sketches of the contour maps of these functions, showing the level curves of each function at 5, 4, 3, 2, 1, $\frac{1}{2}$, and $\frac{1}{4}$.

31. $px = q$ and $qy = p^2$

32. $pqx = 4$ and $p^2qy = 16$

10.3 Partial derivatives

The discussion of the differentiation of functions of several variables is reduced to the one-dimensional case by treating such a function as a function of one variable at a time and holding the others fixed. This leads to the concept of a *partial derivative*. We first define the partial derivative of a function of two variables.

Definition of the partial derivative of a function of two variables

Let f be a function of two variables, x and y. The **partial derivative of f with respect to x** is that function, denoted by $D_x f$, such that its function value at any point (x, y) in the domain of f is given by

$$D_x f(x, y) = \lim_{\Delta x \to 0} \frac{f(x + \Delta x, y) - f(x, y)}{\Delta x} \qquad (1)$$

if this limit exists. Similarly, the **partial derivative of f with respect to y** is that function, denoted by $D_y f$, such that its function value at any point (x, y) in the domain of f is given by

$$D_y f(x, y) = \lim_{\Delta y \to 0} \frac{f(x, y + \Delta y) - f(x, y)}{\Delta y} \qquad (2)$$

if this limit exists.

The process of computing a partial derivative is called **partial differentiation**.

$D_x f$ is read as "D sub x of f," and this denotes the partial derivative function. $D_x f(x, y)$ is read as "D sub x of f of x and y," and this denotes the partial derivative function value at the point (x, y). Other notations for the partial derivative function $D_x f$ are f_x and $\partial f/\partial x$. Other notations for the partial derivative function value $D_x f(x, y)$ are $f_x(x, y)$ and $\partial f(x, y)/\partial x$. Similarly, other notations for $D_y f$ are f_y and $\partial f/\partial y$; other notations for $D_y f(x, y)$ are $f_y(x, y)$ and $\partial f(x, y)/\partial y$. If $z = f(x, y)$, we can write $\partial z/\partial x$ for $D_x f(x, y)$. A partial derivative cannot be thought of as a ratio of ∂z and ∂x because neither of these symbols has a separate meaning. The notation dy/dx can be regarded as the quotient of two differentials when y is a function of the single variable x, but there is not a similar interpretation for $\partial z/\partial x$.

EXAMPLE 1   Given
$$f(x, y) = 3x^2 - 2xy + y^2$$
find $D_x f(x, y)$ and $D_y f(x, y)$ by applying the definition.

Solution

$$D_x f(x, y) = \lim_{\Delta x \to 0} \frac{f(x + \Delta x, y) - f(x, y)}{\Delta x}$$

$$= \lim_{\Delta x \to 0} \frac{3(x + \Delta x)^2 - 2(x + \Delta x)y + y^2 - (3x^2 - 2xy + y^2)}{\Delta x}$$

$$= \lim_{\Delta x \to 0} \frac{3x^2 + 6x\,\Delta x + 3(\Delta x)^2 - 2xy - 2y\,\Delta x + y^2 - 3x^2 + 2xy - y^2}{\Delta x}$$

$$= \lim_{\Delta x \to 0} \frac{6x\,\Delta x + 3(\Delta x)^2 - 2y\,\Delta x}{\Delta x}$$

$$= \lim_{\Delta x \to 0} (6x + 3\,\Delta x - 2y)$$

$$= 6x - 2y$$

$$D_y f(x, y) = \lim_{\Delta y \to 0} \frac{f(x, y + \Delta y) - f(x, y)}{\Delta y}$$

$$= \lim_{\Delta y \to 0} \frac{3x^2 - 2x(y + \Delta y) + (y + \Delta y)^2 - (3x^2 - 2xy + y^2)}{\Delta y}$$

$$= \lim_{\Delta y \to 0} \frac{3x^2 - 2xy - 2x\,\Delta y + y^2 + 2y\,\Delta y + (\Delta y)^2 - 3x^2 + 2xy - y^2}{\Delta y}$$

$$= \lim_{\Delta y \to 0} \frac{-2x\,\Delta y + 2y\,\Delta y + (\Delta y)^2}{\Delta y}$$

$$= \lim_{\Delta y \to 0} (-2x + 2y + \Delta y)$$

$$= -2x + 2y$$

Comparing the definition of a partial derivative with that of an ordinary derivative (in Sec. 2.5), we see that $D_x f(x, y)$ is the ordinary derivative of f if f is considered as a function of one variable x (that is, y is held constant), and $D_y f(x, y)$ is the ordinary derivative of f if f is considered as a function of one variable y (and x is held constant). So the results in Example 1 can be obtained more easily by applying theorems for ordinary differentiation if y is considered constant when finding $D_x f(x, y)$ and if x is considered constant when finding $D_y f(x, y)$. The following example illustrates this.

EXAMPLE 2 Given

$$f(x, y) = 3x^3 - 4x^2 y + 3xy^2 + 7x - 8y$$

find $D_x f(x, y)$ and $D_y f(x, y)$.

Solution Treating f as a function of x and holding y constant we have

$$D_x f(x, y) = 9x^2 - 8xy + 3y^2 + 7$$

Considering f as a function of y and holding x constant we have

$$D_y f(x, y) = -4x^2 + 6xy - 8$$

Geometric interpretations of the partial derivatives of a function of two variables are similar to those of a function of one variable. The graph of a function f of two variables is a surface having equation $z = f(x, y)$. If y is held constant (say, $y = y_0$), then $z = f(x, y_0)$ is an equation of the trace of this surface in the plane $y = y_0$. The curve can be represented by the two equations

$$y = y_0 \quad \text{and} \quad z = f(x, y) \tag{3}$$

because the curve is the intersection of these two surfaces.

Then $D_x f(x_0, y_0)$ is the slope of the tangent line to the curve given by Eqs. (3) at the point $P_0(x_0, y_0, f(x_0, y_0))$ in the plane $y = y_0$. In an analogous fashion, $D_y f(x_0, y_0)$ represents the slope of the tangent line to the curve having equations

$$x = x_0 \quad \text{and} \quad z = f(x, y)$$

at the point P_0 in the plane $x = x_0$. Figure 10.3.1 (a) and (b) shows the portions of the curves and the tangent lines.

EXAMPLE 3 Find the slope of the tangent line to the curve of intersection of the surface $z = \frac{1}{2}\sqrt{24 - x^2 - 2y^2}$ with the plane $y = 2$ at the point $(2, 2, \sqrt{3})$.

Solution The required slope is the value of $\partial z / \partial x$ at the point $(2, 2, \sqrt{3})$.

$$\frac{\partial z}{\partial x} = \frac{-x}{2\sqrt{24 - x^2 - 2y^2}}$$

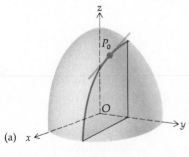

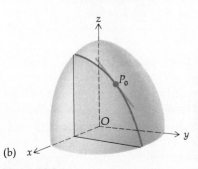

(a)

(b)

Figure 10.3.1

So at $(2, 2, \sqrt{3})$,

$$\frac{\partial z}{\partial x} = \frac{-2}{2\sqrt{12}} = -\frac{1}{2\sqrt{3}}$$

Because every derivative is a measure of a rate of change, a partial derivative can be so interpreted. If f is a function of the two variables x and y, the partial derivative of f with respect to x at the point $P_0(x_0, y_0)$ gives the instantaneous rate of change, at P_0, of $f(x, y)$ per unit change in x (x alone varies and y is held fixed at y_0). Similarly, the partial derivative of f with respect to y at P_0 gives the instantaneous rate of change, at P_0, of $f(x, y)$ per unit change in y.

● ILLUSTRATION 1

Suppose that the production cost of a certain commodity is dependent upon two inputs: the cost of labor and the cost of materials. In particular, if z dollars is the production cost, x dollars is the hourly cost of labor, and y dollars is the cost per pound of materials, then

$$z = 500 + 40x + 7y$$

Because

$$\frac{\partial z}{\partial x} = 40$$

it follows that when the cost of materials remains fixed, an increase of \$1 in the hourly cost of labor results in an increase of \$40 in the production cost. Because

$$\frac{\partial z}{\partial y} = 7$$

then when the cost of labor remains fixed, an increase of \$1 in the cost per pound of materials causes an increase of \$7 in the production cost.   ●

EXAMPLE 4   For a certain retail outlet it is determined that if x is the number of daily television spot announcements, y is the number of minutes in the length of each spot, and z is the number of sales per day, then

$$z = 2xy^2 + x^2 + 9000$$

Suppose that at the present time there are 12 one-minute spot announcements each day. (a) Find the instantaneous rate of change of z per unit change in x if y remains fixed at 1. (b) Use the result of part (a) to approximate the change in the daily sales if the number of daily one-minute spot announcements is increased by 25 percent. (c) Find the instantaneous rate of change of z per unit change in y if x remains fixed at 12. (d) Use the result of part (c) to approximate the change in the daily sales if the length of each of the 12 spots is increased by 25 percent.

Solution $z = 2xy^2 + x^2 + 9000$.

(a) $\dfrac{\partial z}{\partial x} = 2y^2 + 2x$

When $x = 12$ and $y = 1$, $\partial z/\partial x = 26$, which is the answer required.

(b) An increase of 25 percent of 12 is an increase of 3. From the result of part (a), when x is increased by 3 (and y remains fixed), an approximate increase in z is $3 \cdot 26 = 78$. We conclude, then, that if the number of daily one-minute spot announcements is increased from 12 to 15, the increase in daily sales is approximately 78.

(c) $\dfrac{\partial z}{\partial y} = 4xy$

When $x = 12$ and $y = 1$, $\partial z/\partial y = 48$, which is the instantaneous rate of change of z per unit change in y at $x = 12$, $y = 1$, if x remains fixed at 12.

(d) If y is increased by one-fourth (25 percent of 1) and x is held fixed, then from the result of part (c) the change in z is approximately $\frac{1}{4} \cdot 48 = 12$. Hence the increase in daily sales is approximately 12 if the length of each of the 12 spots is increased from 1 min to $1\frac{1}{4}$ min.

The concept of partial derivative can be extended to functions of any number of variables. In particular, if f is a function of the three variables x, y, and z, then the partial derivatives of f are given by

$$D_x f(x, y, z) = \lim_{\Delta x \to 0} \frac{f(x + \Delta x, y, z) - f(x, y, z)}{\Delta x}$$

$$D_y f(x, y, z) = \lim_{\Delta y \to 0} \frac{f(x, y + \Delta y, z) - f(x, y, z)}{\Delta y}$$

and

$$D_z f(x, y, z) = \lim_{\Delta z \to 0} \frac{f(x, y, z + \Delta z) - f(x, y, z)}{\Delta z}$$

if these limits exist.

EXAMPLE 5 Given

$$f(x, y, z) = x^2 y + yz^2 + z^3$$

verify that

$$x f_x(x, y, z) + y f_y(x, y, z) + z f_z(x, y, z) = 3f(x, y, z)$$

Solution Holding y and z constant we get

$$f_x(x, y, z) = 2xy$$

Holding x and z constant we obtain

$$f_y(x, y, z) = x^2 + z^2$$

Holding x and y constant we have

$$f_z(x, y, z) = 2yz + 3z^2$$

Therefore

$$\begin{aligned} xf_x(x, y, z) + yf_y(x, y, z) + zf_z(x, y, z) &= x(2xy) + y(x^2 + z^2) + z(2yz + 3z^2) \\ &= 2x^2y + x^2y + yz^2 + 2yz^2 + 3z^3 \\ &= 3(x^2y + yz^2 + z^3) \\ &= 3f(x, y, z) \end{aligned}$$

If f is a function of two variables x and y, then in general $D_x f$ and $D_y f$ are also functions of two variables. And if the partial derivatives of these functions exist, they are called second partial derivatives of f. In contrast, $D_x f$ and $D_y f$ are called first partial derivatives of f. There are four second partial derivatives of a function of two variables. If f is a function of the two variables x and y, the notations

$$D_{xy}f \qquad f_{xy} \qquad \frac{\partial^2 f}{\partial y\, \partial x}$$

all denote the second partial derivative of f, which is obtained by first partial-differentiating f with respect to x and then partial-differentiating the result with respect to y. Observe that in the subscript notation the order of partial differentiation is from left to right; in the symbolism $\partial^2 f/\partial y\, \partial x$ the order is from right to left. The notations

$$D_{xx}f \qquad f_{xx} \qquad \frac{\partial^2 f}{\partial x^2}$$

all indicate the second partial derivative of f, which is obtained by partial-differentiating twice with respect to x. The other two partial derivatives are written in an analogous way as

$$D_{yx}f \qquad f_{yx} \qquad \frac{\partial^2 f}{\partial x\, \partial y}$$

and

$$D_{yy}f \qquad f_{yy} \qquad \frac{\partial^2 f}{\partial y^2}$$

The notations for higher-order partial derivatives are similar. For example,

$$D_{xxy} \qquad f_{xxy} \qquad \frac{\partial^3 f}{\partial y\, \partial x\, \partial x} \qquad \frac{\partial^3 f}{\partial y\, \partial x^2}$$

all stand for the third partial derivative of f, which is obtained by partial-differentiating twice with respect to x and then once with respect to y. Again observe that when using subscripts, the order of partial differentiation is from left to right, and in the notation $\partial^3 f/\partial y\, \partial x\, \partial x$ the order is from right to left.

EXAMPLE 6 Given

$$f(x, y) = y^2 e^x + \ln xy$$

find (a) $D_{xx}f(x, y)$; (b) $D_{xy}f(x, y)$; and (c) $\dfrac{\partial^3 f}{\partial x \, \partial y^2}$.

Solution

$$D_x f(x, y) = y^2 e^x + \frac{1}{xy}(y) = y^2 e^x + \frac{1}{x}$$

Therefore

(a) $D_{xx}f(x, y) = y^2 e^x - \dfrac{1}{x^2}$

and

(b) $D_{xy}f(x, y) = 2ye^x$

(c) To find $\partial^3 f/\partial x \, \partial y^2$ we partial-differentiate twice with respect to y and then once with respect to x. This gives

$$\frac{\partial f}{\partial y} = 2ye^x + \frac{1}{y} \qquad \frac{\partial^2 f}{\partial y^2} = 2e^x - \frac{1}{y^2} \qquad \frac{\partial^3 f}{\partial x \, \partial y^2} = 2e^x$$

EXAMPLE 7 Given

$$f(x, y, z) = \ln(xy + 2z)$$

find $D_{xzy}f(x, y, z)$.

Solution

$$D_x f(x, y, z) = \frac{y}{xy + 2z}$$

$$D_{xz} f(x, y, z) = \frac{-2y}{(xy + 2z)^2}$$

$$D_{xzy} f(x, y, z) = \frac{-2}{(xy + 2z)^2} + \frac{4xy}{(xy + 2z)^3}$$

EXAMPLE 8 Given

$$f(x, y) = 4x^3 y - 3ye^{xy}$$

find (a) $D_{xy}f(x, y)$ and (b) $D_{yx}f(x, y)$.

Solution

(a) $D_x f(x, y) = 12x^2 y - 3y^2 e^{xy}$

$D_{xy}f(x, y) = 12x^2 - 6ye^{xy} - 3xy^2 e^{xy}$

(b) $D_y f(x, y) = 4x^3 - 3e^{xy} - 3xye^{xy}$

$$D_{yx} f(x, y) = 12x^2 - 3ye^{xy} - 3ye^{xy} - 3xy^2 e^{xy}$$
$$= 12x^2 - 6ye^{xy} - 3xy^2 e^{xy}$$

Observe from the results that for the function of Example 8 the "mixed" partial derivatives $D_{xy} f(x, y)$ and $D_{yx} f(x, y)$ are equal. So for this particular function, when finding the second partial derivative with respect to x and then y, the order of differentiation is immaterial. This condition holds for many functions and for nearly all of the functions that occur in practice. It will hold for all of the functions discussed in this book.

Exercises 10.3

In Exercises 1 through 6, apply only the definition to find the partial derivative.

1. $f(x, y) = 6x + 3y - 7; D_x f(x, y)$

2. $f(x, y) = 4x^2 - 3xy; D_x f(x, y)$

3. $f(x, y) = 3xy + 6x - y^2; D_y f(x, y)$

4. $f(x, y) = xy^2 - 5y + 6; D_y f(x, y)$

5. $f(x, y) = \sqrt{2x + 3y}; D_x f(x, y)$

6. $f(x, y) = \dfrac{x + 2y}{x^2 - y}; D_y f(x, y)$

In Exercises 7 through 22, find the indicated partial derivative by holding all but one of the variables constant and applying theorems for ordinary differentiation.

7. $V = \pi r^2 h; \dfrac{\partial V}{\partial r}$

8. $f(x, y) = x^2 y - 3xy^2 + 4x; D_x f(x, y)$

9. $g(x, y) = x^4 - 2x^2 y + 3xy^2 - y^4; D_y g(x, y)$

10. $A = 100(1 + i)^{-t}; \dfrac{\partial A}{\partial t}$

11. $f(s, t) = t + \sqrt{s^2 + t^2}; f_t(s, t)$

12. $g(x, y) = x\sqrt{y^2 - x^2}; g_x(x, y)$

13. $z = ye^{y/x}; \dfrac{\partial z}{\partial x}$

14. $z = y \ln \dfrac{y}{x}; \dfrac{\partial z}{\partial y}$

15. $f(x, y, z) = x^2 y - 3xy^2 + 2yz; D_y f(x, y, z)$

16. $f(x, y, z) = 4x^2 y^2 - 8xyz + 9x^3 z - z^4; D_x f(x, y, z)$

17. $g(x, y, z) = (x^2 + y^2 + z^2)^{-1/2}; g_z(x, y, z)$

18. $w = xyz + \ln(xyz); \dfrac{\partial w}{\partial z}$

19. $u = e^{rst} + \ln \dfrac{rs}{t}; \dfrac{\partial u}{\partial s}$

20. $g(x, y, z) = \ln \sqrt{x^2 + y^2 + z^2}; g_y(x, y, z)$

21. $f(x, y, z, r, t) = xyr + yzt + yrt + zrt; D_r f(x, y, z, r, t)$

22. $g(r, s, t, u, v, w) = 3r^2 st + st^2 v - 2tuv^2 - tvw + 3uw^2; D_v g(r, s, t, u, v, w)$

23. Given $u = e^{r/t} + \ln \dfrac{t}{r}$, verify $t \dfrac{\partial u}{\partial t} + r \dfrac{\partial u}{\partial r} = 0$.

24. Given $w = x^2 y + y^2 z + z^2 x$, verify $\dfrac{\partial w}{\partial x} + \dfrac{\partial w}{\partial y} + \dfrac{\partial w}{\partial z} = (x + y + z)^2$.

25. Given $u = x^3 + y^3 + z^3 - 3xyz$, verify $x \dfrac{\partial u}{\partial x} + y \dfrac{\partial u}{\partial y} + z \dfrac{\partial u}{\partial z} = 3u$.

26. Given $u = e^{x/y} + e^{y/z} + e^{z/x}$, verify $x \dfrac{\partial u}{\partial x} + y \dfrac{\partial u}{\partial y} + z \dfrac{\partial u}{\partial z} = 0$.

In Exercises 27 through 32, do each of the following: (a) Find $D_{xx}f(x, y)$; (b) find $D_{yy}f(x, y)$; (c) show that $D_{xy}f(x, y) = D_{yx}f(x, y)$.

27. $f(x, y) = 2x^3 - 3x^2y + xy^2$
28. $f(x, y) = \dfrac{x^2}{y} - \dfrac{y}{x^2}$
29. $f(x, y) = \dfrac{x + y}{x - y}$

30. $f(x, y) = e^{2x} \ln y$
31. $f(x, y) = x \ln y + y \ln x$
32. $f(x, y) = xe^y - ye^x$

In Exercises 33 through 36, find the indicated partial derivative.

33. $f(x, y, z) = ye^x + ze^y + e^z$; (a) $f_{xz}(x, y, z)$; (b) $f_{yz}(x, y, z)$
34. $g(x, y, z) = \ln(xyz^2)$; (a) $g_{yz}(x, y, z)$; (b) $g_{xy}(x, y, z)$
35. $f(r, s) = r^3s + r^2s^2 - rs^3$; (a) $f_{rsr}(r, s)$; (b) $f_{ssr}(r, s)$
36. $g(r, s, t) = \ln(r^2 + s^2 + t^2)$; (a) $g_{rts}(r, s, t)$; (b) $g_{rss}(r, s, t)$
37. Find the slope of the tangent line to the curve of intersection of the surface $z = x^2 + y^2$ with the plane $y = 1$ at the point $(2, 1, 5)$. *Hint*: Refer to Fig. 10.1.18.
38. Find the slope of the tangent line to the curve of intersection of the sphere $x^2 + y^2 + z^2 = 9$ with the plane $x = 1$ at the point $(1, 2, 2)$.
39. Suppose that on a day when x workers are in the labor force and y machines are used, a manufacturer produces $f(x, y)$ tables, where

$$f(x, y) = x^2 + 4xy + 3y^2$$

with $4 \le x \le 25$ and $3 \le y \le 10$. (a) Find the number of tables produced in one day if 10 workers are present and 5 machines are used. (b) The partial derivative $D_x f(x, y)$ is called the *marginal productivity of labor*. Use this function to determine the approximate number of additional tables that can be produced in one day if the number of workers is increased from 10 to 11 and the number of machines remains fixed at 5. (c) The partial derivative $D_y f(x, y)$ is called the *marginal productivity of machines*. Use this function to determine the approximate number of additional tables that can be produced in one day if the number of machines is increased from 5 to 6 and the number of workers remains fixed at 10.

40. Suppose that x is the number of tens of thousands of dollars in the inventory carried in a store, y is the number of clerks in the store, P is the number of dollars in the weekly profit of the store, and

$$P = 3000 + 240y + 20y(x - 2y) - 10(x - 12)^2$$

where $15 \le x \le 25$ and $5 \le y \le 12$. At present the inventory is \$180,000 and there are 8 clerks. (a) Find the instantaneous rate of change of P per unit change in x if y remains fixed at 8. (b) Use the result of part (a) to find the approximate change in the weekly profit if the inventory changes from \$180,000 to \$200,000 and the number of clerks remains fixed at 8. (c) Find the instantaneous rate of change of P per unit change in y if x remains fixed at 18. (d) Use the result of part (c) to find the approximate change in the weekly profit if the number of clerks is increased from 8 to 10 and the inventory remains fixed at \$180,000.

41. From Eq. (8) in Sec. 8.8 we know that if V dollars is the present value of an ordinary annuity of equal payments of \$100 per year for t years at an interest rate of $100i$ percent per year, then

$$V = 100 \left[\dfrac{1 - (1 + i)^{-t}}{i} \right]$$

(a) Find the instantaneous rate of change of V per unit change in i if t remains fixed at 8. (b) Use the result of part (a) to find the approximate change in the present value if the interest rate changes from 10 percent to 11 percent and the time remains fixed at 8 years. (c) Find the instantaneous rate of change of V per unit change in t if i remains fixed at 0.10. (d) Use the result of part (c) to find the approximate change in the present value if the time is decreased from 8 to 7 years and the interest rate remains fixed at 10 percent.

42. According to the *ideal gas law* for a confined gas, if P pounds per square unit is the pressure, V cubic units is the volume, and T degrees is the temperature, we have the formula

$$PV = kT$$

where k is a constant of proportionality. Suppose that the volume of gas in a certain container is 100 in.^{3} and the temperature is $90°$, and $k = 8$. (a) Find the instantaneous rate of change of P per unit change in T if V remains fixed at 100. (b) Use the result of part (a) to approximate the change in the pressure if the temperature is increased to $92°$. (c) Find the instantaneous rate of change of V per unit change in P if T remains fixed at 90. (d) Suppose that the temperature is held constant. Use the result of part (c) to find the approximate change in the volume necessary to produce the same change in the pressure as obtained in part (b).

43. The temperature at any point (x, y) of a flat plate is T degrees, and $T = 54 - \frac{2}{3}x^2 - 4y^2$. If distance is measured in centimeters, find at the point $(3, 1)$ the rate of change of the temperature with respect to the distance moved along the plate in the direction of (a) the positive x axis and (b) the positive y axis.

44. Use the ideal gas law for a confined gas (see Exercise 42) to show that

$$\frac{\partial V}{\partial T} \cdot \frac{\partial T}{\partial P} \cdot \frac{\partial P}{\partial V} = -1$$

45. If S square meters is a person's body surface area, then a formula giving an approximate value of S is

$$S = 2W^{0.4}H^{0.7}$$

where W kg is the person's weight and H meters is the person's height. When $W = 70$ and $H = 1.8$, find (a) $\partial S/\partial W$ and (b) $\partial S/\partial H$. Interpret the results.

10.4 Some applications of partial derivatives to economics

In Sec. 10.2 we discussed the demand functions of two related commodities whose demands depend on the price of each commodity. We now use these functions to define the *partial marginal demand*.

Definition of partial marginal demand

Let p dollars be the price of one unit of x units of a first commodity and q dollars be the price of one unit of y units of a second commodity. Suppose that f and g are the respective demand functions for these two commodities so that

$$x = f(p, q) \quad \text{and} \quad y = g(p, q)$$

Then

(i) $\dfrac{\partial x}{\partial p}$ gives the **partial marginal demand of x with respect to p**;

(ii) $\dfrac{\partial x}{\partial q}$ gives the **partial marginal demand of x with respect to q**;

(iii) $\dfrac{\partial y}{\partial p}$ gives the **partial marginal demand of y with respect to p**;

(iv) $\dfrac{\partial y}{\partial q}$ gives the **partial marginal demand of y with respect to q**.

● ILLUSTRATION 1

The demand equations of Example 3 of Sec. 10.2 are

$$x = -2p + 3q + 12$$

and

$$y = -4q + p + 8$$

where p dollars is the price per unit of x units of one commodity and q dollars is the price per unit of y units of a second commodity. From the definition, the four partial marginal demands are given by

$$\frac{\partial x}{\partial p} = -2 \qquad \frac{\partial x}{\partial q} = 3 \qquad \frac{\partial y}{\partial p} = 1 \qquad \frac{\partial y}{\partial q} = -4$$

We interpret these results. Because $\partial x/\partial p = -2$, it follows that if q is held fixed, then an increase of $1 in the price per unit of the first commodity results in a decrease of 2 units in the demand of the first commodity. Because $\partial x/\partial q = 3$, then if p is held fixed, an increase of $1 in the price per unit of the second commodity results in an increase of 3 units in the demand of the first commodity. Because $\partial y/\partial p = 1$, then if q is held fixed, an increase of $1 in the price per unit of the first commodity results in an increase of 1 unit in the demand of the second commodity. Because $\partial y/\partial q = -4$, then if p is held fixed, an increase of $1 in the price per unit of the second commodity results in a decrease of 4 units in the demand of the second commodity. ●

EXAMPLE 1 Suppose that x units of commodity A and y units of commodity B are demanded when the unit price of A is p dollars and the unit price of B is q dollars. The demand equations are

$$x = 4q^2 - 5pq \qquad (1)$$

and

$$y = 7p^2 - 3pq \qquad (2)$$

(a) Find the amount demanded of each commodity when the price of A is $40 and the price of B is $60. (b) Find the four partial marginal demands when $p = 40$ and $q = 60$. (c) Use the results of part (b) to determine how the amount demanded of each commodity is affected when the price of A is increased from $40 to $41 and the price of B is held fixed at $60. (d) Use the results of part (b) to determine how the amount demanded of each commodity is affected when the price of B is increased from $60 to $61 and the price of A is held fixed at $40.

Solution (a) In Eqs. (1) and (2) let $p = 40$ and $q = 60$ to obtain

$$x = 4(60)^2 - 5(40)(60)$$
$$= 14{,}400 - 12{,}000$$
$$= 2400$$

$$y = 7(40)^2 - 3(40)(60)$$
$$= 11{,}200 - 7{,}200$$
$$= 4000$$

Thus, when the price of A is $40 and the price of B is $60, the amount demanded of A is 2400 units and the amount demanded of B is 4000 units.

(b) To find the four partial marginal demands we apply the definition to the demand equations (1) and (2), and we have

$$\frac{\partial x}{\partial p} = -5q \qquad \frac{\partial x}{\partial q} = 8q - 5p$$

$$\frac{\partial y}{\partial p} = 14p - 3q \qquad \frac{\partial y}{\partial q} = -3p$$

Therefore, when $p = 40$ and $q = 60$, we have

$$\frac{\partial x}{\partial p} = -300 \qquad \frac{\partial x}{\partial q} = 480 - 200$$

$$= 280$$

$$\frac{\partial y}{\partial p} = 560 - 180 \qquad \frac{\partial y}{\partial q} = -120$$

$$= 380$$

(c) When the price of B is held fixed at \$60 and the price of A is increased from \$40 to \$41, then because $\partial x/\partial p = -300$, the amount demanded of A is decreased by 300 units; furthermore, because $\partial y/\partial p = 380$, the amount demanded of B is increased by 380 units.

(d) When the price of A is held fixed at \$40 and the price of B is increased from \$60 to \$61, then because $\partial x/\partial q = 280$, the amount demanded of A is increased by 280 units; also, because $\partial y/\partial q = -120$, the amount demanded of B is decreased by 120 units.

Consider again the demand equations

$$x = f(p, q) \quad \text{and} \quad y = g(p, q) \tag{3}$$

where x units of a first commodity and y units of a second commodity are demanded when p dollars is the price of one unit of the first commodity and q dollars is the price of one unit of the second commodity. In normal circumstances, if the variable q is held constant, x decreases as p increases and x increases as p decreases, and so we conclude that $\partial x/\partial p$ is negative. Similarly, $\partial y/\partial q$ is negative in normal circumstances.

We learned in Sec. 10.2 that two commodities are said to be complementary when a decrease in the demand for one commodity as a result of an increase in its price leads to a decrease in the demand for the other. So when the goods are complementary and q is held constant, both $\partial x/\partial p < 0$ and $\partial y/\partial p < 0$, and when p is held constant, then $\partial x/\partial q < 0$ and $\partial y/\partial q < 0$. Therefore we can conclude that the two commodities are complementary if and only if both $\partial x/\partial q$ and $\partial y/\partial p$ are negative.

When a decrease in the demand for one commodity as a result of an increase in its price leads to an increase in the demand for the other commodity, we learned (also in Sec. 10.2) that the goods are said to be substitutes. Hence, when the goods are substitutes, because $\partial x/\partial p$ is always negative, we conclude that $\partial y/\partial p$ is positive, and because $\partial y/\partial q$ is always negative, it follows that $\partial x/\partial q$ is positive. Consequently, the two commodities are substitutes if and only if $\partial x/\partial q$ and $\partial y/\partial p$ are both positive.

If $\partial x/\partial q$ and $\partial y/\partial p$ have opposite signs, the commodities are neither complementary nor substitutes. For example, if $\partial x/\partial q < 0$ and $\partial y/\partial p > 0$, and because $\partial x/\partial p$ and $\partial y/\partial q$ are always negative (in normal circumstances), we have both $\partial x/\partial q < 0$ and $\partial y/\partial q < 0$. Thus a decrease in the price of the second commodity causes an increase in the demands of both commodities. Because $\partial x/\partial p < 0$ and $\partial y/\partial p > 0$, a decrease in the price of the first commodity causes an increase in the demand of the first commodity and a decrease in the demand of the second commodity.

● ILLUSTRATION 2

The demand equations of Illustration 1 are

$$x = -2p + 3q + 12 \quad \text{and} \quad y = -4q + p + 8$$

Because

$$\frac{\partial x}{\partial q} = 3 > 0 \quad \text{and} \quad \frac{\partial y}{\partial p} = 1 > 0$$

the two commodities are substitutes. ●

● ILLUSTRATION 3

The demand equations of Example 4 of Sec. 10.2 are

$$x = \frac{8}{pq} \quad \text{and} \quad y = \frac{12}{pq}$$

Because

$$\frac{\partial x}{\partial q} = -\frac{8}{pq^2} < 0 \quad \text{and} \quad \frac{\partial y}{\partial p} = -\frac{12}{p^2 q} < 0$$

the two commodities are complementary. ●

The definition of **partial elasticity of demand** when the demand is a function of two prices is analogous to the definition of price elasticity of demand for a function of one variable, as given in Sec. 5.4. So for the demand equations (3), the partial elasticity of demand of x with respect to p, which we shall denote by Ex/Ep, is the relative change in x per unit relative change in p when q is held constant. Hence

$$\frac{Ex}{Ep} = \lim_{\Delta p \to 0} \left[\frac{f(p + \Delta p, q) - f(p, q)}{f(p, q)} \div \frac{\Delta p}{p} \right]$$

$$= \lim_{\Delta p \to 0} \left[\frac{p}{f(p, q)} \cdot \frac{f(p + \Delta p, q) - f(p, q)}{\Delta p} \right]$$

$$= \frac{p}{f(p, q)} \lim_{\Delta p \to 0} \frac{f(p + \Delta p, q) - f(p, q)}{\Delta p}$$

$$= \frac{p}{f(p, q)} f_p(p, q)$$

provided, of course, that $f_p(p, q)$ exists. Replacing $f(p, q)$ by x and using the notation $\partial x/\partial p$ in place of $f_p(p, q)$ we may write

$$\frac{Ex}{Ep} = \frac{p}{x} \cdot \frac{\partial x}{\partial p} \tag{4}$$

For the demands represented by x and y in equations (3) there are three other partial elasticities of demand given by

$$\frac{Ex}{Eq} = \frac{q}{x} \cdot \frac{\partial x}{\partial q} \qquad \frac{Ey}{Ep} = \frac{p}{y} \cdot \frac{\partial y}{\partial p} \qquad \frac{Ey}{Eq} = \frac{q}{y} \cdot \frac{\partial y}{\partial q} \tag{5}$$

EXAMPLE 2 Suppose that x represents the demand for butter and y represents the demand for margarine when the price per pound of butter is p cents and the price per pound of margarine is q cents. Furthermore, the demand equations are

$$x = p^{-0.2}q^{0.3} \quad \text{and} \quad y = p^{0.5}q^{-1.2}$$

Show that butter and margarine are substitutes, and find the four partial elasticities of demand. Interpret the results.

Solution

$$\frac{\partial x}{\partial p} = -0.2p^{-1.2}q^{0.3} \qquad \frac{\partial x}{\partial q} = 0.3p^{-0.2}q^{-0.7}$$

$$\frac{\partial y}{\partial p} = 0.5p^{-0.5}q^{-1.2} \qquad \frac{\partial y}{\partial q} = -1.2p^{0.5}q^{-2.2}$$

Because $\partial x/\partial q$ and $\partial y/\partial p$ are both positive, butter and margarine are substitutes.

From Eqs. (4) and (5) we have

$$\frac{Ex}{Ep} = \frac{p}{x} \cdot \frac{\partial x}{\partial p} = \frac{p}{p^{-0.2}q^{0.3}} (-0.2p^{-1.2}q^{0.3}) = -0.2$$

$$\frac{Ex}{Eq} = \frac{q}{x} \cdot \frac{\partial x}{\partial q} = \frac{q}{p^{-0.2}q^{0.3}} (0.3p^{-0.2}q^{-0.7}) = 0.3$$

$$\frac{Ey}{Ep} = \frac{p}{y} \cdot \frac{\partial y}{\partial p} = \frac{p}{p^{0.5}q^{-1.2}} (0.5p^{-0.5}q^{-1.2}) = 0.5$$

$$\frac{Ey}{Eq} = \frac{q}{y} \cdot \frac{\partial y}{\partial q} = \frac{q}{p^{0.5}q^{-1.2}} (-1.2p^{0.5}q^{-2.2}) = -1.2$$

From the values of Ex/Ep and Ey/Ep we can conclude that if the price of margarine is held constant and the price of butter is increased by 1 percent, the demand for butter decreases by 0.2 percent and the demand for margarine increases by 0.5 percent. Similarly, from the values of Ex/Eq and Ey/Eq it follows that in holding the price of butter constant, an increase of 1 percent in the price of margarine causes an increase in the demand for butter of 0.3 percent and a decrease in the demand for margarine of 1.2 percent.

Exercises 10.4

In Exercises 1 through 8, demand equations for two related commodities are given. In each exercise, find the four partial marginal demands. Determine if the commodities are complementary or substitutes.

1. $x = 14 - p - 2q$, $y = 17 - 2p - q$

2. $x = 5 - 2p + q$, $y = 6 + 3p - q$

3. $x = p^{-0.4}q^{0.5}$, $y = p^{0.4}q^{-1.5}$

4. $x = p^{0.5}q^{-1.3}$, $y = p^{-0.6}q^{0.6}$

5. $x = 2^{-p-q}$, $y = 3^{-pq}$

6. $x = 5e^{q-p}$, $y = 3e^{p-q}$

7. $x = \dfrac{q^2}{p}$, $y = \dfrac{p}{q}$

8. $x = \dfrac{1}{pq}$, $y = \dfrac{1}{p^2q}$

In Exercises 9 and 10, for the given demand equations suppose that x units of the first commodity and y units of the second commodity are demanded when the prices per unit are p dollars and q dollars, respectively. Use the partial marginal demands to determine how the amount demanded of each commodity is affected in each of the following cases: (a) q is held fixed and the price of the first commodity is increased by \$1; (b) p is held fixed and the price of the second commodity is increased by \$1; (c) q is held fixed and the price of the first commodity is decreased by \$1; (d) p is held fixed and the price of the second commodity is decreased by \$1.

9. $x = 12 - 4p - 3q$, $y = 15 - 2p - q$

10. $x = 8 - 2p + q$, $y = 16 + 3p - 5q$

In Exercises 11 through 18, demand equations for two related commodities are given. Do each of the following: (a) find the four partial elasticities of demand; (b) at $p = 1$ and $q = 2$, if p is increased by 1 percent and q is held constant, find the percent changes in x and y; and (c) at $p = 1$ and $q = 2$, if q is increased by 1 percent and p is held constant, find the percent changes in x and y; (d) at $p = 1$ and $q = 2$, if p is decreased by 1 percent and q is held constant, find the percent changes in x and y; (e) at $p = 1$ and $q = 2$, if q is decreased by 1 percent and p is held constant, find the percent changes in x and y.

11. The demand equations of Exercise 1.

12. The demand equations of Exercise 2.

13. The demand equations of Exercise 3.

14. The demand equations of Exercise 4.

15. The demand equations of Exercise 5.

16. The demand equations of Exercise 6.

17. The demand equations of Exercise 7.

18. The demand equations of Exercise 8.

19. The demand equations for two commodities A and B are

$$x = 5q^2 - 2pq \quad \text{and} \quad y = 7p^2 - 6pq$$

where x units of A and y units of B are demanded when the unit prices of A and B are, respectively, p dollars and q dollars. (a) Find the amount demanded of each commodity when the price of A is \$10 per unit and the price of B is \$8 per unit. (b) Find the four partial marginal demands when $p = 10$ and $q = 8$. (c) Use the results of part (b) to determine how the amount demanded of each commodity is affected when the price of A is increased from \$10 to \$11 and the price of B is held fixed at \$8. (d) Use the results of part (b) to determine how the amount demanded of each commodity is affected when the price of B is increased from \$8 to \$9 and the price of A is held fixed at \$10.

20. Prove that the commodities of Example 1 are substitutes. *Hint:* Find inequalities involving p and q from the fact that $x \geq 0$ and $y \geq 0$. Then use these inequalities to prove that $\partial x/\partial q > 0$ and $\partial y/\partial p > 0$.

21. Prove that the commodities of Exercise 19 are substitutes. See the hint for Exercise 20.

22. When the price of an umbrella is p dollars and the price of a raincoat is q dollars, x umbrellas and y raincoats are demanded. The respective demand equations are

$$x = 4e^{-p/100q} \quad \text{and} \quad y = 8e^{-q/200p}$$

(a) Show that the two commodities are substitutes. (b) Find the four partial elasticities of demand. Suppose that the price of an umbrella is \$10 and the price of a raincoat is \$30. Find the percent changes in the demands for umbrellas and raincoats if (c) the price of an umbrella is decreased by 1 percent, and (d) the price of a raincoat is decreased by 1 percent.

23. Suppose that x neckties are demanded and y dress shirts are demanded when the price of a necktie is p dollars and the price of a dress shirt is q dollars. The respective demand equations are

$$x = p^{-0.5}q^{-0.2} \quad \text{and} \quad y = p^{-1.3}q^{-0.8}$$

(a) Show that the two commodities are complementary. (b) Find the four partial elasticities of demand. Find the percent changes in the demands for neckties and dress shirts if (c) the price of a necktie is increased by 1 percent and the price of a dress shirt remains constant, and (d) the price of a dress shirt is increased by 1 percent and the price of a necktie remains constant.

10.5 Limits, continuity, and extrema of functions of two variables

When discussing extrema of functions of two variables, we need the concept of continuity of such a function. We first define an *open disk*.

Definition of an
open disk

If (x_0, y_0) is a point in R^2 and r is a positive number, then the **open disk** $B((x_0, y_0); r)$ consists of all points in the interior region bounded by the circle having its center at (x_0, y_0) and radius r.

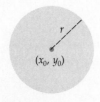

open disk $B((x_0, y_0); r)$

Figure 10.5.1

Figure 10.5.1 shows a circular disk $B((x_0, y_0); r)$. Observe that points on the circumference of the circle are not part of the open disk.

We are now in a position to define what is meant by the limit of a function of two variables.

Definition of the limit of a
function of two variables

Let f be a function of two variables that is defined on some open disk $B((x_0, y_0); r)$ except possibly at the point (x_0, y_0) itself. Then the **limit of $f(x, y)$ as (x, y) approaches (x_0, y_0) is L**, written as

$$\lim_{(x,y)\to(x_0,y_0)} f(x, y) = L$$

if $|f(x, y) - L|$ can be made as small as we please by making the distance between the points (x, y) and (x_0, y_0) small enough but greater than zero.

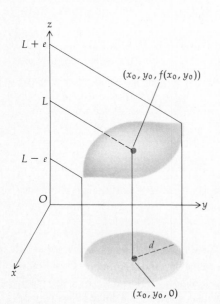

Figure 10.5.2

Another way of stating the above definition is as follows: The function values $f(x, y)$ approach a limit L as the point (x, y) approaches the point (x_0, y_0) if the absolute value of the difference between $f(x, y)$ and L can be made arbitrarily small by taking the point (x, y) sufficiently close to (x_0, y_0) but not equal to (x_0, y_0). Note that in the definition nothing is said about the function value at the point (x_0, y_0); that is, it is not necessary that the function be defined at (x_0, y_0) in order for $\lim_{(x,y)\to(x_0,y_0)} f(x, y)$ to exist.

A geometric interpretation of the definition of the limit of a function of two variables is shown in Fig. 10.5.2. The portion above the open disk $B((x_0, y_0); d)$ of the surface having equation $z = f(x, y)$ is shown. We see that $f(x, y)$ on the z axis will lie between $L - e$ and $L + e$ (that is, $f(x, y)$ will be within e units of L) whenever the point (x, y) in the xy plane is in the open disk $B((x_0, y_0); d)$ (that is, whenever (x, y) is within d units of

(x_0, y_0)). Another way of stating this is that $f(x, y)$ on the z axis can be restricted to lie between $L - e$ and $L + e$ by restricting the point (x, y) in the xy plane to be in the open disk $B((x_0, y_0); d)$.

The limit theorems of Chapter 2, with minor modifications, apply to functions of two variables. We use these theorems without restating them.

● ILLUSTRATION 1

By applying the limit theorems on sums and products we have

$$\lim_{(x,y)\to(-2,1)} (x^3 + 2x^2y - y^2 + 2) = (-2)^3 + 2(-2)^2(1) - (1)^2 + 2$$

$$= 1 \qquad ●$$

EXAMPLE 1 Find

$$\lim_{(x,y)\to(0,0)} \frac{x^2y^2 + y^4 - 4x^2 - 4y^2}{x^2y^2 + x^4 + 2x^2 + 2y^2}$$

Solution

$$\lim_{(x,y)\to(0,0)} \frac{x^2y^2 + y^4 - 4x^2 - 4y^2}{x^2y^2 + x^4 + 2x^2 + 2y^2} = \lim_{(x,y)\to(0,0)} \frac{y^2(x^2 + y^2) - 4(x^2 + y^2)}{x^2(x^2 + y^2) + 2(x^2 + y^2)}$$

$$= \lim_{(x,y)\to(0,0)} \frac{(x^2 + y^2)(y^2 - 4)}{(x^2 + y^2)(x^2 + 2)}$$

$$= \lim_{(x,y)\to(0,0)} \frac{y^2 - 4}{x^2 + 2}$$

$$= \frac{\displaystyle\lim_{(x,y)\to(0,0)} (y^2 - 4)}{\displaystyle\lim_{(x,y)\to(0,0)} (x^2 + 2)}$$

$$= \frac{-4}{2}$$

$$= -2$$

Definition of a function of two variables continuous at a point

The function f of two variables x and y is said to be **continuous at the point (x_0, y_0)** if and only if the following three conditions are satisfied:

(i) $f(x_0, y_0)$ exists.

(ii) $\displaystyle\lim_{(x,y)\to(x_0,y_0)} f(x, y)$ exists.

(iii) $\displaystyle\lim_{(x,y)\to(x_0,y_0)} f(x, y) = f(x_0, y_0)$.

The theorems about continuity for functions of a single variable can be extended to functions of two variables.

> **Theorem 10.5.1**   A polynomial function of two variables is continuous at every point in R^2.

● ILLUSTRATION 2

The function of Illustration 1, with function values

$$x^3 + 2x^2y - y^2 + 2$$

is continuous at every point in R^2 because it is a polynomial function. ●

> **Theorem 10.5.2**   A rational function of two variables is continuous at every point in its domain.

● ILLUSTRATION 3

The function with function values

$$\frac{6x^2 + xy - y^2}{9x^2 - y^2}$$

is a rational function, and therefore it is continuous at every point (x, y) in R^2 except those on the lines $y = 3x$ and $y = -3x$. ●

An important application of the derivative of a function of a single variable is in the study of extreme values of a function, which leads to a variety of problems involving maximum and minimum. We discussed this in Chapter 4, where we had theorems involving the first and second derivatives, which enabled us to determine relative maximum and minimum values of a function of one variable. In extending the theory to functions of two variables we see that it is similar to the one variable case; however, more complications arise.

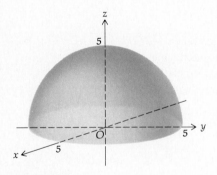

Figure 10.5.3

Definition of relative maximum value of a function of two variables

> The function f of two variables is said to have a **relative maximum value** at the point (x_0, y_0) if there exists an open disk $B((x_0, y_0); r)$ such that $f(x_0, y_0) \geq f(x, y)$ for all (x, y) in B.

● ILLUSTRATION 4

In Fig. 10.5.3 there is the graph of the function f defined by

$$f(x, y) = \sqrt{25 - x^2 - y^2}$$

Let B be any open disk $((0, 0); r)$ for which $r \leq 5$. From the above definition it follows that f has a relative maximum value of 5 at the point where $x = 0$ and $y = 0$. ●

Definition of relative minimum value of a function of two variables	The function f of two variables is said to have a **relative minimum value** at the point (x_0, y_0) if there exists an open disk $B((x_0, y_0); r)$ such that $f(x_0, y_0) \leq f(x, y)$ for all (x, y) in B.

● ILLUSTRATION 5

In Fig. 10.5.4 we have a sketch of the graph of the function g for which

$$g(x, y) = x^2 + y^2$$

Let B be any open disk $((0, 0); r)$. Then from the definition, g has a relative minimum value of 0 at the origin. ●

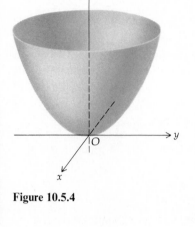

Figure 10.5.4

Analogous to Theorem 4.1.1 for functions of a single variable there is the following one for functions of two variables.

Theorem 10.5.3 If $f(x, y)$ exists at all points in some open disk $B((x_0, y_0); r)$, and if f has a relative extremum at (x_0, y_0), then if $f_x(x_0, y_0)$ and $f_y(x_0, y_0)$ exist,

$$f_x(x_0, y_0) = f_y(x_0, y_0) = 0$$

The proof of Theorem 10.5.3 is too advanced for this text. We do, however, give the following informal geometric argument.

Let f be a function satisfying the hypothesis of Theorem 10.5.3, and let f have a relative maximum value at (x_0, y_0). Consider the curve of intersection of the plane $y = y_0$ with the surface $z = f(x, y)$. Refer to Fig. 10.5.5. This curve is represented by the equations

$$y = y_0 \quad \text{and} \quad z = f(x, y) \tag{1}$$

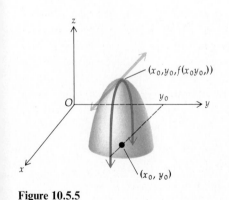

Figure 10.5.5

Because f has a relative maximum value at (x_0, y_0), it follows that the curve having Eqs. (1) has a horizontal tangent line in the plane $y = y_0$ at $(x_0, y_0, f(x_0, y_0))$. The slope of this tangent line is $f_x(x_0, y_0)$, and so $f_x(x_0, y_0) = 0$. In a similar way we can consider the curve of intersection of the plane $x = x_0$ with the surface $z = f(x, y)$ and conclude that $f_y(x_0, y_0) = 0$. An analogous discussion can be given if f has a relative minimum value at (x_0, y_0).

Definition of a critical point	A point (x_0, y_0) for which both $f_x(x_0, y_0) = 0$ and $f_y(x_0, y_0) = 0$ is called a **critical point**.

Theorem 10.5.3 states that a necessary condition for a function of two variables to have a relative extremum at a point, where its first partial derivatives exist, is that this point be a critical point. It is possible for a

function of two variables to have a relative extremum at a point at which the partial derivatives do not exist, but we do not consider this situation here. Furthermore, the vanishing of the first partial derivatives of a function of two variables is not a sufficient condition for the function to have a relative extremum at the point. Such a situation occurs at a **saddle point**.

● ILLUSTRATION 6

A simple example of a function that has a saddle point is the one defined by

$$f(x, y) = y^2 - x^2$$

For this function $f_x(x, y) = -2x$ and $f_y(x, y) = 2y$. Both $f_x(0, 0)$ and $f_y(0, 0)$ equal zero. A sketch of the graph of the function is shown in Fig. 10.5.6, and we see that it is saddle-shaped at points close to the origin. It is apparent that this function f does not satisfy either the definition of a relative maximum value or a relative minimum value when $(x_0, y_0) = (0, 0)$.   ●

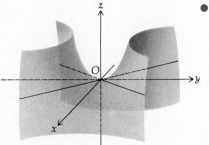

Figure 10.5.6

The basic test for determining relative maxima and minima for functions of two variables is the second-derivative test, which is given in the next theorem.

Second-derivative test

Theorem 10.5.4   Let f be a function of two variables such that f and its first- and second-order partial derivatives are continuous on some open disk $B((a, b); r)$. Suppose further that $f_x(a, b) = f_y(a, b) = 0$. Then

(i) f has a relative minimum value at (a, b) if

$$f_{xx}(a, b)f_{yy}(a, b) - f_{xy}{}^2(a, b) > 0 \quad \text{and} \quad f_{xx}(a, b) > 0$$

(ii) f has a relative maximum value at (a, b) if

$$f_{xx}(a, b)f_{yy}(a, b) - f_{xy}{}^2(a, b) > 0 \quad \text{and} \quad f_{xx}(a, b) < 0$$

(iii) $f(a, b)$ is not a relative extremum if

$$f_{xx}(a, b)f_{yy}(a, b) - f_{xy}{}^2(a, b) < 0$$

(iv) We can make no conclusion if

$$f_{xx}(a, b)f_{yy}(a, b) - f_{xy}{}^2(a, b) = 0$$

Once again we have a theorem for which the proof is too advanced for this book. The following two illustrations show the application of the second-derivative test to two functions previously considered.

● ILLUSTRATION 7

In Illustration 5 we had the function g defined by

$$g(x, y) = x^2 + y^2$$

Partial-differentiating, we obtain

$$g_x(x, y) = 2x \quad \text{and} \quad g_y(x, y) = 2y$$

Setting $g_x(x, y) = 0$ and $g_y(x, y) = 0$ we obtain $x = 0$ and $y = 0$. Thus the one and only critical point of f is $(0, 0)$. To apply the second-derivative test we compute the second partial derivatives of g, and we obtain

$$g_{xx}(x, y) = 2 \qquad g_{yy}(x, y) = 2 \qquad g_{xy}(x, y) = 0$$

Because

$$g_{xx}(0, 0) = 2 > 0$$

and

$$g_{xx}(0, 0)g_{yy}(0, 0) - g_{xy}{}^2(0, 0) = 2 \cdot 2 - 0 = 4 > 0$$

it follows from Theorem 10.5.4(i) that g has a relative minimum value at $(0, 0)$. This agrees with the conclusion in Illustration 5.   ●

● ILLUSTRATION 8

In Illustration 6 we had the function f defined by

$$f(x, y) = y^2 - x^2$$

with $f_x(x, y) = -2x$, $f_y(x, y) = 2y$, and $f_x(0, 0) = f_y(0, 0) = 0$. Therefore $(0, 0)$ is the one and only critical point of f. We compute the second partial derivatives of f.

$$f_{xx}(x, y) = -2 \qquad f_{yy}(x, y) = 2 \qquad f_{xy}(x, y) = 0$$

Because

$$f_{xx}(0, 0)f_{yy}(0, 0) - f_{xy}{}^2(0, 0) = (-2) \cdot 2 - 0 = -4 < 0$$

it follows from Theorem 10.5.4(iii) that $f(0, 0)$ is not a relative extremum. In Illustration 6 we showed that the origin is a saddle point of f.   ●

EXAMPLE 2   If $f(x, y) = 2x^4 + y^2 - x^2 - 2y$, determine the relative extrema of f if there are any.

Solution   To apply the second-derivative test we find the first and second partial derivatives of f.

$$f(x, y) = 2x^4 + y^2 - x^2 - 2y$$

$$f_x(x, y) = 8x^3 - 2x \qquad f_y(x, y) = 2y - 2$$

$$f_{xx}(x, y) = 24x^2 - 2 \qquad f_{yy}(x, y) = 2 \qquad f_{xy}(x, y) = 0$$

Setting $f_x(x, y) = 0$ we get $x = -\frac{1}{2}, x = 0$, and $x = \frac{1}{2}$. Setting $f_y(x, y) = 0$ we obtain $y = 1$. Therefore f_x and f_y are both 0 at the points $(-\frac{1}{2}, 1)$, $(0, 1)$, and $(\frac{1}{2}, 1)$, and these are the critical points of f. The results of applying the second-derivative test at these points are summarized in Table 10.5.1.

Table 10.5.1

Critical point	f_{xx}	f_{yy}	f_{xy}	$f_{xx}f_{yy} - f_{xy}{}^2$	Conclusion
$(-\frac{1}{2}, 1)$	4	2	0	8	f has a relative minimum value
$(0, 1)$	-2	2	0	-4	f does not have a relative extremum
$(\frac{1}{2}, 1)$	4	2	0	8	f has a relative minimum value

At the point $(-\frac{1}{2}, 1)$ $f_{xx} > 0$ and $f_{xx}f_{yy} - f_{xy}{}^2 > 0$; thus from Theorem 10.5.4(i), f has a relative minimum value at $(-\frac{1}{2}, 1)$. At $(0, 1)$, $f_{xx}f_{yy} - f_{xy}{}^2 < 0$; so from Theorem 10.5.4(iii), f does not have a relative extremum at $(0, 1)$. Because $f_{xx} > 0$ and $f_{xx}f_{yy} - f_{xy}{}^2 > 0$ at $(\frac{1}{2}, 1)$, f has a relative minimum value there by Theorem 10.5.4(i).

Because $f(-\frac{1}{2}, 1) = -\frac{9}{8}$ and $f(\frac{1}{2}, 1) = -\frac{9}{8}$, we conclude that f has a relative minimum value of $-\frac{9}{8}$ at each of the points $(-\frac{1}{2}, 1)$ and $(\frac{1}{2}, 1)$.

Applications of extrema of functions of two variables are given in the next section.

Exercises 10.5

In Exercises 1 through 14, find the given limit by the use of limit theorems.

1. $\lim\limits_{(x,y)\to(3,2)} (3x - 4y)$

2. $\lim\limits_{(x,y)\to(1,4)} (5x - 3y)$

3. $\lim\limits_{(x,y)\to(1,1)} (x^2 + y^2)$

4. $\lim\limits_{(x,y)\to(5,3)} (2x^2 - y^2)$

5. $\lim\limits_{(x,y)\to(-2,-4)} (x^2 + 2x - y)$

6. $\lim\limits_{(x,y)\to(3,-1)} (x^2 + y^2 - 4x + 2y)$

7. $\lim\limits_{(x,y)\to(-2,2)} y\sqrt[3]{x^3 + 4y}$

8. $\lim\limits_{(x,y)\to(5,2)} \sqrt{\dfrac{x^2 + 12y}{x - y^2}}$

9. $\lim\limits_{(x,y)\to(0,0)} \dfrac{x^4 - y^4}{x^2 + y^2}$

10. $\lim\limits_{(x,y)\to(1,e)} \ln\dfrac{y}{x}$

11. $\lim\limits_{(x,y)\to(1,1)} \dfrac{e^x + e^y}{e^{-x} + e^{-y}}$

12. $\lim\limits_{(x,y)\to(0,0)} \dfrac{e^x + e^y}{e^{-x} + e^{-y}}$

13. $\lim\limits_{(x,y)\to(0,0)} \dfrac{y^3 + x^2y + 3x^2 + 3y^2}{x^3 + xy^2 - 3x^2 - 3y^2}$

14. $\lim\limits_{(x,y)\to(0,0)} \dfrac{x^3 + y^3 + x^2y + xy^2}{x^2 + y^2}$

In Exercises 15 through 20, determine all points in R^2 for which the given function is continuous.

15. $f(x, y) = x^2y^2(x + y)^2$

16. $f(x, y) = (2x - 3)^2(y^2 + 1)^3$

17. $f(x, y) = \dfrac{x^4 - y^4}{x^2 - y^2}$

18. $f(x, y) = \left(\dfrac{1}{x^2 - 9} - \dfrac{1}{y}\right)^{1/3}$

19. $f(x, y) = \ln x - \ln y$

20. $f(x, y) = \ln(9 - x^2 - y^2) + \ln(x^2 + y^2 - 1)$

In Exercises 21 through 32, determine the relative extrema of f if there are any.

21. $f(x, y) = 2x^2 + y^2 - 8x - 2y + 14$

22. $f(x, y) = x^2 + y^2 - 2x - y + 1$

23. $f(x, y) = x^3 + y^2 - 6x^2 + y - 1$

24. $f(x, y) = x^2 - 4xy + y^3 + 4y$

25. $f(x, y) = \dfrac{1}{x} - \dfrac{64}{y} + xy$

26. $f(x, y) = 18x^2 - 32y^2 - 36x - 128y - 110$

27. $f(x, y) = 4xy^2 - 2x^2y - x$

28. $f(x, y) = x^3 + y^3 - 18xy$

29. $f(x, y) = \dfrac{2}{xy} + x^2 + y^2$

30. $f(x, y) = \dfrac{2x + 2y + 1}{x^2 + y^2 + 1}$

31. $f(x, y) = e^{xy}$

32. $f(x, y) = x^3 + x^2 + y^2 - xy + 8$

10.6 Applications of extrema of functions of two variables

Before giving some applications, we define absolute extrema of functions of two variables.

<table>
<tr><td>Definition of absolute maximum value of a function of two variables</td><td>The function f of two variables is said to have an **absolute maximum value** on its domain D in the xy plane if there is some point (x_0, y_0) in D such that $f(x_0, y_0) \geq f(x, y)$ for all (x, y) in D. In such a case $f(x_0, y_0)$ is the absolute maximum value of f on D.</td></tr>
</table>

<table>
<tr><td>Definition of absolute minimum value of a function of two variables</td><td>The function f of two variables is said to have an **absolute minimum value** on its domain D in the xy plane if there is some point (x_0, y_0) in D such that $f(x_0, y_0) \leq f(x, y)$ for all (x, y) in D. In such a case $f(x_0, y_0)$ is the absolute minimum value of f on D.</td></tr>
</table>

● ILLUSTRATION 1

In Illustration 1 of Sec. 10.2, when x dollars is the selling price of the first kind of sweater, y dollars is the selling price of the second kind of sweater, and $P(x, y)$ dollars is the weekly gross profit from the sales of the two kinds of sweaters, we have, from Eq. (2) of Sec. 10.2,

$$P(x, y) = 50xy - 50x^2 - 25y^2 + 3950x + 650y - 148,000 \qquad (1)$$

Each of the variables x and y is in the interval $(0, +\infty)$. To find the values of x and y for which the profit is greatest we first find the relative extrema of P by the second-derivative test. Differentiating, we get

$$\frac{\partial P}{\partial x} = 50y - 100x + 3950 \qquad \frac{\partial P}{\partial y} = 50x - 50y + 650$$

$$\frac{\partial^2 P}{\partial x^2} = -100 \qquad \frac{\partial^2 P}{\partial y \, \partial x} = 50 \qquad \frac{\partial^2 P}{\partial y^2} = -50$$

Setting $\partial P/\partial x = 0$ and $\partial P/\partial y = 0$ and solving simultaneously we have

$$-2x + y + 79 = 0$$

$$x - y + 13 = 0$$

from which it follows that $x = 92$ and $y = 105$. For all values of x and y,

$$\frac{\partial^2 P}{\partial x^2} = -100 < 0$$

and

$$\frac{\partial^2 P}{\partial x^2} \cdot \frac{\partial^2 P}{\partial y^2} - \left(\frac{\partial^2 P}{\partial y\,\partial x}\right)^2 = (-100)(-50) - (50)^2 = 2500 > 0$$

From Theorem 10.5.4(ii), P has a relative maximum value when $x = 92$ and $y = 105$. Furthermore, there is only one relative extremum for P. Also, x and y are both in the interval $(0, +\infty)$, and from Eq. (1) we observe that $P(x, y)$ is negative when x and y are either close to zero or very large. We therefore conclude that the relative maximum value of P is an absolute maximum value. When $x = 92$ and $y = 105$, we obtain, from (1), $P(92, 105) = 67,825$. Thus the greatest weekly gross profit is $67,825 when the selling price of the first kind of sweater is $92 and the selling price of the second kind of sweater is $105.   ●

EXAMPLE 1   What should be the dimensions of a rectangular box, without a top and having a volume of 32 cubic feet, if the least amount of material is to be used in its manufacture?

Solution   Let x ft be the length of the base of the box, y ft be the width of the base of the box, z ft be the depth of the box, and S ft^2 be the surface area of the box. Figure 10.6.1 shows the box.

Each of the variables x, y, and z is in the interval $(0, +\infty)$. We have the equations

$$S = xy + 2xz + 2yz \quad \text{and} \quad xyz = 32$$

Solving the second equation for z in terms of x and y we get $z = 32/xy$, and substituting this into the first equation gives

$$S(x, y) = xy + \frac{64}{y} + \frac{64}{x} \tag{2}$$

Differentiating we obtain

$$\frac{\partial S}{\partial x} = y - \frac{64}{x^2} \qquad \frac{\partial S}{\partial y} = x - \frac{64}{y^2}$$

$$\frac{\partial^2 S}{\partial x^2} = \frac{128}{x^3} \qquad \frac{\partial^2 S}{\partial y\,\partial x} = 1 \qquad \frac{\partial^2 S}{\partial y^2} = \frac{128}{y^3}$$

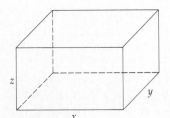

Figure 10.6.1

Setting $\partial S / \partial x = 0$ and $\partial S / \partial y = 0$ and solving simultaneously we get

$$x^2 y - 64 = 0$$

$$xy^2 - 64 = 0$$

From these two equations we obtain $x = 4$ and $y = 4$. For these values of x and y,

$$\frac{\partial^2 S}{\partial x^2} = \frac{128}{64} = 2 > 0 \quad \text{and} \quad \frac{\partial^2 S}{\partial x^2} \cdot \frac{\partial^2 S}{\partial y^2} - \left(\frac{\partial^2 S}{\partial y\, \partial x}\right)^2 = \frac{128}{64} \cdot \frac{128}{64} - 1 = 3 > 0$$

From Theorem 10.5.4(i) it follows that S has a relative minimum value when $x = 4$ and $y = 4$. Recall that x and y are both in the interval $(0, +\infty)$, and notice from Eq. (2) that S is very large when x and y are either close to zero or very large. Furthermore, there is only one relative extremum for S. Thus we conclude that the relative minimum value of S is an absolute minimum value of S. When $x = 4$ and $y = 4$, $z = \frac{32}{16} = 2$. Hence the box should have a square base of side 4 ft and a depth of 2 ft.

For functions of a single variable we had the following theorem (4.1.2), called the extreme-value theorem: If the function f is continuous on the closed interval $[a, b]$, then f has an absolute maximum value and an absolute minimum value on $[a, b]$. We learned that an absolute extremum of a function continuous on a closed interval must be either a relative extremum or a function value at an endpoint of the interval. There is a corresponding situation for functions of two variables. In the statement of the extreme-value theorem for functions of two variables we refer to a *closed region* in the xy plane. By a **closed region** we mean that the region includes its **boundary**. In the following illustration we give some closed regions and identify the boundary of each region.

● ILLUSTRATION 2

(a) A closed disk is an open disk together with the circumference of the disk. Thus a closed disk is a closed region and the circumference is the boundary. See Fig. 10.6.2(a).

(b) The sides of a triangle together with the region enclosed by the triangle is a closed region. The boundary consists of the sides of the triangle. See Fig. 10.6.2(b).

(c) The edges of a rectangle together with the region enclosed by the rectangle is a closed region. The boundary consists of the edges. See Fig. 10.6.2(c). ●

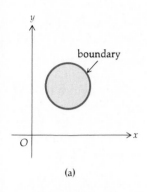

(a)

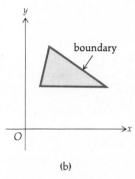

(b)

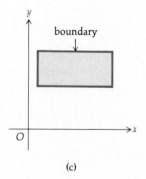

(c)

Figure 10.6.2

The extreme-value theorem for functions of two variables

Theorem 10.6.1 Let R be a closed region in the xy plane, and let f be a function of two variables that is continuous on R. Then there is at least one point in R where f has an absolute maximum value and at least one point in R where f has an absolute minimum value.

The proof of Theorem 10.6.1 is omitted because it is beyond the scope of this book.

If f is a function satisfying Theorem 10.6.1, and if both $f_x(x, y)$ and $f_y(x, y)$ exist at all points of R, then the absolute extrema of f occur at either a point (x_0, y_0), where $f_x(x_0, y_0) = f_y(x_0, y_0) = 0$, or at a point on the boundary of R.

EXAMPLE 2 A manufacturer who is a monopolist makes two types of lamps. From experience the manufacturer has determined that if x lamps of the first type and y lamps of the second type are made, they can be sold for $(100 - 2x)$ dollars each and $(125 - 3y)$ dollars each, respectively. The cost of manufacturing x lamps of the first type and y lamps of the second type is $(12x + 11y + 4xy)$ dollars. How many lamps of each type should be produced to realize the greatest profit, and what is the greatest profit?

Solution The number of dollars in the revenue received from the lamps of the first type is $x(100 - 2x)$, and the number of dollars in the revenue received from the lamps of the second type is $y(125 - 3y)$. Hence if $f(x, y)$ dollars is the manufacturer's profit,

$$f(x, y) = x(100 - 2x) + y(125 - 3y) - (12x + 11y + 4xy)$$

$$f(x, y) = 88x + 114y - 2x^2 - 3y^2 - 4xy \tag{3}$$

Because x and y both represent the number of lamps, we require that $x \geq 0$ and $y \geq 0$ and allow x and y to be any nonnegative real numbers. Furthermore, $(100 - 2x)$ dollars is the selling price of lamps of the first type. Thus we require that $100 - 2x \geq 0$ or, equivalently, $x \leq 50$. Similarly, because $(125 - 3y)$ dollars is the selling price of lamps of the second type, we require that $y \leq \frac{125}{3}$. Therefore the domain of f is the closed region defined by the set

$$\{(x, y) \mid 0 \leq x \leq 50 \text{ and } 0 \leq y \leq \tfrac{125}{3}\} \tag{4}$$

This region is rectangular and is shown in Fig. 10.6.3. The boundary of the region consists of the edges of the rectangle. Because f is a polynomial function, it is continuous everywhere. Hence f is continuous on the closed region defined by set (4); so the extreme-value theorem can be applied. The critical points of f are found by determining where $f_x(x, y) = 0$ and $f_y(x, y) = 0$.

$$f_x(x, y) = 88 - 4x - 4y$$

$$f_y(x, y) = 114 - 6y - 4x$$

Setting $f_x(x, y) = 0$ and $f_y(x, y) = 0$ we have

$$x + y = 22$$

$$2x + 3y = 57$$

Solving these two equations simultaneously we obtain $x = 9$ and $y = 13$. To apply the second-derivative test we find the second partial derivatives.

$$f_{xx}(x, y) = -4 \qquad f_{yy}(x, y) = -6 \qquad f_{xy}(x, y) = -4$$

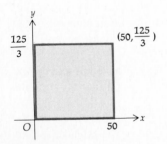

Figure 10.6.3

At the point (9, 13),

$$f_{xx}(9, 13) = -4 < 0$$

$$f_{xx}(9, 13)f_{yy}(9, 13) - f_{xy}{}^2(9, 13) = (-4)(-6) - (-4)^2$$
$$= 8 > 0$$

It follows, then, by Theorem 10.5.4(ii) that f has a relative maximum value at (9, 13).

From (3),

$$f(x, y) = x(88 - 2x) + y(114 - 3y) - 4xy \tag{5}$$

Thus

$$f(9, 13) = 9(70) + 13(75) - 468$$
$$= 1137$$

The absolute maximum value of f must occur at either (9, 13) or on the boundary of the domain of f. Let us compare $f(9, 13)$ with the function values on the boundary.

For that part of the boundary on the x axis with $x \in [0, 50]$ we have the function values computed from (5) as follows:

$$f(x, 0) = 88x - 2x^2$$

Let

$$g(x) = 88x - 2x^2 \qquad x \in [0, 50]$$

Then

$$g'(x) = 88 - 4x \quad \text{and} \quad g''(x) = -4$$

Because $g'(22) = 0$ and $g''(22) < 0$, g has a relative maximum value of 968 at $x = 22$. Furthermore, $g(0) = 0$ and $g(50) < 0$. Because $f(9, 13) = 1137 > 968$, the absolute maximum value of f does not occur on the x axis.

For that part of the boundary on the y axis with $y \in [0, \frac{125}{3}]$, from (5),

$$f(0, y) = 114y - 3y^2$$

Let

$$h(y) = 114y - 3y^2 \qquad y \in [0, \tfrac{125}{3}]$$

Then

$$h'(y) = 114 - 6y \quad \text{and} \quad h''(y) = -6$$

Because $h'(19) = 0$ and $h''(19) < 0$, h has a relative maximum value of 1083 at $y = 19$. Moreover, $h(0) = 0$ and $h(\frac{125}{3}) < 0$. Because $f(9, 13) = 1137 > 1083$, the absolute maximum value of f does not occur on the y axis.

We now consider the part of the boundary on the line $x = 50$ with $y \in [0, \frac{125}{3}]$. From (5),

$$f(50, y) = y(114 - 3y) - 600 - 200y \tag{6}$$

$$f(0, y) = y(114 - 3y) \tag{7}$$

By comparing (6) and (7),

$$f(50, y) < f(0, y) \tag{8}$$

Because $f(9, 13) > f(0, y)$ for all y in $[0, \frac{125}{3}]$, then from inequality (8),

$$f(9, 13) > f(50, y) \qquad \text{for } y \in [0, \tfrac{125}{3}]$$

Hence the absolute maximum value of f does not occur on the line $x = 50$.

Finally, we have that part of the boundary on the line $y = \frac{125}{3}$ with $x \in [0, 50]$. From (5),

$$f(x, \tfrac{125}{3}) = x(88 - 2x) - \tfrac{1375}{3} - \tfrac{500}{3}x \tag{9}$$

$$f(x, 0) = x(88 - 2x) \tag{10}$$

From (9) and (10) it follows that $f(x, \frac{125}{3}) < f(x, 0)$. Therefore, because $f(9, 13) > f(x, 0)$ for all x in $[0, 50]$, we can conclude that it is also greater than $f(x, \frac{125}{3})$ for all x in $[0, 50]$. Thus the absolute maximum value cannot occur on the line $y = \frac{125}{3}$.

Therefore the absolute maximum value of f is not on the boundary, and thus it is at the point $(9, 13)$. We conclude, then, that 9 lamps of the first type and 13 lamps of the second type should be produced for the greatest profit of $1137.

If the total cost of producing x units of one commodity and y units of another commodity is given by $C(x, y)$, then C is called a **joint-cost function**. The partial derivatives of C are called **marginal cost functions**.

Suppose that a monopolist produces two related commodities whose demand equations are $x = f(p, q)$ and $y = g(p, q)$ and the joint-cost function is C. Because the revenue from the two commodities is given by $px + qy$, then if S dollars is the profit,

$$S = px + qy - C(x, y)$$

To determine the greatest profit that can be earned we first use the demand equations to express S in terms of either p and q or x and y alone. The following example shows the procedure.

EXAMPLE 3 A monopolist produces two commodities that are substitutes and have demand equations

$$x = 8 - p + q$$

and

$$y = 9 + p - 5q$$

where $1000x$ units of the first commodity are demanded if the price is p dollars per unit and $1000y$ units of the second commodity are demanded if the price is q dollars per unit. It costs $4 to produce each unit of the first commodity and $2 to produce each unit of the second commodity. Find the amounts of output and the corresponding prices of the two commodities for the monopolist to have the greatest total profit.

Solution When $1000x$ units of the first commodity and $1000y$ units of the second commodity are produced and sold, the number of dollars in the total revenue is $1000px + 1000qy$, and the number of dollars in the total cost of production is $4000x + 2000y$. Thus the number of dollars in the total profit is

$$1000px + 1000qy - (4000x + 2000y)$$
$$= 1000p(8 - p + q) + 1000q(9 + p - 5q)$$
$$- 4000(8 - p + q) - 2000(9 + p - 5q)$$
$$= 1000(-p^2 + 2pq - 5q^2 + 10p + 15q - 50)$$

Therefore, if $S(p, q)$ dollars is the total profit,

$$S(p, q) = 1000(-p^2 + 2pq - 5q^2 + 10p + 15q - 50) \qquad (11)$$

Because x, y, p, and q must be nonnegative,

$$8 - p + q \geq 0 \qquad 9 + p - 5q \geq 0 \qquad p \geq 0 \qquad q \geq 0$$

From these inequalities we determine that the domain of the function S is the closed region shaded in Fig. 10.6.4. Because S is a polynomial function, it is continuous on its closed-region domain. Therefore the extreme-value theorem can be applied. We find the critical points of S by determining where $\partial S/\partial p$ and $\partial S/\partial q$ are zero.

$$\frac{\partial S}{\partial p} = 1000(-2p + 2q + 10) \qquad \frac{\partial S}{\partial q} = 1000(2p - 10q + 15)$$

Setting $\partial S/\partial p = 0$ and $\partial S/\partial q = 0$ we have

$$-2p + 2q + 10 = 0$$
$$2p - 10q + 15 = 0$$

From these two equations we obtain

$$p = \tfrac{65}{8} \quad \text{and} \quad q = \tfrac{25}{8}$$

Hence $(\tfrac{65}{8}, \tfrac{25}{8})$ is a critical point. Because

$$\frac{\partial^2 S}{\partial p^2} = -2000 \qquad \frac{\partial^2 S}{\partial q^2} = -10{,}000 \qquad \frac{\partial^2 S}{\partial q\, \partial p} = 2000$$

we have

$$\frac{\partial^2 S}{\partial p^2} \cdot \frac{\partial^2 S}{\partial q^2} - \left(\frac{\partial^2 S}{\partial q\, \partial p}\right)^2 = (-2000)(-10{,}000) - (2000)^2 > 0$$

Also, $\partial^2 S/\partial p^2 < 0$, and so from Theorem 10.5.4(ii) we conclude that S has a relative maximum value at $(\tfrac{65}{8}, \tfrac{25}{8})$. From (11) we obtain $S(\tfrac{65}{8}, \tfrac{25}{8}) = 14{,}062.5$. The absolute maximum value of S must occur at either $(\tfrac{65}{8}, \tfrac{25}{8})$ or on the boundary of S.

We now consider the function values $S(p, q)$ on the boundary. For points on the p axis we have, from (11),

$$S(p, 0) = -1000(p^2 - 10p + 50) \qquad (12)$$

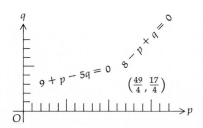

Figure 10.6.4

The expression $p^2 - 10p + 50$ is always positive. This fact can be shown by considering the function f for which $f(p) = p^2 - 10p + 50$. The graph of f is always above the p axis. Therefore, because $p^2 - 10p + 50 > 0$ for all p, then, from (12), $S(p, 0) < 0$ for all p. Hence the absolute maximum value of S cannot occur on the p axis.

For points on the q axis we have, from (11),

$$S(0, q) = -5000(q^2 - 3q + 10)$$

Because the expression $q^2 - 3q + 10$ is always positive (the graph of the function g, for which $g(q) = q^2 - 3q + 10$, is always above the q axis), it follows that $S(0, q) < 0$ for all q. Thus the absolute maximum value of S cannot occur on the q axis.

We find the function values of S for points on the line $9 + p - 5q = 0$ by letting $p = 5q - 9$ on the right-hand side of (11). We obtain

$$S(5q - 9, q) = 1000[-(5q - 9)^2 + 2q(5q - 9) - 5q^2$$
$$+ 10(5q - 9) + 15q - 50]$$

$$S(5q - 9, q) = 1000(-20q^2 + 137q - 221) \qquad (13)$$

We wish to determine the largest value that can be had by the right-hand side of (13). To do this, consider the function h defined by

$$h(q) = -20q^2 + 137q - 221$$

Because

$$h'(q) = -40q + 137 \quad \text{and} \quad h''(q) = -40$$

it follows from the second-derivative test for functions of a single variable that h has an absolute maximum value when $q = \frac{137}{40}$. We substitute this value of q into the right-hand side of (13) and obtain

$$1000[-20(\tfrac{137}{40})^2 + 137(\tfrac{137}{40}) - 221] = 13{,}612.5$$

Therefore $S(5q - 9, q) \le 13{,}612.5$ for all q. Because $S(\frac{65}{8}, \frac{25}{8}) = 14{,}062.5$ and $13{,}612.5 < 14{,}062.5$, we conclude that the absolute maximum value of S cannot occur on the line $9 + p - 5q = 0$.

In a similar way we can show that the absolute maximum value of S cannot occur on the line $8 - p + q = 0$. We let $q = p - 8$ on the right-hand side of (11) and get

$$S(p, p - 8) = 1000(-4p^2 + 89p - 490)$$

We determine that $S(p, p - 8)$ has its largest value of 5062.5 when $p = \frac{89}{8}$. Thus $S(p, p - 8) \le 5062.5 < 14{,}062.5$ for all p.

We have therefore shown that the absolute maximum value of S cannot occur on the boundary of its closed-region domain. Hence the absolute maximum value of S occurs at the point $(\frac{65}{8}, \frac{25}{8})$. From the demand equations we find that when $p = \frac{65}{8}$ and $q = \frac{25}{8}$, $x = 3$ and $y = \frac{3}{2}$.

Thus the greatest total profit of \$14,062.50 is attained when 3000 units of the first commodity are produced and sold at \$8.12$\frac{1}{2}$ per unit and 1500 units of the second commodity are produced and sold at \$3.12$\frac{1}{2}$ per unit.

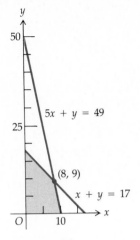

Figure 10.6.5

● ILLUSTRATION 3

In the preceding example, if the demand equations are solved for p and q in terms of x and y, we obtain

$$p = \tfrac{1}{4}(49 - 5x - y) \quad \text{and} \quad q = \tfrac{1}{4}(17 - x - y)$$

Because q and p must be nonnegative as well as x and y, it follows that

$$17 - x - y \geq 0 \qquad 49 - 5x - y \geq 0 \qquad x \geq 0 \qquad y \geq 0$$

From these four inequalities we determine that the permissible values of x and y occur in the closed region shaded in Fig. 10.6.5. The problem can be solved by considering x and y as the independent variables. You are asked to do this in Exercise 8.

●

In Sec. 10.2 we referred to a production function where the independent variables can be any of several inputs, such as the number of machines used in production, the number of person-hours available, the amount of working capital to be had, the quantity of materials used, and the amount of land available. Consider now a production function f of two variables, where the amounts of the inputs are given by x and y, and where z gives the amount of the output; then $z = f(x, y)$. Suppose that the prices of the two inputs are a dollars and b dollars per unit, respectively, and that the price of the output is c dollars, where a, b, and c are constants. This situation could occur if there were so many producers in the market that a change in the output of any particular producer would not affect the price of the commodity. Such a market is called a **perfectly competitive market**. If P dollars is the total profit, and because the total profit is obtained by subtracting the total cost from the total revenue, then

$$P = cz - (ax + by)$$

and because $z = f(x, y)$,

$$P = cf(x, y) - ax - by$$

It is, of course, desired to maximize P. This is illustrated by an example.

EXAMPLE 4 Suppose that the production of a certain commodity depends on two inputs. The amounts of these are given by $100x$ and $100y$, whose prices are, respectively, \$7 and \$4. The amount of the output is given by $100z$, the price per unit of which is \$9. Furthermore, the production function f has the function values

$$f(x, y) = \frac{x}{3} + \frac{y}{3} + 5 - \frac{1}{x} - \frac{1}{y}$$

Determine the greatest profit.

Solution If P dollars is the profit,

$$P = 9(100z) - 7(100x) - 4(100y) \tag{14}$$

Letting $z = f(x, y)$ we get

$$P = 900\left(\frac{x}{3} + \frac{y}{3} + 5 - \frac{1}{x} - \frac{1}{y}\right) - 700x - 400y$$

$$= 4500 - \frac{900}{x} - \frac{900}{y} - 400x - 100y$$

x and y are both in the interval $(0, +\infty)$. Hence

$$\frac{\partial P}{\partial x} = \frac{900}{x^2} - 400 \quad \text{and} \quad \frac{\partial P}{\partial y} = \frac{900}{y^2} - 100$$

Also,

$$\frac{\partial^2 P}{\partial x^2} = -\frac{1800}{x^3} \qquad \frac{\partial^2 P}{\partial y^2} = -\frac{1800}{y^3} \qquad \frac{\partial^2 P}{\partial y\, \partial x} = 0$$

Setting $\partial P/\partial x = 0$ and $\partial P/\partial y = 0$ we have

$$\frac{900}{x^2} - 400 = 0 \quad \text{and} \quad \frac{900}{y^2} - 100 = 0$$

from which we obtain $x = \frac{3}{2}$ and $y = 3$ (the negative result is rejected because x and y must be positive). At $(\frac{3}{2}, 3)$,

$$\frac{\partial^2 P}{\partial x^2} \cdot \frac{\partial^2 P}{\partial y^2} - \left(\frac{\partial^2 P}{\partial y\, \partial x}\right)^2 = \left(-\frac{1800}{\frac{27}{8}}\right)\left(-\frac{1800}{27}\right) - (0)^2 > 0$$

From the above and the fact that at $(\frac{3}{2}, 3)$, $\partial^2 P/\partial x^2 < 0$, it follows from Theorem 10.5.4(ii) that P has a relative maximum value at $(\frac{3}{2}, 3)$. Because x and y are both in the interval $(0, +\infty)$ and P is a negative number when x and y are either close to zero or very large, we conclude that the relative maximum value of P is an absolute maximum value. Because $z = f(x, y)$, the value of z at $(\frac{3}{2}, 3)$ is $f(\frac{3}{2}, 3) = \frac{1}{2} + 1 + 5 - \frac{2}{3} - \frac{1}{3} = \frac{11}{2}$. Hence, from (14),

$$P_{\max} = 900 \cdot \tfrac{11}{2} - 700 \cdot \tfrac{3}{2} - 400 \cdot 3 = 2700$$

The greatest profit, then, is $2700.

Exercises 10.6

1. Find the three positive numbers whose sum is 24 such that their product is as great as possible.
2. Find the three positive numbers whose product is 24 such that their sum is as small as possible.
3. Suppose that when the production of a particular commodity requires x machine-hours and y person-hours, the cost of production is given by $f(x, y)$, where $f(x, y) = 2x^3 - 6xy + y^2 + 500$. Determine the number of machine-hours and the number of person-hours needed to produce the commodity at the least cost.
4. A rectangular box without a top is to be made at a cost of $10 for the material. The cost of the material for the bottom of the box is 15 cents per square foot, and the material for the sides costs 30 cents per square foot. Find the dimensions of the box of greatest volume that can be made.
5. A closed rectangular box to contain 16 ft^3 is to be made of three kinds of material. The cost of the material for the top and bottom is 18 cents per square foot, the cost of the material for the front and back is 16 cents per square foot, and the cost

of the material for the other two sides is 12 cents per square foot. Find the dimensions of the box such that the total cost of the materials is a minimum.

6. A rectangular box without a top is to be made from 300 ft^2 of material. What should be the dimensions for the box to have the greatest possible volume?

7. The demand equations for two commodities that are produced by a monopolist are

$$x = 6 - 2p + q \quad \text{and} \quad y = 7 + p - q$$

where $100x$ is the quantity of the first commodity demanded if the price is p dollars per unit and $100y$ is the quantity of the second commodity demanded if the price is q dollars per unit. Show that the two commodities are substitutes. If it costs $2 to produce each unit of the first commodity and $3 to produce each unit of the second commodity, find the quantities demanded and the prices of the two commodities in order to have the greatest total profit. Take p and q as the independent variables.

8. Solve Example 3 of this section by taking x and y as the independent variables.

9. Solve Exercise 7 by taking x and y as the independent variables.

10. A monopolist produces staplers and staples having demand equations

$$x = \frac{10}{pq} \quad \text{and} \quad y = \frac{20}{pq}$$

where $1000x$ staplers are demanded if the price is p dollars per stapler and $1000y$ boxes of staples are demanded if the price per box of staples is q dollars. It costs $2 to produce each stapler and $1 to produce each box of staples. Determine the price of each commodity in order to have the greatest total profit.

11. If the demand equations in Exercise 10 are

$$x = 11 - 2p - 2q \quad \text{and} \quad y = 19 - 2p - 3q$$

show that to have the greatest total profit the staplers should be free and the staples should be expensive.

12. A monopolist produces two commodities A and B, and their demand equations are

$$x = 16 - 3p - 2q \quad \text{and} \quad y = 11 - 2p - 2q$$

where $100x$ units of A and $100y$ units of B are demanded when the unit prices of A and B are, respectively, p dollars and q dollars. Show that the two commodities are complementary. If the cost of production of each unit of commodity A is $1 and the cost of production of each unit of commodity B is $3, find the quantities demanded and the price of each commodity in order to have the greatest total profit.

13. The production function f for a certain commodity has function values

$$f(x, y) = x + \tfrac{5}{2}y - \tfrac{1}{8}x^2 - \tfrac{1}{4}y^2 - \tfrac{9}{8}$$

The amounts of the two inputs are given by $100x$ and $100y$, whose prices per unit are, respectively, $4 and $8, and the amount of the output is given by $100z$, whose price per unit is $16. Determine the greatest total profit.

14. A particular commodity has a production function f defined by

$$f(x, y) = 4 - \frac{8}{xy}$$

The amounts of the two inputs are given by $100x$ and $100y$, whose prices per unit are, respectively, $10 and $5, and the amount of the output is given by $100z$, whose price per unit is $20. Determine the greatest total profit.

15. A manufacturing plant has two classifications for its workers, A and B. Class A workers earn $14 per run, and class B workers earn $13 per run. For a certain production run it is determined that in addition to the salaries of the workers, if x class A workers and y class B workers are used, the number of dollars in the cost of the run is $y^3 + x^2 - 8xy + 600$. How many workers of each class should be used so that the total cost of the run is a minimum if at least three workers of each class are required for a run?

16. Suppose that t hr after the injection of x mg of adrenalin the response is R units, and

$$R = te^{-t}(c - x)x$$

where c is a positive constant. What values of x and t will cause the maximum response?

17. An injection of x mg of drug A and y mg of drug B cause a response of R units, and

$$R = x^2 y^3 (c - x - y)$$

where c is a positive constant. What quantity of each drug will cause the maximum response?

18. Suppose that t degrees is the temperature at any point (x, y) on the sphere $x^2 + y^2 + z^2 = 4$, and $t = 100xy^2z$. Find the points on the sphere where the temperature is the greatest and also the points where the temperature is the least. Also find the temperature at these points.

19. Determine the relative dimensions of a rectangular box, without a top, to be made from a given amount of material in order for the box to have the greatest possible volume.

20. Determine the relative dimensions of a rectangular box, without a top and having a specific volume, if the least amount of material is to be used in its manufacture.

10.7 Lagrange multipliers

In the solution of Example 4 in Sec. 10.6 we maximized the function P for which

$$P(x, y, z) = 900z - 700x - 400y \qquad (1)$$

subject to the condition that x, y, and z satisfy the equation

$$z = \frac{x}{3} + \frac{y}{3} + 5 - \frac{1}{x} - \frac{1}{y} \qquad (2)$$

Compare this with Example 2 in Sec. 10.5, in which we found the relative extrema of the function f for which

$$f(x, y) = 2x^4 + y^2 - x^2 - 2y$$

These are essentially two different kinds of problems, because in the first case we had an additional condition, called a **constraint** (or **side condition**). Such a problem is called one in **constrained extrema**; that of the second type is called a problem in **free extrema**.

The solution of Example 4 of Sec. 10.6 involved obtaining a function of the two variables x and y by replacing z in the first equation by its value from the second equation. Another method that can be used to solve this example is due to Joseph Lagrange, and it is known as the method of **Lagrange multipliers**. The theory behind this method involves theorems known as implicit function theorems, which are studied in advanced calculus. Hence a proof is not given here. The procedure is outlined and illustrated by examples.

Suppose that we wish to find the critical points of a function f of the three variables x, y, and z, subject to the constraint $g(x, y, z) = 0$. We introduce a new variable, usually denoted by λ, and form the auxiliary function F for which

$$F(x, y, z, \lambda) = f(x, y, z) + \lambda g(x, y, z)$$

The problem, then, becomes one of finding the critical points of the function F of the four variables x, y, z, and λ. The values of x, y, and z that give the extrema of f, subject to the constraint g, are among these critical points.

EXAMPLE 1 Solve Example 1 of Sec. 10.6 by the method of Lagrange multipliers.

Solution Using the variables x, y, and z as defined in the solution of Example 1 of Sec. 10.6 we have

$$S = f(x, y, z) = xy + 2xz + 2yz$$

and

$$g(x, y, z) = xyz - 32$$

We wish to minimize the function f subject to the constraint that

$$g(x, y, z) = 0$$

Let

$$F(x, y, z, \lambda) = f(x, y, z) + \lambda g(x, y, z)$$
$$= xy + 2xz + 2yz + \lambda(xyz - 32)$$

Finding the four partial derivatives F_x, F_y, F_z, and F_λ, and setting the function values equal to zero, we have

$$F_x(x, y, z, \lambda) = y + 2z + \lambda yz = 0 \tag{3}$$

$$F_y(x, y, z, \lambda) = x + 2z + \lambda xz = 0 \tag{4}$$

$$F_z(x, y, z, \lambda) = 2x + 2y + \lambda xy = 0 \tag{5}$$

$$F_\lambda(x, y, z, \lambda) = xyz - 32 = 0 \tag{6}$$

Subtracting corresponding members of (4) from those of (3) we obtain

$$y - x + \lambda z(y - x) = 0$$

$$(y - x)(1 + \lambda z) = 0$$

giving the two equations

$$y = x \tag{7}$$

$$\lambda = -\frac{1}{z} \tag{8}$$

Substituting from (8) into (4) we get $x + 2z - x = 0$, giving $z = 0$, which is impossible, because z is in the interval $(0, +\infty)$. Substituting from (7) into (5) gives

$$2x + 2x + \lambda x^2 = 0$$

$$x(4 + \lambda x) = 0$$

and because $x \neq 0$,

$$\lambda = -\frac{4}{x}$$

If, in (4), $\lambda = -4/x$,

$$x + 2z - \frac{4}{x}(xz) = 0$$

$$x + 2z - 4z = 0$$

$$z = \frac{x}{2} \qquad (9)$$

Substituting from (7) and (9) into (6) we get $\frac{1}{2}x^3 - 32 = 0$ from which it follows that $x = 4$. From (7) and (9) we obtain $y = 4$ and $z = 2$. These results agree with those found in the solution of Example 1 in Sec. 10.6.

Observe in the solution that the equation $F_\lambda(x, y, z, \lambda) = 0$ is the same constraint given by the equation $xyz = 32$.

A disadvantage of the method of Lagrange multipliers is that the procedure gives only the critical points of a function; it does not indicate whether there is a relative maximum value, a relative minimum value, or neither. However, usually we can determine from the conditions of the problem which situation prevails.

EXAMPLE 2 Solve Example 4 of Sec. 10.6 by the method of Lagrange multipliers.

Solution We wish to maximize the function P defined by

$$P(x, y, z) = 900z - 700x - 400y$$

subject to the constraint given by the equation $z = x/3 + y/3 + 5 - 1/x - 1/y$, which we can write as

$$g(x, y, z) = \frac{1}{x} + \frac{1}{y} + z - \frac{x}{3} - \frac{y}{3} - 5 = 0$$

Let

$$F(x, y, z, \lambda) = P(x, y, z) + \lambda g(x, y, z)$$

$$= 900z - 700x - 400y + \lambda\left(\frac{1}{x} + \frac{1}{y} + z - \frac{x}{3} - \frac{y}{3} - 5\right)$$

We find the four partial derivatives F_x, F_y, F_z, and F_λ and set them equal to zero.

$$F_x(x, y, z, \lambda) = -700 - \frac{\lambda}{x^2} - \frac{\lambda}{3} = 0$$

$$F_y(x, y, z, \lambda) = -400 - \frac{\lambda}{y^2} - \frac{\lambda}{3} = 0$$

$$F_z(x, y, z, \lambda) = 900 + \lambda = 0$$

$$F_\lambda(x, y, z, \lambda) = \frac{1}{x} + \frac{1}{y} + z - \frac{x}{3} - \frac{y}{3} - 5 = 0$$

Solving these equations simultaneously we obtain

$$\lambda = -900 \qquad x = \tfrac{3}{2} \qquad y = 3 \qquad z = \tfrac{11}{2}$$

The values of x, y, and z agree with those found previously, and $P(\tfrac{3}{2}, 3, \tfrac{11}{2}) = 2700$. P is shown to have an absolute maximum value of 2700 in the same way as before.

In the next example there is an economic situation involving a **utility function** that measures the satisfaction of quantities of various commodities. A function value of a utility function is called a **utility index**, and it describes numerically the preference of an individual for the commodities.

EXAMPLE 3 Suppose that U is a utility function for which

$$U(x, y, z) = xyz$$

where x, y, and z represent the number of units of commodities A, B, and C, respectively, that are consumed weekly bv a particular person. Assume that $2, $3, and $4 are the unit prices of A, B, and C, respectively, and that the total weekly expense for the commodities is budgeted at $90. How many units of each commodity should be purchased in a week to maximize the consumer's utility index?

Solution We wish to determine the values of x, y, and z that maximize $U(x, y, z)$ subject to the budget constraint

$$2x + 3y + 4z = 90$$

Each of the variables x, y, and z is in the interval $[0, +\infty)$. Let

$$g(x, y, z) = 2x + 3y + 4z - 90$$

and

$$F(x, y, z, \lambda) = U(x, y, z) + \lambda g(x, y, z)$$
$$= xyz + \lambda(2x + 3y + 4z - 90)$$

We find F_x, F_y, F_z, and F_λ and set them equal to zero.

$$F_x(x, y, z, \lambda) = yz + 2\lambda = 0 \tag{10}$$

$$F_y(x, y, z, \lambda) = xz + 3\lambda = 0 \tag{11}$$

$$F_z(x, y, z, \lambda) = xy + 4\lambda = 0 \tag{12}$$

$$F_\lambda(x, y, z, \lambda) = 2x + 3y + 4z - 90 = 0 \tag{13}$$

From (10) and (11),

$$\frac{yz}{xz} = \frac{-2\lambda}{-3\lambda}$$

$$y = \tfrac{2}{3}x \tag{14}$$

From (10) and (12),

$$\frac{yz}{xy} = \frac{-2\lambda}{-4\lambda}$$

$$z = \tfrac{1}{2}x \qquad\qquad\qquad\qquad\qquad (15)$$

We substitute from (14) and (15) into (13), and we have

$$2x + 3(\tfrac{2}{3}x) + 4(\tfrac{1}{2}x) - 90 = 0$$

$$2x + 2x + 2x = 90$$

$$6x = 90$$

$$x = 15$$

Therefore $y = \tfrac{2}{3}(15) = 10$ and $z = \tfrac{1}{2}(15) = \tfrac{15}{2}$. With these values of x, y, and z we obtain

$$U(15, 10, \tfrac{15}{2}) = 15 \cdot 10 \cdot \tfrac{15}{2}$$

$$= 1125$$

It is apparent that this is the maximum utility index. Therefore the numbers of units of the three commodities that should be purchased in a week are 15, 10, and $\tfrac{15}{2}$.

In some applications of functions of several variables to problems in economics, the Lagrange multiplier λ is related to marginal concepts, in particular marginal cost and marginal utility. For details you should consult references in mathematical economics.

If several constraints are imposed, the method of Lagrange multipliers can be extended by using several multipliers. In particular, if we wish to find critical points of the function having function values $f(x, y, z)$ subject to the two side conditions $g(x, y, z) = 0$ and $h(x, y, z) = 0$, we find the critical points of the function F of the five variables x, y, z, λ, and μ for which

$$F(x, y, z, \lambda, \mu) = f(x, y, z) + \lambda g(x, y, z) + \mu h(x, y, z)$$

The following example shows the method.

EXAMPLE 4   Find the relative extrema of the function f if

$$f(x, y, z) = xz + yz$$

with the two constraints $x^2 + z^2 = 2$ and $yz = 2$.

Solution   We form the function F for which

$$F(x, y, z, \lambda, \mu) = xz + yz + \lambda(x^2 + z^2 - 2) + \mu(yz - 2)$$

Finding the five partial derivatives and setting them equal to zero we have

$$F_x(x, y, z, \lambda, \mu) = z + 2\lambda x = 0 \qquad\qquad (16)$$

$$F_y(x, y, z, \lambda, \mu) = z + \mu z = 0 \qquad\qquad (17)$$

$$F_z(x, y, z, \lambda, \mu) = x + y + 2\lambda z + \mu y = 0 \tag{18}$$

$$F_\lambda(x, y, z, \lambda, \mu) = x^2 + z^2 - 2 = 0 \tag{19}$$

$$F_\mu(x, y, z, \lambda, \mu) = yz - 2 = 0 \tag{20}$$

From (17) we obtain $\mu = -1$ or $z = 0$. We reject $z = 0$ because this contradicts (20). From (16) we obtain

$$\lambda = -\frac{z}{2x} \tag{21}$$

Substituting from (21) and letting $\mu = -1$ in (18) we get

$$x + y - \frac{z^2}{x} - y = 0$$

and so

$$x^2 = z^2 \tag{22}$$

Substituting from (22) into (19) we have $2x^2 - 2 = 0$, or $x^2 = 1$. This gives two values for x, namely 1 and -1; and for each of these values of x we get, from (22), the two values 1 and -1 for z. Obtaining the corresponding values for y from (20) we have four sets of solutions for the five Eqs. (16) through (20). These solutions are

$x = 1$	$y = 2$	$z = 1$	$\lambda = -\frac{1}{2}$	$\mu = -1$
$x = 1$	$y = -2$	$z = -1$	$\lambda = \frac{1}{2}$	$\mu = -1$
$x = -1$	$y = 2$	$z = 1$	$\lambda = \frac{1}{2}$	$\mu = -1$
$x = -1$	$y = -2$	$z = -1$	$\lambda = -\frac{1}{2}$	$\mu = -1$

The first and fourth sets of solutions give $f(x, y, z) = 3$, and the second and third sets of solutions give $f(x, y, z) = 1$. Hence f has a relative maximum value of 3 and a relative minimum value of 1.

Exercises 10.7

In Exercises 1 through 4, use the method of Lagrange multipliers to find the critical points of the given function subject to the given constraint.

1. $f(x, y) = 25 - x^2 - y^2$ with constraint $x^2 + y^2 - 4y = 0$
2. $f(x, y) = 4x^2 + 2y^2 + 5$ with constraint $x^2 + y^2 - 2y = 0$
3. $f(x, y, z) = x^2 + y^2 + z^2$ with constraint $3x - 2y + z - 4 = 0$
4. $f(x, y, z) = x^2 + y^2 + z^2$ with constraint $y^2 - x^2 = 1$

In Exercises 5 through 8, use the method of Lagrange multipliers to find the relative extrema of f subject to the indicated constraint. Also find the points at which the extrema occur. Assume that the relative extrema exist.

5. $f(x, y) = x^2 + y$ with constraint $x^2 + y^2 = 9$
6. $f(x, y) = x^2 y$ with constraint $x^2 + 8y^2 = 24$
7. $f(x, y, z) = xyz$ with constraint $x^2 + 2y^2 + 4z^2 = 4$
8. $f(x, y, z) = y^3 + xz^2$ with constraint $x^2 + y^2 + z^2 = 1$

In Exercises 9 and 10, find the minimum value of f subject to the indicated constraint. Assume that the minimum value exists.

9. $f(x, y, z) = x^2 + y^2 + z^2$ with constraint $xyz = 1$ **10.** $f(x, y, z) = xyz$ with constraint $x^2 + y^2 + z^2 = 1$

In Exercises 11 and 12, find the maximum value of f subject to the indicated constraint. Assume that the maximum value exists.

11. $f(x, y, z) = x + y + z$ with constraint $x^2 + y^2 + z^2 = 9$ **12.** $f(x, y, z) = xyz$ with constraint $2xy + 3xz + yz = 72$

13. Use the method of Lagrange multipliers to find a relative minimum function value of f if $f(x, y, z) = x^2 + y^2 + z^2$ with the two constraints $x + 2y + 3z = 6$ and $x - y - z = -1$.

14. Use the method of Lagrange multipliers to find a relative minimum function value of f if $f(x, y, z) = x^2 + y^2 + z^2$ with the two constraints $x + y + 2z = 1$ and $3x - 2y + z = -4$.

15. Use the method of Lagrange multipliers to find a relative maximum function value of f if $f(x, y, z) = xyz$ with the two constraints $x + y + z = 4$ and $x - y - z = 3$.

16. Use the method of Lagrange multipliers to find a relative maximum function value of f if $f(x, y, z) = x^3 + y^3 + z^3$ with the two constraints $x + y + z = 1$ and $x + y - z = 0$.

In Exercises 17 through 24, use the method of Lagrange multipliers to solve the indicated exercise of Exercises 10.6.

17. Exercise 1   **18.** Exercise 2   **19.** Exercise 5   **20.** Exercise 6

21. Exercise 13   **22.** Exercise 14   **23.** Exercise 19   **24.** Exercise 18

25. A circular disk is in the shape of the region bounded by the circle $x^2 + y^2 = 1$. If T degrees is the temperature at any point (x, y) of the disk and $T = 2x^2 + y^2 - y$, find the hottest and coldest points on the disk.

26. Find the minimum value of the function f for which $f(x, y) = x^2 + 4y^2 + 16z^2$ with the constraint (a) $xyz = 1$; (b) $xy = 1$; (c) $x = 1$.

27. Solve Example 3 if $U(x, y, z) = e^{x^2yz}$.   **28.** Solve Example 3 if $U(x, y, z) = x^2y^3z$.

29. In Example 3, suppose that the utility function involves five commodities A, B, C, D, and E. Furthermore, suppose that x units of A, y units of B, z units of C, s units of D, and t units of E are consumed weekly, and the unit prices of A, B, C, D, and E are, respectively, \$2, \$3, \$4, \$1, and \$5. If $U(x, y, z, s, t) = xyzst$ and the total weekly expense for the commodities is to be \$150, how many units of each commodity should be purchased in a week to maximize the consumer's utility index?

30. A company has three factories, each manufacturing the same product. If factory A produces x units, factory B produces y units, and factory C produces z units, their respective manufacturing costs are $(3x^2 + 200)$ dollars, $(y^2 + 400)$ dollars, and $(2z^2 + 300)$ dollars. If an order for 1100 units is to be filled, how should the production be distributed among the three factories to minimize the total manufacturing cost?

10.8 The method of least squares

Suppose that we wish to find a mathematical model for some data given by a set of points $(x_1, y_1), (x_2, y_2), \ldots, (x_n, y_n)$. In particular, y_i may be the number of dollars in a manufacturer's weekly profit when x_i is the number of units sold in a week, or y_i could be a company's total annual sales when x_i years have elapsed since the start of the company. The number of new cases of a certain disease could be y_i when x_i is the number of days since the outbreak of an epidemic of the disease. The desired model is a relationship involving x and y that can be used to make future predictions. Such a relationship is afforded by a line that "fits" the data.

Table 10.8.1

● ILLUSTRATION 1

Table 10.8.1 gives a company's total annual receipts from sales during its first 4 years of operation where x is the number of years of operation and y is the number of millions of dollars in annual sales.

x	1	2	3	4
y	5	8	7	12

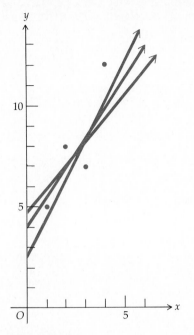

Figure 10.8.1

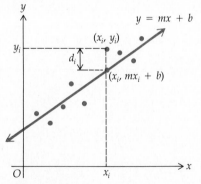

Figure 10.8.2

Suppose that one wanted to make a prediction of the fifth year's receipts from this information. The four points (x_i, y_i) are plotted in Fig. 10.8.1. There are various lines that one could consider "fit" the four points, and three such lines are shown in the figure. Each line gives a different estimate of the receipts for the fifth year. ●

As observed in Illustration 1, there are often several lines that "fit" a set of data points. We would like to obtain the "line of best fit." To arrive at a suitable definition for such a line we first indicate how well a particular line fits a set of data points by measuring the vertical distances from the points to the line. For instance, in Fig. 10.8.2, there are n data points, and the line $y = mx + b$. The point (x_i, y_i) is the ith data point, and corresponding to it on the line is the point $(x_i, mx_i + b)$. The **derivation** (or **error**) between the ith data point and the line is defined to be d_i, where

$$d_i = y_i - (mx_i + b) \qquad (1)$$

Note that d_i is positive if the point (x_i, y_i) is above the line and negative if it is below the line. If the line is to fit the data points well, it may seem reasonable to require that

$$d_1 + d_2 + \ldots + d_n$$

be small. However, such a requirement would allow for the situation in which the sum of some positive deviations and negative deviations would be small while the points were not close to the line. To avoid this possibility we consider the sum of the squares of the deviations

$$d_1{}^2 + d_2{}^2 + \ldots + d_n{}^2 = \sum_{i=1}^{n} d_i{}^2$$

Observe that $\sum_{i=1}^{n} d_i{}^2$ is never negative and is zero only if each d_i is zero, in which case all the data points lie on the line. We shall take as the line of best fit the one for which $\sum_{i=1}^{n} d_i{}^2$ is an absolute minimum. This line is called the **regression line**, and the process for finding it is called the **method of least squares**.

● ILLUSTRATION 2

We apply the method of least squares to find the regression line for the data points of Illustration 1. These points are $(1, 5)$, $(2, 8)$, $(3, 7)$, and $(4, 12)$. Let the regression line have the equation

$$y = mx + b$$

We wish to find m and b. From (1), the deviations between the data points and the line are

$$d_1 = 5 - [m(1) + b] = 5 - m - b$$
$$d_2 = 8 - [m(2) + b] = 8 - 2m - b$$
$$d_3 = 7 - [m(3) + b] = 7 - 3m - b$$
$$d_4 = 12 - [m(4) + b] = 12 - 4m - b$$

Thus

$$\sum_{i=1}^{4} d_i^2 = (5 - m - b)^2 + (8 - 2m - b)^2 + (7 - 3m - b)^2 + (12 - 4m - b)^2$$

Observe that $\sum_{i=1}^{4} d_i^2$ is a function of m and b. Denote this function by f so that

$$f(m, b) = (5 - m - b)^2 + (8 - 2m - b)^2 + (7 - 3m - b)^2 + (12 - 4m - b)^2 \qquad (2)$$

We apply the second-derivative test to find the values of m and b that give the absolute minimum value of f.

$$f_m(m, b)$$
$$= 2(5 - m - b)(-1) + 2(8 - 2m - b)(-2)$$
$$\qquad\qquad + 2(7 - 3m - b)(-3) + 2(12 - 4m - b)(-4)$$
$$= 2(-5 + m + b - 16 + 4m + 2b - 21 + 9m + 3b - 48 + 16m + 4b)$$
$$= 2(-90 + 30m + 10b)$$

$$f_b(m, b)$$
$$= 2(5 - m - b)(-1) + 2(8 - 2m - b)(-1)$$
$$\qquad\qquad + 2(7 - 3m - b)(-1) + 2(12 - 4m - b)(-1)$$
$$= 2(-5 + m + b - 8 + 2m + b - 7 + 3m + b - 12 + 4m + b)$$
$$= 2(-32 + 10m + 4b)$$

$$f_{mm}(m, b) = 60 \qquad f_{mb}(m, b) = 20 \qquad f_{bb}(m, b) = 8 \qquad f_{bm}(m, b) = 20$$

Setting $f_m(m, b) = 0$ and $f_b(m, b) = 0$ we have

$$-9 + 3m + b = 0$$
$$-16 + 5m + 2b = 0$$

Solving these two equations simultaneously we obtain $m = 2$ and $b = 3$. For all values of m and b,

$$f_{mm}(m, b) = 60 > 0$$

and

$$f_{mm}(m, b) \cdot f_{bb}(m, b) - f_{mb}^2(m, b) = (60)(8) - (20)^2 = 80 > 0$$

Therefore, from Theorem 10.5.4(i), f has a relative minimum value when $m = 2$ and $b = 3$. Furthermore, there is only one relative extremum for f. Also, m and b are both in the interval $(-\infty, +\infty)$, and from (2) we observe that $f(m, b)$ is large when either the absolute value of m or the absolute value of b is large. Thus we conclude that the relative minimum value of f is an absolute minimum value. Hence the regression line is

$$y = 2x + 3 \qquad (3)$$

Figure 10.8.3 shows the four points and this line. From (3), when $x = 5$, $y = 13$. Therefore the fifth year's receipts are estimated to be \$13 million. ●

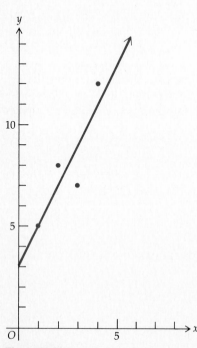

Figure 10.8.3

We now give the general procedure for using the method of least squares to find the regression line $y = mx + b$ for a set of n data points (x_1, y_1), $(x_2, y_2), \ldots, (x_n, y_n)$. The sum of the squares of the deviations between the points and the line is

$$\sum_{i=1}^{n} d_i^2 = \sum_{i=1}^{n} [y_i - (mx_i + b)]^2$$

Remember that x_i and y_i are constants and m and b are variables, so that

$$f(m, b) = \sum_{i=1}^{n} (y_i - mx_i - b)^2$$

We wish to find the values of m and b that make $f(m, b)$ an absolute minimum. We first find the partial derivatives $f_m(m, b)$ and $f_b(m, b)$.

$$f_m(m, b) = \sum_{i=1}^{n} \frac{\partial}{\partial m} [(y_i - mx_i - b)^2]$$

$$= \sum_{i=1}^{n} 2(y_i - mx_i - b)(-x_i)$$

$$= 2 \sum_{i=1}^{n} (-x_i y_i + mx_i^2 + bx_i)$$

$$= 2 \left[-\sum_{i=1}^{n} x_i y_i + m \sum_{i=1}^{n} x_i^2 + b \sum_{i=1}^{n} x_i \right]$$

$$f_b(m, b) = \sum_{i=1}^{n} \frac{\partial}{\partial b} [(y_i - mx_i - b)^2]$$

$$= \sum_{i=1}^{n} 2(y_i - mx_i - b)(-1)$$

$$= 2 \sum_{i=1}^{n} (-y_i + mx_i + b)$$

$$= 2 \left[-\sum_{i=1}^{n} y_i + m \sum_{i=1}^{n} x_i + nb \right]$$

Setting $f_m(m, b) = 0$ and $f_b(m, b) = 0$ we obtain

$$\left(\sum_{i=1}^{n} x_i^2\right) m + \left(\sum_{i=1}^{n} x_i\right) b = \sum_{i=1}^{n} x_i y_i \tag{4}$$

$$\left(\sum_{i=1}^{n} x_i\right) m + nb = \sum_{i=1}^{n} y_i \tag{5}$$

Equations (4) and (5) are to be solved for m and b. These values of m and b make $f(m, b)$ an absolute minimum. The verification that we obtain an absolute minimum value is the same as given in Illustration 2.

When applying the method of least squares, the simultaneous equations (4) and (5) are to be found. A convenient way of computing the coefficients in these equations is shown in the following example.

EXAMPLE 1   Find the regression line for the data in Table 10.8.2.

Table 10.8.2

x	1	3	5	7	9	11
y	3	5	6	5	7	8

Solution   To find the values of m and b that give the regression line $y = mx + b$, we first obtain the pair of simultaneous equations (4) and (5). The computation of the coefficients is shown in Table 10.8.3. From the table,

$$\sum_{i=1}^{6} x_i = 36 \qquad \sum_{i=1}^{6} y_i = 34 \qquad \sum_{i=1}^{6} x_i^2 = 286 \qquad \sum_{i=1}^{6} x_i y_i = 234$$

Substituting these values in (4) and (5) and taking $n = 6$ we have

$$286m + 36b = 234$$

$$36m + 6b = 34$$

or, equivalently,

$$143m + 18b = 117$$

$$108m + 18b = 102$$

Table 10.8.3

x_i	y_i	x_i^2	$x_i y_i$
1	3	1	3
3	5	9	15
5	6	25	30
7	5	49	35
9	7	81	63
11	8	121	88
$\sum$ 36	34	286	234

Eliminating b we get $35m = 15$; thus $m = \frac{3}{7}$. Substituting this value of m in one of the equations gives $b = \frac{65}{21}$. Therefore the regression line has the equation

$$y = \tfrac{3}{7}x + \tfrac{65}{21}$$

$$9x - 21y + 65 = 0$$

EXAMPLE 2   A rare antique was purchased in 1965 for $1200. Its value was $1800 in 1970, $2500 in 1975, and $3500 in 1980. Use the method of least squares to estimate the value of the antique in 1990.

Table 10.8.4

x	0	1	2	3
y	1200	1800	2500	3500

Solution   To find a regression line $y = mx + b$, we let x be the number of 5-year periods since 1965 and let y dollars be the value of the antique $5x$ years since 1965. Thus we have the data points given in Table 10.8.4.

Table 10.8.5

x_i	y_i	x_i^2	$x_i y_i$
0	1200	0	0
1	1800	1	1800
2	2500	4	5000
3	3500	9	10,500
Σ 6	9000	14	17,300

To determine m and b we first obtain the simultaneous equations (4) and (5) from the computation in Table 10.8.5. From the table,

$$\sum_{i=1}^{4} x_i = 6 \qquad \sum_{i=1}^{4} y_i = 9000 \qquad \sum_{i=1}^{4} x_i^2 = 14 \qquad \sum_{i=1}^{4} x_i y_i = 17{,}300$$

With these values and $n = 4$, we obtain from (4) and (5) the equations

$$14m + 6b = 17{,}300$$
$$6m + 4b = 9{,}000$$

or, equivalently,

$$14m + 6b = 17{,}300$$
$$9m + 6b = 13{,}500$$

Eliminating b gives $5m = 3800$; therefore $m = 760$. With this value of m in one of the equations we get $b = 1110$. Hence the regression line has the equation

$$y = 760x + 1110$$

For the year 1990, $x = 5$. For this value of x we have

$$y = 760(5) + 1110 = 4910$$

Thus in 1990 the value of the antique is estimated to be $4910.

Table 10.8.6

x	1	2	3	4	5
y	20	24	30	35	42

EXAMPLE 3   In Table 10.8.6, x days have elapsed since the outbreak of a particular disease, and y is the number of new cases of the disease on the xth day. (a) Find the regression line for the data points (x_i, y_i). (b) Use the regression line to estimate the number of new cases of the disease on the sixth day.

Solution   (a) The required line has the equation $y = mx + b$. To determine m and b we find the pair of simultaneous equations (4) and (5) from the computation in Table 10.8.7. From the table,

$$\sum_{i=1}^{5} x_i = 15 \qquad \sum_{i=1}^{5} y_i = 151 \qquad \sum_{i=1}^{5} x_i^2 = 55 \qquad \sum_{i=1}^{5} x_i y_i = 508$$

Table 10.8.7

x_i	y_i	x_i^2	$x_i y_i$
1	20	1	20
2	24	4	48
3	30	9	90
4	35	16	140
5	42	25	210
Σ 15	151	55	508

From (4) and (5) with these values and $n = 5$, we obtain the equations

$$55m + 15b = 508$$
$$15m + 5b = 151$$

or, equivalently,

$$55m + 15b = 508$$
$$45m + 15b = 453$$

We eliminate b by subtracting corresponding members of each equation, and we obtain $m = 5.5$. Substituting this value of m in one of the equations gives $b = 13.7$. Therefore the regression line has the equation

$$y = 5.5x + 13.7$$

(b) From the equation of the regression line, when $x = 6$, then $y = 46.7$. Therefore, on the sixth day of the epidemic 47 new cases are estimated.

EXAMPLE 4   A monopolist wants to determine the selling price of a particular commodity in order to maximize the daily profits. A linear demand equation was obtained by testing the commodity in five similar regions. The test results are shown in Table 10.8.8, where u units are demanded per day when p dollars is the price per unit. If the monopolist's daily overhead is $1200 and in addition it costs $1.50 to manufacture each unit, what should be the selling price to yield the greatest daily profit?

Table 10.8.8

	Region A	Region B	Region C	Region D	Region E
p	5.00	5.25	5.50	6.00	6.25
u	920	860	820	730	700

Solution   As the demand curve, we take the regression line for the data in Table 10.8.8. To obtain an equation of this line we first find the equations (4) and (5) from the computation in Table 10.8.9.

Table 10.8.9

p_i	u_i	p_i^2	$p_i u_i$
5.00	920	25.00	4600
5.25	860	27.56	4515
5.50	820	30.25	4510
6.00	730	36.00	4380
6.25	700	39.06	4375
$\sum$ 28.00	4030	157.87	22,380

From the table,

$$\sum_{i=1}^{5} p_i = 28.00 \qquad \sum_{i=1}^{5} u_i = 4030 \qquad \sum_{i=1}^{5} p_i^2 = 157.87 \qquad \sum_{i=1}^{5} p_i u_i = 22,380$$

Therefore the simultaneous equations in m and b are

$$157.87m + 28.00b = 22,380$$

$$28.00m + 5.00b = 4030$$

From the second of these equations we get $b = 806 - 5.60m$. Substituting this value of b into the first equation gives

$$157.87m + 22,568 - 156.8m = 22,380$$

$$m = -175.7$$

With this value of m we obtain $b = 1790$. Therefore the regression line has the equation

$$u = -176p + 1790$$

which we take as the monopolist's demand equation for the commodity. Because the total revenue is the product of the price per unit and the number of units demanded, then if $R(p)$ dollars is the total daily revenue,

$$R(p) = p(-176p + 1790)$$

$$R(p) = -176p^2 + 1790p \tag{6}$$

The total cost is the sum of the overhead and the cost of manufacturing u units. The daily overhead is \$1200, and it costs \$1.50 to manufacture each unit. Therefore, if $C(p)$ is the total daily cost,

$$C(p) = 1200 + 1.50(-176p + 1790)$$

$$C(p) = 3885 - 264p \tag{7}$$

If $S(p)$ dollars is the total daily profit, $S(p) = R(p) - C(p)$. Thus, from (6) and (7), we have

$$S(p) = -176p^2 + 1790p - (3885 - 264p)$$

$$S(p) = -176p^2 + 2054p - 3885$$

We wish to find the value of p that makes $S(p)$ an absolute maximum.

$$S'(p) = -352p + 2054$$

$$S''(p) = -352$$

Setting $S'(p) = 0$ we get

$$-352p + 2054 = 0$$

$$p = 5.84$$

Because $S''(5.84) < 0$, it follows that S has a relative maximum value when $p = 5.84$. This is an absolute maximum value because S is continuous everywhere and S has only one relative extremum. Therefore the selling price should be \$5.84 per unit in order to yield the greatest daily profit.

Exercises 10.8

In Exercises 1 through 6, find the regression line for the given set of data points. Make a figure showing the data points and the regression line.

1. $(2, 0), (4, -1), (6, 2)$
2. $(1, 1), (2, 3), (3, 4)$
3. $(0, 2), (1, 1), (2, 3), (3, 2)$
4. $(1, -2), (3, 0), (5, 1), (7, 4)$
5. $(-2, 6), (-1, 4), (0, 1), (1, -1), (2, -2)$
6. $(0, 1), (2, 3), (4, 3), (6, 5), (8, 6)$

7. An early abstract painting was sold by the artist in 1915 for $100. Because of its historical importance its value has increased over the years. Its value was $4600 in 1935, $11,000 in 1955, and $20,000 in 1975. Use the method of least squares to estimate the value of the painting in 1995.

8. A 1980 model car was sold as a used car in 1981 for $6800. Its value was $6200 in 1982, $5700 in 1983, and $5400 in 1984. Use the method of least squares to determine its value in 1985.

9. A motion picture has been playing at Cinema One for 5 weeks, and the weekly attendance (to the nearest 100) for each week is given in the table.

Week Number	1	2	3	4	5
Attendance	5000	4500	4100	3900	3500

(a) Use the regression line for the data in the table to determine the expected attendance for the sixth week. (b) The film will move over to the smaller Cinema Two when the weekly attendance drops below 2250. How many weeks is the film expected to play in Cinema One?

10. The following table gives a manufacturer's monthly production and profit for the first 5 months of the year, where x thousands of units were produced and y thousands of dollars was the profit.

	Jan.	Feb.	Mar.	April	May
x	65	72	82	90	100
y	30	35	42	48	60

If the production for June is to be 105,000 units, use the regression line for the data in the table to estimate that month's profit.

11. The following table gives data for five patients having a particular surgical operation at a certain hospital, where x years is the patient's age and y days is the length of time the patient remained in the hospital while recuperating after the surgery.

	Patient A	Patient B	Patient C	Patient D	Patient E
x	54	46	40	36	30
y	15	12	9	10	8

(a) Find an equation of the regression line for the data in the table. (b) Use the regression line to estimate the length of stay in the hospital for a 42-year-old person having the surgery.

12. The score on a student's entrance examination was used to predict the student's grade-point average at the end of the freshman year. The following table gives the data for six students, where x is the test score and y is the grade-point average.

	Student A	Student B	Student C	Student D	Student E	Student F
x	92	81	73	98	79	85
y	3.4	2.7	3.1	3.8	2.2	3.0

(a) Find an equation of the regression line for the data in the table. (b) Use the regression line to estimate a student's grade-point average at the end of the freshman year if the student had a score of 88 on the college entrance examination.

13. Five joggers were given examinations to determine their maximum oxygen uptake, a measure used to denote a person's cardiovascular fitness. The results are given in the following table, where x is the number of seconds in the jogger's best time for a mile run and y is the number of milliliters per minute per kilogram of body weight in the jogger's maximum oxygen uptake.

	Jogger A	Jogger B	Jogger C	Jogger D	Jogger E
x	300.5	350.6	407.3	326.2	512.8
y	350.2	325.8	375.6	418.5	400.2

(a) Find an equation of the regression line for the data in the table. (b) Use the regression line to estimate a jogger's maximum oxygen uptake if the jogger's best time for a mile run is 340.4 sec.

14. Five trees had their sap analyzed for the amount of a plant hormone that causes the detachment of leaves. For the trees in the following table, when x micrograms (μg) of plant hormone were released, y leaves were detached.

Oak tree	Maple tree	Birch tree	Pine tree	Locust tree
28	57	38	75	82
208	350	300	620	719

(a) Find an equation of the regression line for the data in the table. (b) Use the regression line to estimate the number of leaves detached from another kind of tree when 100 μg of plant hormone are released.

15. In deserts, water is a factor that sharply limits plant activity. In the following table x is the number of millimeters of precipitation per year for six different regions, and y is the number of kilograms per hectare in the net photosynthate produced.

	Region A	Region B	Region C	Region D	Region E	Region F
x	100	200	400	500	600	650
y	1000	1900	3200	4400	5800	6400

(a) Find an equation of the regression line for the data in the table. (b) Use the regression line to estimate the net photosynthate produced in a region having an annual precipitation of 300 mm.

16. In the following table, for five healthy children, w kg is the body weight and y mm of mercury is the mean arterial pressure (average of systolic and diastolic blood pressures).

	Child A	Child B	Child C	Child D	Child E
w	20	30	35	40	50
y	70	85	90	96	100

(a) An equation that "fits" the data in this table is

$$y = m(\ln w) + b$$

To find such an equation let $x = \ln w$, and use the method of least squares for the points (x_i, y_i). (b) Use the result of part (a) to estimate the mean arterial pressure of a healthy child of weight 45 kg.

17. In order to set a selling price for a certain cosmetic, a monopolist test-marketed the cosmetic in four cities of the same size and at different prices. The following table gives the results, where u units were demanded per day when p dollars was the price per unit.

	City A	City B	City C	City D
p	8.50	9.00	9.50	10.00
u	1160	1080	980	740

From the data in the table a linear demand equation was obtained by the method of least squares. If the monopolist must pay \$5 to produce each unit and the daily overhead is \$900, what should be the selling price to maximize the daily profit?

Review Exercises for Chapter 10

1. Draw a sketch of the graph of $x = 3$ in R^1, R^2, and R^3.
2. Draw a sketch of the set of points satisfying the simultaneous equations $x = 6$ and $y = 3$ in R^2 and R^3.

In Exercises 3 and 4, draw a sketch of the given plane.

3. $2x + 10y + 5z = 10$ 4. $2x + 3y - 12 = 0$

In Exercises 5 through 7, draw a sketch showing as a shaded region in R^2 the set of points in the domain of f.

5. $f(x, y) = \sqrt{x^2 + y^2 - 1}$ 6. $f(x, y) = \dfrac{\sqrt{4 - x^2 - y^2}}{x}$ 7. $f(x, y) = \ln(xy - 4)$

8. Determine the domain of g if $g(x, y, z) = \ln(x^2 + y^2 + z^2 - 4)$.
9. The production function for a certain commodity is f, where $f(x, y) = 4x^{1/2}y$, and x and y give the amounts of two inputs. Draw a contour map of f showing the constant product curves at 16, 8, 4, and 2.
10. The temperature at a point (x, y) of a flat metal plate is t degrees, and $t = x^2 + 2y$. Draw the isothermals for $t = 0, 2, 4, 6,$ and 8.

In Exercises 11 through 20, find the indicated partial derivatives.

11. $f(x, y) = 2x^2y - 3xy^2 + 4x - 2y$; (a) $D_x f(x, y)$; (b) $D_y f(x, y)$; (c) $D_{xx} f(x, y)$; (d) $D_{yy} f(x, y)$; (e) $D_{xy} f(x, y)$; (f) $D_{yx} f(x, y)$
12. $f(x, y) = (4x^2 - 2y)^3$; (a) $D_x f(x, y)$; (b) $D_y f(x, y)$; (c) $D_{xx} f(x, y)$; (d) $D_{yy} f(x, y)$; (e) $D_{xy} f(x, y)$; (f) $D_{yx} f(x, y)$
13. $f(r, s) = re^{2rs}$; (a) $D_r f(r, s)$; (b) $D_s f(r, s)$; (c) $D_{rs} f(r, s)$; (d) $D_{sr} f(r, s)$

14. $f(x, y) = e^{x/y} + \ln \dfrac{x}{y}$; (a) $D_x f(x, y)$; (b) $D_y f(x, y)$; (c) $D_{xy} f(x, y)$; (d) $D_{yx} f(x, y)$

15. $f(x, y) = \ln\sqrt{x^2 + y^2}$; (a) $D_x f(x, y)$; (b) $D_{xx} f(x, y)$; (c) $D_{xy} f(x, y)$
16. $f(u, v) = ve^{u^2v}$; (a) $D_u f(u, v)$; (b) $D_v f(u, v)$; (c) $D_{uv} f(u, v)$; (d) $D_{vu} f(u, v)$

17. $f(x, y, z) = \dfrac{x}{x^2 + y^2 + z}$; (a) $D_x f(x, y, z)$; (b) $D_y f(x, y, z)$; (c) $D_z f(x, y, z)$

18. $f(x, y, z) = \sqrt{x^2 + 3yz - z^2}$; (a) $D_x f(x, y, z)$; (b) $D_y f(x, y, z)$; (c) $D_z f(x, y, z)$
19. $f(r, s, t) = t^2 e^{4rst}$; (a) $f_r(r, s, t)$; (b) $f_{rt}(r, s, t)$; (c) $f_{rts}(r, s, t)$
20. $f(u, v, w) = \ln(u^2 + 4v^2 - 5w^2)$; (a) $f_{uwv}(u, v, w)$; (b) $f_{uvv}(u, v, w)$

21. Given $u = \ln(x^2 + y^2)$, verify $\dfrac{\partial^2 u}{\partial x^2} + \dfrac{\partial^2 u}{\partial y^2} = 0$.

22. If $w = x^2y - y^2x + y^2z - z^2y + z^2x - x^2z$, show that

$$\frac{\partial w}{\partial x} + \frac{\partial w}{\partial y} + \frac{\partial w}{\partial z} = 0$$

23. If $u = (x^2 + y^2 + z^2)^{-1/2}$, show that

$$\frac{\partial^2 u}{\partial x^2} + \frac{\partial^2 u}{\partial y^2} + \frac{\partial^2 u}{\partial z^2} = 0$$

In Exercises 24 and 25, find the given limit.

24. $\displaystyle\lim_{(x,y)\to(e,0)} \ln\left(\frac{x^2}{y+1}\right)$

25. $\displaystyle\lim_{(x,y)\to(0,0)} \frac{e^{2x} - e^{2y}}{e^x - e^y}$

In Exercises 26 and 27, determine all points in R^2 for which the given function is continuous.

26. $f(x, y) = \dfrac{x^2 + 4y^2}{x^2 - 4y^2}$

27. $f(x, y) = \ln\left(\dfrac{x^2 + y^2 - 4}{16 - x^2 - y^2}\right)$

28. Find the slope of the tangent line to the curve of intersection of the surface $z = 2x^2 + y^2$ with the plane $x = 2$ at the point $(2, -1, 9)$.

29. If $f(x, y)$ units are produced by x workers and y machines, then $D_x f(x, y)$ is called the *marginal productivity of labor* and $D_y f(x, y)$ is called the *marginal productivity of machines*. Suppose that

$$f(x, y) = x^2 + 6xy + 3y^2$$

where $5 \le x \le 30$ and $4 \le y \le 12$. (a) Find the number of units produced in a day when the labor force for that day consists of 15 workers, and 8 machines are used. (b) Use the marginal productivity of labor to determine the approximate number of additional units that can be produced in 1 day if the labor force is increased from 15 to 16 and the number of machines remains fixed at 8. (c) Use the marginal productivity of machines to determine the approximate number of additional units that can be produced in 1 day if the number of machines is increased from 8 to 9 and the number of workers remains fixed at 15.

30. For the demand equations

$$x = 20 - 5p - 4q \quad \text{and} \quad y = 18 - 3p - q$$

suppose that x units of the first commodity and y units of the second commodity are demanded when the prices per unit are p dollars and q dollars, respectively. Use the partial marginal demands to determine how the amount demanded of each commodity is affected in each of the following cases: (a) q is held fixed and the price of the first commodity is increased by \$1; (b) p is held fixed and the price of the second commodity is increased by \$1; (c) q is held fixed and the price of the first commodity is decreased by \$1; (d) p is held fixed and the price of the second commodity is decreased by \$1.

In Exercises 31 and 32, demand equations for two related commodities are given. In each exercise, find the four partial marginal demands. Determine if the commodities are complementary or substitutes.

31. $x = 6 - 3p + 2q,\ y = 2 + 3p - 2q$

32. $x = \dfrac{2p}{q},\ y = \dfrac{3q^2}{p}$

In Exercises 33 and 34, demand equations are given. Do each of the following: (a) find the four partial elasticities of demand; (b) at $p = 3$ and $q = 2$, if p is increased by 1 percent and q is held constant, find the percent changes in x and y; (c) at $p = 3$ and $q = 2$, if q is increased by 1 percent and p is held constant, find the percent changes in x and y; (d) at $p = 3$ and $q = 2$, if p is decreased by 1 percent and q is held constant, find the percent changes in x and y; (e) at $p = 3$ and $q = 2$, if q is decreased by 1 percent and p is held constant, find the percent changes in x and y.

33. The demand equations of Exercise 31.

34. The demand equations of Exercise 32.

35. In a particular community, when the price of product A is p dollars and the price of product B is q dollars, x units of A and y units of B are demanded, and the respective demand equations are

$$x = 100e^{-p/10q} \quad \text{and} \quad y = 200e^{-q/20p}$$

(a) Show that the two products are substitutes. (b) Find the four partial elasticities of demand. Suppose that the price of A is $40 and the price of B is $30. Find the percent changes in the demands for the two products if (c) the price of A is decreased by 1 percent and the price of B remains constant, and (d) the price of B is decreased by 1 percent and the price of A remains constant.

In Exercises 36 and 37, determine the relative extrema of f, if there are any.

36. $f(x, y) = 2x^2 - 3xy + 2y^2 + 10x - 11y$ **37.** $f(x, y) = x^3 + y^3 + 3xy$

In Exercises 38 and 39, use the method of Lagrange multipliers to find the critical point(s) of the given function subject to the indicated constraint. Determine if the function has a relative maximum or a relative minimum value at any critical point.

38. $f(x, y) = 5 + x^2 - y^2$ with constraint $x^2 - 2y^2 = 5$ **39.** $f(x, y, z) = x^2 + y^2 + z^2$ with constraint $x^2 - y^2 = 1$

40. Find three numbers whose sum is 100 and the sum of whose squares is least.

41. A manufacturer produces daily x units of commodity A and y units of commodity B. If $P(x, y)$ dollars is the daily profit from their sale, and

$$P(x, y) = 33x + 66y + xy - x^2 - 3y^2$$

how many units of each commodity should be produced daily for the manufacturer to receive the greatest profit?

42. A rectangular crate without a top is to have a surface area of 216 ft^2. What are the dimensions for a crate of greatest volume?

43. For the crate of Exercise 42, suppose instead of the surface area being 216 ft^2, the sum of the lengths of the edges is 216 ft. What then are the dimensions for a crate of greatest volume?

44. The temperature is t degrees at any point (x, y) of the curve $4x^2 + 12y^2 = 1$, and $t = 4x^2 + 24y^2 - 2x$. Find the points on the curve where the temperature is the greatest and where it is the least. Also find the temperature at these points.

45. If $f(x, y, z) = xyz$, find the maximum and minimum values of f subject to the constraints $y = z$ and $x^2 + y^2 = 1$.

46. Find the regression line for the data points $(0, 4)$, $(1, 3)$, $(2, 4)$, $(3, -2)$, and $(4, -1)$. Make a figure showing the data points and the regression line.

47. In the following table, a patient's systolic blood pressure and corresponding heart rate are given, where x mm of mercury is the systolic blood pressure and y beats per minute is the heart rate.

	Patient A	Patient B	Patient C	Patient D	Patient E	Patient F
x	110	117	133	146	115	127
y	70	74	80	65	60	77

(a) Find an equation of the regression line for the data in the table. (b) Use the regression line to estimate a patient's heart rate if the systolic blood pressure is 85 mm of mercury.

48. A breakfast cereal was test-sold in four cities of the same size at different prices; and the results are shown in the table, where x cents was the price per box and y thousands of cases were sold per week.

	City A	City B	City C	City D
x	130	140	150	160
y	100	85	75	63

(a) Find an equation of the regression line for the data in the table. Use the regression line in part (a) as the demand curve to estimate the weekly sales if the price per box is (b) $1.20 and (c) $1.70.

APPENDIX

TRIGONOMETRIC FUNCTIONS

A.1 The sine and cosine functions

The functions considered in this text have been algebraic, exponential, or logarithmic. The trigonometric functions are another type of function; they are used to describe repetitive events, such as business cycles, wave motions, vibrations, and biological rhythms. Because they are related to angles and their measure, the trigonometric functions also have applications in physics, engineering, navigation, and architecture.

In geometry an **angle** is defined as the union of two rays called the **sides**, having a common endpoint called the **vertex**. Any angle is congruent to some angle having its vertex at the origin and one side, called the **initial side**, lying on the positive side of the x axis. Such an angle is said to be in **standard position**. Figure A.1.1 shows an angle AOB in standard position with OA as the initial side. The other side, OB, is called the **terminal side**. The angle AOB can be formed by rotating the side OA to the side OB, and under such a rotation the point A moves along the circumference of a circle, having its center at O and radius $|\overline{OA}|$, to the point B.

In dealing with problems involving angles of triangles, the measurement of an angle is usually given in degrees. However, in the calculus we are concerned with trigonometric functions of real numbers, and these functions are defined in terms of *radian measure*.

The length of an arc of a circle is used to define the radian measure of an angle.

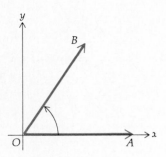

Figure A.1.1

Definition of radian measure

Let AOB be an angle in standard position, and $|\overline{OA}| = 1$. If s units is the length of the arc of the circle traveled by point A as the initial side OA is rotated to the terminal side OB, the **radian measure**, t, of angle AOB is given by

$$t = s \qquad \text{if the rotation is counterclockwise}$$

and

$$t = -s \qquad \text{if the rotation is clockwise}$$

● ILLUSTRATION 1

From the fact that the measure of the length of the unit circle's circumference is 2π, the radian measures of the angles in Fig. A.1.2 (a), (b), (c), (d), (e), and (f) are determined. They are $\frac{1}{2}\pi$, $\frac{1}{4}\pi$, $-\frac{1}{2}\pi$, $\frac{3}{2}\pi$, $-\frac{3}{4}\pi$, and $\frac{7}{4}\pi$, respectively.   ●

In the definition of radian measure, it is possible that there may be more than one complete revolution in the rotation of OA.

● ILLUSTRATION 2

Figure A.1.3(a) shows an angle whose radian measure is $\frac{5}{2}\pi$, and Fig. A.1.3(b) shows one whose radian measure is $-\frac{13}{4}\pi$.   ●

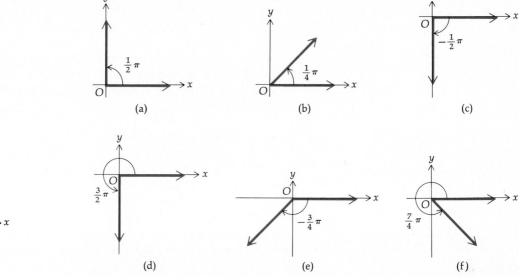

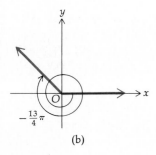

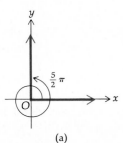

Figure A.1.2

An angle formed by one complete revolution so that OA is coincident with OB has degree measure of 360 and radian measure of 2π. Hence there is the following correspondence between degree measure and radian measure (where the symbol $\sim$ indicates that the given measurements are for the same or congruent angles):

$$360° \sim 2\pi \text{ rad}$$

$$180° \sim \pi \text{ rad}$$

From this it follows that

$$1° \sim \tfrac{1}{180}\pi \text{ rad}$$

and

$$1 \text{ rad} \sim \frac{180°}{\pi} \approx 57°18'$$

From this correspondence the measurement of an angle can be converted from one system of units to the other.

● ILLUSTRATION 3

$$162° \sim 162 \cdot \tfrac{1}{180}\pi \text{ rad} \qquad \tfrac{5}{12}\pi \text{ rad} \sim \tfrac{5}{12}\pi \cdot \frac{180°}{\pi}$$

$$162° \sim \tfrac{9}{10}\pi \text{ rad} \qquad \tfrac{5}{12}\pi \text{ rad} \sim 75° \qquad\qquad ●$$

Table A.1.1 gives the corresponding degree and radian measures of certain angles.

Figure A.1.3

Table A.1.1

Degree measure	Radian measure
30	$\tfrac{1}{6}\pi$
45	$\tfrac{1}{4}\pi$
60	$\tfrac{1}{3}\pi$
90	$\tfrac{1}{2}\pi$
120	$\tfrac{2}{3}\pi$
135	$\tfrac{3}{4}\pi$
150	$\tfrac{5}{6}\pi$
180	π
270	$\tfrac{3}{2}\pi$
360	2π

We now define the *sine* and *cosine* functions of any real number.

Definition of the sine and
cosine functions

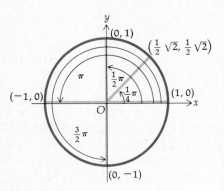

Figure A.1.4

Table A.1.2

x	$\sin x$	$\cos x$
0	0	1
$\frac{1}{6}\pi$	$\frac{1}{2}$	$\frac{1}{2}\sqrt{3}$
$\frac{1}{4}\pi$	$\frac{1}{2}\sqrt{2}$	$\frac{1}{2}\sqrt{2}$
$\frac{1}{3}\pi$	$\frac{1}{2}\sqrt{3}$	$\frac{1}{2}$
$\frac{1}{2}\pi$	1	0
$\frac{2}{3}\pi$	$\frac{1}{2}\sqrt{3}$	$-\frac{1}{2}$
$\frac{3}{4}\pi$	$\frac{1}{2}\sqrt{2}$	$-\frac{1}{2}\sqrt{2}$
$\frac{5}{6}\pi$	$\frac{1}{2}$	$-\frac{1}{2}\sqrt{3}$
π	0	-1
$\frac{3}{2}\pi$	-1	0
2π	0	1

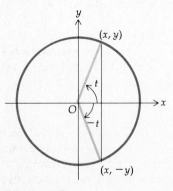

Figure A.1.5

Suppose that t is a real number. Place an angle, having radian measure t, in standard position, and let point P be at the intersection of the terminal side of the angle with the unit circle having its center at the origin. If P is the point (x, y), then the **cosine function** is defined by

$$\cos t = x$$

and the **sine function** is defined by

$$\sin t = y$$

From the above definition it is seen that $\sin t$ and $\cos t$ are defined for any value of t. Hence the domain of the sine and cosine functions is the set of all real numbers. The largest value either function may have is 1, and the smallest value is -1. It will be shown later that the sine and cosine functions take on all values between -1 and 1, and from this it follows that the range of the two functions is $[-1, 1]$.

For certain values of t, the cosine and sine are easily obtained from a figure. From Fig. A.1.4 we see that $\cos 0 = 1$ and $\sin 0 = 0$, $\cos \frac{1}{4}\pi = \frac{1}{2}\sqrt{2}$ and $\sin \frac{1}{4}\pi = \frac{1}{2}\sqrt{2}$, $\cos \frac{1}{2}\pi = 0$ and $\sin \frac{1}{2}\pi = 1$, $\cos \pi = -1$ and $\sin \pi = 0$, and $\cos \frac{3}{2}\pi = 0$ and $\sin \frac{3}{2}\pi = -1$. Table A.1.2 gives these values and some others that are frequently used.

An equation of the unit circle having its center at the origin is $x^2 + y^2 = 1$. Because $x = \cos t$ and $y = \sin t$, it follows that

$$\cos^2 t + \sin^2 t = 1 \tag{1}$$

Note that $\cos^2 t$ and $\sin^2 t$ stand for $(\cos t)^2$ and $(\sin t)^2$. Equation (1) is an identity because it is valid for any real number t.

Figures A.1.5 and A.1.6 show angles having a negative radian measure of $-t$ and corresponding angles having a positive radian measure of t. From these figures observe that

$$\cos(-t) = \cos t \quad \text{and} \quad \sin(-t) = -\sin t \tag{2}$$

These equations hold for any real number t because the points where the terminal sides of the angles (having radian measures t and $-t$) intersect the unit circle have equal abscissas and ordinates that differ only in sign. Hence Eqs. (2) are identities.

From the definition of the sine and cosine functions the following identities are obtained.

$$\cos(t + 2\pi) = \cos t \quad \text{and} \quad \sin(t + 2\pi) = \sin t \tag{3}$$

The property of cosine and sine stated by Eqs. (3) is called *periodicity*, which is now defined.

Definition of periodic function

A function f is said to be **periodic** with period $p \neq 0$ if whenever x is in the domain of f, then $x + p$ is also in the domain of f, and

$$f(x + p) = f(x)$$

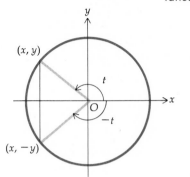

Figure A.1.6

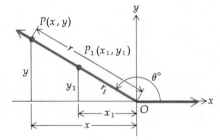

Figure A.1.7

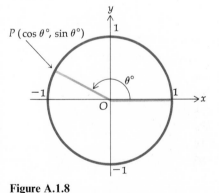

Figure A.1.8

Definition of the sine and cosine of angle measurements

From the above definition and Eqs. (3) it is seen that the sine and cosine functions are periodic with period 2π; that is, whenever the value of the independent variable t is increased by 2π, the value of each of the functions is repeated. It is because of the periodicity of the sine and cosine that these functions have important applications in connection with periodically repetitive phenomena.

When referring to the trigonometric functions with a domain of angle measurements, we use the notation $\theta°$ to denote the measurement of an angle if its degree measure is θ. For example, $45°$ is the measurement of an angle whose degree measure is 45 or, equivalently, whose radian measure is $\frac{1}{4}\pi$.

Consider an angle of $\theta°$ in standard position on a rectangular cartesian coordinate system. Choose any point P, excluding the vertex, on the terminal side of the angle, and let its abscissa be x, its ordinate be y, and $|\overline{OP}| = r$. Refer to Fig. A.1.7. The ratios x/y and y/r are independent of the choice of P, because if the point P_1 is chosen instead of P, we see by Fig. A.1.7 that $x/r = x_1/r_1$ and $y/r = y_1/r_1$. Because the position of the terminal side depends on the angle, these two ratios are functions of the measurement of the angle, and we define

$$\cos \theta° = \frac{x}{r} \quad \text{and} \quad \sin \theta° = \frac{y}{r}$$

Because any point P (other than the origin) may be chosen on the terminal side, we could choose the point for which $r = 1$; this is the point where the terminal side intersects the unit circle $x^2 + y^2 = 1$ (see Fig. A.1.8). Then $\cos \theta°$ is the abscissa of the point and $\sin \theta°$ is the ordinate of the point. This gives the analogy between the sine and cosine of real numbers and those of angle measurements stated in the next definition.

If α degrees and x radians are measurements for the same angle, then

$$\sin \alpha° = \sin x \quad \text{and} \quad \cos \alpha° = \cos x$$

The value of either the sine or cosine function may be found by a table such as Table 9 in the back of the book or by a hand calculator having sin and cos keys. The values are usually given for either degree or radian measure.

When obtaining the derivative of the sine function,

$$\lim_{t \to 0} \frac{\sin t}{t}$$

appears, and we need to know its value. Observe that if $f(t) = \dfrac{\sin t}{t}$, $f(0)$ is not defined. However, the following theorem states that $\lim_{t \to 0} f(t)$ exists and is equal to 1.

Theorem A.1.1

$$\lim_{t \to 0} \frac{\sin t}{t} = 1$$

The proof of this theorem is too advanced for this text. However, its validity should seem plausible by referring to Table A.1.3.

There is another limit that we need when obtaining the derivative of the sine function. However, before discussing this limit we need to prove that the sine function and the cosine function are continuous at 0.

To show that the sine function is continuous at 0 we test the three conditions for continuity at a number.

(i) $\sin 0 = 0$

(ii) $\displaystyle\lim_{t \to 0} \sin t = \lim_{t \to 0} \frac{\sin t}{t} \cdot t = \lim_{t \to 0} \frac{\sin t}{t} \cdot \lim_{t \to 0} t = 1 \cdot 0 = 0$

(iii) $\displaystyle\lim_{t \to 0} \sin t = \sin 0$

Therefore the sine function is continuous at 0.

We now show that the cosine function is continuous at 0.

(i) $\cos 0 = 1$

(ii) $\displaystyle\lim_{t \to 0} \cos t = \lim_{t \to 0} \sqrt{1 - \sin^2 t}$

$$= \sqrt{\lim_{t \to 0}(1 - \sin^2 t)} = \sqrt{1 - 0} = 1$$

Note: We can replace $\cos t$ by $\sqrt{1 - \sin^2 t}$, which follows from (1), since $\cos t > 0$ when $0 < t < \frac{1}{2}\pi$ and when $-\frac{1}{2}\pi < t < 0$.

Table A.1.3

$\alpha°$	$4°$	$3°$	$2°$	$1°$	$-1°$	$-2°$	$-3°$	$-4°$
x radians	0.06981	0.05236	0.03491	0.017453	-0.017453	-0.03491	-0.05236	-0.06981
$\sin x$	0.06976	0.05234	0.03490	0.017452	-0.017452	-0.03490	-0.05234	-0.06976
$\dfrac{\sin x}{x}$	0.99928	0.99962	0.99971	0.99994	0.99994	0.99971	0.99962	0.99928

(iii) $\lim\limits_{t \to 0} \cos t = \cos 0$

Therefore the cosine function is continuous at 0.

Theorem A.1.2

$$\lim_{t \to 0} \frac{1 - \cos t}{t} = 0$$

Proof

$$\lim_{t \to 0} \frac{1 - \cos t}{t} = \lim_{t \to 0} \frac{(1 - \cos t)(1 + \cos t)}{t(1 + \cos t)}$$

$$= \lim_{t \to 0} \frac{1 - \cos^2 t}{t(1 + \cos t)}$$

$$= \lim_{t \to 0} \frac{\sin^2 t}{t(1 + \cos t)}$$

$$= \lim_{t \to 0} \frac{\sin t}{t} \cdot \lim_{t \to 0} \frac{\sin t}{1 + \cos t}$$

By Theorem A.1.1,

$$\lim_{t \to 0} \frac{\sin t}{t} = 1$$

and because the sine and cosine functions are continuous at 0 it follows that

$$\lim_{t \to 0} \frac{\sin t}{1 + \cos t} = \frac{0}{1 + 1} = 0$$

Therefore

$$\lim_{t \to 0} \frac{1 - \cos t}{t} = 1 \cdot 0 = 0 \qquad \blacksquare$$

To show that the sine function has a derivative we apply the following formula from trigonometry:

$$\sin(a + b) = \sin a \cos b + \cos a \sin b \tag{4}$$

and make use of Theorems A.1.1 and A.1.2.

Let f be the function defined by

$$f(x) = \sin x$$

From the definition of a derivative,

$$f'(x) = \lim_{\Delta x \to 0} \frac{f(x + \Delta x) - f(x)}{\Delta x}$$

$$= \lim_{\Delta x \to 0} \frac{\sin(x + \Delta x) - \sin x}{\Delta x}$$

Formula (4) for $\sin(x + \Delta x)$ is used to obtain

$$f'(x) = \lim_{\Delta x \to 0} \frac{\sin x \cos(\Delta x) + \cos x \sin(\Delta x) - \sin x}{\Delta x}$$

$$= \lim_{\Delta x \to 0} \frac{\sin x [\cos(\Delta x) - 1]}{\Delta x} + \lim_{\Delta x \to 0} \frac{\cos x \sin(\Delta x)}{\Delta x}$$

$$= -\lim_{\Delta x \to 0} \frac{1 - \cos(\Delta x)}{\Delta x} \left(\lim_{\Delta x \to 0} \sin x \right) + \left(\lim_{\Delta x \to 0} \cos x \right) \lim_{\Delta x \to 0} \frac{\sin(\Delta x)}{\Delta x} \quad (5)$$

From Theorem A.1.2,

$$\lim_{\Delta x \to 0} \frac{1 - \cos(\Delta x)}{\Delta x} = 0 \tag{6}$$

and from Theorem A.1.1,

$$\lim_{\Delta x \to 0} \frac{\sin(\Delta x)}{\Delta x} = 1 \tag{7}$$

Substituting from (6) and (7) into (5) we get

$$f'(x) = -0 \cdot \sin x + \cos x \cdot 1$$

$$= \cos x$$

From this result and the chain rule we have the following theorem.

Theorem A.1.3   If u is a differentiable function of x,

$$D_x(\sin u) = \cos u \, D_x u$$

EXAMPLE 1   Given $y = \sin 3x^2$, find $\dfrac{dy}{dx}$.

Solution   From Theorem A.1.3,

$$\frac{dy}{dx} = (\cos 3x^2)D_x(3x^2)$$

$$= 6x \cos 3x^2$$

From (4), with $a = \frac{1}{2}\pi$ and $b = -x$, we get

$$\sin(\tfrac{1}{2}\pi - x) = \sin \tfrac{1}{2}\pi \cos(-x) + \cos \tfrac{1}{2}\pi \sin(-x)$$

From (2), $\cos(-x) = \cos x$, and $\sin(-x) = -\sin x$. Also $\sin \frac{1}{2}\pi = 1$, and $\cos \frac{1}{2}\pi = 0$. Thus

$$\sin(\tfrac{1}{2}\pi - x) = \cos x \tag{8}$$

If in (8) we let $\frac{1}{2}\pi - x = t$, then $x = \frac{1}{2}\pi - t$, and we obtain

$$\cos(\tfrac{1}{2}\pi - t) = \sin t \tag{9}$$

The derivative of the cosine function is found by making use of identities (8) and (9) and Theorem A.1.3.

$$D_x(\cos x) = D_x[\sin(\tfrac{1}{2}\pi - x)]$$
$$= \cos(\tfrac{1}{2}\pi - x)D_x(\tfrac{1}{2}\pi - x)$$
$$= (\sin x)(-1)$$
$$= -\sin x$$

From this result and the chain rule we have the following theorem.

Theorem A.1.4   If u is a differentiable function of x,

$$D_x(\cos u) = -\sin u \, D_x u$$

EXAMPLE 2   Given $F(t) = \cos e^{2t}$, find $F'(t)$.

Solution   From Theorem A.1.4,

$$F'(t) = (-\sin e^{2t})D_t(e^{2t})$$
$$= -2e^{2t}\sin e^{2t}$$

EXAMPLE 3   Given

$$f(x) = \frac{\sin x}{1 - 2\cos x}$$

find $f'(x)$.

Solution

$$f'(x) = \frac{(1 - 2\cos x)D_x(\sin x) - \sin x \cdot D_x(1 - 2\cos x)}{(1 - 2\cos x)^2}$$

$$= \frac{(1 - 2\cos x)(\cos x) - \sin x(2\sin x)}{(1 - 2\cos x)^2}$$

$$= \frac{\cos x - 2(\cos^2 x + \sin^2 x)}{(1 - 2\cos x)^2}$$

$$= \frac{\cos x - 2}{(1 - 2\cos x)^2}$$

Because $D_x(\sin x) = \cos x$ and $\cos x$ exists for all values of x, the sine function is differentiable everywhere and therefore continuous everywhere. Similarly, the cosine function is differentiable and continuous everywhere. Because the largest value either function may have is 1 and the smallest value is -1, and because they are both continuous everywhere, the range of each of these functions is $[-1, 1]$.

The graph of the sine function is now discussed. Let

$$f(x) = \sin x$$

Then

$$f'(x) = \cos x$$

To determine the relative extrema, we set $f'(x) = 0$ and get $x = \frac{1}{2}\pi + n\pi$, where $n = 0, \pm1, \pm2, \ldots$. At these values of x, $\sin x$ is either $+1$ or -1; and these are the largest and smallest values that $\sin x$ assumes. The graph intersects the x axis at those points where $\sin x = 0$, that is, at the points where $x = n\pi$ and n is any integer. From this information we draw a sketch of the graph of the sine function. It is shown in Fig. A.1.9.

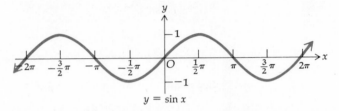

Figure A.1.9

For the graph of the cosine function, use the identity

$$\cos x = \sin(\tfrac{1}{2}\pi + x)$$

which follows from (4). Hence the graph of the cosine function is obtained from the graph of the sine function by translating the y axis $\frac{1}{2}\pi$ units to the right. See Fig. A.1.10.

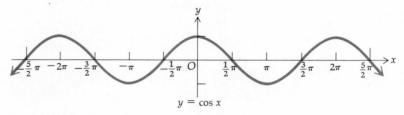

Figure A.1.10

The theorems for the indefinite integral of the sine and cosine functions follow immediately from the corresponding ones for differentiation.

Theorem A.1.5

$$\int \sin u \, du = -\cos u + C$$

Proof

$$D_u(-\cos u) = -(-\sin u) = \sin u \qquad \blacksquare$$

Theorem A.1.6

$$\int \cos u \, du = \sin u + C$$

Proof

$$D_u(\sin u) = \cos u \qquad \blacksquare$$

EXAMPLE 4   Find $\displaystyle\int 3 \sin 2x \, dx$.

Solution   Let $u = 2x$. Then $du = 2\, dx$. Thus

$$\int 3 \sin 2x \, dx = \frac{3}{2} \int \sin 2x(2\, dx)$$

$$= \frac{3}{2} \int \sin u \, du$$

$$= \tfrac{3}{2}(-\cos u) + C$$

$$= -\tfrac{3}{2} \cos 2x + C$$

EXAMPLE 5   Find

$$\int \frac{\cos \sqrt{x}}{\sqrt{x}} \, dx$$

Solution   Let $u = \sqrt{x}$. Then

$$du = \frac{1}{2\sqrt{x}} \, dx$$

Therefore

$$\int \frac{\cos \sqrt{x}}{\sqrt{x}} \, dx = 2 \int \cos \sqrt{x} \left(\frac{1}{2\sqrt{x}} \, dx \right)$$

$$= 2 \int \cos u \, du$$

$$= 2 \sin u + C$$

$$= 2 \sin \sqrt{x} + C$$

EXAMPLE 6 Find $\displaystyle\int x \sin x \, dx$.

Solution We use integration by parts. Let $u = x$ and $dv = \sin x \, dx$. Then

$$du = dx \quad \text{and} \quad v = -\cos x$$

Therefore

$$\int x \sin x \, dx = -x \cos x + \int \cos x \, dx$$

$$= -x \cos x + \sin x + C$$

The next example shows a procedure that can be used to find the indefinite integral of an odd-numbered power of the sine or cosine.

EXAMPLE 7 Find $\displaystyle\int \cos^3 x \, dx$.

Solution

$$\int \cos^3 x \, dx = \int \cos^2 x (\cos x \, dx)$$

From (1), $\cos^2 x = 1 - \sin^2 x$. Therefore

$$\int \cos^3 x \, dx = \int (1 - \sin^2 x)(\cos x \, dx)$$

$$= \int \cos x \, dx - \int \sin^2 x (\cos x \, dx)$$

$$= \sin x - \tfrac{1}{3} \sin^3 x + C$$

Exercises A.1

In Exercises 1 and 2, find the equivalent radian measurement.

1. (a) $60°$; (b) $135°$; (c) $210°$; (d) $-150°$; (e) $20°$; (f) $450°$; (g) $-75°$; (h) $100°$
2. (a) $45°$; (b) $120°$; (c) $240°$; (d) $-225°$; (e) $15°$; (f) $540°$; (g) $-48°$; (h) $2°$

In Exercises 3 and 4 find the equivalent degree measurement.

3. (a) $\tfrac{1}{4}\pi$ rad; (b) $\tfrac{2}{3}\pi$ rad; (c) $\tfrac{11}{6}\pi$ rad; (d) $-\tfrac{1}{2}\pi$ rad; (e) $\tfrac{1}{2}$ rad; (f) 3π rad; (g) -2 rad; (h) $\tfrac{1}{12}\pi$ rad
4. (a) $\tfrac{1}{6}\pi$ rad; (b) $\tfrac{4}{3}\pi$ rad; (c) $\tfrac{3}{4}\pi$ rad; (d) -5π rad; (e) $\tfrac{1}{3}$ rad; (f) -5 rad; (g) $\tfrac{11}{12}\pi$ rad; (h) 0.2 rad

In Exercises 5 and 6, find the indicated value.

5. (a) $\sin \tfrac{1}{6}\pi$; (b) $\cos \tfrac{1}{4}\pi$; (c) $\sin(-\tfrac{3}{2}\pi)$; (d) $\cos(-\tfrac{2}{3}\pi)$; (e) $\cos \tfrac{5}{6}\pi$; (f) $\sin(-\tfrac{5}{4}\pi)$; (g) $\cos 3\pi$; (h) $\sin(-5\pi)$
6. (a) $\cos \tfrac{1}{3}\pi$; (b) $\sin \tfrac{1}{4}\pi$; (c) $\cos(-\tfrac{1}{2}\pi)$; (d) $\sin(-2\pi)$; (e) $\sin \tfrac{4}{3}\pi$; (f) $\cos(-\tfrac{1}{6}\pi)$; (g) $\sin 7\pi$; (h) $\cos(-\tfrac{5}{2}\pi)$

In Exercises 7 and 8, find all values of t in the interval $[0, 2\pi)$ for which the given equation is satisfied.

7. (a) $\sin t = \tfrac{1}{2}\sqrt{3}$; (b) $\cos t = -\tfrac{1}{2}\sqrt{2}$; (c) $\cos t = 0$; (d) $\sin t = -\tfrac{1}{2}$
8. (a) $\cos t = -\tfrac{1}{2}$; (b) $\sin t = \tfrac{1}{2}\sqrt{2}$; (c) $\sin t = -1$; (d) $\cos t = -\tfrac{1}{2}\sqrt{3}$

In Exercises 9 through 20, find the derivative of the given function.

9. $f(x) = 3 \sin x$

10. $g(x) = \sin x + \cos x$

11. $g(x) = x \sin x + \cos x$

12. $f(y) = 3 \sin y - y \cos y$

13. $f(x) = 12 \sin 3x \cos 4x$

14. $f(x) = \cos^2 x^2$

15. $f(t) = \dfrac{\sin t}{1 - \cos t}$

16. $f(x) = \dfrac{\sin x - 1}{\cos x + 1}$

17. $f(x) = (\sin^2 x - x^2)^3$

18. $F(x) = 4 \cos(\sin 3x)$

19. $h(y) = \cos \sqrt{y^2 + 1}$

20. $g(x) = \sqrt{x} \sin \sqrt{\dfrac{1}{x}}$

In Exercises 21 through 24, find $f'(a)$ for the given value of a.

21. $f(x) = \dfrac{\cos x}{x}; a = \frac{1}{2}\pi$

22. $f(x) = x^2 \cos x - \sin x; a = 0$

23. $f(x) = \dfrac{2 \cos x - 1}{\sin x}; a = \frac{2}{3}\pi$

24. $f(x) = \dfrac{\sin x}{\cos x - \sin x}; a = \frac{3}{4}\pi$

In Exercises 25 through 36, evaluate the indefinite integral.

25. $\displaystyle\int \sin 4x \, dx$

26. $\displaystyle\int \frac{1}{2} \cos 6x \, dx$

27. $\displaystyle\int \cos x(2 + \sin x)^3 \, dx$

28. $\displaystyle\int \frac{\sin x}{(1 + \cos x)^2} \, dx$

29. $\displaystyle\int 2 \sin x \sqrt[3]{1 + \cos x} \, dx$

30. $\displaystyle\int \frac{\frac{1}{2} \cos \frac{1}{4}x}{\sqrt{\sin \frac{1}{4}x}} \, dx$

31. $\displaystyle\int x \cos 2x \, dx$

32. $\displaystyle\int x^2 \sin 3x \, dx$

33. $\displaystyle\int \sin^3 x \, dx$

34. $\displaystyle\int \sin^2 x \cos^3 x \, dx$

35. $\displaystyle\int \sqrt{\cos z} \sin^3 z \, dz$

36. $\displaystyle\int \sin^5 t \, dt$

In Exercises 37 and 38, evaluate the definite integral.

37. $\displaystyle\int_0^{\pi/3} \sin^3 t \cos^2 t \, dt$

38. $\displaystyle\int_{-\pi}^{\pi} z^2 \cos 2z \, dz$

39. Find the area of the region bounded by one arch of the sine curve.

40. Find the average value of the cosine function on the closed interval $[\frac{1}{3}\pi, \frac{1}{2}\pi]$.

41. A company that sells men's overcoats starts its fiscal year on July 1. For the three fiscal years beginning on July 1, 1981, the profit from sales was given approximately by

$$P(t) = 20{,}000(1 - \cos \tfrac{1}{6}\pi t) \qquad 0 \le t \le 36$$

where $P(t)$ dollars per month is the profit t months since July 1, 1981. (a) Draw a sketch of the graph of P. Find the profit per month on (b) October 1, 1981; (c) January 1, 1982; (d) April 1, 1982; (e) July 1, 1982.

42. Suppose that on a particular day in a certain city the Fahrenheit temperature is $f(t)$ degrees t hours since midnight, where

$$f(t) = 60 - 15 \sin \tfrac{1}{12}\pi(8 - t) \qquad 0 \le t \le 24$$

(a) Draw a sketch of the graph of f. Find the temperature at (b) 12 midnight; (c) 8 A.M.; (d) 12 noon; (e) 2 P.M.; (f) 6 P.M.

43. For the company of Exercise 41, determine the total profit for the three fiscal years beginning on July 1, 1981, by finding the area of the region bounded by the graph of P and the t axis.

44. On the particular day in the city of Exercise 42, find the average temperature between 8 A.M. and 6 P.M.

45. If a body of weight W lb is dragged along a horizontal floor by means of a force of magnitude F lb and directed at an angle of θ radians with the plane of the floor, then F is given by the equation

$$F = \frac{kW}{k \sin \theta + \cos \theta}$$

where k is a constant and is called the coefficient of friction. If $k = 0.5$, find the instantaneous rate of change of F with respect to θ when (a) $\theta = \frac{1}{4}\pi$; (b) $\theta = \frac{1}{2}\pi$.

46. For the body of Exercise 45, find $\cos \theta$ when F is least if $0 \leq \theta \leq \frac{1}{2}\pi$.

47. If R ft is the range of a projectile, then

$$R = \frac{v_0{}^2 \sin 2\theta}{g} \qquad 0 \leq \theta \leq \frac{1}{2}\pi$$

where v_0 ft/sec is the initial velocity, g ft/sec^2 is the acceleration due to gravity, and θ is the radian measure of the angle that the gun makes with the horizontal. Find the value of θ that makes the range a maximum.

48. The cross section of a trough has the shape of an inverted isosceles triangle. Find the size of the vertex angle for which the area of the triangle is a maximum, and thus the capacity of the trough is a maximum.

A.2 The tangent, cotangent, secant, and cosecant functions

In addition to the sine and cosine there are four other trigonometric functions. They are defined in terms of the sine and cosine.

Definition of the tangent, cotangent, secant, and cosecant functions

The **tangent** and **secant** functions are defined by

$$\tan x = \frac{\sin x}{\cos x} \qquad \sec x = \frac{1}{\cos x}$$

for all real numbers x for which $\cos x \neq 0$.

The **cotangent** and **cosecant** functions are defined by

$$\cot x = \frac{\cos x}{\sin x} \qquad \csc x = \frac{1}{\sin x}$$

for all real numbers x for which $\sin x \neq 0$.

The tangent and secant functions are not defined when $\cos x = 0$ which occurs when x is $\frac{1}{2}\pi$, $\frac{3}{2}\pi$, or $\frac{1}{2}\pi + n\pi$, where n is a positive or negative integer or zero. Therefore the domain of the tangent and secant functions is the set of all real numbers except numbers of the form $\frac{1}{2}\pi + n\pi$, where n is any integer. Similarly, because $\cot x$ and $\csc x$ are not defined when $\sin x = 0$, the domain of the cotangent and cosecant functions is the set of all real numbers except numbers of the form $n\pi$, where n is any integer.

Because the sine and cosine functions are continuous at all real numbers, it follows that the tangent, cotangent, secant, and cosecant functions are continuous at all numbers in their domain.

By using the identity

$$\cos^2 x + \sin^2 x = 1 \tag{1}$$

and the above definition, we obtain two other important identities. If on both sides of (1) we divide by $\cos^2 x$ when $\cos x \neq 0$, we get

$$\frac{\cos^2 x}{\cos^2 x} + \frac{\sin^2 x}{\cos^2 x} = \frac{1}{\cos^2 x}$$

and because $\sin x/\cos x = \tan x$ and $1/\cos x = \sec x$, we have the identity

$$1 + \tan^2 x = \sec^2 x \tag{2}$$

By dividing on both sides of (1) by $\sin^2 x$ when $\sin x \neq 0$, we obtain in a similar way the identity

$$\cot^2 x + 1 = \csc^2 x \tag{3}$$

Three other important identities that follow immediately from the definition are

$$\sin x \csc x = 1 \tag{4}$$

$$\cos x \sec x = 1 \tag{5}$$

and

$$\tan x \cot x = 1 \tag{6}$$

The derivatives of the tangent, cotangent, secant, and cosecant functions are obtained from those of the sine and cosine functions and theorems on differentiation.

$$D_x(\tan x) = D_x\left(\frac{\sin x}{\cos x}\right) = \frac{\cos x \cdot D_x(\sin x) - \sin x \cdot D_x(\cos x)}{\cos^2 x}$$

$$= \frac{(\cos x)(\cos x) - (\sin x)(-\sin x)}{\cos^2 x}$$

$$= \frac{\cos^2 x + \sin^2 x}{\cos^2 x}$$

$$= \frac{1}{\cos^2 x}$$

$$= \sec^2 x \tag{7}$$

From (7) and the chain rule we have the following theorem.

Theorem A.2.1   If u is a differentiable function of x,

$$D_x(\tan u) = \sec^2 u \, D_x u$$

EXAMPLE 1   Find $f'(x)$ if $f(x) = 2 \tan \frac{1}{2}x - x$.

Solution $f'(x) = 2 \sec^2 \frac{1}{2}x(\frac{1}{2}) - 1 = \sec^2 \frac{1}{2}x - 1$. If we use identity (2), this simplifies to

$$f'(x) = \tan^2 \frac{1}{2}x$$

The derivative of the cotangent function is obtained by a method analogous to that for the tangent function. The result is

$$D_x(\cot x) = -\csc^2 x \tag{8}$$

The derivation of (8) is left as an exercise (see Exercise 1). The next theorem follows from (8) and the chain rule.

Theorem A.2.2   If u is a differentiable function of x,

$$D_x(\cot u) = -\csc^2 u \, D_x u$$

We now proceed to find the derivative of the secant function.

$$
\begin{aligned}
D_x(\sec x) &= D_x[(\cos x)^{-1}] \\
&= -1(\cos x)^{-2}(-\sin x) \\
&= \frac{1}{\cos^2 x} \cdot \sin x \\
&= \frac{1}{\cos x} \cdot \frac{\sin x}{\cos x} \\
&= \sec x \tan x \tag{9}
\end{aligned}
$$

By applying (9) and the chain rule, the following theorem is obtained.

Theorem A.2.3   If u is a differentiable function of x,

$$D_x(\sec u) = \sec u \tan u \, D_x u$$

EXAMPLE 2   Find $f'(x)$ if $f(x) = \sec^4 3x$.

Solution

$$
\begin{aligned}
f'(x) &= 4 \sec^3 3x \cdot D_x(\sec 3x) \\
&= 4 \sec^3 3x(\sec 3x \tan 3x)(3) \\
&= 12 \sec^4 3x \tan 3x
\end{aligned}
$$

In a manner similar to that for the secant, the formula for the derivative of the cosecant function may be derived, and we obtain

$$D_x(\csc x) = -\csc x \cot x \tag{10}$$

The derivation of (10) is left as an exercise (see Exercise 2). We obtain the following theorem from (10) and the chain rule.

Theorem A.2.4 If u is a differentiable function of x,

$$D_x(\csc u) = -\csc u \cot u\, D_x u$$

EXAMPLE 3 Find $f'(x)$ if $f(x) = \cot x \csc x$.

Solution

$$f'(x) = \cot x \cdot D_x(\csc x) + \csc x \cdot D_x(\cot x)$$
$$= \cot x(-\csc x \cot x) + \csc x(-\csc^2 x)$$
$$= -\csc x \cot^2 x - \csc^3 x$$

EXAMPLE 4 An airplane is flying west at 500 ft/sec at an altitude of 4000 ft. The airplane is in a vertical plane with a searchlight on the ground. If the light is to be kept on the plane, how fast is the searchlight revolving when the airplane is due east of the searchlight at an airline distance of 2000 ft?

Solution Refer to Fig. A.2.1. The searchlight is at point L, and at a particular instant of time the airplane is at point P.

Let t be the number of seconds in the time; x be the number of feet due east in the airline distance of the airplane from the searchlight at time t sec; and θ be the number of radians in the angle of elevation of the airplane at the searchlight at time t sec.

We are given $D_t x = -500$, and we wish to find $D_t \theta$ when $x = 2000$.

$$\tan \theta = \frac{4000}{x} \tag{11}$$

Differentiating on both sides of (11) with respect to t we obtain

$$\sec^2 \theta\, D_t\theta = -\frac{4000}{x^2}\, D_t x$$

Substituting $D_t x = -500$ in the above and dividing by $\sec^2 \theta$ gives

$$D_t\theta = \frac{2{,}000{,}000}{x^2 \sec^2 \theta} \tag{12}$$

When $x = 2000$, $\tan \theta = 2$. Therefore $\sec^2 \theta = 1 + \tan^2 \theta = 5$. Substituting these values into (12) we have, when $x = 2000$,

$$D_t\theta = \frac{2{,}000{,}000}{4{,}000{,}000(5)} = \frac{1}{10}$$

We conclude that at the given instant the measurement of the angle is increasing at the rate of $\frac{1}{10}$ rad/sec, and this is how fast the searchlight is revolving.

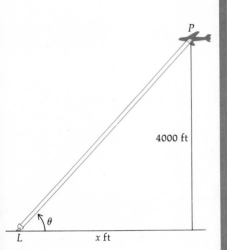

Figure A.2.1

A formula for the indefinite integral of the tangent function is derived as follows. Because

$$\int \tan u \, du = \int \frac{\sin u}{\cos u} \, du$$

we let $v = \cos u$ and $dv = -\sin u \, du$, and we obtain

$$\int \tan u \, du = -\int \frac{dv}{v} = -\ln |v| + C$$

Hence

$$\int \tan u \, du = -\ln |\cos u| + C$$

Because $-\ln |\cos u| = \ln |(\cos u)^{-1}| = \ln |\sec u|$, we have the following theorem.

Theorem A.2.5

$$\int \tan u \, du = \ln |\sec u| + C$$

EXAMPLE 5   Find $\int \tan 3x \, dx$.

Solution

$$\int \tan 3x \, dx = \frac{1}{3} \int \tan 3x (3 \, dx)$$
$$= \tfrac{1}{3} \ln |\sec 3x| + C$$

The theorem giving the indefinite integral of the cotangent function is proved in a way similar to the proof of Theorem A.2.5 (see Exercise 27).

Theorem A.2.6

$$\int \cot u \, du = \ln |\sin u| + C$$

To obtain the formula for $\int \sec u \, du$, we multiply the numerator and denominator of the integrand by $(\sec u + \tan u)$, and we have

$$\int \sec u \, du = \int \frac{\sec u(\sec u + \tan u)}{\sec u + \tan u} \, du = \int \frac{(\sec^2 u + \sec u \tan u)}{\sec u + \tan u} \, du$$

Let $v = \sec u + \tan u$. Then $dv = (\sec u \tan u + \sec^2 u)du$; so we have

$$\int \sec u\; du = \int \frac{dv}{v} = \ln |v| + C$$

We have proved the following theorem.

Theorem A.2.7

$$\int \sec u\; du = \ln |\sec u + \tan u| + C$$

The formula for $\int \csc u\; du$ is derived by multiplying the numerator and denominator of the integrand by $(\csc u - \cot u)$ and proceeding as above. The derivation is left as an exercise (see Exercise 28). The formula is given in the next theorem.

Theorem A.2.8

$$\int \csc u\; du = \ln |\csc u - \cot u| + C$$

EXAMPLE 6 Find

$$\int \frac{dx}{\sin 2x}$$

Solution

$$\int \frac{dx}{\sin 2x} = \int \csc 2x\; dx = \frac{1}{2} \int \csc 2x\; (2\; dx)$$

$$= \tfrac{1}{2} \ln |\csc 2x - \cot 2x| + C$$

The indefinite integral formulas in the next two theorems follow immediately from the corresponding differentiation formulas.

Theorem A.2.9

$$\int \sec^2 u\; du = \tan u + C$$

Theorem A.2.10

$$\int \sec u \tan u \, du = \sec u + C$$

There are similar theorems involving the cotangent and cosecant.

Theorem A.2.11

$$\int \csc^2 u \, du = -\cot u + C$$

Proof

$$D_u(-\cot u) = -(-\csc^2 u) = \csc^2 u \qquad \blacksquare$$

Theorem A.2.12

$$\int \csc u \cot u \, du = -\csc u + C$$

Proof

$$D_u(-\csc u) = -(-\csc u \cot u) = \csc u \cot u \qquad \blacksquare$$

EXAMPLE 7   Find

$$\int \frac{2 + 3 \cos u}{\sin^2 u} \, du$$

Solution

$$\int \frac{2 + 3 \cos u}{\sin^2 u} \, du = 2 \int \frac{1}{\sin^2 u} \, du + 3 \int \frac{1}{\sin u} \cdot \frac{\cos u}{\sin u} \, du$$

$$= 2 \int \csc^2 u \, du + 3 \int \csc u \cot u \, du$$

$$= -2 \cot u - 3 \csc u + C$$

Exercises A.2

1. Derive: $D_x(\cot x) = -\csc^2 x$.

2. Derive: $D_x(\csc x) = -\csc x \cot x$.

In Exercises 3 through 26, find the derivative of the given function.

3. $f(x) = \csc 4x$

4. $g(x) = 3 \cot 4x$

5. $h(x) = \tan^2 x$

6. $f(x) = \sqrt{\sec x}$

7. $f(x) = \sec x^2$

8. $g(x) = \ln \csc^2 x$

9. $g(x) = 3 \cot e^x$

10. $f(w) = 2 \tan(w^2 - 1)$

11. $F(x) = \ln |\sec 2x|$

12. $h(x) = \ln |\cot \frac{1}{2}x|$

13. $G(r) = \sqrt{\cot 3r}$

14. $f(x) = \sin x \tan x$

15. $g(x) = \sec^2 x \tan^2 x$

16. $h(t) = \sec^2 2t - \tan^2 2t$

17. $f(t) = \csc(t^3 + 1)$

18. $f(x) = e^{\sec 5x}$

19. $f(x) = 2^{\csc 3x}$

20. $g(t) = 2 \sec \sqrt{t}$

21. $H(t) = \cot^4 t - \csc^4 t$

22. $F(x) = \dfrac{\cot^2 2x}{1 + x^2}$

23. $f(x) = \dfrac{\tan x}{\cos x - 4}$

24. $G(x) = (\tan x)^x; \tan x > 0$

25. $F(x) = (\sin x)^{\tan x}; \sin x > 0$

26. $f(x) = \ln |\sec 5x + \tan 5x|$

27. Derive the formula $\int \cot u \, du = \ln |\sin u| + C$.

28. Derive the formula $\int \csc u \, du = \ln |\csc u - \cot u| + C$.

In Exercises 29 through 42, evaluate the indefinite integral.

29. $\displaystyle\int \sec^2 5x \, dx$

30. $\displaystyle\int \cot(3x + 1) \, dx$

31. $\displaystyle\int \tan 2w \, dw$

32. $\displaystyle\int \csc^2 4t \, dt$

33. $\displaystyle\int \csc 3x \cot 3x \, dx$

34. $\displaystyle\int \sec 3y \, dy$

35. $\displaystyle\int \csc 10t \, dt$

36. $\displaystyle\int e^x \csc^2 e^x \, dx$

37. $\displaystyle\int x \csc^2 5x^2 \, dx$

38. $\displaystyle\int \dfrac{\tan \sqrt{x}}{\sqrt{x}} \, dx$

39. $\displaystyle\int e^t \cot e^t \, dt$

40. $\displaystyle\int \sec x \tan x \tan(\sec x) \, dx$

41. $\displaystyle\int y \sec 3y^2 \, dy$

42. $\displaystyle\int t^2 \sec^2 t^3 \, dt$

In Exercises 43 and 44, evaluate the definite integral.

43. $\displaystyle\int_{\pi/16}^{\pi/12} \tan 4x \, dx$

44. $\displaystyle\int_{\pi/8}^{\pi/4} 3 \csc 2x \, dx$

45. Find the average value of the tangent function on the closed interval $[0, \frac{1}{4}\pi]$.

46. We can integrate $\int \sec^2 x \tan x \, dx$ in two ways as follows:

$$\int \sec^2 x \tan x \, dx = \int \tan x(\sec^2 x \, dx) = \tfrac{1}{2} \tan^2 x + C$$

and

$$\int \sec^2 x \tan x \, dx = \int \sec x(\sec x \tan x \, dx) = \tfrac{1}{2} \sec^2 x + C$$

Explain the difference in the appearance of the two answers.

47. An airplane is flying at a constant speed at an altitude of 10,000 ft on a line that will take it directly over an observer on the ground. At a given instant the observer notes that the angle of elevation of the airplane is $\frac{1}{3}\pi$ radians and is increasing at the rate of $\frac{1}{60}$ rad/sec. Find the speed of the airplane.

48. A radar antenna is located on a ship 4 mi from a straight shore, and it is rotating at 32 rpm. How fast is the radar beam moving along the shoreline when the beam makes an angle of 45° with the shore?

TABLES

Table 1 Powers and roots

n	n^2	$\sqrt{n}$	n^3	$\sqrt[3]{n}$	n	n^2	$\sqrt{n}$	n^3	$\sqrt[3]{n}$
1	1	1.000	1	1.000	51	2,601	7.141	132,651	3.708
2	4	1.414	8	1.260	52	2,704	7.211	140,608	3.732
3	9	1.732	27	1.442	53	2,809	7.280	148,877	3.756
4	16	2.000	64	1.587	54	2,916	7.348	157,464	3.780
5	25	2.236	125	1.710	55	3,025	7.416	166,375	3.803
6	36	2.449	216	1.817	56	3,136	7.483	175,616	3.826
7	49	2.646	343	1.913	57	3,249	7.550	185,193	3.848
8	64	2.828	512	2.000	58	3,364	7.616	195,112	3.871
9	81	3.000	729	2.080	59	3,481	7.681	205,379	3.893
10	100	3.162	1,000	2.154	60	3,600	7.746	216,000	3.915
11	121	3.317	1,331	2.224	61	3,721	7.810	226,981	3.936
12	144	3.464	1,728	2.289	62	3,844	7.874	238,328	3.958
13	169	3.606	2,197	2.351	63	3,969	7.937	250,047	3.979
14	196	3.742	2,744	2.410	64	4,096	8.000	262,144	4.000
15	225	3.873	3,375	2.466	65	4,225	8.062	274,625	4.021
16	256	4.000	4,096	2.520	66	4,356	8.124	287,496	4.041
17	289	4.123	4,913	2.571	67	4,489	8.185	300,763	4.062
18	324	4.243	5,832	2.621	68	4,624	8.246	314,432	4.082
19	361	4.359	6,859	2.668	69	4,761	8.307	328,509	4.102
20	400	4.472	8,000	2.714	70	4,900	8.367	343,000	4.121
21	441	4.583	9,261	2.759	71	5,041	8.426	357,911	4.141
22	484	4.690	10,648	2.802	72	5,184	8.485	373,248	4.160
23	529	4.796	12,167	2.844	73	5,329	8.544	389,017	4.179
24	576	4.899	13,824	2.884	74	5,476	8.602	405,224	4.198
25	625	5.000	15,625	2.924	75	5,625	8.660	421,875	4.217
26	676	5.099	17,576	2.962	76	5,776	8.718	438,976	4.236
27	729	5.196	19,683	3.000	77	5,929	8.775	456,533	4.254
28	784	5.291	21,952	3.037	78	6,084	8.832	474,552	4.273
29	841	5.385	24,389	3.072	79	6,241	8.888	493,039	4.291
30	900	5.477	27,000	3.107	80	6,400	8.944	512,000	4.309
31	961	5.568	29,791	3.141	81	6,561	9.000	531,441	4.327
32	1,024	5.657	32,768	3.175	82	6,724	9.055	551,368	4.344
33	1,089	5.745	35,937	3.208	83	6,889	9.110	571,787	4.362
34	1,156	5.831	39,304	3.240	84	7,056	9.165	592,704	4.380
35	1,225	5.916	42,875	3.271	85	7,225	9.220	614,125	4.397
36	1,296	6.000	46,656	3.302	86	7,396	9.274	636,056	4.414
37	1,369	6.083	50,653	3.332	87	7,569	9.327	658,503	4.431
38	1,444	6.164	54,872	3.362	88	7,744	9.381	681,472	4.448
39	1,521	6.245	59,319	3.391	89	7,921	9.434	704,969	4.465
40	1,600	6.325	64,000	3.420	90	8,100	9.487	729,000	4.481
41	1,681	6.403	68,921	3.448	91	8,281	9.539	753,571	4.498
42	1,764	6.481	74,088	3.476	92	8,464	9.592	778,688	4.514
43	1,849	6.557	79,507	3.503	93	8,649	9.643	804,357	4.531
44	1,936	6.633	85,184	3.530	94	8,836	9.695	830,584	4.547
45	2,025	6.708	91,125	3.557	95	9,025	9.747	857,375	4.563
46	2,116	6.782	97,336	3.583	96	9,216	9.798	884,736	4.579
47	2,209	6.856	103,823	3.609	97	9,409	9.849	912,673	4.595
48	2,304	6.928	110,592	3.634	98	9,604	9.899	941,192	4.610
49	2,401	7.000	117,649	3.659	99	9,801	9.950	970,299	4.626
50	2,500	7.071	125,000	3.684	100	10,000	10.000	1,000,000	4.642

Table 2 Common logarithms

N	0	1	2	3	4	5	6	7	8	9
10	0000	0043	0086	0128	0170	0212	0253	0294	0334	0374
11	0414	0453	0492	0531	0569	0607	0645	0682	0719	0755
12	0792	0828	0864	0899	0934	0969	1004	1038	1072	1106
13	1139	1173	1206	1239	1271	1303	1335	1367	1399	1430
14	1461	1492	1523	1553	1584	1614	1644	1673	1703	1732
15	1761	1790	1818	1847	1875	1903	1931	1959	1987	2014
16	2041	2068	2095	2122	2148	2175	2201	2227	2253	2279
17	2304	2330	2355	2380	2405	2430	2455	2480	2504	2529
18	2553	2577	2601	2625	2648	2672	2695	2718	2742	2765
19	2788	2810	2833	2856	2878	2900	2923	2945	2967	2989
20	3010	3032	3054	3075	3096	3118	3139	3160	3181	3201
21	3222	3243	3263	3284	3304	3324	3345	3365	3385	3404
22	3424	3444	3464	3483	3502	3522	3541	3560	3579	3598
23	3617	3636	3655	3674	3692	3711	3729	3747	3766	3784
24	3802	3820	3838	3856	3874	3892	3909	3927	3945	3962
25	3979	3997	4014	4031	4048	4065	4082	4099	4116	4133
26	4150	4166	4183	4200	4216	4232	4249	4265	4281	4298
27	4314	4330	4346	4362	4378	4393	4409	4425	4440	4456
28	4472	4487	4502	4518	4533	4548	4564	4579	4594	4609
29	4624	4639	4654	4669	4683	4698	4713	4728	4742	4757
30	4771	4786	4800	4814	4829	4843	4857	4871	4886	4900
31	4914	4928	4942	4955	4969	4983	4997	5011	5024	5038
32	5051	5065	5079	5092	5105	5119	5132	5145	5159	5172
33	5185	5198	5211	5224	5237	5250	5263	5276	5289	5302
34	5315	5328	5340	5353	5366	5378	5391	5403	5416	5428
35	5441	5453	5465	5478	5490	5502	5514	5527	5539	5551
36	5563	5575	5587	5599	5611	5623	5635	5647	5658	5670
37	5682	5694	5705	5717	5729	5740	5752	5763	5775	5786
38	5798	5809	5821	5832	5843	5855	5866	5877	5888	5899
39	5911	5922	5933	5944	5955	5966	5977	5988	5999	6010
40	6021	6031	6042	6053	6064	6075	6085	6096	6107	6117
41	6128	6138	6149	6160	6170	6180	6191	6201	6212	6222
42	6232	6243	6253	6263	6274	6284	6294	6304	6314	6325
43	6335	6345	6355	6365	6375	6385	6395	6405	6415	6425
44	6435	6444	6454	6464	6474	6484	6493	6503	6513	6522
45	6532	6542	6551	6561	6571	6580	6590	6599	6609	6618
46	6628	6637	6646	6656	6665	6675	6684	6693	6702	6712
47	6721	6730	6739	6749	6758	6767	6776	6785	6794	6803
48	6812	6821	6830	6839	6848	6857	6866	6875	6884	6893
49	6902	6911	6920	6928	6937	6946	6955	6964	6972	6981
50	6990	6998	7007	7016	7024	7033	7042	7050	7059	7067
51	7076	7084	7093	7101	7110	7118	7126	7135	7143	7152
52	7160	7168	7177	7185	7193	7202	7210	7218	7226	7235
53	7243	7251	7259	7267	7275	7284	7292	7300	7308	7316
54	7324	7332	7340	7348	7356	7364	7372	7380	7388	7396

Table 2 (*Continued*)

N	0	1	2	3	4	5	6	7	8	9
55	7404	7412	7419	7427	7435	7443	7451	7459	7466	7474
56	7482	7490	7497	7505	7513	7520	7528	7536	7543	7551
57	7559	7566	7574	7582	7589	7597	7604	7612	7619	7627
58	7634	7642	7649	7657	7664	7672	7679	7686	7694	7701
59	7709	7716	7723	7731	7738	7745	7752	7760	7767	7774
60	7782	7789	7796	7803	7810	7818	7825	7832	7839	7846
61	7853	7860	7868	7875	7882	7889	7896	7903	7910	7917
62	7924	7931	7938	7945	7952	7959	7966	7973	7980	7987
63	7993	8000	8007	8014	8021	8028	8035	8041	8048	8055
64	8062	8069	8075	8082	8089	8096	8102	8109	8116	8122
65	8129	8136	8142	8149	8156	8162	8169	8176	8182	8189
66	8195	8202	8209	8215	8222	8228	8235	8241	8248	8254
67	8261	8267	8274	8280	8287	8293	8299	8306	8312	8319
68	8325	8331	8338	8344	8351	8357	8363	8370	8376	8382
69	8388	8395	8401	8407	8414	8420	8426	8432	8439	8445
70	8451	8457	8463	8470	8476	8482	8488	8494	8500	8506
71	8513	8519	8525	8531	8537	8543	8549	8555	8561	8567
72	8573	8579	8585	8591	8597	8603	8609	8615	8621	8627
73	8633	8639	8645	8651	8657	8663	8669	8675	8681	8686
74	8692	8698	8704	8710	8716	8722	8727	8733	8739	8745
75	8751	8756	8762	8768	8774	8779	8785	8791	8797	8802
76	8808	8814	8820	8825	8831	8837	8842	8848	8854	8859
77	8865	8871	8876	8882	8887	8893	8899	8904	8910	8915
78	8921	8927	8932	8938	8943	8949	8954	8960	8965	8971
79	8976	8982	8987	8993	8998	9004	9009	9015	9020	9025
80	9031	9036	9042	9047	9053	9058	9063	9069	9074	9079
81	9085	9090	9096	9101	9106	9112	9117	9122	9128	9133
82	9138	9143	9149	9154	9159	9165	9170	9175	9180	9186
83	9191	9196	9201	9206	9212	9217	9222	9227	9232	9238
84	9243	9248	9253	9258	9263	9269	9274	9279	9284	9289
85	9294	9299	9304	9309	9315	9320	9325	9330	9335	9340
86	9345	9350	9355	9360	9365	9370	9375	9380	9385	9390
87	9395	9400	9405	9410	9415	9420	9425	9430	9435	9440
88	9445	9450	9455	9460	9465	9469	9474	9479	9484	9489
89	9494	9499	9504	9509	9513	9518	9523	9528	9533	9538
90	9542	9547	9552	9557	9562	9566	9571	9576	9581	9586
91	9590	9595	9600	9605	9609	9614	9619	9624	9628	9633
92	9638	9643	9647	9652	9657	9661	9666	9671	9675	9680
93	9685	9689	9694	9699	9703	9708	9713	9717	9722	9727
94	9731	9736	9741	9745	9750	9754	9759	9763	9768	9773
95	9777	9782	9786	9791	9795	9800	9805	9809	9814	9818
96	9823	9827	9832	9836	9841	9845	9850	9854	9859	9863
97	9868	9872	9877	9881	9886	9890	9894	9899	9903	9908
98	9912	9917	9921	9926	9930	9934	9939	9943	9948	9952
99	9956	9961	9965	9969	9974	9978	9983	9987	9991	9996

Table 3 Natural logarithms

N	0	1	2	3	4	5	6	7	8	9
1.0	0000	0100	0198	0296	0392	0488	0583	0677	0770	0862
1.1	0953	1044	1133	1222	1310	1398	1484	1570	1655	1740
1.2	1823	1906	1989	2070	2151	2231	2311	2390	2469	2546
1.3	2624	2700	2776	2852	2927	3001	3075	3148	3221	3293
1.4	3365	3436	3507	3577	3646	3716	3784	3853	3920	3988
1.5	4055	4121	4187	4253	4318	4383	4447	4511	4574	4637
1.6	4700	4762	4824	4886	4947	5008	5068	5128	5188	5247
1.7	5306	5365	5423	5481	5539	5596	5653	5710	5766	5822
1.8	5878	5933	5988	6043	6098	6152	6206	6259	6313	6366
1.9	6419	6471	6523	6575	6627	6678	6729	6780	6831	6881
2.0	6931	6981	7031	7080	7129	7178	7227	7275	7324	7372
2.1	7419	7467	7514	7561	7608	7655	7701	7747	7793	7839
2.2	7885	7930	7975	8020	8065	8109	8154	8198	8242	8286
2.3	8329	8372	8416	8459	8502	8544	8587	8629	8671	8713
2.4	8755	8796	8838	8879	8920	8961	9002	9042	9083	9123
2.5	9163	9203	9243	9282	9322	9361	9400	9439	9478	9517
2.6	9555	9594	9632	9670	9708	9746	9783	9821	9858	9895
2.7	9933	9969	*0006	*0043	*0080	*0116	*0152	*0188	*0225	*0260
2.8	1.0296	0332	0367	0403	0438	0473	0508	0543	0578	0613
2.9	0647	0682	0716	0750	0784	0818	0852	0886	0919	0953
3.0	1.0986	1019	1053	1086	1119	1151	1184	1217	1249	1282
3.1	1314	1346	1378	1410	1442	1474	1506	1537	1569	1600
3.2	1632	1663	1694	1725	1756	1787	1817	1848	1878	1909
3.3	1939	1969	2000	2030	2060	2090	2119	2149	2179	2208
3.4	2238	2267	2296	2326	2355	2384	2413	2442	2470	2499
3.5	1.2528	2556	2585	2613	2641	2669	2698	2726	2754	2782
3.6	2809	2837	2865	2892	2920	2947	2975	3002	3029	3056
3.7	3083	3110	3137	3164	3191	3218	3244	3271	3297	3324
3.8	3350	3376	3403	3429	3455	3481	3507	3533	3558	3584
3.9	3610	3635	3661	3686	3712	3737	3762	3788	3813	3838
4.0	1.3863	3888	3913	3938	3962	3987	4012	4036	4061	4085
4.1	4110	4134	4159	4183	4207	4231	4255	4279	4303	4327
4.2	4351	4375	4398	4422	4446	4469	4493	4516	4540	4563
4.3	4586	4609	4633	4656	4679	4702	4725	4748	4770	4793
4.4	4816	4839	4861	4884	4907	4929	4951	4974	4996	5019
4.5	1.5041	5063	5085	5107	5129	5151	5173	5195	5217	5239
4.6	5261	5282	5304	5326	5347	5369	5390	5412	5433	5454
4.7	5476	5497	5518	5539	5560	5581	5602	5623	5644	5665
4.8	5686	5707	5728	5748	5769	5790	5810	5831	5851	5872
4.9	5892	5913	5933	5953	5974	5994	6014	6034	6054	6074
5.0	1.6094	6114	6134	6154	6174	6194	6214	6233	6253	6273
5.1	6292	6312	6332	6351	6371	6390	6409	6429	6448	6467
5.2	6487	6506	6525	6544	6563	6582	6601	6620	6639	6658
5.3	6677	6696	6715	6734	6752	6771	6790	6808	6827	6845
5.4	6864	6882	6901	6919	6938	6956	6974	6993	7011	7029

Table 3 (*Continued*)

N	0	1	2	3	4	5	6	7	8	9
5.5	1.7047	7066	7084	7102	7120	7138	7156	7174	7192	7210
5.6	7228	7246	7263	7281	7299	7317	7334	7352	7370	7387
5.7	7405	7422	7440	7457	7475	7492	7509	7527	7544	7561
5.8	7579	7596	7613	7630	7647	7664	7681	7699	7716	7733
5.9	7750	7766	7783	7800	7817	7834	7851	7867	7884	7901
6.0	1.7918	7934	7951	7967	7984	8001	8017	8034	8050	8066
6.1	8083	8099	8116	8132	8148	8165	8181	8197	8213	8229
6.2	8245	8262	8278	8294	8310	8326	8342	8358	8374	8390
6.3	8405	8421	8437	8453	8469	8485	8500	8516	8532	8547
6.4	8563	8579	8594	8610	8625	8641	8656	8672	8687	8703
6.5	1.8718	8733	8749	8764	8779	8795	8810	8825	8840	8856
6.6	8871	8886	8901	8916	8931	8946	8961	8976	8991	9006
6.7	9021	9036	9051	9066	9081	9095	9110	9125	9140	9155
6.8	9169	9184	9199	9213	9228	9242	9257	9272	9286	9301
6.9	9315	9330	9344	9359	9373	9387	9402	9416	9430	9445
7.0	1.9459	9473	9488	9502	9516	9530	9544	9559	9573	9587
7.1	9601	9615	9629	9643	9657	9671	9685	9699	9713	9727
7.2	9741	9755	9769	9782	9796	9810	9824	9838	9851	9865
7.3	9879	9892	9906	9920	9933	9947	9961	9974	9988	*0001
7.4	2.0015	0028	0042	0055	0069	0082	0096	0109	0122	0136
7.5	2.0149	0162	0176	0189	0202	0215	0229	0242	0255	0268
7.6	0281	0295	0308	0321	0334	0347	0360	0373	0386	0399
7.7	0412	0425	0438	0451	0464	0477	0490	0503	0516	0528
7.8	0541	0554	0567	0580	0592	0605	0618	0630	0643	0656
7.9	0669	0681	0694	0707	0719	0732	0744	0757	0769	0782
8.0	2.0794	0807	0819	0832	0844	0857	0869	0882	0894	0906
8.1	0919	0931	0943	0956	0968	0980	0992	1005	1017	1029
8.2	1041	1054	1066	1078	1090	1102	1114	1126	1138	1150
8.3	1163	1175	1187	1199	1211	1223	1235	1247	1258	1270
8.4	1282	1294	1306	1318	1330	1342	1353	1365	1377	1389
8.5	2.1401	1412	1424	1436	1448	1459	1471	1483	1494	1506
8.6	1518	1529	1541	1552	1564	1576	1587	1599	1610	1622
8.7	1633	1645	1656	1668	1679	1691	1702	1713	1725	1736
8.8	1748	1759	1770	1782	1793	1804	1815	1827	1838	1849
8.9	1861	1872	1883	1894	1905	1917	1928	1939	1950	1961
9.0	2.1972	1983	1994	2006	2017	2028	2039	2050	2061	2072
9.1	2083	2094	2105	2116	2127	2138	2148	2159	2170	2181
9.2	2192	2203	2214	2225	2235	2246	2257	2268	2279	2289
9.3	2300	2311	2322	2332	2343	2354	2364	2375	2386	2396
9.4	2407	2418	2428	2439	2450	2460	2471	2481	2492	2502
9.5	2.2513	2523	2534	2544	2555	2565	2576	2586	2597	2607
9.6	2618	2628	2638	2649	2659	2670	2680	2690	2701	2711
9.7	2721	2732	2742	2752	2762	2773	2783	2793	2803	2814
9.8	2824	2834	2844	2854	2865	2875	2885	2895	2905	2915
9.9	2925	2935	2946	2956	2966	2976	2986	2996	3006	3016

Use ln 10 = 2.30259 to find logarithms of numbers greater than 10 or less than 1. *Example:* ln 220 = ln 2.2 + 2 ln 10 = 0.7885 + 2(2.30259) = 5.3937.

Table 4 Exponential functions

x	e^x	$\log_{10}(e^x)$	e^{-x}	x	e^x	$\log_{10}(e^x)$	e^{-x}
0.00	1.0000	0.00000	1.000000	**0.50**	1.6487	0.21715	0.606531
0.01	1.0101	.00434	0.990050	0.51	1.6653	.22149	.600496
0.02	1.0202	.00869	.980199	0.52	1.6820	.22583	.594521
0.03	1.0305	.01303	.970446	0.53	1.6989	.23018	.588605
0.04	1.0408	.01737	.960789	0.54	1.7160	.23452	.582748
0.05	1.0513	0.02171	0.951229	**0.55**	1.7333	0.23886	0.576950
0.06	1.0618	.02606	.941765	0.56	1.7507	.24320	.571209
0.07	1.0725	.03040	.932394	0.57	1.7683	.24755	.565525
0.08	1.0833	.03474	.923116	0.58	1.7860	.25189	.559898
0.09	1.0942	.03909	.913931	0.59	1.8040	.35623	.554327
0.10	1.1052	0.04343	0.904837	**0.60**	1.8221	0.26058	0.548812
0.11	1.1163	.04777	.895834	0.61	1.8404	.26492	.543351
0.12	1.1275	.05212	.886920	0.62	1.8589	.26926	.537944
0.13	1.1388	.05646	.878095	0.63	1.8776	.27361	.532592
0.14	1.1503	.06080	.869358	0.64	1.8965	.27795	.527292
0.15	1.1618	0.06514	0.860708	**0.65**	1.9155	0.28229	0.522046
0.16	1.1735	.06949	.852144	0.66	1.9348	.28663	.516851
0.17	1.1853	.07383	.843665	0.67	1.9542	.29098	.511709
0.18	1.1972	.07817	.835270	0.68	1.9739	.29532	.506617
0.19	1.2092	.08252	.826959	0.69	1.9937	.29966	.501576
0.20	1.2214	0.08686	0.818731	**0.70**	2.0138	0.30401	0.496585
0.21	1.2337	.09120	.810584	0.71	2.0340	.30835	.491644
0.22	1.2461	.09554	.802519	0.72	2.0544	.31269	.486752
0.23	1.2586	.09989	.794534	0.73	2.0751	.31703	.481909
0.24	1.2712	.10423	.786628	0.74	2.0959	.32138	.477114
0.25	1.2840	0.10857	0.778801	**0.75**	2.1170	0.32572	0.472367
0.26	1.2969	.11292	.771052	0.76	2.1383	.33006	.467666
0.27	1.3100	.11726	.763379	0.77	2.1598	.33441	.463013
0.28	1.3231	.12160	.755784	0.78	2.1815	.33875	.458406
0.29	1.3364	.12595	.748264	0.79	2.2034	.34309	.453845
0.30	1.3499	0.13029	0.740818	**0.80**	2.2255	0.34744	0.449329
0.31	1.3634	.13463	.733447	0.81	2.2479	.35178	.444858
0.32	1.3771	.13897	.726149	0.82	2.2705	.35612	.440432
0.33	1.3910	.14332	.718924	0.83	2.2933	.36046	.436049
0.34	1.4049	.14766	.711770	0.84	2.3164	.36481	.431711
0.35	1.4191	0.15200	0.704688	**0.85**	2.3396	0.36915	0.427415
0.36	1.4333	.15635	.697676	0.86	2.3632	.37349	.423162
0.37	1.4477	.16069	.690734	0.87	2.3869	.37784	.418952
0.38	1.4623	.16503	.683861	0.88	2.4109	.38218	.414783
0.39	1.4770	.16937	.677057	0.89	2.4351	.38652	.410656
0.40	1.4918	0.17372	0.670320	**0.90**	2.4596	0.39087	0.406570
0.41	1.5068	.17806	.663650	0.91	2.4843	.39521	.402524
0.42	1.5220	.18240	.657047	0.92	2.5093	.39955	.398519
0.43	1.5373	.18675	.650509	0.93	2.5345	.40389	.394554
0.44	1.5527	.19109	.644036	0.94	2.5600	.40824	.390628
0.45	1.5683	0.19543	0.637628	**0.95**	2.5857	0.41258	0.386741
0.46	1.5841	.19978	.631284	0.96	2.6117	.41692	.382893
0.47	1.6000	.20412	.625002	0.97	2.6379	.42127	.379083
0.48	1.6161	.20846	.618783	0.98	2.6645	.42561	.375311
0.49	1.6323	.21280	.612626	0.99	2.6912	.42995	.371577
0.50	1.6487	0.21715	0.606531	**1.00**	2.7183	0.43429	0.367879

Table 4 (*Continued*)

x	e^x	$\log_{10}(e^x)$	e^{-x}	x	e^x	$\log_{10}(e^x)$	e^{-x}
1.00	2.7183	0.43429	0.367879	**1.50**	4.4817	0.65144	0.223130
1.01	2.7456	.43864	.364219	1.51	4.5267	.65578	.220910
1.02	2.7732	.44298	.360595	1.52	4.5722	.66013	.218712
1.03	2.8011	.44732	.357007	1.53	4.6182	.66447	.216536
1.04	2.8292	.45167	.353455	1.54	4.6646	.66881	.214381
1.05	2.8577	0.45601	0.349938	**1.55**	4.7115	0.67316	0.212248
1.06	2.8864	.46035	.346456	1.56	4.7588	.67750	.210136
1.07	2.9154	.46470	.343009	1.57	4.8066	.68184	.208045
1.08	2.9447	.46904	.339596	1.58	4.8550	.68619	.205975
1.09	2.9743	.47338	.336216	1.59	4.9037	.69053	.203926
1.10	3.0042	0.47772	0.332871	**1.60**	4.9530	0.69487	0.201897
1.11	3.0344	.48207	.329559	1.61	5.0028	.69921	.199888
1.12	3.0649	.48641	.326280	1.62	5.0531	.70356	.197899
1.13	3.0957	.49075	.323033	1.63	5.1039	.70790	.195930
1.14	3.1268	.49510	.319819	1.64	5.1552	.71224	.193980
1.15	3.1582	0.49944	0.316637	**1.65**	5.2070	0.71659	0.192050
1.16	3.1899	.50378	.313486	1.66	5.2593	.72093	.190139
1.17	3.2220	.50812	.310367	1.67	5.3122	.72527	.188247
1.18	3.2544	.51247	.307279	1.68	5.3656	.72961	.186374
1.19	3.2871	.51681	.304221	1.69	5.4195	.73396	.184520
1.20	3.3201	0.52115	0.301194	**1.70**	5.4739	0.73830	0.182684
1.21	3.3535	.52550	.298197	1.71	5.5290	.74264	.180866
1.22	3.3872	.52984	.295230	1.72	5.5845	.74699	.179066
1.23	3.4212	.53418	.292293	1.73	5.6407	.75133	.177284
1.24	3.4556	.53853	.289384	1.74	5.6973	.75567	.175520
1.25	3.4903	0.54287	0.286505	**1.75**	5.7546	0.76002	0.173774
1.26	3.5254	.54721	.283654	1.76	5.8124	.76436	.172045
1.27	3.5609	.55155	.280832	1.77	5.8709	.76870	.170333
1.28	3.5966	.55590	.278037	1.78	5.9299	.77304	.168638
1.29	3.6328	.56024	.275271	1.79	5.9895	.77739	.166960
1.30	3.6693	0.56458	0.272532	**1.80**	6.0496	0.78173	0.165299
1.31	3.7062	.56893	.269820	1.81	6.1104	.78607	.163654
1.32	3.7434	.57327	.267135	1.82	6.1719	.79042	.162026
1.33	3.7810	.57761	.264477	1.83	6.2339	.79476	.160414
1.34	3.8190	.58195	.261846	1.84	6.2965	.79910	.158817
1.35	3.8574	0.58630	0.259240	**1.85**	6.3598	0.80344	0.157237
1.36	3.8962	.59064	.256661	1.86	6.4237	.80779	.155673
1.37	3.9354	.59498	.254107	1.87	6.4483	.81213	.154124
1.38	3.9749	.59933	.251579	1.88	6.5535	.81647	.152590
1.39	4.0149	.60367	.249075	1.89	6.6194	.82082	.151072
1.40	4.0552	0.60801	0.246597	**1.90**	6.6859	0.82516	0.149569
1.41	4.0960	.61236	.244143	1.91	6.7531	.82950	.148080
1.42	4.1371	.61670	.241714	1.92	6.8210	.83385	.146607
1.43	4.1787	.62104	.239309	1.93	6.8895	.83819	.145148
1.44	4.2207	.62538	.236928	1.94	6.9588	.84253	.143704
1.45	4.2631	0.62973	0.234570	**1.95**	7.0287	0.84687	0.142274
1.46	4.3060	.63407	.232236	1.96	7.0993	.85122	.140858
1.47	4.3492	.63841	.229925	1.97	7.1707	.85556	.139457
1.48	4.3929	.64276	.227638	1.98	7.2427	.85990	.138069
1.49	4.4371	.64710	.225373	1.99	7.3155	.86425	.136695
1.50	4.4817	0.65144	0.223130	**2.00**	7.3891	0.86859	0.135335

Table 4 (*Continued*)

x	e^x	$\log_{10}(e^x)$	e^{-x}	x	e^x	$\log_{10}(e^x)$	e^{-x}
2.00	7.3891	0.86859	0.135335	**2.50**	12.182	1.08574	0.082085
2.01	7.4633	.87293	.133989	2.51	12.305	1.09008	.081268
2.02	7.5383	.87727	.132655	2.52	12.429	1.09442	.080460
2.03	7.6141	.88162	.131336	2.53	12.554	1.09877	.079659
2.04	7.6906	.88596	.130029	2.54	12.680	1.10311	.078866
2.05	7.7679	0.89030	0.128735	**2.55**	12.807	1.10745	0.078082
2.06	7.8460	.89465	.127454	2.56	12.936	1.11179	.077305
2.07	7.9248	.89899	.126186	2.57	13.066	1.11614	.076536
2.08	8.0045	.90333	.124930	2.58	13.197	1.12048	.075774
2.09	8.0849	.90756	.123687	2.59	13.330	1.12482	.075020
2.10	8.1662	0.91202	0.122456	**2.60**	13.464	1.12917	0.074274
2.11	8.2482	.91636	.121238	2.61	13.599	1.13351	.073535
2.12	8.3311	.92070	.120032	2.62	13.736	1.13785	.072803
2.13	8.4149	.92505	.118837	2.63	13.874	1.14219	.072078
2.14	8.4994	.92939	.117655	2.64	14.013	1.14654	.071361
2.15	8.5849	0.93373	0.116484	**2.65**	14.154	1.15088	0.070651
2.16	8.6711	.93808	.115325	2.66	14.296	1.15522	.069948
2.17	8.7583	.94242	.114178	2.67	14.440	1.15957	.069252
2.18	8.8463	.94676	.113042	2.68	14.585	1.16391	.068563
2.19	8.9352	.95110	.111917	2.69	14.732	1.16825	.067881
2.20	9.0250	0.95545	0.110803	**2.70**	14.880	1.17260	0.067206
2.21	9.1157	.95979	.109701	2.71	15.029	1.17694	.066537
2.22	9.2073	.96413	.108609	2.72	15.180	1.18128	.065875
2.23	9.2999	.96848	.107528	2.73	15.333	1.18562	.065219
2.24	9.3933	.97282	.106459	2.74	15.487	1.18997	.064570
2.25	9.4877	0.97716	0.105399	**2.75**	15.643	1.19431	0.063928
2.26	9.5831	.98151	.104350	2.76	15.800	1.19865	.063292
2.27	9.6794	.98585	.103312	2.77	15.959	1.20300	.062662
2.28	9.7767	.99019	.102284	2.78	16.119	1.20734	.062039
2.29	9.8749	.99453	.101266	2.79	16.281	1.21168	.061421
2.30	9.9742	0.99888	0.100259	**2.80**	16.445	1.21602	0.060810
2.31	10.074	1.00322	.099261	2.81	16.610	1.22037	.060205
2.32	10.176	1.00756	.098274	2.82	16.777	1.22471	.059606
2.33	10.278	1.01191	.097296	2.83	16.945	1.22905	.059013
2.34	10.381	1.01625	.096328	2.84	17.116	1.23340	.058426
2.35	10.486	1.02059	0.095369	**2.85**	17.288	1.23774	0.057844
2.36	10.591	1.02493	.094420	2.86	17.462	1.24208	.057269
2.37	10.697	1.02928	.093481	2.87	17.637	1.24643	.056699
2.38	10.805	1.03362	.092551	2.88	17.814	1.25077	.056135
2.39	10.913	1.03796	.091630	2.89	17.993	1.25511	.055576
2.40	11.023	1.04231	0.090718	**2.90**	18.174	1.25945	0.055023
2.41	11.134	1.04665	.089815	2.91	18.357	1.26380	.054476
2.42	11.246	1.05099	.088922	2.92	18.541	1.26814	.053934
2.43	11.359	1.05534	.088037	2.93	18.728	1.27248	.053397
2.44	11.473	1.05968	.087161	2.94	18.916	1.27683	.052866
2.45	11.588	1.06402	0.086294	**2.95**	19.106	1.28117	0.052340
2.46	11.705	1.06836	.085435	2.96	19.298	1.28551	.051819
2.47	11.822	1.07271	.084585	2.97	19.492	1.28985	.051303
2.48	11.941	1.07705	.083743	2.98	19.688	1.29420	.050793
2.49	12.061	1.08139	.082910	2.99	19.886	1.29854	.050287
2.50	12.182	1.08574	0.082085	**3.00**	20.086	1.30288	0.049787

Table 4 (*Continued*)

x	e^x	$\log_{10}(e^x)$	e^{-x}	x	e^x	$\log_{10}(e^x)$	e^{-x}
3.00	20.086	1.30288	0.049787	**3.50**	33.115	1.52003	0.030197
3.01	20.287	1.30723	.049292	3.51	33.448	1.52437	.029897
3.02	20.491	1.31157	.048801	3.52	33.784	1.52872	.029599
3.03	20.697	1.31591	.048316	3.53	34.124	1.53306	.029305
3.04	20.905	1.32026	.047835	3.54	34.467	1.53740	.029013
3.05	21.115	1.32460	0.047359	**3.55**	34.813	1.54175	0.028725
3.06	21.328	1.32894	.046888	3.56	35.163	1.54609	.028439
3.07	21.542	1.33328	.046421	3.57	35.517	1.55043	.028156
3.08	21.758	1.33763	.045959	3.58	35.874	1.55477	.027876
3.09	21.977	1.34197	.045502	3.59	36.234	1.55912	.027598
3.10	22.198	1.34631	0.045049	**3.60**	36.598	1.56346	0.027324
3.11	22.421	1.35066	.044601	3.61	36.966	1.56780	.027052
3.12	22.646	1.35500	.044157	3.62	37.338	1.57215	.026783
3.13	22.874	1.35934	.043718	3.63	37.713	1.57649	.026516
3.14	23.104	1.36368	.043283	3.64	38.092	1.58083	.026252
3.15	23.336	1.36803	0.042852	**3.65**	38.475	1.58517	0.025991
3.16	23.571	1.37237	.042426	3.66	38.861	1.58952	.025733
3.17	23.807	1.36671	.042004	3.67	39.252	1.59386	.025476
3.18	24.047	1.38106	.041586	3.68	39.646	1.59820	.025223
3.19	24.288	1.38540	.041172	3.69	40.045	1.60255	.024972
3.20	24.533	1.38974	0.040764	**3.70**	40.447	1.60689	0.024724
3.21	24.779	1.39409	.040357	3.71	40.854	1.61123	.024478
3.22	25.028	1.39843	.039955	3.72	41.264	1.61558	.024234
3.23	25.280	1.40277	.039557	3.73	41.679	1.61992	.023993
3.24	25.534	1.40711	.039164	3.74	42.098	1.62426	.023754
3.25	25.790	1.41146	0.038774	**3.75**	42.521	1.62860	0.023518
3.26	26.050	1.41580	.038388	3.76	42.948	1.63295	.023284
3.27	26.311	1.42014	.038006	3.77	43.380	1.63729	.023052
3.28	26.576	1.42449	.037628	3.78	43.816	1.64163	.022823
3.29	26.843	1.42883	.037254	3.79	44.256	1.64598	.022596
3.30	27.113	1.44317	0.036883	**3.80**	44.701	1.65032	0.022371
3.31	27.385	1.43751	.036516	3.81	45.150	1.65466	.022148
3.32	27.660	1.44186	.036153	3.82	45.604	1.65900	.021928
3.33	27.938	1.44620	.035793	3.83	46.063	1.66335	.021710
3.34	28.219	1.45054	.035437	3.84	46.525	1.66769	.021494
3.35	28.503	1.45489	0.035084	**3.85**	46.993	1.67203	0.021280
3.36	28.789	1.45923	.034735	3.86	47.465	1.67638	.021068
3.37	29.079	1.46357	.034390	3.87	47.942	1.68072	.020858
3.38	29.371	1.46792	.034047	3.88	48.424	1.68506	.020651
3.39	29.666	1.47226	.033709	3.89	48.911	1.68941	.020445
3.40	29.964	1.47660	0.033373	**3.90**	49.402	1.69375	0.020242
3.41	30.265	1.48094	.033041	3.91	49.899	1.69809	.020041
3.42	30.569	1.48529	.032712	3.92	50.400	1.70243	.019840
3.43	30.877	1.48963	.032387	3.93	50.907	1.70678	.019644
3.44	31.187	1.49397	.032065	3.94	51.419	1.71112	.019448
3.45	31.500	1.49832	0.031746	**3.95**	51.935	1.71546	0.019255
3.46	31.817	1.50266	.031430	3.96	52.457	1.71981	.019063
3.47	32.137	1.50700	.031117	3.97	52.985	1.72415	.018873
3.48	32.460	1.51134	.030807	3.98	53.517	1.72849	.018686
3.49	32.786	1.51569	.030501	3.99	54.055	1.73283	.018500
3.50	33.115	1.52003	0.030197	**4.00**	54.598	1.73718	0.018316

Table 4 (*Continued*)

x	e^x	$\log_{10}(e^x)$	e^{-x}	x	e^x	$\log_{10}(e^x)$	e^{-x}
4.00	54.598	1.73718	0.018316	**4.50**	90.017	1.95433	0.011109
4.01	55.147	1.74152	.018133	4.51	90.922	1.95867	.010998
4.02	55.701	1.74586	.017953	4.52	91.836	1.96301	.010889
4.03	56.261	1.75021	.017774	4.53	92.759	1.96735	.010781
4.04	56.826	1.75455	.017597	4.54	93.691	1.97170	.010673
4.05	57.397	1.75889	0.017422	**4.55**	94.632	1.97604	0.010567
4.06	57.974	1.76324	.017249	4.56	95.583	1.98038	.010462
4.07	58.577	1.76758	.017077	4.57	96.544	1.98473	.010358
4.08	59.145	1.77192	.016907	4.58	97.514	1.98907	.010255
4.09	59.740	1.77626	.016739	4.59	98.494	1.99341	.010153
4.10	60.340	1.78061	0.016573	**4.60**	99.484	1.99775	0.010052
4.11	60.947	1.78495	.016408	4.61	100.48	2.00210	.009952
4.12	61.559	1.78929	.016245	4.62	101.49	2.00644	.009853
4.13	62.178	1.79364	.016083	4.63	102.51	2.01078	.009755
4.14	62.803	1.79798	.015923	4.64	103.54	2.01513	.009658
4.15	63.434	1.80232	0.015764	**4.65**	104.58	2.01947	0.009562
4.16	64.072	1.80667	.015608	4.66	105.64	2.02381	.009466
4.17	64.715	1.81101	.015452	4.67	106.70	2.02816	.009372
4.18	65.366	1.81535	.015299	4.68	107.77	2.03250	.009279
4.19	66.023	1.81969	.015146	4.69	108.85	2.03684	.009187
4.20	66.686	1.82404	0.014996	**4.70**	109.95	2.04118	0.009095
4.21	67.357	1.82838	.014846	4.71	111.05	2.04553	.009005
4.22	68.033	1.83272	.014699	4.72	112.17	2.04987	.008915
4.23	68.717	1.83707	.014552	4.73	113.30	2.05421	.008826
4.24	69.408	1.84141	.014408	4.74	114.43	2.05856	.008739
4.25	70.105	1.84575	0.014264	**4.75**	115.58	2.06290	0.008652
4.26	70.810	1.85009	.014122	4.76	116.75	2.06724	.008566
4.27	71.522	1.85444	.013982	4.77	117.92	2.07158	.008480
4.28	72.240	1.85878	.013843	4.78	119.10	2.07593	.008396
4.29	72.966	1.86312	.013705	4.79	120.30	2.08027	.008312
4.30	73.700	1.86747	0.013569	**4.80**	121.51	2.08461	0.008230
4.31	74.440	1.87181	.013434	4.81	122.73	2.08896	.008148
4.32	75.189	1.87615	.013300	4.82	123.97	2.09330	.008067
4.33	75.944	1.88050	.013168	4.83	125.21	2.09764	.007987
4.34	76.708	1.88484	.013037	4.84	126.47	2.10199	.007907
4.35	77.478	1.88918	0.012907	**4.85**	127.74	2.10633	0.007828
4.36	78.257	1.89352	.012778	4.86	129.02	2.11067	.007750
4.37	79.044	1.89787	.012651	4.87	130.32	2.11501	.007673
4.38	79.838	1.90221	.012525	4.88	131.63	2.11936	.007597
4.39	80.640	1.90655	.012401	4.89	132.95	2.12370	.007521
4.40	81.451	1.91090	0.012277	**4.90**	134.29	2.12804	0.007477
4.41	82.269	1.91524	.012155	4.91	135.64	2.13239	.007372
4.42	83.096	1.91958	.012034	4.92	137.00	2.13673	.007299
4.43	83.931	1.92392	.011914	4.93	138.38	2.14107	.007227
4.44	84.775	1.92827	.011796	4.94	139.77	2.14541	.007155
4.45	85.627	1.93261	0.011679	**4.95**	141.17	2.14976	0.007083
4.46	86.488	1.93695	.011562	4.96	142.59	2.15410	.007013
4.47	87.357	1.94130	.011447	4.97	144.03	2.15844	.006943
4.48	88.235	1.94564	.011333	4.98	145.47	2.16279	.006874
4.49	89.121	1.94998	.011221	4.99	146.94	2.16713	.006806
4.50	90.017	1.95433	0.011109	**5.00**	148.41	2.17147	0.006738

Table 4 (*Continued*)

x	e^x	$\log_{10}(e^x)$	e^{-x}	x	e^x	$\log_{10}(e^x)$	e^{-x}
5.00	148.41	2.17147	0.006738	**5.50**	244.69	2.38862	0.0040868
5.01	149.90	2.17582	.006671	5.55	257.24	2.41033	.0038875
5.02	151.41	2.18016	.006605	5.60	270.43	2.43205	.0036979
5.03	152.93	2.18450	.006539	5.65	284.29	2.45376	.0035175
5.04	154.47	2.18884	.006474	5.70	298.87	2.47548	.0033460
5.05	156.02	2.19319	0.006409	**5.75**	314.19	2.49719	0.0031828
5.06	157.59	2.19753	.006346	5.80	330.30	2.51891	.0030276
5.07	159.17	2.20187	.006282	5.85	347.23	2.54062	.0028799
5.08	160.77	2.20622	.006220	5.90	365.04	2.56234	.0027394
5.09	162.39	2.21056	.006158	5.95	383.75	2.58405	.0026058
5.10	164.02	2.21490	0.006097	**6.00**	403.43	2.60577	0.0024788
5.11	165.67	2.21924	.006036	6.05	424.11	2.62748	.0023579
5.12	167.34	2.22359	.005976	6.10	445.86	2.64920	.0022429
5.13	169.02	2.22793	.005917	6.15	468.72	2.67091	.0021335
5.14	170.72	2.23227	.005858	6.20	492.75	2.69263	.0020294
5.15	172.43	2.23662	0.005799	**6.25**	518.01	2.71434	0.0019305
5.16	174.16	2.24096	.005742	6.30	544.57	2.73606	.0018363
5.17	175.91	2.24530	.005685	6.35	572.49	2.75777	.0017467
5.18	177.68	2.24965	.005628	6.40	601.85	2.77948	.0016616
5.19	179.47	2.25399	.005572	6.45	632.70	2.80120	.0015805
5.20	181.27	2.25833	0.005517	**6.50**	665.14	2.82291	0.0015034
5.21	183.09	2.26267	.005462	6.55	699.24	2.84463	.0014301
5.22	184.93	2.26702	.005407	6.60	735.10	2.86634	.0013604
5.23	186.79	2.27136	.005354	6.65	772.78	2.88806	.0012940
5.24	188.67	2.27570	.005300	6.70	812.41	2.90977	.0012309
5.25	190.57	2.28005	0.005248	**6.75**	854.06	2.93149	0.0011709
5.26	192.48	2.28439	.005195	6.80	897.85	2.95320	.0011138
5.27	194.42	2.28873	.005144	6.85	943.88	2.97492	.0010595
5.28	196.37	2.29307	.005092	6.90	992.27	2.99663	.0010078
5.29	198.34	2.29742	.005042	6.95	1043.1	3.01835	.0009586
5.30	200.34	2.30176	0.004992	**7.00**	1096.6	3.04006	0.0009119
5.31	202.35	2.30610	.004942	7.05	1152.9	3.06178	.0008674
5.32	204.38	2.31045	.004893	7.10	1212.0	3.08349	.0008251
5.33	206.44	2.31479	.004844	7.15	1274.1	3.10521	.0007849
5.34	208.51	2.31913	.004796	7.20	1339.4	3.12692	.0007466
5.35	210.61	2.32348	0.004748	**7.25**	1408.1	3.14863	0.0007102
5.36	212.72	2.32782	.004701	7.30	1480.3	3.17035	.0006755
5.37	214.86	2.33216	.004654	7.35	1556.2	3.19206	.0006426
5.38	217.02	2.33650	.004608	7.40	1636.0	3.21378	.0006113
5.39	219.20	2.34085	.004562	7.45	1719.9	3.23549	.0005814
5.40	221.41	2.34519	0.004517	**7.50**	1808.0	3.25721	0.0005531
5.41	223.63	2.34953	.004472	7.55	1900.7	3.27892	.0005261
5.42	225.88	2.35388	.004427	7.60	1998.2	3.30064	.0005005
5.43	228.15	2.35822	.004383	7.65	2100.6	3.32235	.0004760
5.44	230.44	2.36256	.004339	7.70	2208.3	3.34407	.0004528
5.45	232.76	2.36690	0.004296	**7.75**	2321.6	3.36578	0.0004307
5.46	235.10	2.37125	.004254	7.80	2440.6	3.38750	.0004097
5.47	237.46	2.37559	.004211	7.85	2565.7	3.40921	.0003898
5.48	239.85	2.37993	.004169	7.90	2697.3	3.43093	.0003707
5.49	242.26	2.38428	.004128	7.95	2835.6	3.45264	.0003527
5.50	244.69	2.38862	0.004087	**8.00**	2981.0	3.47436	0.0003355

Table 4 (*Continued*)

x	e^x	$\log_{10}(e^x)$	e^{-x}	x	e^x	$\log_{10}(e^x)$	e^{-x}
8.00	2981.0	3.47436	0.0003355	**9.00**	8103.1	3.90865	0.0001234
8.05	3133.8	3.49607	.0003191	9.05	8518.5	3.93037	.0001174
8.10	3294.5	3.51779	.0003035	9.10	8955.3	3.95208	.0001117
8.15	3463.4	3.53950	.0002887	9.15	9414.4	3.97379	.0001062
8.20	3641.0	3.56121	.0002747	9.20	9897.1	3.99551	.0001010
8.25	3827.6	3.58293	0.0002613	**9.25**	10405	4.01722	0.0000961
8.30	4023.9	3.60464	.0002485	9.30	10938	4.03894	.0000914
8.35	4230.2	3.62636	.0002364	9.35	11499	4.06065	.0000870
8.40	4447.1	3.64807	.0002249	9.40	12088	4.08237	.0000827
8.45	4675.1	3.66979	.0002139	9.45	12708	4.10408	.0000787
8.50	4914.8	3.69150	0.0002036	**9.50**	13360	4.12580	0.0000749
8.55	5166.8	3.71322	.0001935	9.55	14045	4.14751	.0000712
8.60	5431.7	3.73493	.0001841	9.60	14765	4.16923	.0000677
8.65	5710.0	3.75665	.0001751	9.65	15522	4.19094	.0000644
8.70	6002.9	3.77836	.0001666	9.70	16318	4.21266	.0000613
8.75	6310.7	3.80008	0.0001585	**9.75**	17154	4.23437	0.0000583
8.80	6634.2	3.82179	.0001507	9.80	18034	4.25609	.0000555
8.85	6974.4	3.84351	.0001434	9.85	18958	4.27780	.0000527
8.90	7332.0	3.86522	.0001364	9.90	19930	4.29952	.0000502
8.95	7707.9	3.88694	.0001297	9.95	20952	4.32123	0.0000477
9.00	8103.1	3.90865	0.0001234	**10.00**	22026	4.34294	0.0000454

Table 5 Amount of $1 n periods hence at compound interest at rate j per period: $(1 + j)^n$

n	$\frac{1}{2}\%$	1%	2%	3%	4%	5%	6%	8%	10%
1	1.0050	1.0100	1.0200	1.0300	1.0400	1.0500	1.0600	1.0800	1.1000
2	1.0100	1.0201	1.0404	1.0609	1.0816	1.1025	1.1236	1.1664	1.2100
3	1.0151	1.0303	1.0612	1.0927	1.1249	1.1576	1.1910	1.2597	1.3310
4	1.0201	1.0406	1.0824	1.1255	1.1699	1.2155	1.2625	1.3605	1.4641
5	1.0253	1.0510	1.1041	1.1593	1.2167	1.2763	1.3382	1.4693	1.6105
6	1.0304	1.0615	1.1262	1.1941	1.2653	1.3401	1.4185	1.5869	1.7716
7	1.0355	1.0721	1.1487	1.2299	1.3159	1.4071	1.5036	1.7148	1.9487
8	1.0407	1.0829	1.1717	1.2668	1.3686	1.4775	1.5938	1.8510	2.1436
9	1.0459	1.0937	1.1951	1.3048	1.4233	1.5513	1.6895	1.9990	2.3579
10	1.0511	1.1046	1.2190	1.3439	1.4802	1.6289	1.7908	2.1590	2.5937
11	1.0564	1.1157	1.2434	1.3842	1.5395	1.7103	1.8983	2.3316	2.8531
12	1.0617	1.1268	1.2682	1.4258	1.6010	1.7959	2.0122	2.5182	3.1384
13	1.0670	1.1381	1.2936	1.4685	1.6651	1.8856	2.1329	2.7196	3.4523
14	1.0723	1.1495	1.3195	1.5126	1.7317	1.9799	2.2609	2.9372	3.7975
15	1.0777	1.1610	1.3459	1.5580	1.8009	2.0789	2.3966	3.1723	4.1772
16	1.0831	1.1726	1.3728	1.6047	1.8730	2.1829	2.5404	3.4260	4.5950
17	1.0885	1.1843	1.4002	1.6528	1.9479	2.2920	2.6928	3.7000	5.0545
18	1.0939	1.1961	1.4282	1.7024	2.0258	2.4066	2.8543	3.9960	5.5599
19	1.0994	1.2081	1.4568	1.7535	2.1068	2.5270	3.0256	4.3157	6.1159
20	1.1049	1.2202	1.4859	1.8061	2.1911	2.6533	3.2071	4.6610	6.7275
21	1.1104	1.2324	1.5157	1.8603	2.2788	22.7860	3.3996	5.8338	7.4002
22	1.1160	1.2447	1.5460	1.9161	2.3699	2.9253	3.6035	5.4368	8.1403
23	1.1216	1.2572	1.5769	1.9736	2.4647	3.0715	3.8197	5.8715	8.9543
24	1.1272	1.2697	1.6084	2.0328	2.5633	3.2251	4.0489	6.3412	9.8497
25	1.1328	1.2824	2.6406	2.0938	2.6658	3.3864	4.2919	6.8485	10.8347
26	1.1385	1.2953	1.6734	2.1566	2.7725	3.5557	4.5494	7.3964	11.9182
27	1.1442	1.3082	1.7069	2.2213	2.8834	3.7335	4.8223	7.9881	13.1010
28	1.1499	1.3213	1.7410	2.2879	2.9987	3.9201	5.1117	8.6271	14.4210
29	1.1556	1.3345	1.7758	2.3566	3.1187	4.1161	5.4184	9.3173	15.8631
30	1.1614	1.3478	1.8114	2.4273	3.2334	4.3219	5.7435	10.0627	17.4494
31	1.1672	1.3613	1.8476	2.5001	3.3731	4.5380	6.0881	10.8677	19.1943
32	1.1730	1.3749	1.8845	2.5751	3.5081	4.7649	6.4534	11.7371	21.1138
33	1.1789	1.3887	1.9222	2.6523	3.6484	5.0032	6.8406	12.6760	23.2252
34	1.1848	1.4026	1.9607	2.7319	3.7943	5.2533	7.2510	13.6901	25.5477
35	1.1907	1.4166	1.9999	2.8139	3.9461	5.5160	7.6861	14.7853	28.1024
36	1.1967	1.4308	2.0399	2.8983	4.1039	5.7918	8.1473	15.9682	30.9127
37	1.2027	1.4451	2.0807	2.9852	4.2681	6.0814	8.6361	17.2456	34.0039
38	1.2087	1.4595	2.1223	3.0748	4.4388	6.3855	9.1543	18.6253	37.4043
39	1.2147	1.4741	2.1647	3.1670	4.6164	6.7048	9.7035	20.1153	41.1448
40	1.2208	1.4889	2.2080	3.2620	4.8010	7.0400	10.2857	21.7245	45.2593
41	1.2269	1.5038	2.2522	3.3599	4.9931	7.3920	10.9029	23.4625	49.7852
42	1.2330	1.5188	2.2972	3.4607	5.1928	7.7616	11.5570	25.3395	54.7637
43	1.2392	1.5340	2.3432	3.5645	5.4005	8.1497	12.2505	27.3666	60.2401
44	1.2454	1.5493	2.3901	3.6715	5.6665	8.5572	12.9855	29.5560	66.2641
45	1.2516	1.5648	2.4379	3.7816	5.8412	8.9850	13.7647	31.9204	72.8905
46	1.2579	1.5805	2.4866	3.8950	6.0748	9.4343	14.5905	34.4741	80.1795
47	1.2642	1.5963	2.5363	4.0119	6.3178	9.9060	15.4659	37.2320	88.1975
48	1.2705	1.6122	2.5871	4.1323	6.5705	10.4013	16.3938	40.2106	97.0172
49	1.2768	1.6283	2.6388	4.2562	6.8333	10.9213	17.3775	43.4274	106.7190
50	1.2832	1.6446	2.6916	4.3839	7.1067	11.4674	18.4202	46.9016	117.3909

Table 6 Present value of $1 to be received n periods hence at compound interest at rate j per period: $(1 + j)^{-n}$

n	½%	1%	2%	3%	4%	5%	6%	8%	10%
1	0.9950	0.9901	0.9804	0.9709	0.9615	0.9524	0.9434	0.9259	0.9091
2	0.9901	0.9803	0.9612	0.9226	0.9426	0.9070	0.8900	0.8573	0.8264
3	0.9851	0.9706	0.9423	0.9151	0.8890	0.8238	0.8396	0.7938	0.7513
4	0.9802	0.9610	0.9238	0.8885	0.8548	0.8227	0.7921	0.7350	0.6380
5	0.9754	0.9515	0.9057	0.8626	0.8219	0.7835	0.7473	0.6806	0.6209
6	0.9705	0.9420	0.8880	0.8375	0.7903	0.7462	0.7050	0.6302	0.5645
7	0.9657	0.9327	0.8706	0.8131	0.7599	0.7107	0.6651	0.5835	0.5132
8	0.9609	0.9235	0.8535	0.7894	0.7307	0.6768	0.2674	0.5403	0.4665
9	0.9561	0.9143	0.8368	0.7664	0.7026	0.6446	0.5919	0.5002	0.4241
10	0.9513	0.9053	0.8203	0.7441	0.6756	0.6139	0.5584	0.4632	0.3855
11	0.9466	0.8963	0.8043	0.7224	0.6496	0.5847	0.5268	0.4289	0.3505
12	0.9419	0.8874	0.7885	0.7014	0.6246	0.5568	0.4970	0.3971	0.3186
13	0.9372	0.8787	0.7730	0.6810	0.6006	0.5303	0.4688	0.3677	0.2897
14	0.9326	0.8700	0.7579	0.6611	0.5775	0.5051	0.4423	0.3405	0.2633
15	0.9279	0.8613	0.7430	0.6419	0.5553	0.4810	0.4173	0.3152	0.2394
16	0.9233	0.8528	0.7284	0.6232	0.5339	0.4581	0.3936	0.2919	0.2176
17	0.9187	0.8444	0.7142	0.6050	0.5134	0.4363	0.3714	0.2703	0.1978
18	0.9141	0.8360	0.7002	0.5874	0.4936	0.4155	0.3503	0.2502	0.1799
19	0.9096	0.8277	0.6864	0.5703	0.4746	0.3957	0.3305	0.2317	0.1635
20	0.9051	0.8195	0.6730	0.5537	0.4564	0.3769	0.3118	0.2145	0.1486
21	0.9006	0.8114	0.6598	0.5375	0.4388	0.3589	0.2942	0.1987	0.1351
22	0.8961	0.8304	0.6468	0.5219	0.4220	0.3418	0.2775	0.1839	0.1228
23	0.8916	0.7954	0.6342	0.5067	0.4057	0.3256	0.2618	0.1703	0.1117
24	0.8872	0.7876	0.6217	0.4919	0.3901	0.3101	0.2470	0.1577	0.1051
25	0.8828	0.7798	0.6095	0.4776	0.3715	0.2953	0.2330	0.1460	0.0923
26	0.8784	0.7720	0.5976	0.4637	0.3607	0.2812	0.2198	0.1352	0.0839
27	0.8740	0.7644	0.5859	0.4502	0.3468	0.2678	0.2074	0.1252	0.0763
28	0.8697	0.7568	0.5744	0.4371	0.3335	0.2551	0.1956	0.1159	0.0693
29	0.8653	0.7493	0.5631	0.4243	0.3207	0.2429	0.1846	0.1073	0.0630
30	0.8610	0.7419	0.5521	0.4120	0.3083	0.2314	0.1741	0.0994	0.0573
31	0.8567	8.7346	0.5412	0.4000	0.2965	0.2204	0.1643	0.0920	0.0521
32	0.8525	0.7273	0.5306	0.3883	0.2851	0.2099	0.1880	0.0852	0.0474
33	0.8482	0.7201	0.5202	0.3770	0.2741	0.1999	0.1462	0.0789	0.0431
34	0.8440	0.7130	0.5100	0.3660	0.2636	0.1904	0.1397	0.0730	0.0391
35	0.8398	0.7059	0.5000	0.3554	0.2534	0.1813	0.1301	0.0676	0.0356
36	0.8356	0.6989	0.4902	0.3450	0.2437	0.1727	0.1227	0.0626	0.0323
37	0.8315	0.6920	0.4806	0.3350	0.2343	0.1644	0.1158	0.0580	0.0294
38	0.8274	0.6852	0.4712	0.3252	0.2253	0.1566	0.1092	0.0537	0.0267
39	0.8232	0.6784	0.4619	0.3158	0.2166	0.1491	0.1031	0.0497	0.0243
40	0.8191	0.6717	0.4529	0.3066	0.2083	0.1420	0.0972	0.0460	0.0221
41	0.8151	0.6650	0.4440	0.2976	0.2003	0.1353	0.0912	0.0426	0.0201
42	0.8110	0.6584	0.4353	0.2890	0.1926	0.1288	0.0865	0.0395	0.0183
43	0.8070	0.6519	0.4268	0.2805	0.1852	0.1227	0.0816	0.0365	0.0166
44	0.8030	0.6454	0.4184	0.2724	0.1780	0.1169	0.0770	0.0338	0.0151
45	0.7990	0.6391	0.4102	0.2644	0.1712	0.1113	0.0727	0.0313	0.0137
46	0.7950	0.6327	0.4022	0.2567	0.1646	0.1060	0.0685	0.0290	0.0125
47	0.7910	0.6265	0.3943	0.2493	0.1583	0.1009	0.0647	0.0269	0.0113
48	0.7871	0.6203	0.3865	0.2420	0.1522	0.0961	0.0610	0.0249	0.0103
49	0.7832	0.6141	0.3790	0.2350	0.1463	0.0916	0.0575	0.0230	0.0094
50	0.7793	0.6080	0.3715	0.2281	0.1407	0.0872	0.0543	0.0213	0.0085

Table 7 Amount of an annuity of $1 per period, n periods hence, at rate j per period: $[(1 + j)^n - 1]/j$

n	$\frac{1}{2}\%$	1%	2%	3%	4%	5%	6%	8%	10%
1	1.0000	1.0000	1.0000	1.0000	1.0000	1.0000	1.0000	1.0000	1.0000
2	2.0050	2.0100	2.0200	2.0300	2.0400	2.0500	2.0600	2.0800	2.1000
3	3.0150	3.0301	3.0604	3.0909	3.1216	3.1525	3.1836	3.2464	3.3100
4	4.0301	4.0604	4.1216	4.1836	4.2465	4.3101	4.3746	4.5061	4.6410
5	5.0503	5.1010	5.2040	5.3091	5.4163	5.5256	5.6371	5.8666	6.1051
6	6.0755	6.1520	6.3081	6.4684	6.6330	6.8019	6.9753	7.3359	7.7156
7	7.1059	7.2135	7.4343	7.6625	7.8983	8.1420	8.3938	8.9228	9.4872
8	8.1414	8.2857	8.5830	8.8923	9.2142	9.5491	9.8975	10.6366	11.4359
9	9.1821	9.3685	9.7546	10.1591	10.5828	11.0266	11.4913	12.4876	13.5795
10	10.2280	10.4622	10.9497	11.4639	12.0061	12.5779	13.1808	14.4855	15.9374
11	11.2792	11.5668	12.1687	12.8078	13.4864	14.2068	14.9716	16.6455	18.5312
12	12.3356	12.6825	13.4121	14.1920	15.0258	15.9171	16.8699	18.9771	21.3843
13	13.3972	13.8093	14.6803	15.6178	16.6268	17.7130	18.8821	21.4953	24.5227
14	14.4642	14.9474	15.9739	17.0863	18.2919	19.5986	21.0151	24.2149	27.9750
15	15.5365	16.0969	17.2934	18.5989	20.0236	21.5786	23.2760	27.1521	31.7725
16	16.6142	17.2579	18.6393	20.1569	21.8245	23.6575	25.6725	30.3243	35.9497
17	17.6973	18.4304	20.0121	21.7616	23.6975	25.8404	28.2129	33.7502	40.5447
18	18.7858	19.6147	21.4123	23.4144	25.6454	28.1324	30.9057	37.4502	45.5992
19	19.8797	20.8109	22.8406	25.1169	27.6712	30.5390	33.7600	41.4463	51.1591
20	20.9791	22.0190	24.2974	26.8704	29.7781	33.0660	36.7856	45.7620	57.2750
21	22.0840	23.2392	25.7833	28.6765	31.9692	35.7193	39.9927	50.4229	64.0025
22	23.1944	24.4716	27.2990	30.5368	34.2480	38.5052	43.3923	55.4568	71.4027
23	24.3104	25.7163	28.8450	32.4529	36.6179	41.4305	46.9958	60.8933	79.5430
24	25.4320	26.9735	30.4219	34.4265	39.0826	44.5020	50.8156	66.7648	88.4973
25	26.5591	28.2432	32.0303	36.4593	41.6459	47.7271	54.8645	73.1059	98.3471
26	27.6919	29.5256	33.6709	38.5530	44.3117	51.1135	59.1564	79.9544	109.1818
27	28.8304	30.8209	35.3443	40.7096	47.0842	54.6691	63.7058	87.3508	121.0999
28	29.9745	32.1291	37.0512	42.9309	49.9676	58.4026	68.5281	95.3388	134.2099
29	31.1244	33.4504	38.7922	45.2189	52.9663	62.3227	73.6398	103.9659	148.6309
30	32.2800	34.7849	40.5681	47.5754	56.0849	66.4388	79.0582	113.2832	164.4940
31	33.4414	36.1327	42.3794	50.0027	59.3283	70.7608	84.8017	123.3459	181.9434
32	34.6086	37.4941	44.2270	82.5028	62.7015	75.2988	90.8898	134.2135	201.1378
33	35.7817	38.8690	46.1116	55.0278	66.2095	80.0638	97.3432	145.9506	222.2515
34	36.9606	40.2577	48.0338	57.7302	69.8579	85.0670	104.1838	158.6267	245.4767
35	38.1454	41.6603	49.9945	60.4621	73.6522	90.3203	111.4348	172.3168	271.0244
36	39.3361	43.0769	51.9944	63.2759	77.5983	95.8363	119.1209	187.1021	299.1268
37	40.5328	44.5076	54.0343	66.1742	81.7022	101.6281	127.2681	203.0703	330.0395
38	41.7355	45.9527	86.1149	69.1594	85.9703	107.7095	135.9042	220.3159	364.0434
39	42.9441	47.4123	58.2372	72.2342	90.4091	114.0950	143.0585	238.9412	401.4478
40	44.1589	48.8864	60.4020	75.4013	95.0255	120.7998	154.7620	259.0565	442.5926
41	45.3796	50.3752	62.6100	78.6633	99.8265	127.8398	165.0477	280.7810	487.8518
42	46.6065	51.8790	64.8622	82.0232	104.8196	135.2318	175.9505	304.2435	537.6370
43	47.8396	53.3978	67.1595	85.4839	110.0124	142.9933	187.5076	329.5830	592.4007
44	49.0788	54.9318	69.5027	89.0484	115.4129	151.1430	199.7580	356.9496	652.6408
45	50.3242	56.4811	71.8927	92.7199	121.0294	159.7002	212.7435	386.5056	718.9049
46	51.5758	58.0459	74.3306	96.5018	126.8706	168.6852	226.5081	418.4261	791.7953
47	52.8337	59.6263	76.8172	100.3965	132.9454	178.1194	241.0986	452.9002	871.9749
48	54.0978	61.2226	79.3539	104.4084	139.2632	188.0254	256.5645	490.1322	960.1723
49	55.3683	62.8348	81.9406	108.5406	145.8337	198.4267	272.9584	530.3427	1057.1896
50	56.6452	64.4632	84.5794	112.7969	152.6671	209.3480	290.3359	573.7702	1163.9085

Table 8 Present value of an annuity of \$1 per period for n periods at interest rate j per period: $[1 - (1 + j)^{-n}]/j$

n	$\frac{1}{2}\%$	1%	2%	3%	4%	5%	6%	8%	10%
1	0.9950	0.9901	0.9804	0.9709	0.9615	0.9524	0.9434	0.9259	0.9091
2	1.9851	1.9704	1.9416	1.9135	1.8861	1.8594	1.8334	1.7833	1.7355
3	2.9702	2.9410	2.8839	2.8286	2.7751	2.7232	2.6730	2.5771	2.4869
4	3.9505	3.9020	3.8077	3.7171	3.6299	3.5460	3.4651	3.3121	3.1699
5	4.9259	4.8534	4.7135	4.4797	4.4518	4.3295	4.2124	3.9927	3.7908
6	5.8964	5.7955	5.6014	5.4172	5.2421	5.0757	4.9173	4.6229	4.3553
7	6.8621	6.4720	6.2303	6.2303	6.0021	5.7864	5.5824	5.2064	4.8684
8	7.8230	7.6517	7.3255	7.0197	6.7327	6.4632	6.2098	5.7466	5.3349
9	8.7791	8.5660	8.1622	7.7861	7.4353	7.1078	6.8017	6.2469	5.7590
10	9.7304	9.4713	8.9826	8.5302	8.1109	7.7217	7.3601	6.7101	6.1446
11	10.6770	10.3676	9.7868	9.2526	8.7605	8.3064	7.8869	7.1390	6.4951
12	11.6189	11.2551	10.5753	9.9540	9.3851	8.3838	8.3838	7.5361	6.8137
13	12.5562	12.1337	11.3484	10.6350	9.9856	9.3936	8.8527	7.9038	7.1034
14	13.4887	13.0037	12.1062	11.2961	10.5631	9.8986	9.2950	8.2442	7.3667
15	14.4166	13.8651	12.8493	11.9379	11.1184	10.3797	9.7122	8.5595	7.6061
16	15.3399	14.7179	13.5777	12.5611	11.6523	10.8378	10.1059	8.8514	7.8237
17	16.2586	15.5623	14.2919	13.1661	12.1657	11.2741	10.4773	9.1216	8.0216
18	17.1728	16.3983	14.9920	13.7535	12.6593	11.6896	10.8267	9.3719	8.2014
19	18.0824	17.2260	15.6785	14.3238	13.1339	12.0853	11.1581	9.6036	8.3649
20	18.9874	18.0456	16.3514	14.8775	13.5903	12.4622	11.4699	9.8181	8.5136
21	19.8880	18.8570	17.0112	15.4150	14.0292	12.8212	11.7641	10.0168	8.6487
22	20.7841	19.6604	17.6580	15.9369	14.4511	13.1630	12.0416	10.2007	8.7715
23	21.6757	20.4558	18.2922	16.4436	14.8568	13.4886	12.3034	10.3711	8.8832
24	22.5629	21.2434	18.9139	16.9355	15.2470	13.7986	12.5504	10.5288	8.9847
25	23.4456	22.0232	19.5235	17.4131	15.6221	14.0939	12.7834	10.6748	9.0770
26	24.3240	22.7952	20.1201	17.8768	15.9828	14.3752	13.0032	10.8100	9.1609
27	25.1980	23.5596	20.7069	18.3270	16.3296	14.6430	13.2105	10.9352	9.2372
28	26.0677	24.3164	21.2813	18.7641	16.6631	14.8981	13.4062	11.0511	9.3066
29	26.9330	25.0658	21.8444	19.1885	16.9837	15.1411	13.5907	11.1584	9.3696
30	27.7941	25.8077	22.3965	19.6004	17.2920	15.3725	13.7648	11.2578	9.4269
31	28.6508	26.5423	22.9377	20.0004	17.5885	15.5928	13.9291	11.3498	9.4790
32	29.5033	27.2696	23.4683	20.3888	17.8736	15.8027	14.0840	11.4350	9.5264
33	30.3515	27.9897	23.9886	80.7658	18.1476	16.0025	14.2302	11.5139	9.5694
34	31.1956	28.7027	24.4986	21.1318	18.4112	16.1929	14.3681	11.5869	9.6086
35	32.0354	29.4086	24.9986	21.4872	18.6646	16.3742	14.4982	11.6546	9.6442
36	32.8710	30.1075	25.4888	21.8323	18.9083	16.5469	14.6210	11.7172	9.6765
37	33.7025	30.7995	25.9695	22.1672	19.1426	16.7113	14.7368	11.7752	9.7059
38	34.5299	31.4847	26.4406	22.4925	19.3679	16.8679	14.8460	11.8289	9.7327
39	35.3531	32.1630	26.9026	22.8082	19.5845	17.0170	14.9491	11.8786	9.7570
40	36.1722	32.8347	27.3555	23.1148	19.7928	17.1591	15.0463	11.9246	9.7791
41	36.9873	33.4997	27.7995	23.4124	19.9931	17.2944	15.1380	11.9672	9.7991
42	37.7983	34.1581	28.2348	23.7014	20.1856	17.4232	15.2245	12.0067	9.8174
43	38.6053	34.8100	28.6616	23.9819	20.3708	17.5459	15.3062	12.0432	9.8340
44	39.4082	38.4555	29.0800	24.2543	20.5488	17.6628	15.3832	12.0771	9.8491
45	40.2072	36.0945	29.4902	24.5187	20.7200	17.7741	15.4558	12.1084	9.8628
46	41.0022	36.7272	29.8923	24.7754	20.8847	17.8801	15.5244	12.1374	9.8753
47	41.7932	37.3537	30.2866	25.0247	21.0429	17.9810	15.5890	12.1643	9.8866
48	42.5803	37.9740	30.6731	25.2667	21.1951	18.0772	15.6500	12.1891	9.8969
49	43.3635	38.5881	31.0521	25.5017	21.3415	18.1687	15.7076	12.2122	9.9063
50	44.1428	39.1961	31.4236	25.7298	21.4822	18.2559	15.7619	12.2335	9.9148

Table 9 Trigonometric Functions

Degrees	Radians	Sin	Cos	Tan	Cot		
0	0.0000	0.0000	1.0000	0.0000		1.5708	90
1	0.0175	0.0175	0.9998	0.0175	57.290	1.5533	89
2	0.0349	0.0349	0.9994	0.0349	28.636	1.5359	88
3	0.0524	0.0523	0.9986	0.0524	19.081	1.5184	87
4	0.0698	0.0698	0.9976	0.0699	14.301	1.5010	86
5	0.0873	0.0872	0.9962	0.0875	11.430	1.4835	85
6	0.1047	0.1045	0.9945	0.1051	9.5144	1.4661	84
7	0.1222	0.1219	0.9925	0.1228	8.1443	1.4486	83
8	0.1396	0.1392	0.9903	0.1405	7.1154	1.4312	82
9	0.1571	0.1564	0.9877	0.1584	6.3138	1.4137	81
10	0.1745	0.1736	0.9848	0.1763	5.6713	1.3963	80
11	0.1920	0.1908	0.9816	0.1944	5.1446	1.3788	79
12	0.2094	0.2079	0.9781	0.2126	4.7046	1.3614	78
13	0.2269	0.2250	0.9744	0.2309	4.3315	1.3439	77
14	0.2443	0.2419	0.9703	0.2493	4.0108	1.3265	76
15	0.2618	0.2588	0.9659	0.2679	3.7321	1.3090	75
16	0.2793	0.2756	0.9613	0.2867	3.4874	1.2915	74
17	0.2967	0.2924	0.9563	0.3057	3.2709	1.2741	73
18	0.3142	0.3090	0.9511	0.3249	3.0777	1.2566	72
19	0.3316	0.3256	0.9455	0.3443	2.9042	1.2392	71
20	0.3491	0.3420	0.9397	0.3640	2.7475	1.2217	70
21	0.3665	0.3584	0.9336	0.3839	2.6051	1.2043	69
22	0.3840	0.3746	0.9272	0.4040	2.4751	1.1868	68
23	0.4014	0.3907	0.9205	0.4245	2.3559	1.1694	67
24	0.4189	0.4067	0.9135	0.4452	2.2460	1.1519	66
25	0.4363	0.4226	0.9063	0.4663	2.1445	1.1345	65
26	0.4538	0.4384	0.8988	0.4877	2.0503	1.1170	64
27	0.4712	0.4540	0.8910	0.5095	1.9626	1.0996	63
28	0.4887	0.4695	0.8829	0.5317	1.8807	1.0821	62
29	0.5061	0.4848	0.8746	0.5543	1.8040	1.0647	61
30	0.5236	0.5000	0.8660	0.5774	1.7321	1.0472	60
31	0.5411	0.5150	0.8572	0.6009	1.6643	1.0297	59
32	0.5585	0.5299	0.8480	0.6249	1.6003	1.0123	58
33	0.5760	0.5446	0.8387	0.6494	1.5399	0.9948	57
34	0.5934	0.5592	0.8290	0.6745	1.4826	0.9774	56
35	0.6109	0.5736	0.8192	0.7002	1.4281	0.9599	55
36	0.6283	0.5878	0.8090	0.7265	1.3764	0.9425	54
37	0.6458	0.6018	0.7986	0.7536	1.3270	0.9250	53
38	0.6632	0.6157	0.7880	0.7813	1.2799	0.9076	52
39	0.6807	0.6293	0.7771	0.8098	1.2349	0.8901	51
40	0.6981	0.6428	0.7660	0.8391	1.1918	0.8727	50
41	0.7156	0.6561	0.7547	0.8693	1.1504	0.8552	49
42	0.7330	0.6691	0.7431	0.9004	1.1106	0.8378	48
43	0.7505	0.6820	0.7314	0.9325	1.0724	0.8203	47
44	0.7679	0.6947	0.7193	0.9657	1.0355	0.8029	46
45	0.7854	0.7071	0.7071	1.0000	1.0000	0.7854	45
		Cos	Sin	Cot	Tan	Radians	Degrees

ANSWERS TO ODD-NUMBERED EXERCISES

EXERCISES 1.1 (Page 9)

Sketches of the graphs for Exercises 1 through 25 appear in Figs. EX 1.1-1 through EX 1.1-25.
27. (a) $y = 0$; (b) $x = 0$; (c) $xy = 0$

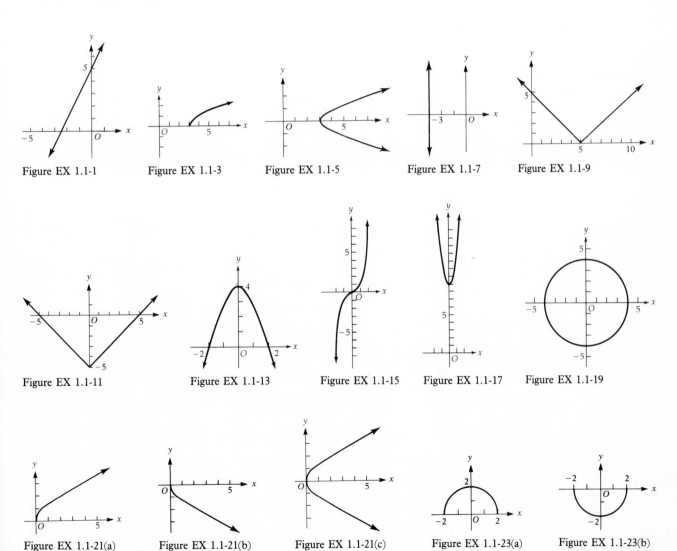

Figure EX 1.1-1   Figure EX 1.1-3   Figure EX 1.1-5   Figure EX 1.1-7   Figure EX 1.1-9

Figure EX 1.1-11   Figure EX 1.1-13   Figure EX 1.1-15   Figure EX 1.1-17   Figure EX 1.1-19

Figure EX 1.1-21(a)   Figure EX 1.1-21(b)   Figure EX 1.1-21(c)   Figure EX 1.1-23(a)   Figure EX 1.1-23(b)

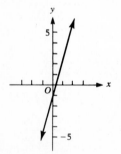

Figure EX 1.1-23(c)

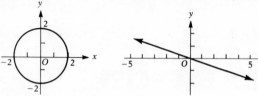

Figure EX 1.1-25(a)

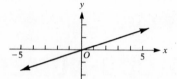

Figure EX 1.1-25(b)

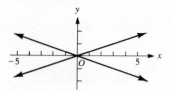

Figure EX 1.1-25(c)

EXERCISES 1.2 (Page 18)

1. -1 **3.** $-\frac{1}{7}$ **5.** 0 **7.** $4x - y - 11 = 0$ **9.** $y = 0$ **11.** $2x - 3y - 7 = 0$ **13.** $4x + 3y + 12 = 0$
15. $y = -7$ **17.** $x = y$ **19.** $\frac{2}{3}$ **21.** 0 **23.** $y = -\frac{3}{2}x - \frac{1}{2}$ **25.** $m_1 = -\frac{3}{5}$, $m_2 = \frac{5}{3}$, and $m_1 m_2 = -1$ **27.** (a) yes;
(b) no   **29.** (a) no; (b) yes   **31.** $(-4, 7)$ **33.** $\$7200$ **35.** (a) $y = 25x + 3000$ **37.** (a) $\$600$; (b) $y = 30x + 600$

EXERCISES 1.3 (Page 25)

Sketches of the graphs appear in Figs. EX 1.3-1 through EX 1.3-31.   **1.** domain: $(-\infty, +\infty)$; range: $(-\infty, +\infty)$ **3.** domain: $(-\infty, +\infty)$;
range: $[-6, +\infty)$ **5.** domain: $[-1, +\infty)$; range: $[0, +\infty)$ **7.** domain: $[\frac{4}{3}, +\infty)$; range: $[0, +\infty)$ **9.** domain: $(-\infty, -2] \cup [2, +\infty)$;
range: $[0, +\infty)$ **11.** domain: $(-\infty, +\infty)$; range: $[0, +\infty)$ **13.** domain: $\{x \mid x \neq 1\}$; range: $\{y \mid y \neq -2\}$ **15.** domain: $(-\infty, +\infty)$; range:
$\{-2, 2\}$ **17.** domain: $(-\infty, +\infty)$; range: $\{y \mid y \neq 3\}$ **19.** domain: $(-\infty, +\infty)$; range: $[-4, +\infty)$ **21.** domain: $(-\infty, +\infty)$; range:
$(-\infty, -2) \cup [1, +\infty)$ **23.** domain: $\{x \mid x \neq -5 \text{ and } x \neq -1\}$; range: $\{y \mid y \neq -7 \text{ and } y \neq -3\}$ **25.** domain: $[-3, 3]$; range: $[0, 3]$
27. domain: $\{x \mid x \neq 2\}$; range: $[0, +\infty)$ **29.** domain: $\{x \mid x \neq -5\}$; range: $[-6, +\infty)$ **31.** domain: $(-\infty, +\infty)$; range: $[1, +\infty)$

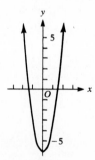

Figure EX 1.3-1

Figure EX 1.3-3

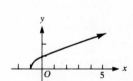

Figure EX 1.3-5

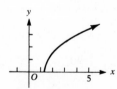

Figure EX 1.3-7

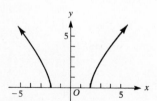

Figure EX 1.3-9

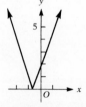

Figure EX 1.3-11

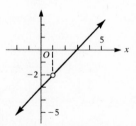

Figure EX 1.3-13

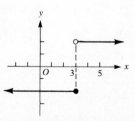

Figure EX 1.3-15

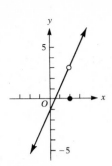

Figure EX 1.3-17

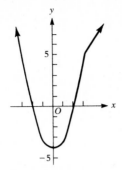

Figure EX 1.3-19

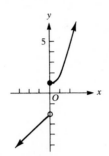

Figure EX 1.3-21

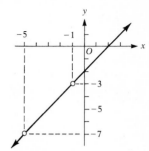

Figure EX 1.3-23

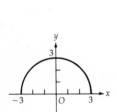

Figure EX 1.3-25

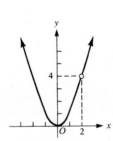

Figure EX 1.3-27

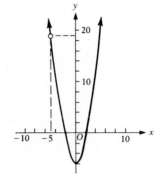

Figure EX 1.3-29

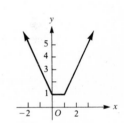

Figure EX 1.3-31

EXERCISES 1.4 (Page 33)

1. (a) 5; (b) -5; (c) -1; (d) $2a + 1$; (e) $2x + 1$; (f) $4x - 1$; (g) $4x - 2$; (h) $2x + 2h - 1$; (i) $2x + 2h - 2$; (j) 2 **3.** (a) -5; (b) -6;
(c) -3; (d) 30; (e) $2h^2 + 9h + 4$; (f) $8x^4 + 10x^2 - 3$; (g) $2x^4 - 7x^2$; (h) $2x^2 + (4h + 5)x + (2h^2 + 5h - 3)$; (i) $2x^2 + 5x + (2h^2 + 5h - 6)$;

(j) $4x + 2h + 5$ **5.** (a) $\frac{1}{2}$; (b) -1; (c) $\frac{2}{x}$; (d) $\frac{1 - x}{1 + x}$; (e) $\frac{2}{x^2 + 1}$; (f) $\frac{4}{x^2 + 2x + 1}$; (g) $\frac{-2}{(x + h + 1)(x + 1)}$ **7.** (a) 1; (b) $\sqrt{11}$; (c) 2;

(d) $3\sqrt{7}$; (e) $\sqrt{4x + 9}$; (f) $\frac{2}{\sqrt{2x + 2h + 3} + \sqrt{2x + 3}}$ **9.** (a) $x^2 - 6$, domain: $(-\infty, +\infty)$; (b) $x^2 - 10x + 4$, domain: $(-\infty, +\infty)$

11. (a) $\frac{1 + x}{1 - x}$, domain: $\{x \mid x \neq 0, x \neq 1\}$; (b) $\frac{x - 1}{x + 1}$, domain: $\{x \mid x \neq -1, x \neq 1\}$ **13.** (a) $\sqrt{x^2 - 1}$, domain: $(-\infty, -1] \cup [1, +\infty)$;

(b) $x - 1$, domain: $[0, +\infty)$ **15.** (a) $\frac{x - 2}{2x - 2}$, domain: $\{x \mid x \neq 1, x \neq 2\}$; (b) $-\frac{1}{2x + 1}$, domain: $\{x \mid x \neq -1, x \neq -\frac{1}{2}\}$

17. (a) $f(x) = 40x - 140{,}000$; (b) \$780{,}000; (c) 3500 **19.** (a) $f(x) = \frac{5}{9}(x - 32)$; (b) $35°$ **21.** (a) $f(x) = 45x$; (b) \$675

23. (a) $f(x) = \frac{3}{2}\sqrt{\frac{x}{2}}$; (b) $\frac{3}{2}$ sec **25.** (a) $f(x) = \frac{32(10)^8}{x^2}$; (b) 165 lb **27.** (a) $f(x) = \frac{9x}{490{,}000}(5000 - x)$; (b) 17.6 people per day

EXERCISES 1.5 (Page 38)

1. (a) $P(x) = -2x^2 + 380x - 12{,}000$; (b) \$5600 **3.** (a) $V(x) = 4x^3 - 46x^2 + 120x$; (b) $[0, 4]$ **5.** (a) $A(x) = 120x - x^2$;

(b) $[0, 120]$ **7.** (a) $A(x) = 120x - \frac{1}{2}x^2$; (b) $[0, 240]$ **9.** (a) $A(x) = 96x - \frac{6}{5}x^2$; (b) $[0, 80]$ **11.** (a) $C(x) = 48x + \frac{97{,}200}{x}$;

(b) $(0, +\infty)$ **13.**   (a) $S(x) = 6x^2 + \dfrac{768}{x}$; (b) $(0, +\infty)$ **15.**   (a) $A(x) = 3x + \dfrac{48}{x} + 30$; (b) $(0, +\infty)$

17. (a) $f(x) = \begin{cases} 600x & \text{if } 0 \le x \le 20 \\ 900x - 15x^2 & \text{if } 20 < x \le 60 \end{cases}$; (b) $[0, 60]$ **19.**   (a) $R(x) = 400x - \frac{1}{2}x^2$; (b) $[0, 800]$ **21.**   (a) $f(x) = kx(900{,}000 - x)$;
(b) $[0, 900{,}000]$

EXERCISES 1.6 (Page 44)

1. supply curve   **3.** demand curve   **5.** neither   **7.** neither   **9.** supply curve   **11.** (b) \$7.50; (c) 5   **13.** (b) \$1.74; (c) 5
15. (b) \$3   **17.** \$2   **19.** (a) $x_E = \frac{39}{5}$; $p_E = \frac{18}{5}$ **21.** (a) $x_E = 2$; $p_E = 5$ **23.** (a) $x_E = 1$; $p_E = 7$ **25.** (a) $x_E = 4$; $p_E = 3$
27. $x + 1000p = 34{,}000$ **29.** $x - 800p + 54{,}000 = 0$

REVIEW EXERCISES FOR CHAPTER 1 (Page 45)

Sketches of the graphs for Exercises 1 through 9 appear in Figs. REV 1-1 through REV 1-9.   **13.** (a) -2; (b) $2x + y - 3 = 0$ **15.** $\frac{3}{4}$
17. $5x + 2y - 19 = 0$ **19.** domain: $[-\frac{5}{2}, +\infty)$; range: $[0, +\infty)$; see Fig. REV 1-19   **21.** domain: $\{x \mid x \ne 2\}$; range: $\{y \mid y \ne 5\}$; see Fig.
REV 1-21   **23.** (a) 7; (b) 35; (c) $3x^4 + x^2 + 5$; (d) $-9x^4 + 6x^3 - 31x^2 + 10x - 25$; (e) $6x + 3h - 1$ **25.** (a) $16x^2 - 24x + 5$, domain:
$(-\infty, +\infty)$; (b) $4x^2 - 19$, domain: $(-\infty, +\infty)$ **27.** (a) $x - 4$, domain: $[-5, +\infty)$; (b) $\sqrt{x^2 - 4}$, domain: $(-\infty, -2] \cup [2, +\infty)$
29. (b) \$5   **31.** (a) $x_E = 5$, $p_E = 12$ **33.** \$700,000   **35.** (a) $f(t) = 16t^2$; (b) 100 ft   **37.** (a) $V(x) = 4x^3 - 80x^2 + 400x$;
(b) $[0, 10]$ **39.** (a) $f(x) = \begin{cases} 15x & \text{if } 0 \le x \le 150 \\ 22.5x - 0.05x^2 & \text{if } 150 < x \le 250 \end{cases}$; (b) $[0, 250]$ **41.** (a) $S(x) = x^2 + \dfrac{128}{x}$; (b) $(0, +\infty)$

43. (a) $A(x) = 8x + \dfrac{200}{x} + 82$; (b) $(0, +\infty)$ **45.** $p = -\frac{1}{500}x + 50$

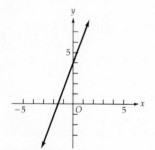

Figure REV 1-1

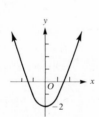

Figure REV 1-3

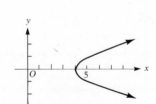

Figure REV 1-5

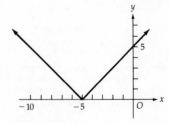

Figure REV 1-7

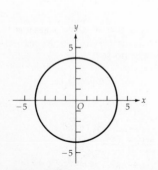

Figure REV 1-9

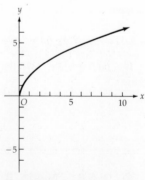

Figure REV 1-11(a)

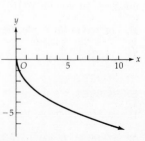

Figure REV 1-11(b)

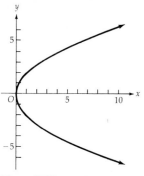

Figure REV 1-11(c)

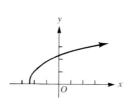

Figure REV 1-19

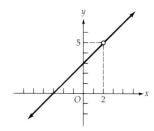

Figure REV 1-21

EXERCISES 2.1 (Page 57)

1. 14　**3.** -2　**5.** 7　**7.** -62　**9.** 5　**11.** $-\frac{1}{22}$　**13.** -2　**15.** $\frac{1}{4}$　**17.** $\frac{5}{2}$　**19.** 1　**21.** 3　**23.** $\frac{3}{2}$
25. $\frac{1}{5}\sqrt{30}$　**27.** $\frac{1}{6}$　**29.** $\frac{1}{4}$　**37.** (a) 0

EXERCISES 2.2 (Page 67)

1. (a) -3; (b) 2; (c) does not exist　**3.** (a) 8; (b) 0; (c) does not exist　**5.** (a) 4; (b) 4; (c) 4　**7.** (a) 5; (b) 5; (c) 5　**9.** (a) 0; (b) 0; (c) 0　**11.** (a) 0; (b) 0; (c) 0　**13.** (a) 1; (b) -1; (c) does not exist　**15.** (a) 0; (b) 1; (c) does not exist; (d) 1; (e) 0; (f) does not exist　**17.** $+\infty$　**19.** $+\infty$　**21.** $-\infty$　**23.** $-\infty$　**25.** $+\infty$　**27.** $+\infty$　**29.** $+\infty$　**31.** $-\infty$　**33.** $-\infty$　**35.** $+\infty$
37. (b) 40; (c) 35; (d) 140; (e) 130

EXERCISES 2.3 (Page 77)

1. -3; $f(-3)$ does not exist　**3.** -3; $\lim_{x \to -3} g(x) \neq g(-3)$　**5.** 4; $h(4)$ does not exist　**7.** 4; $\lim_{x \to 4} f(x)$ does not exist　**9.** -2, 2;
$F(-2)$ and $F(2)$ do not exist　**11.** -2, 2; $G(-2)$ and $G(2)$ do not exist　**13.** 0; $\lim_{x \to 0} f(x)$ does not exist　**15.** 2; $\lim_{t \to 2} g(t) \neq g(2)$
17. 0; $\lim_{x \to 0} g(x)$ does not exist　**19.** 0; $f(0)$ does not exist　**21.** all real numbers　**23.** all real numbers　**25.** all real numbers
except 3　**27.** all real numbers except -1 and 1　**29.** all real numbers except -2 and 2　**35.** C is discontinuous at 50 and 200
because $\lim_{x \to 50} C(x)$ and $\lim_{x \to 200} C(x)$ do not exist.　**37.** (a) $V(x) = 4x^3 - 46x^2 + 120x$　**39.** (a) $A(x) = 120x - x^2$

41. (a) $f(x) = \begin{cases} 600x & \text{if } 0 \le x \le 20 \\ 900x - 15x^2 & \text{if } 20 < x \le 60 \end{cases}$　**43.** (a) $f(x) = kx(900{,}000 - x)$

EXERCISES 2.4 (Page 84)

1. $m(x_1) = -2x_1$; horizontal tangent at $(0, 9)$　**3.** $m(x_1) = -4x_1 + 4$; horizontal tangent at $(1, 2)$　**5.** $m(x_1) = 3x_1^2$; horizontal tangent at
$(0, 1)$　**7.** $m(x_1) = 3x_1^2 - 3$; horizontal tangent at $(1, -2)$ and $(-1, 2)$　**9.** $m(x_1) = 3x_1^2 - 8x_1 + 4$; horizontal tangent at $(\frac{2}{3}, -\frac{22}{27})$ and
$(2, -2)$　**11.** $8x + y + 9 = 0$　**13.** $x - 6y + 13 = 0$　**15.** $2x + 3y - 12 = 0$　**17.** $8x - y - 5 = 0$
19. $4x + 4y - 11 = 0$　**21.** $x + 4y - 4 = 0$; $x + 4y + 4 = 0$　**23.** (a) $P(x) = -2x^2 + 380x - 12{,}000$; (b) \$95

EXERCISES 2.5 (Page 91)

1. 7　**3.** 0　**5.** $-4x$　**7.** $8x + 5$　**9.** $-3x^2$　**11.** $\dfrac{1}{2\sqrt{x}}$　**13.** $-\dfrac{1}{(x + 1)^2}$　**15.** $-\dfrac{1}{(2x + 1)^{3/2}}$　**17.** -6

19. $-\frac{3}{128}$　**21.** $-\frac{1}{4}$　**23.** $2x - 2x^{-3}$　**25.** $\dfrac{-3}{(2x - 3)^2}$　**27.** $-\dfrac{8}{x^3} + 3 - 2x$　**29.** $\dfrac{-x}{\sqrt{4 - x^2}}$　**31.** (a) $\dfrac{1}{3(x - 1)^{2/3}}$; (b) no

EXERCISES 2.6 (Page 100)

1. 7   **3.** $-2 - 2x$ **5.** $3x^2 - 6x + 5$ **7.** $x^7 - 4x^3$ **9.** $t^3 - t$ **11.** $4\pi r^2$ **13.** $2x + 3 - \dfrac{2}{x^3}$ **15.** $16x^3 + \dfrac{1}{x^5}$

17. $-\dfrac{6}{x^3} - \dfrac{20}{x^5}$ **19.** $2x^{-1/2} - 3x^{-3/2}$ **21.** $\dfrac{1}{3\sqrt[3]{t^2}} - \dfrac{1}{3t\sqrt[3]{t}}$ **23.** $3\sqrt{3}s^2 - 2\sqrt{3}s$ **25.** $70x^6 + 60x^4 - 15x^2 - 6$

27. $-18y^2(7 - 3y^3)$ **29.** $10x^4 - 24x^3 + 12x^2 + 2x - 3$ **31.** $\dfrac{-1}{(x-1)^2}$ **33.** $\dfrac{-4(x+1)}{(x-1)^3}$ **35.** $\dfrac{5 - 10x^2}{(1 + 2x^2)^2}$ **37.** $\dfrac{48x^2}{(x^3 + 8)^2}$

39. $\dfrac{6x^2 + 60x + 6}{(x+5)^2}$ **41.** $\dfrac{1}{\sqrt{x}(\sqrt{x} + 1)^2}$ **43.** $5x + 2y - 1 = 0$ **45.** (a) $(1, -2)$ **47.** $(-2, -2)$ and $(2, 6)$

REVIEW EXERCISES FOR CHAPTER 2 (Page 101)

1. 9   **3.** -6 **5.** $\frac{5}{8}$ **7.** $-\infty$ **9.** $\dfrac{1}{8\sqrt{2}}$ **11.** discontinuous at -2; $f(-2)$ does not exist   **13.** discontinuous at -2;

$\lim\limits_{x\to -2} g(x)$ does not exist   **17.** $15x^2 - 14x + 2$ **19.** $-\dfrac{8}{x^3} + \dfrac{12}{x^5}$ **21.** $60t^4 - 39t^2 - 6t - 4$ **23.** $\dfrac{-6x^2}{(x^3 - 1)^2}$

25. $6x^2 + 6x^{-4}$ **27.** $\dfrac{x^2 - 2x}{(x-1)^2}$ **29.** $8x - 2$ **31.** $-\frac{3}{49}$ **33.** $3x + y - 5 = 0$ **35.** (a) $V(x) = 4x^3 - 80x^2 + 400x$

37. (a) $f(x) = \begin{cases} 15x & \text{if } 0 \le x \le 150 \\ 22.5x - 0.05x^2 & \text{if } 150 < x \le 250 \end{cases}$ **39.** (a) $P(x) = -x^2 + 220x - 4000$; (b) \$110   **41.** (a) $\dfrac{1}{3(x-3)^{2/3}}$; (b) no

EXERCISES 3.1 (Page 110)

1. (a) $C'(x) = 30 + 2x$; (b) \$110; (c) \$111   **3.** (a) \$0.50; (b) 25   **5.** (a) 46 cents; (b) 20 cents; (c) 0.435   **7.** (a) \$6.20; (b) \$5.20;

(c) 0.839   **9.** (a) $Q(x) = x + 6 + \dfrac{12}{x}$; (b) $C'(x) = 2x + 6$ **11.** (a) $R'(x) = 200 - \frac{2}{3}x$; (b) \$180; (c) \$179.67

13. (a) $R(x) = 3x - \frac{3}{4}x^2$; (b) $R'(x) = 3 - \frac{3}{2}x$ **15.** (a) $2x + 3p = 9$; (b) $R'(x) = 3 - \frac{4}{3}x$

EXERCISES 3.2 (Page 117)

1. (a) 8.6; (b) 8.3; (c) 8.1; (d) 8   **3.** (a) -20 units per dollar increase in the price; (b) -19 units per dollar increase in the price
5. (a) 369,000 bulbs per one cent increase in the price; (b) 363,000 bulbs per one cent increase in the price   **7.** (a) increase of 1640 units;
(b) increase of 1638 units   **9.** increase of 5 units   **11.** (a) \$30,000 per year; (b) 10 percent   **13.** (a) 20 units per \$100 increase in budget;
(b) 10 units per \$100 increase in budget   **15.** (a) 920 people per year; (b) 1400 people per year; (c) 6.1 percent; (d) 6.4 percent
17. (a) -2.9 degrees per hour; (b) -3 degrees per hour   **19.** (a) 10 nonsense words per hour; (b) 5 nonsense words per hour
21. -2 **23.** (a) 496 ft/sec; (b) 17.5 sec

EXERCISES 3.3 (Page 124)

1. $6(2x + 1)^2$ **3.** $8(x + 2)(x^2 + 4x - 5)^3$ **5.** $2(2t^4 - 7t^3 + 2t - 1)(8t^3 - 21t^2 + 2)$ **7.** $\dfrac{4x}{\sqrt{1 + 4x^2}}$ **9.** $\dfrac{-2}{(5 - 3x)^{1/3}}$

11. $\dfrac{y}{(25 - y^2)^{3/2}}$ **13.** $\dfrac{17}{2(3r + 1)^{3/2}(2r - 5)^{1/2}}$ **15.** $\dfrac{t - 1}{t\sqrt{2t}}$ **17.** $\dfrac{-2x(2x^2 + 1)(6x^3 + 9x - 4)}{(3x^3 + 1)^3}$

19. $2(s^2 + 1)^2(2s + 5)(8s^2 + 15s + 2)$ **21.** $-2(2x - 5)^{-2}(4x + 3)^{-3}(12x - 17)$

23. $(2x - 9)(x^3 + 4x - 5)^2(22x^3 - 81x^2 + 40x - 128)$ **25.** $\dfrac{1}{x^2\sqrt{x^2 - 1}}$ **27.** $\dfrac{3(y - 3)}{(2y - 3)^{3/2}}$ **29.** $\dfrac{-1}{4\sqrt{9 - x}\sqrt{9 + \sqrt{9 - x}}}$

31. $\dfrac{-4(1 - 3w)^3}{[(1 - 3w)^4 + 4]^{2/3}}$ **33.** $4x - 5y + 9 = 0$ **35.** $18x + y + 28 = 0$ **37.** (a) $C'(x) = 4 + \dfrac{3x}{\sqrt{3x^2 + 24}}$; (b) \$5.67

39. \$44.27   **41.** 100   **43.** -8 units per week   **45.** 329 predators per week

EXERCISES 3.4 (Page 129)

1. $-\dfrac{x}{y}$ **3.** $\dfrac{8y - 3x^2}{3y^2 - 8y}$ **5.** $-\dfrac{y^2}{x^2}$ **7.** $-y^{1/2}x^{-1/2}$ **9.** $\dfrac{x - xy^2}{x^2y - y}$ **11.** $\dfrac{y + 4\sqrt{xy}}{\sqrt{x} - x}$ **13.** $\dfrac{y^{2/3}(1 + y^{1/3})}{x^{2/3}(24y^{5/3} - x^{1/3})}$ **15.** $\dfrac{3x^2 - 4y}{4x - 3y^2}$

17. $\dfrac{3x^2 - y^3}{x^3 - 6xy}$ **19.** $\dfrac{x^3 + 8y^3}{4x^3 - 3x^2y}$ **21.** $2x + y = 4$ **23.** $-\dfrac{1}{p + 2}$ **25.** $-\dfrac{x^2}{p^2}$ **27.** (a) $f_1(x) = 2\sqrt{x - 2}$, domain: $x \geq 2$;

$f_2(x) = -2\sqrt{x - 2}$, domain: $x \geq 2$; (d) $f_1'(x) = (x - 2)^{-1/2}$, domain: $x > 2$; $f_2'(x) = -(x - 2)^{-1/2}$, domain: $x > 2$; (e) $\dfrac{2}{y}$; (f) $x - y - 1 = 0$;

$x + y - 1 = 0$ **29.** (a) $f_1(x) = \sqrt{x^2 - 9}$, domain: $|x| \geq 3$; $f_2(x) = -\sqrt{x^2 - 9}$, domain: $|x| \geq 3$; (d) $f_1'(x) = x(x^2 - 9)^{-1/2}$, domain: $|x| > 3$;

$f_2'(x) = -x(x^2 - 9)^{-1/2}$, domain: $|x| > 3$; (e) $\dfrac{x}{y}$; (f) $5x + 4y + 9 = 0$

EXERCISES 3.5 (Page 134)

1. -3 **3.** -8 **5.** $-\frac{27}{2}$ **7.** $-\frac{3}{4}$ **9.** \$1020 per week **11.** decreasing at the rate of 55 shirts per week **13.** increasing at
the rate of 2 units per week **15.** (a) increasing at the rate of \$500 per week; (b) increasing at the rate of \$1050 per week; (c) increasing at

the rate of \$550 per week **17.** $\dfrac{1}{2\pi}$ ft/min **19.** 0.001π cm³/day **21.** 0.004π cm²/day **25.** $\frac{8}{3}$ ft/sec **27.** $\frac{10}{3}$ ft/sec

29. decreasing at the rate of 1800 lb/ft² per minute **31.** 128π cm²/sec **33.** $\dfrac{6}{25\pi}$ m/min

REVIEW EXERCISES FOR CHAPTER 3 (Page 136)

1. $6(3x - 1)(3x^2 - 2x + 1)^2$ **3.** $\dfrac{t + 2}{\sqrt{t^2 + 4t - 3}}$ **5.** $\dfrac{-10x(3x^2 + 4)^9(15x^7 + 28x^5 - 6)}{(x^7 + 1)^{11}}$ **7.** $\dfrac{x(x^2 - 1)^{1/2}(4x^2 - 13)}{(x^2 - 4)^{1/2}}$

9. $\dfrac{8x}{3y^2 - 8y}$ **11.** $-\dfrac{y^{1/3}}{x^{1/3}}$ **13.** $2x + 3x[x^3 + (x^4 + x)^2]^2(8x^6 + 10x^3 + 3x + 2)$ **15.** $5x - 4y - 6 = 0$

17. (a) $C'(x) = 2x + 40$; (b) \$80; (c) \$81 **19.** (a) \$130; (b) \$60; (c) 0.46 **21.** (a) $R'(x) = 100 - \frac{1}{3}x$; (b) \$95; (c) \$94.83
23. (a) increase of 3312 calculators per \$1 increase in the price; (b) increase of 3212.5 calculators per \$1 increase in the price **25.** (a) 340
people per year; (b) 1.62 percent **27.** decreasing at the rate of 242 bars per week **29.** (a) increasing at the rate of \$10,400 per year;
(b) increasing at the rate of \$800 per year **31.** decreasing at the rate of 5 units per week **33.** decreasing at the rate of 0.31 square
centimeters per day **35.** 9.6 ft/sec **37.** 648 fish per week

EXERCISES 4.1 (Page 146)

1. $-5, \frac{1}{3}$ **3.** $-3, -1, 1$ **5.** $0, 2$ **7.** $-2, 0, 2$ **9.** no critical numbers **11.** no critical numbers **13.** abs min:
$f(2) = -2$ **15.** no absolute extrema **17.** abs min: $f(-3) = 0$ **19.** abs min: $f(5) = 1$ **21.** abs min: $f(4) = 1$ **23.** abs max:
$f(5) = 2$ **25.** abs min: $g(-3) = -46$; abs max: $g(-1) = -10$ **27.** abs min: $f(-2) = 0$; abs max: $f(-4) = 144$ **29.** abs min:
$f(2) = 0$; abs max: $f(3) = 25$ **31.** abs min: $f(-1) = -1$; abs max: $f(2) = \frac{1}{2}$ **33.** abs min: $f(-1) = 0$; abs max: $f(1) = \sqrt[3]{4}$

EXERCISES 4.2 (Page 150)

1. $\frac{1}{2}$ **3.** $-10, 10$ **5.** $\frac{5}{3}$ in. **7.** 60 m by 60 m **9.** 60 m by 120 m **11.** from A to P to C, where P is 8 km down the beach
from B **13.** from A to P to C, where P is 4 km down the river from B **15.** 48 ft by 40 ft **17.** 30 **19.** 400
21. 450,000 **23.** A, 7 machines; B, 8 machines **25.** $\frac{2}{3}k$

EXERCISES 4.3 (Page 156)

1. (a) and (b) $f(2) = -5$, rel min; (c) $[2, +\infty)$; (d) $(-\infty, 2]$ **3.** (a) and (b) $f(-\frac{1}{3}) = \frac{5}{27}$, rel max; $f(1) = -1$, rel min; (c) $(-\infty, -\frac{1}{3}]$,
$[1, +\infty)$; (d) $[-\frac{1}{3}, 1]$ **5.** (a) and (b) $f(0) = 2$, rel max; $f(3) = -25$, rel min; (c) $(-\infty, 0]$, $[3, +\infty)$; (d) $[0, 3]$ **7.** (a) and (b) $f(0) = 0$,
rel min; $f(1) = \frac{1}{4}$, rel max; (c) $[0, 1]$, $[2, +\infty)$; (d) $(-\infty, 0]$, $[1, 2]$ **9.** (a) and (b) $f(-2) = 46$, rel max; $f(2) = -50$, rel min; (c) $(-\infty, -2]$,
$[2, +\infty)$; (d) $[-2, 2]$ **11.** (a) and (b) no relative extrema; (c) $(0, +\infty)$; (d) nowhere **13.** (a) and (b) $f(2) = 4$, rel max; (c) $(-\infty, 2]$;
(d) $[2, 3]$ **15.** (a) and (b) $f(\frac{1}{5}) = \frac{3456}{3125}$, rel max; $f(1) = 0$, rel min; (c) $(-\infty, \frac{1}{5}]$; $[1, +\infty)$; (d) $[\frac{1}{5}, 1]$ **17.** (a) and (b) $f(4) = 2$, rel max;
(c) $(-\infty, 4]$; (d) $[4, +\infty)$ **19.** (a) and (b) $f(\frac{1}{8}) = -\frac{1}{4}$, rel min; (c) $[\frac{1}{8}, +\infty)$; (d) $(-\infty, \frac{1}{8}]$ **21.** (a) and (b) $f(-1) = 0$, rel max;
$f(1) = -\sqrt[3]{4}$, rel min; (c) $(-\infty, -1]$, $[1, +\infty)$; (d) $[-1, 1]$ **23.** $a = -3, b = 7$ **25.** $a = -2, b = 9, c = -12, d = 7$

EXERCISES 4.4 (Page 162)

1. $f'(x) = 5x^4 - 6x^2 + 1; f''(x) = 20x^3 - 12x$ **3.** $g'(s) = 8s^3 - 12s^2 + 7; g''(s) = 24s^2 - 24s$ **5.** $F'(x) = \frac{5}{2}x^{3/2} - 5;$

$F''(x) = \frac{15}{4}x^{1/2}$ **7.** $f'(x) = x(x^2 + 1)^{-1/2}; f''(x) = (x^2 + 1)^{-3/2}$ **9.** $g'(x) = \dfrac{8x}{(x^2 + 4)^2}; g''(x) = \dfrac{32 - 24x^2}{(x^2 + 4)^3}$ **11.** $24x$

13. $\frac{5}{8}(\frac{21}{2}x^{-1/2} + 3x^{-3/2} - \frac{3}{2}x^{-7/2})$ **15.** $1152(2x - 1)^{-5}$ **17.** $\dfrac{18 + 6x}{(1 - x)^5}$ **23.** $-\dfrac{3a^4x^2}{y^7}$ **25.** 24 **27.** $f(\frac{1}{3}) = \frac{2}{3},$ rel min

29. $f(-1) = -11,$ rel min; $f(\frac{3}{2}) = \frac{81}{4},$ rel max **31.** $g(0) = 3,$ rel max; $g(2) = \frac{5}{3},$ rel min **33.** $f(4) = 0,$ rel min
35. $h(-\frac{3}{4}) = -\frac{99}{256},$ rel min; $h(0) = 0,$ rel max; $h(1) = -\frac{5}{8},$ rel min **37.** $f(1) = 8,$ rel min **39.** $f(-\frac{1}{2}) = -\frac{27}{16},$ rel min
41. $h(-2) = -2,$ rel min **43.** $F(27) = 9,$ rel max **45.** (a) 76 cents, 77 cents, 79 cents, 81 cents, 81 cents

EXERCISES 4.5 (Page 169)

1. $f(0) = 0,$ abs min **3.** $F(-1) = 7,$ abs max **5.** no absolute extrema **7.** $f(\frac{1}{4}) = \frac{3}{4},$ abs min **9.** $g(-1) = 0,$ abs min
11. $f(0) = 0,$ abs min; $f(\sqrt{2}) = \frac{1}{18}\sqrt{3},$ abs max **13.** $f(1) = -6,$ abs min **15.** $F(3) = 75,$ abs max **17.** 45 m by 60 m
19. width, 4 in.; length, 12 in.; depth, 6 in. **21.** 6 in. by 9 in. **23.** height, $4\sqrt[3]{4}$ in.; radius, $\sqrt[3]{4}$ in. **25.** 1 month; 7.5 percent
27. radius of semicircle, $\dfrac{32}{4 + \pi}$ ft; height of rectangle, $\dfrac{32}{4 + \pi}$ ft **29.** 90 km/hr **31.** 120 **33.** 12 orders of 1000 cases each

35. $\sqrt{\dfrac{RS}{2K}}$ orders of $\sqrt{\dfrac{2KR}{S}}$ cases each

REVIEW EXERCISES FOR CHAPTER 4 (Page 171)

1. $f'(x) = 12x^3 - 6x^2 + 14x - 5; f''(x) = 36x^2 - 12x + 14$ **3.** $g'(x) = 5(x + 2)^{-2}; g''(x) = -10(x + 2)^{-3}$ **5.** $-3(3 - 2x)^{-5/2}$
7. abs min: $f(5) = 0$ **9.** abs min: $f(1) = -\frac{1}{2};$ abs max: $f(2) = 64$ **11.** abs min: $f(-1) = f(1) = 0;$ abs max: $f(2) = 9$ **13.** no

absolute extrema **15.** (a) and (b) $f(2) = 10,$ rel max; $f(4) = 6,$ rel min; (c) $(-\infty, 2], [4, +\infty);$ (d) $[2, 4]$ **17.** (a) and (b) $f(\frac{8}{5}) = \dfrac{839,808}{3125},$

rel max; $f(4) = 0,$ rel min; (c) $(-\infty, \frac{8}{5}], [4, +\infty);$ (d) $[\frac{8}{5}, 4]$ **19.** (a) and (b) $f(\sqrt[3]{2}) = \dfrac{3}{\sqrt[3]{4}},$ rel min; (c) $(-\infty, 0), [\sqrt[3]{2}, +\infty);$ (d) $(0, \sqrt[3]{2}]$

21. $f(-2) = 6,$ rel max; $f(0) = 2,$ rel min **23.** $g(2) = 0,$ rel min **25.** $f(2) = 3,$ rel min **27.** $G(\frac{1}{2}) = -\frac{27}{16},$ rel min
29. 6, 6 **33.** $\frac{10}{3}$ in. **35.** 225 **37.** side of base, 4 ft; depth, 2 ft **39.** 9 m by 18 m **41.** 120 **43.** 1500

EXERCISES 5.1 (Page 180)

1. concave downward for $x < 0;$ concave upward for $x > 0;$ (0, 0) pt. of infl. **3.** concave downward for $x < -\frac{1}{2};$ concave upward for
$x > -\frac{1}{2};$ $(-\frac{1}{2}, 5)$ pt. of infl. **5.** concave upward everywhere; no pts. of infl. **7.** concave downward for $x < -1$ and $0 < x < 1;$ concave
upward for $-1 < x < 0$ and $x > 1;$ (0, 0) pt. of infl. **9.** concave upward for $x < 2;$ concave downward for $x > 2;$ (2, 0) pt. of infl.
11. concave upward for $x < -4;$ concave downward for $x > -4;$ no pts. of infl. (Sketches of the graphs for Exercises 13–25 appear in Figs.
EX 5.1-13 through EX 5.1-25.) **27.** 11 A.M. **29.** (b) $(2, 78\frac{2}{3})$ **31.** $a = 4, b = -12, c = 10$

Figure EX 5.1-13

Figure EX 5.1-15

Figure EX 5.1-17

Figure EX 5.1-19

Figure EX 5.1-21

(c, f(c))

Figure EX 5.1-23

Figure EX 5.1-25

EXERCISES 5.2 (Page 185)

(Sketches of the graphs appear in Figs. EX 5.2-1 through EX 5.2-23.)   **1.** $f(-1) = 5$, rel max; $f(1) = -3$, rel min; $(0, 1)$, pt. of infl.; f increasing on $(-\infty, -1]$ and $[1, +\infty)$; f decreasing on $[-1, 1]$; graph concave downward for $x < 0$; graph concave upward for $x > 0$
3. $f(\frac{3}{2}) = -\frac{27}{16}$, rel min; $(0, 0)$, $(1, -1)$, pts. of infl.; f increasing on $[\frac{3}{2}, +\infty)$; f decreasing on $(-\infty, \frac{3}{2}]$; graph concave upward for $x < 0$ and $x > 1$; graph concave downward for $0 < x < 1$ **5.** $f(-3) = 5$, rel max; $f(-\frac{1}{3}) = -\frac{121}{27}$, rel min; $(-\frac{5}{3}, \frac{7}{27})$, pt. of infl.; f increasing on $(-\infty, -3]$ and $[-\frac{1}{3}, +\infty)$; f decreasing on $[-3, -\frac{1}{3}]$; graph concave downward for $x < -\frac{5}{3}$; graph concave upward for $x > -\frac{5}{3}$ **7.** $f(0) = 1$, rel min; $(\frac{1}{2}, \frac{23}{16})$, $(1, 2)$, pts. of infl.; f decreasing on $(-\infty, 0]$; f increasing on $[0, +\infty)$; graph concave upward for $x < \frac{1}{2}$ and $x > 1$; graph concave downward for $\frac{1}{2} < x < 1$ **9.** $f(-1) = \frac{7}{12}$, rel min; $f(0) = 1$, rel max; $f(2) = -\frac{5}{3}$, rel min; pts. of infl. at $x = \frac{1}{3}(1 \pm \sqrt{7})$; f decreasing on $(-\infty, -1]$ and $[0, 2]$; f increasing on $[-1, 0]$ and $[2, +\infty)$; graph concave upward for $x < \frac{1}{3}(1 - \sqrt{7})$ and $x > \frac{1}{3}(1 + \sqrt{7})$; graph concave downward for $\frac{1}{3}(1 - \sqrt{7}) < x < \frac{1}{3}(1 + \sqrt{7})$ **11.** $f(0) = 0$, rel min; no pts. of infl.; f decreasing on $(-\infty, 0]$; f increasing on $[0, +\infty)$; graph concave upward everywhere   **13.** $f(\frac{4}{5}) = \dfrac{26,244}{3,125}$, rel max; $f(2) = 0$, rel min; pts. of infl. at $(-1, 0)$ and $x = \frac{1}{10}(8 \pm 3\sqrt{6})$; f increasing on

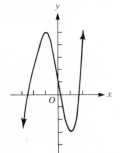

Figure EX 5.2-1

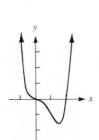

Figure EX 5.2-3

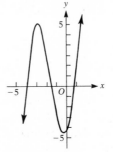

Figure EX 5.2-5

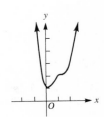

Figure EX 5.2-7

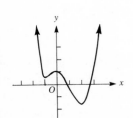

Figure EX 5.2-9

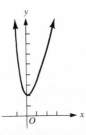

Figure EX 5.2-11

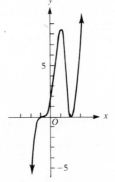

Figure EX 5.2-13

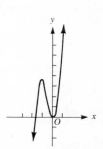

Figure EX 5.2-15

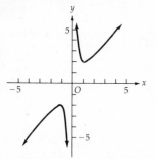

Figure EX 5.2-17

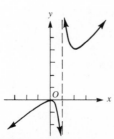

Figure EX 5.2-19

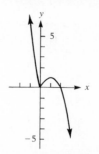

Figure EX 5.2-21

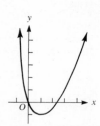

Figure EX 5.2-23

$(-\infty, \frac{4}{5}]$ and $[2, +\infty)$; f increasing on $[\frac{4}{5}, 2]$; graph concave downward for $x < -1$ and $\frac{1}{10}(8 - 3\sqrt{6}) < x < \frac{1}{10}(8 + 3\sqrt{6})$; graph concave upward for $-1 < x < \frac{1}{10}(8 - 3\sqrt{6})$ and $x > \frac{1}{10}(8 + 3\sqrt{6})$ **15.** $f(-\frac{4}{3}) = \frac{256}{81}$, rel max; $f(0) = 0$, rel min; $(-1, 2)$, pt. of infl.; f increasing on $(-\infty, -\frac{4}{3}]$ and $[0, +\infty)$; f decreasing on $[-\frac{4}{3}, 0]$; graph concave downward for $x < -1$; graph concave upward for $x > -1$ **17.** $f(-1) = -2$, rel max; $f(1) = 2$, rel min; no pts. of infl.; f increasing on $(-\infty, -1]$ and $[1, +\infty)$; f decreasing on $[-1, 0)$ and $(0, 1]$; graph concave downward for $x < 0$; graph concave upward for $x > 0$ **19.** $f(0) = 0$, rel max; $f(2) = 4$, rel min; no pts. of infl.; f increasing on $(-\infty, 0]$ and $[2, +\infty)$; f decreasing on $[0, 1)$ and $(1, 2]$; graph concave downward for $x < 1$; graph concave upward for $x > 1$ **21.** $f(0) = 0$, rel min; $f(1) = 1$, rel max; f decreasing on $(-\infty, 0)$ and $[1, +\infty)$; f increasing on $(0, 1]$; graph concave downward for $x < 0$ and $x > 0$ **23.** $f(1) = -1$, rel min; f decreasing on $(-\infty, 1]$; f increasing on $[1, +\infty)$; graph concave upward for all x

EXERCISES 5.3 (Page 191)

1. (a) $Q(x) = x + 4 + \dfrac{8}{x}$; (b) $C'(x) = 2x + 4$; (c) \$9.66   **3.** (a) $Q(x) = 3x - 6 + \dfrac{4}{x}$; (b) $C'(x) = 6x - 6$; (c) $[1, +\infty)$; (d) \$0.93

5. (a) $[2, +\infty)$; (b) $C'(x) = x^2 - 4x + 5$; (c) decreasing on $[0, 2]$ and increasing on $[2, +\infty)$; (d) graph concave downward for $0 < x < 2$, graph concave upward for $x > 2$, pt. of infl. at $(2, \frac{20}{3})$, infl. tangent: $3x - 3y + 14 = 0$ **7.** (a) $C(x) = 3x + 400$; (b) $Q(x) = 3 + \dfrac{400}{x}$;

(c) $C'(x) = 3$; (e) 953   **9.** (a) $R(x) = 100x\sqrt{36 - x^2}$; (b) $R'(x) = \dfrac{3600 - 200x^2}{\sqrt{36 - x^2}}$; (c) \$1800   **11.** (a) $p = 200 - \frac{1}{3}x$;

(b) $R'(x) = 200 - \frac{2}{3}x$; (c) \$30,000   **13.** (a) $px = 30 + 50\sqrt{x + 1}$; (b) $R'(x) = \dfrac{25}{\sqrt{x + 1}}$; (c) \$280

EXERCISES 5.4 (Page 198)

1. (a) 18.2 percent; (b) -1.82; (c) -1.78 **3.** (a) 2.44 percent; (b) -0.488; (c) -0.5 **5.** (a) -0.5; (b) approximate increase of 3 percent   **7.** (a) -1.14; (b) approximate decrease of 3.42 percent   **9.** (a) -2.625; (b) 1.52 percent   **11.** (a) $0 < x < 13\frac{1}{3}$; (b) $x = 13\frac{1}{3}$;

(c) $13\frac{1}{3} < x \leq 20$; (d) $R(x) = \dfrac{x}{2}\sqrt{20 - x}$, $R'(x) = \dfrac{40 - 3x}{4\sqrt{20 - x}}$ **13.** (a) $0 < x < 30$; (b) $x = 30$; (c) $30 < x \leq 45$; (d) $R(x) = 20x\sqrt{45 - x}$,

$R'(x) = \dfrac{900 - 30x}{\sqrt{45 - x}}$ **15.** at any price an increase of 1 percent in the price will cause a decrease of n percent in the demand

EXERCISES 5.5 (Page 204)

1. (a) $P(x) = -2x^2 + 120x - 300$; (c) $R'(x) = 140 - 2x$, $C'(x) = 2x + 20$; (d) \$1500   **3.** 80; \$3400

5. (a) $R'(x) = \dfrac{60\sqrt{x - 100} - 3x + 200}{10\sqrt{x - 100}}$, $C'(x) = 2$; (b) 200   **7.** (a) 1875; (b) 62.5 cents; (c) \$503.12   **9.** (a) 1625; (b) 67.5 cents;

(c) \$328.12   **11.** 37.5 cents   **13.** (a) $x \in [1, 10]$; (b) $R'(x) = \dfrac{6\sqrt{x - 1} - 3x + 2}{\sqrt{x - 1}}$, $C'(x) = 2$; (c) 2   **15.** (a) \$8.67; (b) \$9.18

17. (a) \$49; (b) \$64; (c) $64 - $49 = $15 > $9

REVIEW EXERCISES FOR CHAPTER 5 (Page 205)

1. concave downward for $x < -1$; concave upward for $x > -1$; $(-1, 0)$, pt. of infl. **3.** concave upward for $x < 1$; concave downward for $x > 1$; $(1, 0)$, pt. of infl. **5.** $f(-2) = 0$, rel max; $f(0) = -4$, rel min; $(-1, -2)$, pt. of infl.; f increasing on $(-\infty, -2]$ and $[0, +\infty)$; f decreasing on $[-2, 0]$; graph concave downward for $x < -1$; graph concave upward for $x > -1$ **7.** no relative extrema; $(3, 1)$, pt. of infl.; f increasing on $(-\infty, +\infty)$; graph concave upward for $3 < x < +\infty$; graph concave downward for $-\infty < x < 3$ **9.** $f(0) = 0$, rel max; $f(6) = 12$, rel min; no pts. of infl.; f increasing on $(-\infty, 0]$ and $[6, +\infty)$; f decreasing on $[0, 3)$ and $(3, 6]$; graph concave downward for $x < 3$; graph concave upward for $x > 3$ **11.** (a) $Q(x) = 2x + 4 + \dfrac{32}{x}$; (b) $C'(x) = 4x + 4$; (c) \$20 **13.** (a) $R(x) = x\sqrt{16 - x}$; (b) $R'(x) = \dfrac{32 - 3x}{2\sqrt{16 - x}}$ **15.** (a) $p = 1350 - \frac{1}{2}x^2$; (b) $R'(x) = 1350 - \frac{3}{2}x^2$; (c) \$27,000 **17.** (a) -0.667; (b) approximate decrease of 1.33 percent **19.** (a) $0 < x < 133\frac{1}{3}$; (b) $x = 133\frac{1}{3}$; (c) $133\frac{1}{3} < x \le 200$ **21.** 135 **23.** (a) 2500; (b) \$1.15; (c) \$1375 **25.** 75 cents **27.** \$14.88 **29.** $a = -1$, $b = 6$

EXERCISES 6.1 (Page 214)

3. (a) $2x\,\Delta x + (\Delta x)^2$; (b) $2x\,\Delta x$; (c) $(\Delta x)^2$ **5.** (a) $(-3 - 4x)\,\Delta x - 2(\Delta x)^2$; (b) $(-3 - 4x)\,\Delta x$; (c) $-2(\Delta x)^2$ **7.** (a) $\sqrt{x + \Delta x} - \sqrt{x}$; (b) $\dfrac{\Delta x}{2\sqrt{x}}$; (c) $-\dfrac{(\sqrt{x} - \sqrt{x + \Delta x})^2}{2\sqrt{x}}$ **9.** (a) $\dfrac{-2\,\Delta x}{(x - 1)(x + \Delta x - 1)}$; (b) $\dfrac{-2\,\Delta x}{(x - 1)^2}$; (c) $\dfrac{2(\Delta x)^2}{(x - 1)^2(x + \Delta x - 1)}$ **11.** (a) 0.0309; (b) 0.03; (c) 0.0009 **13.** (a) $\frac{1}{42} \approx 0.0238$; (b) $\frac{1}{40} = 0.025$; (c) $-\frac{1}{84} \approx -0.0012$ **15.** (a) -0.875; (b) -1.5; (c) 0.625 **17.** $3(3x^2 - 2x + 1)^2(6x - 2)\,dx$ **19.** $\dfrac{(14x^2 + 18x)\,dx}{3(2x + 3)^{2/3}}$ **21.** $\dfrac{dx}{(x + 1)^{3/2}(x - 1)^{1/2}}$ **23.** 9.06 **25.** 1.96 **27.** (a) 6.75 cm³; (b) 0.3 cm² **29.** $\frac{12}{5}\pi$ m³ **31.** -0.4π cm² **33.** 0.9π cm³ **35.** 10 ft³ **37.** The cost is increased by approximately 66 cents **39.** The cost is decreased by approximately \$1.69

EXERCISES 6.2 (Page 225)

1. $\frac{3}{5}x^5 + C$ **3.** $3t - t^2 + \frac{1}{3}t^3 + C$ **5.** $\frac{8}{5}x^5 + x^4 - 2x^3 - 2x^2 + 5x + C$ **7.** $\frac{2}{5}x^{5/2} + \frac{2}{3}x^{3/2} + C$ **9.** $\frac{2}{5}x^{5/2} - \frac{1}{2}x^2 + C$ **11.** $-\dfrac{1}{x^2} - \dfrac{3}{x} + 5x + C$ **13.** $\frac{2}{5}x^{5/2} + \frac{8}{3}x^{3/2} - 8x^{1/2} + C$ **15.** $\frac{2}{3}x\sqrt{2x} - \sqrt{2x} + C$ **17.** $-\frac{1}{6}(1 - 4y)^{3/2} + C$ **19.** $-\frac{3}{8}(6 - 2x)^{4/3} + C$ **21.** $\frac{1}{3}(x^2 - 9)^{3/2} + C$ **23.** $\frac{2}{9}(x^3 - 1)^{3/2} + C$ **25.** $-\frac{3}{8}(9 - 4x^2)^{5/3} + C$ **27.** $\dfrac{1}{32(1 - 2y^4)^4} + C$ **29.** $\frac{1}{9}\sqrt{3s^2 + 1} + C$ **31.** $\frac{2}{5}(t + 3)^{3/2} - 6(t + 3)^{1/2} + C$ **33.** $-2\left(1 + \dfrac{1}{3x}\right)^{3/2} + C$ **35.** $-\frac{2}{7}(3 - x)^{7/2} + \frac{12}{5}(3 - x)^{5/2} - 6(3 - x)^{3/2} + C$ **37.** $\frac{2}{3}\sqrt{x^3 + 3x^2 + 1} + C$ **39.** $-\dfrac{1}{4(3x^4 + 2x^2 + 1)} + C$ **41.** $\frac{3}{4}(3 - y)^{4/3} - 18(3 - y)^{1/3} + C$ **43.** $\frac{3}{5}(r^{1/3} + 2)^5 + C$ **45.** $\sqrt{x^2 + 4} + \dfrac{4}{\sqrt{x^2 + 4}} + C$ **47.** (a) $2x^4 + 4x^3 + 3x^2 + x + C$; (b) $\frac{1}{8}(2x + 1)^4 + C$

EXERCISES 6.3 (Page 230)

1. $y = 2x^2 - 5x + C$ **3.** $y = x^3 + x^2 - 7x + C$ **5.** $u = 2v + \dfrac{3}{v} + C$ **7.** $y = (2x - 1)^{3/2} + C$ **9.** $3x^2y + Cy + 2 = 0$ **11.** $2\sqrt{1 + y^2} = 3x^2 + C$ **13.** $12y = 5x^4 + 6x^2 + C_1x + C_2$ **15.** $u = \frac{4}{135}(3v + 1)^{5/2} + C_1v + C_2$ **17.** $3y = x^3 - 3x^2 - 12x + 18$ **19.** $x^2 = 4y^2$ **21.** $12y = x^4 + 6x^3 - 10x + 27$ **23.** $u = 3v^4 + 4v^3 + 2v^2 + 2v$ **25.** $y = x^2 - 3x + 2$ **27.** $3y = -2x^3 + 3x^2 + 2x + 6$ **29.** \$325 **31.** \$3.25 million **33.** (a) 2.5 cm²; (b) 1.5 cm² **35.** $3.1\mu m^3$

EXERCISES 6.4 (Page 234)

1. (a) $R(x) = 12x - \frac{3}{2}x^2$; (b) $p = 12 - \frac{3}{2}x$ **3.** $C(x) = \frac{6}{5}\sqrt{5x + 4} + \frac{38}{5}$ **5.** $p(x + 5) = 4x + 22$ **7.** $C(x) = 3x^2 - 17x + 47$ **9.** (a) $R(x) = 16x - x^3$; (b) $p = 16 - x^2$ **11.** $C(x) = x^3 + 4x^2 + 4x + 6$ **13.** \$3375 **15.** (a) \$2.50; (b) \$3 **17.** $C(x) = x^2 + x + 6$ **19.** \$5

REVIEW EXERCISES FOR CHAPTER 6 (Page 235)

1. $\frac{1}{2}x^4 - \frac{1}{3}x^3 + 3x + C$ **3.** $2y^2 + 4y^{3/2} + C$ **5.** $-\dfrac{2}{3x^3} + \dfrac{5}{x} + C$ **7.** $\frac{5}{9}(2 + 3x^2)^{3/2} + C$ **9.** $\frac{2}{3}x\sqrt{3x} + \frac{2}{5}\sqrt{5x} + C$

11. $\frac{1}{3}\sqrt{x^4 + 2x^2} + C$ **13.** $\frac{1}{420}(4x + 3)^{3/2}(30x^2 - 18x + 79) + C$ **15.** $\frac{1}{2}(y^2 - 1)^{-1} = x^{-1} + C$ **17.** $y + y^{-1} = -x + C$
19. $y = \frac{1}{15}(2x - 1)^{5/2} + C_1x + C_2$ **21.** (a) -0.16; (b) -0.64 **23.** 2π in.^{3} **25.** $x^4 - 24x^2 + 36x + 12y - 1 = 0$
27. $y = \frac{4}{15}(x + 4)^{5/2} + \frac{1}{3}(6 - 32\sqrt{2})x - \frac{1}{15}(75 - 128\sqrt{2})$ **29.** $C(x) = \frac{8}{3}\sqrt{x + 1} + \frac{28}{3}$ **31.** (a) $R(x) = \frac{1}{3}x^3 - 8x^2 + 48x$;
(b) $3p = x^2 - 24x + 144$ **33.** \$850   **35.** 4738 units   **37.** \$600   **39.** 1.46 cm^3 **41.** (a) $V = \frac{2}{3}(t + 1)^{3/2} + \frac{1}{3}t^2 + \frac{74}{3}$;
(b) 64 cm^3

EXERCISES 7.1 (Page 249)

1. 51   **3.** $\frac{73}{12}$ **5.** $\frac{63}{4}$ **7.** $\frac{7}{12}$ **9.** 10,880   **11.** 10,400   **13.** $\frac{1}{4}(n^4 + 2n^3 + 3n^2 + 22n)$ **15.** $n^4 - \frac{2}{3}n^3 - 3n^2 - \frac{1}{3}n$
17. 0   **19.** $\frac{2}{5}$ **21.** $-\frac{2}{5}$ **23.** $\frac{7}{3}$ **25.** 0   **27.** $+\infty$ **29.** $\frac{1}{2}$ **31.** $+\infty$ **33.** $\frac{8}{3}$ **35.** $\frac{15}{4}$

EXERCISES 7.2 (Page 260)

1. $\frac{8}{3}$ sq units   **3.** 5 sq units   **5.** $\frac{5}{3}$ sq units   **7.** 9 sq units   **9.** $\frac{3}{2}$ sq units   **11.** $\frac{17}{4}$ sq units   **13.** $\frac{27}{4}$ sq units
15. $\frac{1}{2}m(b^2 - a^2)$ sq units   **17.** $\frac{1}{2}h(b_1 + b_2)$ sq units   **19.** 9 sq units   **21.** 15 sq units   **23.** 9 sq units   **25.** (b) 492
27. (b) 1650 units

EXERCISES 7.3 (Page 270)

1. $\frac{459}{64}$ **3.** $\frac{533}{495}$ **5.** 0.835   **7.** $\frac{1}{3}$ **9.** $\frac{15}{4}$ **11.** 66   **13.** 12   **15.** 36   **17.** $\frac{3}{2}$ **19.** $\frac{3}{16}$ **21.** $\frac{134}{3}$ **23.** -8
25. $\frac{2}{9}(27 - 2\sqrt{2})$ **27.** $2 - \sqrt[3]{2}$ **29.** $\frac{104}{5}$ **31.** $\frac{256}{15}$ **33.** $\frac{6215}{12}$ **35.** $\frac{32}{3}$ sq units   **37.** $\frac{22}{3}$ sq units   **39.** $\frac{52}{3}$ sq units

EXERCISES 7.4 (Page 279)

1. 492   **3.** 1650 units   **5.** 375   **7.** (a) \$96,000; (b) 5 years   **9.** \$50,000   **11.** (a) 1764; (b) 412   **13.** 10th
15. \$57.60   **17.** (a) \$5.40   **21.** $\frac{32}{3}$; $\frac{4}{3}(3 - \sqrt{3})$ **23.** 4800   **25.** 79 words per minute   **27.** \$2.82   **29.** (b) 0.1171
31. 0.2186

EXERCISES 7.5 (Page 288)

1. $\frac{2}{3}$ sq units   **3.** $\frac{1}{6}$ sq units   **5.** $\frac{1}{6}$ sq units   **17.** $\frac{32}{3}$ sq units   **9.** $\frac{32}{3}$ sq units   **11.** $\frac{12}{5}$ sq units   **13.** $\frac{9}{2}$ sq units
15. 72 sq units   **17.** $\frac{64}{3}$ sq units   **19.** $\frac{5}{12}$ sq units   **21.** 12 sq units   **23.** (b) 10; (c) \$80,000

EXERCISES 7.6 (Page 295)

1. \$93.33   **3.** \$120   **5.** \$267   **7.** \$267   **9.** \$360   **11.** \$360   **13.** \$200, consumers' surplus; \$25, producers' surplus
15. \$270, consumers' surplus; \$120, producers' surplus   **17.** \$178

EXERCISES 7.7 (Page 303)

1. $\sqrt{4 + x^2}$ **3.** $\dfrac{1}{\sqrt{x^2 - 1}}$ **5.** $-\dfrac{1}{3 + x^2}$ **7.** $\frac{5}{2}$ **9.** $\frac{29}{2}$ **11.** $\frac{2}{3}\sqrt{2}$ **13.** 40   **15.** $\frac{5}{2}$ sq units   **17.** $25\frac{5}{6}$ sq units
19. $\frac{7}{3}$ sq units   **21.** \$3813.33

REVIEW EXERCISES FOR CHAPTER 7 (Page 304)

1. 3230   **3.** $\frac{1}{2}$ **5.** $+\infty$ **7.** 6   **9.** $-\frac{3}{4}$ **11.** $\frac{4}{3}$ **13.** $\frac{52}{9}$ **15.** 18 sq units   **17.** $\frac{224}{3}$ sq units   **19.** $\frac{1}{3}(40\sqrt{5} - 20)$ sq

units   **21.** $\frac{1}{6}$ sq units   **23.** 36 sq units   **25.** $\frac{256}{15}$ sq units   **27.** $\dfrac{42,304}{175}$ **29.** \$1566.67   **31.** 1190 units   **33.** 20,800

people   **35.** 141   **37.** \$18   **39.** \$1733.33   **41.** \$4, consumers' surplus; \$5.33, producers' surplus

EXERCISES 8.1 (Page 315)

1. (a) \$2060; (b) \$2120; (c) \$2240   **3.** (a) \$1320; (b) \$1360.49; (c) \$1368.57; (d) \$1372.79   **5.** (a) \$620; (b) \$627.20; (c) \$631.24;
(d) \$634.87   **7.** (a) \$849.29; (b) 16.99 percent   **9.** (a) \$867.55; (b) 17.35 percent   **11.** (a) 25.44 percent; (b) 26.82 percent; 27.12

percent   **13.** (a) \$2376.28; (b) \$2361.30   **15.** (a) \$1094.17; (b) 9.42 percent   **17.** (a) \$4631.93; (b) \$4493.29   **19.** \$1750   **21.** \$1120   **23.** 2.7183

EXERCISES 8.2 (Page 325)

19. (a) y^4; (b) y^{-12}; (c) y^8; (d) y^4 **21.** (a) $2^{5\sqrt{3}}$; (b) e^{18} **23.** (a) $f(t) = 200 \cdot 2^{t/10}$; (b) \$12,800   **25.** 22.3 g   **27.** 64,000   **29.** (b) 40; (c) 82; (d) 100   **31.** (a) 144; (b) 4075; (c) 5000   **33.** 50 percent

EXERCISES 8.3 (Page 335)

7. (a) 2; (b) 3; (c) $\frac{1}{2}$; (d) -6 **9.** (a) 0; (b) 1; (c) 2; (d) -1 **11.** (a) -4; (b) $-\frac{4}{3}$; (c) -4; (d) $-\frac{3}{4}$ **13.** (a) $\log_b x + \log_b y + \log_b z$; (b) $4 \ln x + \ln y - 2 \ln z$ **15.** (a) $\frac{1}{3}(\log_b y + 2 \log_b z)$; (b) $\frac{1}{5}(3 \ln x + 4 \ln y - 2 \ln z)$ **17.** (a) $\log_{10} x^4 \sqrt{y}$; (b) $\ln \frac{x^5 \sqrt{y}}{\sqrt[3]{z}}$

19. (a) $\log_b \frac{z\sqrt[3]{x^2}}{by^4}$; (b) $\ln \frac{1}{3}\pi r^2 h$ **21.** (b) \$44   **23.** 6.531 years   **25.** 6.821 years   **27.** 8.664 years   **29.** (a) 9.006 years; (b) 8.751 years   **31.** 17.33   **33.** 1987   **35.** 15.35   **37.** 9   **39.** 6   **41.** 0.4343   **43.** 3.3219

EXERCISES 8.4 (Page 343)

1. (a) $\frac{5}{4 + 5x}$; (b) $\frac{5}{8 + 10x}$ **3.** (a) $\frac{6}{3t + 1}$; (b) $\frac{6 \ln (3t + 1)}{3t + 1}$ **5.** (a) $\frac{12x - 4}{3x^2 - 2x + 1}$; (b) $\frac{(6x - 2) \log_2 e}{3x^2 - 2x + 1}$ **7.** $\frac{\log_{10} e}{t(t + 1)}$

9. $\frac{\log_b e}{2x \sqrt{\log_b x}}$ **11.** $\frac{1 - 2w - w^2}{3(w + 1)(w^2 + 1)}$ **13.** $\frac{1}{y \ln y}$ **15.** $\frac{3x^2}{x^3 + 1}$ **17.** $2x(x + 1)^6(x - 1)^2(6x^2 - 2x - 1)$

19. $\frac{8x^9 - 4x^7 + 15x^2 + 10}{5(x^7 + 1)^{6/5}}$ **21.** $-\frac{xy + y}{xy + x}$ **23.** $-\frac{1}{2} \ln |3 - 2x| + C$ **25.** $\frac{3}{2} \ln (x^2 + 4) + C$ **27.** $\frac{1}{5} \ln |5x^3 - 1| + C$

29. $x^2 + 4 \ln |x^2 - 4| + C$ **31.** $\ln 5$ **33.** $\frac{1}{\ln 4}$ **35.** $px = 12 \ln |x + 2| - 12 \ln 2$ **37.** (a) -0.5; (b) decrease of 5 percent   **39.** $(42 \ln 6 - 35)$ dollars $= \$40.25$ **41.** \$154   **43.** $\ln \left| \ln \frac{a}{y} \right| + kt = C$

EXERCISES 8.5 (Page 349)

1. (a) $5e^{5x}$; (b) $(5 \ln 2)2^{5x}$ **3.** (a) $-6xe^{-3x^2}$; (b) $(-6x \ln b)b^{-3x^2}$ **5.** $\frac{e^t(t - 1)}{t^2}$ **7.** $\frac{4}{(e^x + e^{-x})^2}$ **9.** $2 \ln 10(x - 1)10^{x^2 - 2x}$

11. $\frac{e^{2x} - 1}{e^{2x} + 1}$ **13.** $\frac{e^x(e^y - 1)}{e^y(1 - e^x)}$ **15.** (a) $\frac{1}{2}e^{2x} + C$; (b) $\frac{3^{2x}}{2 \ln 3} + C$ **17.** (a) $-\frac{1}{5}e^{2 - 5x} + C$; (b) $-\frac{10^{2 - 5x}}{5 \ln 10} + C$ **19.** $\frac{1}{6}e^{2x^3} + C$

21. $e^x - e^{-x} + C$ **23.** $\frac{1}{6(1 - 2e^{3x})} + C$ **25.** $\frac{3^t e^t}{\ln 3 + 1} + C$ **27.** $\frac{5^{x^4 + 2x}}{2 \ln 5} + C$ **29.** $\frac{1}{2}(e^3 - e^{-3})$ **31.** $\frac{1}{2}(e^4 - 1)$

33. $\frac{1}{2}x^{\sqrt{x} - 1/2}(2 + \ln x)$ **35.** $(2 \ln x)x^{\ln x - 1}$ **37.** $x^{x^x + x}\left(\ln^2 x + \ln x + \frac{1}{x}\right)$ **39.** 1.63   **41.** 31.54   **43.** \$887 per year

45. (a) 61 sales per day; (b) 2.3 sales per day   **47.** $(9 - \ln 10)$ dollars $= \$6.70$

EXERCISES 8.6 (Page 356)

1. 68.4 years   **3.** 38,720   **5.** 800   **7.** $\left(10 \frac{\ln 3}{\ln 2}\right)$ years $= 15.85$ years   **9.** 12.21 percent   **11.** \$2734   **13.** 16,000

15. 1389   **17.** 43.9 g   **19.** (a) 96 percent; (b) 66 percent   **21.** 11.6 kg   **23.** 15,327 years ago

EXERCISES 8.7 (Page 364)

1. (a) 34.7; (b) 55.5   **3.** 69.9   **5.** (a) 102 sec; (b) 42.1°   **7.** 0.2134   **9.** (a) $Q(x) = \frac{40\, e^{x/4}}{x}$; (b) $C'(x) = 10\, e^{x/4}$; (c) \$27.18

11. (a) \$10,000; (b) \$33,834   **13.** 3 years   **15.** 2 years and 10 months

EXERCISES 8.8 (Page 373)

1. $1842.84 **3.** $331.11 **5.** $185.84 **7.** $75,059.33 **9.** $4739.94 **11.** $791.26 **13.** no, because the annual payment of $52,759 exceeds the yearly income **15.** yes **17.** $1885.64 **19.** 738.80 **21.** $4843.91 **23.** (a) $a\left(\dfrac{1 - e^{-(i+b)T}}{i + b}\right)$

REVIEW EXERCISES FOR CHAPTER 8 (Page 374)

1. $\dfrac{5}{5x + 3}$ **3.** $\dfrac{x}{x^2 + 1}$ **5.** $\dfrac{2}{x} \log_{10} e$ **7.** $2re^{r^2}$ **9.** $2^t t(t \ln 2 + 2)$ **11.** $\dfrac{2}{(e^x + e^{-x})^2}$ **13.** $2x^{2x} (\ln x + 1)$

15. $\frac{3}{2} \ln (1 + e^{2x}) + C$ **17.** $\dfrac{1}{3}\left(e^{3t} + \dfrac{2^{3t}}{\ln 2}\right) + C$ **19.** $\frac{1}{12} \ln (1 + e^{6x^2}) + C$ **21.** $\frac{1}{3}(e^8 - 1)$ **23.** $\frac{3}{2} \ln 2$

25. $-\dfrac{ye^x + e^y + 1}{e^x + xe^y + 1}$ **27.** (a) 16.64 percent; (b) 16.99 percent; (c) 17.35 percent **29.** $14,000 **31.** (a) 5.86 years; (b) 5.78 years **33.** (a) 57.65 mg; (b) 455.8 years **35.** (a) 16; (b) 1686; (c) 9610; (d) 10,000 **37.** (a) $R(x) = 20xe^{-x/200}$;

(b) $R'(x) = 20e^{-x/200}\left(1 - \dfrac{x}{200}\right)$; (c) $1471.52 **39.** (a) 0.148; (b) 0.551; (c) 0.699 **41.** 8212 years **43.** 187,500

45. 8.63 min **47.** (a) 32; (b) 43.5; (c) 3.83 **49.** (a) $5717.94; (b) $524.66 **51.** $6127.70 **53.** 73.7°

EXERCISES 9.1 (Page 383)

1. $\frac{1}{3}xe^{3x} - \frac{1}{9}e^{3x} + C$ **3.** $\dfrac{3^x(x \ln 3 - 1)}{(\ln 3)^2} + C$ **5.** $x \ln x - x + C$ **7.** $\dfrac{x^2(2 \ln x - 1)}{4 \ln 10} + C$ **9.** $\frac{1}{9}x^3(3 \ln x - 1) + C$

11. $x \ln^2 x - 2x \ln x + 2x + C$ **13.** $\dfrac{a^x(x \ln a - 1)}{(\ln a)^2} + C$ **15.** $-\dfrac{e^{-2x}}{4}(2x^2 + 2x + 1) + C$ **17.** $\dfrac{e^x}{x + 1} + C$

19. $-\dfrac{1}{x}(\ln x + 1) + C$ **21.** $-x^2\sqrt{1 - x^2} - \frac{2}{3}(1 - x^2)^{3/2} + C$ **23.** $\frac{2}{9}e^{3\sqrt{x}}(3\sqrt{x} - 1) + C$ **25.** $\dfrac{36}{\ln 3} - \dfrac{36}{\ln^2 3} + \dfrac{16}{\ln^3 3}$

27. $8 \ln 2 - 3$ **29.** $(e^2 + 1)$ sq units **31.** 105 **33.** (a) $(10 \ln \frac{10}{7} - 3)$ dollars = $0.57; (b) $(3 - \frac{1}{2} \ln 7)$ dollars = $2.03
35. $1542 **37.** $1502 **39.** (a) 0.50; (b) 0.11

EXERCISES 9.2 (Page 396)

1. $\frac{1}{4} \ln \left|\dfrac{x - 2}{x + 2}\right| + C$ **3.** $x + \frac{1}{5} \ln \left|\dfrac{C(x - 2)^4}{(x + 3)^9}\right|$ **5.** $\ln \left|\dfrac{C(w + 4)^3}{2w - 1}\right|$ **7.** $\frac{1}{4} \ln \left|\dfrac{Cx^4(2x + 1)^3}{2x - 1}\right|$

9. $\frac{1}{9} \ln \left|\dfrac{x + 3}{x}\right| - \dfrac{1}{3x} + C$ **11.** $2 \ln \left|\dfrac{x + 1}{x}\right| - \dfrac{1}{x} - \dfrac{1}{x + 1} + C$ **13.** $\dfrac{3}{x + 1} + \ln |x + 1| - \frac{1}{2} \ln |2x + 3| + C$

15. $\dfrac{5}{16(z + 2)} - \dfrac{7}{16(z - 2)} + \frac{1}{32} \ln \left|\dfrac{z + 2}{z - 2}\right| + C$ **17.** $4 \ln \frac{4}{3} - \frac{3}{2}$ **19.** $\ln \frac{27}{4} - 2$ **21.** $13 \ln 2 - 4 \ln 5$
23. $(2 - \ln 3)$ sq units **25.** (a) 3000; (b) 16.8 days **27.** (a) 2848; (b) 3.55 hours **29.** $\frac{31}{19}$ **31.** 7.4 lb

EXERCISES 9.3 (Page 408)

1. approx: 0.695; exact: $\ln 2 \approx 0.693$ **3.** approx: 4.250; exact: 4 **5.** 1.106 **7.** 0.881 **9.** 3.689
11. $-0.007 \le \epsilon_T \le -0.001$ **13.** $-0.5 \le \epsilon_T \le 0$ **15.** $0 \le \epsilon_T \le 0.072$ **17.** approx: 4.000; exact: 4 **19.** approx: 0.693; exact:
$\ln 2 \approx 0.693$ **21.** 1.107 **23.** 0 **25.** $-0.0005 \le \epsilon_s \le 0$ **27.** 1.569 **29.** 1.402 **31.** 3.090 **33.** (a) 0.3401;
(b) 0.3414 **35.** $24.78 **37.** $29.79 **39.** 0.9436

EXERCISES 9.4 (Page 414)

1. $\frac{1}{3}x - \frac{2}{9} \ln |2 + 3x| + C$ **3.** $x + \dfrac{36}{6 - x} + 12 \ln |6 - x| + C$ **5.** $\frac{1}{30}(3x - 1)(1 + 2x)^{3/2} + C$

7. $2\sqrt{1 + 2x} + \ln \left|\dfrac{\sqrt{1 + 2x} - 1}{\sqrt{1 + 2x} + 1}\right| + C$ **9.** $\frac{1}{4} \ln \left|\dfrac{x + 2}{x - 2}\right| + C$ **11.** $\ln |x + 3 + \sqrt{x^2 + 6x}| + C$

13. $\sqrt{9x^2 + 4} - 2 \ln \left|\dfrac{2 + \sqrt{9x^2 + 4}}{3x}\right| + C$ **15.** $\sqrt{9 - 4x^2} - 3 \ln \left|\dfrac{3 + \sqrt{9 - 4x^2}}{2x}\right| + C$

17. $e^x(x^4 - 4x^3 + 12x^2 - 24x + 24) + C$ **19.** $\dfrac{e^{4x}}{32}(8x^2 - 4x + 1) + C$ **21.** $\dfrac{x^4}{16}(4\ln 3x - 1) + C$ **23.** $\frac{2}{5}\ln 2$

25. $\frac{15}{2} - 8\ln 2$ **27.** $\frac{32}{5}\ln 2 - \frac{31}{25}$ **29.** $\frac{15}{2} - 8\ln 2$

EXERCISES 9.5 (Page 421)

1. e **3.** 2   **5.** divergent   **7.** 2   **9.** 6   **11.** 1   **13.** 0   **15.** divergent   **17.** divergent   **19.** 2   **21.** $\frac{3}{2}$

23. 0.223   **25.** possible; $\frac{1}{2}\pi$ **27.** possible; 2   **29.** $\dfrac{1000}{0.08 + \ln 2}$ dollars = \$1293.41   **33.** (a) does not exist; (b) 0

REVIEW EXERCISES FOR CHAPTER 9 (Page 422)

1. $-\frac{1}{16}e^{-4x}(4x + 1) + C$ **3.** $-2\sqrt{4 - e^x} + C$ **5.** $\ln|x^5 - x^3| + C$ **7.** $2t + \ln\dfrac{t^2}{(t + 2)^{10}} - \dfrac{15}{t + 2} + C$

9. $x\ln(x + 1) - x + \ln(x + 1) + C$ **11.** $2(\ln 2 - 1)^2$ **13.** $\frac{1}{6} + \ln\frac{3}{2}$ **15.** $\frac{256}{15}$ **17.** $\frac{1}{2} + 2\ln\frac{6}{5}$ **19.** divergent

21. $\frac{1}{2}$ **23.** divergent   **25.** $\dfrac{2}{\ln 3}$ **27.** 2.977   **29.** 2.958   **31.** $\dfrac{x}{2}\sqrt{9x^2 + 1} + \frac{1}{6}\ln|3x + \sqrt{9x^2 + 1}| + C$

33. $\frac{255}{4} - 32\ln 2$ **35.** \$2754.74   **37.** (a) 4824; (b) 8.18 weeks   **39.** (a) $x = 300\left(\dfrac{18^t - 17^t}{3\cdot18^t - 2\cdot17^t}\right)$; (b) 35.94 kg   **41.** 0.202

43. 152,500

EXERCISES 10.1 (Page 431)

1. (b) (7, 2, 0), (0, 0, 3), (0, 2, 0), (0, 2, 3), (7, 0, 3), (7, 0, 0)   **3.** (b) (2, 1, 2), (−1, 3, 2), (−1, 1, 5), (2, 3, 2), (−1, 3, 5), (2, 1, 5)   **5.** (b) (3, −1, 0), (3, 3, 0), (1, 3, 0), (1, 3, 5), (1, −1, 5), (3, −1, 5)   **7.** (0, 0, 0), (14, 0, 0), (14, 16, 0), (0, 16, 0), (14, 0, 10), (14, 16, 10), (0, 16, 10), (0, 0, 10)   **17.** the z axis

EXERCISES 10.2 (Page 440)

1. (a) 2; (b) −13; (c) $3a + 2b - 6$; (d) $3x + 2y - 6$; (e) $6x + 6y - 5$; (f) $3h$; (g) $2h$ **3.** (a) $-\frac{1}{7}$; (b) $\dfrac{x^2 + y^2}{x^2 - y^2}$; (c) $\dfrac{x^2 + 2xy + y^2}{x^2 - 2xy + y^2}$; (d) 0; (e) the set of all points (x, y) in R^2 except those on the line $x = y$ **5.** (a) 1; (b) $\frac{1}{2}\sqrt{16 - 4a^2 - 16b^2 - c^2}$; (c) $\sqrt{4 - x^2 - 2y^2}$; (d) $4x + 4y + 8$; (e) $\{(x, y, z)\,|\,x^2 + y^2 + z^2 \le 4\}$; (f) [0, 2]   **7.** domain: the set of all points (x, y) in R^2 interior to and on the circumference of the circle $x^2 + y^2 = 16$ **9.** domain: the set of all points (x, y) in R^2 exterior to and on the circumference of the circle $x^2 + y^2 = 16$ **11.** domain: the set of all points (x, y) in R^2 for which $x^2 > 4y$ **13.** domain: the set of all points (x, y) in R^2 for which $xy > 1$ **15.** domain: $\{(x, y, z)\,|\,z \ne 0\}$; range: $[0, +\infty)$ **25.** complementary   **27.** neither   **29.** substitutes   **31.** $f(p, q) = q/p$; $g(p, q) = p^2/q$

EXERCISES 10.3 (Page 448)

1. 6   **3.** $3x - 2y$ **5.** $\dfrac{1}{\sqrt{2x + 3y}}$ **7.** $2\pi rh$ **9.** $-2x^2 + 6xy - 4y^3$ **11.** $1 + \dfrac{t}{\sqrt{s^2 + t^2}}$ **13.** $-\dfrac{y^2}{e^2}e^{y/x}$

15. $x^2 - 6xy + 2z$ **17.** $\dfrac{-z}{(x^2 + y^2 + z^2)^{3/2}}$ **19.** $rte^{rst} + \dfrac{1}{s}$ **21.** $xy + yt + zt$ **27.** (a) $12x - 6y$; (b) $2x$; (c) $-6x + 2y$

29. (a) $\dfrac{4y}{(x - y)^3}$; (b) $\dfrac{4x}{(x - y)^3}$; (c) $\dfrac{-2(x + y)}{(x - y)^3}$ **31.** (a) $-\dfrac{y}{x^2}$; (b) $-\dfrac{x}{y^2}$; (c) $\dfrac{x + y}{xy}$ **33.** (a) 0; (b) e^y **35.** (a) $6r + 4s$;

(b) $4r - 6s$ **37.** 4   **39.** (a) 375; (b) 40; (c) 70   **41.** (a) $\dfrac{100}{i^2}\left[\dfrac{1 + 9i}{(1 + i)^9} - 1\right]$; (b) decrease of \$19.42; (c) $1000(1.1)^{-t}\ln(1.1)$;

(d) decrease of \$44.46   **43.** (a) −4 deg/cm; (b) −8 deg/cm   **45.** (a) 0.0943 m²/kg; (b) 6.42 m²/m

EXERCISES 10.4 (Page 455)

1. $\dfrac{\partial x}{\partial p} = -1$, $\dfrac{\partial x}{\partial q} = -2$, $\dfrac{\partial y}{\partial p} = -2$, $\dfrac{\partial y}{\partial q} = -1$; complementary   **3.** $\dfrac{\partial x}{\partial p} = -0.4p^{-1.4}q^{0.5}$, $\dfrac{\partial x}{\partial q} = 0.5p^{-0.4}q^{-0.5}$, $\dfrac{\partial y}{\partial p} = 0.4p^{-0.6}q^{-1.5}$,

$\dfrac{\partial y}{\partial q} = -1.5p^{0.4}q^{-2.5}$; substitutes **5.** $\dfrac{\partial x}{\partial p} = -2^{-p-q}\ln 2$, $\dfrac{\partial x}{\partial q} = -2^{-p-q}\ln 2$, $\dfrac{\partial y}{\partial p} = -q3^{-pq}\ln 3$, $\dfrac{\partial y}{\partial q} = -p3^{-pq}\ln 3$; complementary

7. $\dfrac{\partial x}{\partial p} = -\dfrac{q^2}{p^2}$, $\dfrac{\partial x}{\partial q} = \dfrac{2q}{p}$, $\dfrac{\partial y}{\partial p} = \dfrac{1}{q}$, $\dfrac{\partial y}{\partial q} = -\dfrac{p}{q^2}$; substitutes **9.** (a) There is a decrease of 4 units in the demand of the first commodity and a decrease of 2 units in the demand of the second commodity; (b) there is a decrease of 3 units in the demand of the first commodity and a decrease of 1 unit in the demand of the second commodity; (c) there is an increase of 4 units in the demand of the first commodity and an increase of 2 units in the demand of the second commodity; (d) there is an increase of 3 units in the demand of the first commodity and an increase of 1 unit in the demand of the second commodity. **11.** (a) $\dfrac{Ex}{Ep} = -\dfrac{p}{x}$, $\dfrac{Ex}{Eq} = -\dfrac{2q}{x}$, $\dfrac{Ey}{Ep} = -\dfrac{2p}{y}$, $\dfrac{Ey}{Eq} = -\dfrac{q}{y}$; (b) x decreases by 0.11 percent and y decreases by 0.15 percent; (c) x decreases by 0.44 percent and y decreases by 0.15 percent; (d) x increases by 0.11 percent and y increases by 0.15 percent; (e) x increases by 0.44 percent and y increases by 0.15 percent

13. (a) $\dfrac{Ex}{Ep} = -0.4$, $\dfrac{Ex}{Eq} = 0.5$, $\dfrac{Ey}{Ep} = 0.4$, $\dfrac{Ey}{Eq} = -1.5$; (b) x decreases by 0.4 percent and y increases by 0.4 percent; (c) x increases by 0.5 percent and y decreases by 1.5 percent; (d) x increases by 0.4 percent and y decreases by 0.4 percent; (e) x decreases by 0.5 percent and y increases by 1.5 percent **15.** (a) $\dfrac{Ex}{Ep} = -p\ln 2$, $\dfrac{Ex}{Eq} = -q\ln 2$, $\dfrac{Ey}{Ep} = -pq\ln 3$, $\dfrac{Ey}{Eq} = -pq\ln 3$; (b) x decreases by 0.69 percent and y decreases by 2.20 percent; (c) x decreases by 1.39 percent and y decreases by 2.20 percent; (d) x increases by 0.69 percent and y increases by 2.20 percent; (e) x increases by 1.39 percent and y increases by 2.20 percent **17.** (a) $\dfrac{Ex}{Ep} = -1$, $\dfrac{Ex}{Eq} = 2$, $\dfrac{Ey}{Ep} = 1$, $\dfrac{Ey}{Eq} = -1$; (b) x decreases by 1 percent and y increases by 1 percent; (c) x increases by 2 percent and y decreases by 1 percent; (d) x increases by 1 percent and y decreases by 1 percent; (e) x decreases by 2 percent and y increases by 1 percent **19.** (a) 160 units of A and 220 units of B; (b) $\dfrac{\partial x}{\partial p} = -16$, $\dfrac{\partial x}{\partial q} = 60$, $\dfrac{\partial y}{\partial p} = 92$, $\dfrac{\partial y}{\partial q} = -60$; (c) the amount demanded of A is decreased by 16 units, and the amount demanded of B is increased by 92 units; (d) the amount demanded of A is increased by 60 units; and the amount demanded of B is decreased by 60 units. **23.** (b) $\dfrac{Ex}{Ep} = -0.5$, $\dfrac{Ex}{Eq} = -0.2$, $\dfrac{Ey}{Ep} = -1.3$, $\dfrac{Ey}{Eq} = -0.8$; (c) the demand for neckties decreases by 0.5 percent and the demand for dress shirts decreases by 1.3 percent; (d) the demand for neckties decreases by 0.2 percent and the demand for dress shirts decreases by 0.8 percent.

EXERCISES 10.5 (Page 462)

1. 1 **3.** 2 **5.** 4 **7.** 0 **9.** 0 **11.** e^2 **13.** -1 **15.** continuous at all points in R^2 **17.** continuous at all points in R^2 except those on the lines $y = x$ and $y = -x$ **19.** continuous at all points (x, y) in R^2 for which $x > 0$ and $y > 0$
21. $f(2, 1) = 5$, rel min **23.** $f(4, -\frac{1}{2}) = -\frac{133}{4}$, rel min **25.** $f(-\frac{1}{4}, 16) = -12$, rel max **27.** no relative extrema
29. $f(1, 1) = 4$, rel min; $f(-1, -1) = 4$, rel min **31.** no relative extrema

EXERCISES 10.6 (Page 472)

1. 8, 8, 8 **3.** 3 machine-hours and 9 person-hours **5.** length of the base is $2\frac{2}{3}$ ft; width of the base is 2 ft; depth is 3 ft **7.** 250 units of the first commodity at $7.50 per unit and 300 units of the second commodity at $11.50 per unit **13.** $6400 **15.** 24 class A, 8 class B
17. $\frac{1}{3}c$ mg of drug A and $\frac{1}{2}c$ mg of drug B **19.** $l:w:h = 1:1:\frac{1}{2}$

EXERCISES 10.7 (Page 479)

1. $(0, 0)$ and $(0, 4)$ **3.** $(\frac{6}{7}, -\frac{4}{7}, \frac{2}{7})$ **5.** $f(\pm\frac{1}{2}\sqrt{35}, \frac{1}{2}) = \frac{37}{4}$, rel max; $f(0, -3) = -3$, rel min; $f(0, 3) = 3$, rel min
7. $f(-\frac{2}{3}\sqrt{3}, -\frac{1}{3}\sqrt{6}, -\frac{1}{3}\sqrt{3}) = f(-\frac{2}{3}\sqrt{3}, \frac{1}{3}\sqrt{6}, \frac{1}{3}\sqrt{3}) = f(\frac{2}{3}\sqrt{3}, -\frac{1}{3}\sqrt{6}, \frac{1}{3}\sqrt{3}) = f(\frac{2}{3}\sqrt{3}, \frac{1}{3}\sqrt{6}, -\frac{1}{3}\sqrt{3}) = -\frac{2}{9}\sqrt{6}$, rel min;
$f(\frac{2}{3}\sqrt{3}, \frac{1}{3}\sqrt{6}, \frac{1}{3}\sqrt{3}) = f(-\frac{2}{3}\sqrt{3}, -\frac{1}{3}\sqrt{6}, \frac{1}{3}\sqrt{3}) = f(-\frac{2}{3}\sqrt{3}, \frac{1}{3}\sqrt{6}, -\frac{1}{3}\sqrt{3}) = f(\frac{2}{3}\sqrt{3}, -\frac{1}{3}\sqrt{6}, -\frac{1}{3}\sqrt{3}) = \frac{2}{9}\sqrt{6}$, rel max **9.** 3 **11.** $3\sqrt{3}$
13. $\frac{37}{13}$ **15.** $\frac{7}{32}$ **25.** hottest at $(\pm\frac{1}{2}\sqrt{3}, -\frac{1}{2})$; coldest at $(0, \frac{1}{2})$ **27.** $\frac{45}{2}$ units of A, $\frac{15}{2}$ units of B, $\frac{45}{8}$ units of C **29.** 15 units of A, 10 units of B, $\frac{15}{2}$ units of C, 30 units of D, 6 units of E

EXERCISES 10.8 (Page 487)

1. $3x - 6y - 10 = 0$ **3.** $2x - 10y + 17 = 0$ **5.** $21x + 10y - 16 = 0$ **7.** $25,400   **9.** (a) 3120; (b) 9
11. (a) $81x - 284y - 270 = 0$; (b) 11 days   **13.** (a) $y = 0.1560x + 314.9$; (b) 368.0 milliliters per minute per kilogram
15. (a) $17,070x - 1,743y - 375,900 = 0$; (b) 2722 kilograms per hectare   **17.** $8.94

REVIEW EXERCISES FOR CHAPTER 10 (Page 490)

1. a point in R^1, a line in R^2, a plane in R^3 **5.** the set of all points (x, y) in R^2 exterior to and on the circumference of the circle $x^2 + y^2 = 1$ **7.** the set of all points (x, y) in R^2 for which $xy > 4$ **11.** (a) $4xy - 3y^2 + 4$; (b) $2x^2 - 6xy - 2$; (c) $4y$; (d) $-6x$;

(e) $4x - 6y$; (f) $4x - 6y$ **13.** (a) $e^{2rs}(1 + 2rs)$; (b) $2r^2e^{2rs}$; (c) $4re^{2rs}(1 + rs)$; (d) $4re^{2rs}(1 + rs)$ **15.** (a) $\dfrac{x}{x^2 + y^2}$; (b) $\dfrac{y^2 - x^2}{(x^2 + y^2)^2}$;

(c) $\dfrac{-2xy}{(x^2 + y^2)^2}$ **17.** (a) $\dfrac{y^2 + z^2 - x^2}{(x^2 + y^2 + z^2)^2}$; (b) $\dfrac{-2xy}{(x^2 + y^2 + z^2)^2}$; (c) $\dfrac{-x}{(x^2 + y^2 + z^2)^2}$ **19.** (a) $4st^3e^{4rst}$; (b) $4st^2e^{4rst}(3 + 4rst)$;

(c) $4t^2e^{4rst}(3 + 20rst + 16r^2s^2t^2)$ **25.** 2   **27.** continuous at all points in R^2 that are exterior to the circle $x^2 + y^2 = 4$ and interior to the

circle $x^2 + y^2 = 16$ **29.** (a) 1137; (b) 78; (c) 138   **31.** $\dfrac{\partial x}{\partial p} = -3$, $\dfrac{\partial x}{\partial q} = 2$, $\dfrac{\partial y}{\partial p} = 3$, $\dfrac{\partial y}{\partial q} = -2$; substitutes   **33.** (a) $\dfrac{Ex}{Ep} = -\dfrac{3p}{x}$,

$\dfrac{Ex}{Eq} = \dfrac{2q}{x}$, $\dfrac{Ey}{Ep} = \dfrac{3p}{y}$, $\dfrac{Ey}{Eq} = -\dfrac{2q}{y}$; (b) x decreases by 9 percent and y increases by 1.29 percent; (c) x increases by 4 percent and y decreases by 0.57 percent; (d) x increases by 9 percent and y decreases by 1.29 percent; (e) x decreases by 4 percent and y increases by 0.57 percent

35. (b) $\dfrac{Ex}{Ep} = -\dfrac{p}{10q}$, $\dfrac{Ex}{Eq} = \dfrac{p}{10q}$, $\dfrac{Ey}{Ep} = \dfrac{q}{20p}$, $\dfrac{Ey}{Eq} = \dfrac{-q}{20p}$; (c) the demand for A is increased by 0.13 percent and the demand for B is decreased by 0.04 percent; (d) the demand for A is decreased by 0.13 percent and the demand for B is increased by 0.04 percent
37. $f(-1, -1) = 1$, rel max   **39.** $f(1, 0, 0) = 1$, rel min; $f(-1, 0, 0) = 1$, rel min   **41.** 24 units of A and 15 units of B   **43.** 18 ft
by 18 ft by 18 ft   **45.** $f(\tfrac{1}{3}\sqrt{3}, \tfrac{1}{3}\sqrt{6}, \tfrac{1}{3}\sqrt{6}) = \tfrac{2}{9}\sqrt{3}$, rel max; $f(-\tfrac{1}{3}\sqrt{3}, -\tfrac{1}{3}\sqrt{6}, -\tfrac{1}{3}\sqrt{6}) = -\tfrac{2}{9}\sqrt{3}$, rel min
47. (a) $177x - 2692y + 169,066 = 0$; (b) 68 beats per minute

EXERCISES A.1 (Page A-12)

1. (a) $\tfrac{1}{3}\pi$; (b) $\tfrac{3}{4}\pi$; (c) $\tfrac{7}{6}\pi$; (d) $-\tfrac{5}{8}\pi$; (e) $\tfrac{1}{9}\pi$; (f) $\tfrac{2}{3}\pi$; (g) $-\tfrac{5}{12}\pi$; (h) $\tfrac{5}{9}\pi$ **3.** (a) $45°$; (b) $120°$; (c) $330°$; (d) $-90°$; (e) $28.65°$; (f) $540°$;
(g) $-114.6°$; (h) $15°$ **5.** (a) $\tfrac{1}{2}$; (b) $\tfrac{1}{2}\sqrt{2}$; (c) 1; (d) $-\tfrac{1}{2}$; (e) $-\tfrac{1}{2}\sqrt{3}$; (f) $\tfrac{1}{2}\sqrt{2}$; (g) -1; (h) 0   **7.** (a) $\tfrac{1}{3}\pi, \tfrac{2}{3}\pi$; (b) $\tfrac{3}{4}\pi, \tfrac{5}{4}\pi$; (c) $\tfrac{1}{2}\pi, \tfrac{3}{2}\pi$;

(d) $\tfrac{7}{6}\pi, \tfrac{11}{6}\pi$ **9.** $3 \cos x$ **11.** $x \cos x$ **13.** $36 \cos 3x \cos 4x - 48 \sin 3x \sin 4x$ **15.** $\dfrac{1}{\cos t - 1}$

17. $6(\sin^2 x - x^2)^2(\sin x \cos x - x)$ **19.** $\dfrac{-y \sin\sqrt{y^2 + 1}}{\sqrt{y^2 + 1}}$ **21.** $-\dfrac{2}{\pi}$ **23.** $-\tfrac{10}{3}$ **25.** $-\tfrac{1}{4}\cos 4x + C$

27. $\tfrac{1}{4}(2 + \sin x)^4 + C$ **29.** $-\tfrac{2}{3}(1 + \cos x)^{4/3} + C$ **31.** $\tfrac{1}{2}x \sin^2 x + \tfrac{1}{4}\cos^2 x + C$ **33.** $\tfrac{1}{3}\cos^3 x - \cos x + C$
35. $\tfrac{2}{7}\cos^{7/2}z - \tfrac{2}{3}\cos^{3/2}z + C$ **37.** $\tfrac{47}{480}$ **39.** 2 sq units   **41.** (b) $20,000; (c) $40,000; (d) $20,000; (e) 0   **43.** $720,000
45. (a) $\tfrac{1}{9}\sqrt{2}W$; (b) $2W$ **47.** $\tfrac{1}{4}\pi$

EXERCISES A.2 (Page A-21)

3. $-4 \csc 4x \cot 4x$ **5.** $2 \tan x \sec^2 x$ **7.** $2x \sec x^2 \tan x^2$ **9.** $-3e^x\csc^2 e^x$ **11.** $2 \tan 2x$ **13.** $-\dfrac{3 \csc^2 3r}{2\sqrt{\cot 3r}}$

15. $4 \tan^5 x + 6 \tan^3 x + 2 \tan x$ **17.** $-3t^2 \csc(t^3 + 1) \cot(t^3 + 1)$ **19.** $(-3 \ln 2)2^{\csc 3x}\csc 3x \cot 3x$ **21.** $4 \cot t \csc^2 t$

23. $\dfrac{\sin^2 x \cos x + \cos x - 4}{\cos^2 x(\cos x - 4)^2}$ **25.** $(\sin x)^{\tan x}(\sec^2 x \ln(\sin x) + 1)$ **29.** $\tfrac{1}{5}\tan 5x + C$ **31.** $\tfrac{1}{2}\ln|\sec 2w| + C$

33. $-\tfrac{1}{3}\csc 3x + C$ **35.** $\tfrac{1}{10}\ln|\csc 10t - \cot 10t| + C$ **37.** $-\tfrac{1}{10}\cot 5x^2 + C$ **39.** $\ln|\sin e^t| + C$

41. $\tfrac{1}{6}\ln|\sec 3y^2 + \tan 3y^2| + C$ **43.** $\tfrac{1}{8}\ln 2$ **45.** $\dfrac{2 \ln 2}{\pi}$ **47.** $\tfrac{2000}{9}$ ft/sec

INDEX

83 84 85 86 87 9 8 7 6 5 4 3 2 1

Forms Containing $\sqrt{a^2 - u^2}$

39. $\displaystyle \int \frac{du}{\sqrt{a^2 - u^2}} = \sin^{-1}\frac{u}{a} + C$

40. $\displaystyle \int \sqrt{a^2 - u^2}\, du = \frac{u}{2}\sqrt{a^2 - u^2} + \frac{a^2}{2}\sin^{-1}\frac{u}{a} + C$

41. $\displaystyle \int u^2 \sqrt{a^2 - u^2}\, du = \frac{u}{8}(2u^2 - a^2)\sqrt{a^2 - u^2} + \frac{a^4}{8}\sin^{-1}\frac{u}{a} + C$

42. $\displaystyle \int \frac{\sqrt{a^2 - u^2}\, du}{u} = \sqrt{a^2 - u^2} - a\ln\left|\frac{a + \sqrt{a^2 - u^2}}{u}\right| + C$

43. $\displaystyle \int \frac{\sqrt{a^2 - u^2}\, du}{u^2} = -\frac{\sqrt{a^2 - u^2}}{u} - \sin^{-1}\frac{u}{a} + C$

44. $\displaystyle \int \frac{u^2\, du}{\sqrt{a^2 - u^2}} = -\frac{u}{2}\sqrt{a^2 - u^2} + \frac{a^2}{2}\sin^{-1}\frac{u}{a} + C$

45. $\displaystyle \int \frac{du}{u\sqrt{a^2 - u^2}} = -\frac{1}{a}\ln\left|\frac{a + \sqrt{a^2 - u^2}}{u}\right| + C$

46. $\displaystyle \int \frac{du}{u^2\sqrt{a^2 - u^2}} = -\frac{\sqrt{a^2 - u^2}}{a^2 u} + C$

47. $\displaystyle \int (a^2 - u^2)^{3/2}\, du = -\frac{u}{8}(2u^2 - 5a^2)\sqrt{a^2 - u^2} + \frac{3a^4}{8}\sin^{-1}\frac{u}{a} + C$

48. $\displaystyle \int \frac{du}{(a^2 - u^2)^{3/2}} = \frac{u}{a^2\sqrt{a^2 - u^2}} + C$

Forms Containing Exponential and Logarithmic Functions

49. $\displaystyle \int e^u\, du = e^u + C$

50. $\displaystyle \int a^u\, du = \frac{a^u}{\ln a} + C$

51. $\displaystyle \int ue^u\, du = e^u(u - 1) + C$

52. $\displaystyle \int u^n e^u\, du = u^n e^u - n\int u^{n-1}e^u\, du$

53. $\displaystyle \int u^n a^u\, du = \frac{u^n a^u}{\ln a} - \frac{n}{\ln a}\int u^{n-1}a^u\, du + C$

54. $\displaystyle \int \frac{e^u\, du}{u^n} = -\frac{e^u}{(n-1)u^{n-1}} + \frac{1}{n-1}\int \frac{e^u\, du}{u^{n-1}}$

55. $\displaystyle \int \frac{a^u\, du}{u^n} = -\frac{a^u}{(n-1)u^{n-1}} + \frac{\ln a}{n-1}\int \frac{a^u\, du}{u^{n-1}}$

56. $\displaystyle \int \ln u\, du = u\ln u - u + C$

57. $\displaystyle \int u^n \ln u\, du = \frac{u^{n+1}}{(n+1)^2}[(n+1)\ln u - 1] + C$

58. $\displaystyle \int \frac{du}{u\ln u} = \ln|\ln u| + C$